2nd International Conference on Engineering and Applied Sciences (2nd InCEAS)

IOP Conference Series: Materials Science and Engineering
Volume 771

Yogyakarta, Indonesia
16 November 2019

Editors:

Haryanto
Mohammad Mansoob Khan
Setyawan Widyarto
Wakhyu Dwiono
Anwar Ma'ruf
Mukhlis Prasetyo Aji
Hindayati Mustafidah
Anton Yudhana
Oyas Wahyunggoro
Anton Satria Prabuwono
Gatot Rusbintardjo
Sri Sunarjono

ISBN: 978-1-7138-4709-0
ISSN: 1757-8981

Printed from e-media with permission by:

Curran Associates, Inc.
57 Morehouse Lane
Red Hook, NY 12571

Some format issues inherent in the e-media version may also appear in this print version.

Institute of Physics
Dirac House, Temple Back
Bristol BS1 6BE UK

Phone: 44 1 17 929 7481
Fax: 44 1 17 920 0979

techtracking@iop.org

Additional copies of this publication are available from:

Curran Associates, Inc.
57 Morehouse Lane
Red Hook, NY 12571 USA
Phone: 845-758-0400
Fax: 845-758-2633
Email: curran@proceedings.com
Web: www.proceedings.com

TABLE OF CONTENTS

IOP Conf. Series: Materials Science and Engineering **771** (2020) 011001 doi:10.1088/1757-899X/771/1/011001

PREFACE

The 2nd International Conference on Engineering and Applied Science (INCEAS) to be held in Shantika Premier Hotel Yogyakarta, Indonesia on November 16 2019. The 2nd INCEAS 2019 is hosted by Faculty of Engineering and Science Universitas Muhammadiyah Purwokerto Indonesia. The objective of 2nd INCEAS 2019 is to bring together researchers, scientists, academics, and engineers in the area of Civil Engineering, Chemical Engineering, Electrical Engineering, Informatical Technology Mechanical Engineering, Industrial Engineering and Applied Science from all over the world.

This conference program is highlighted by four invited keynote speakers & plenary speaker:

- Assoc. Prof. Ahmad Ruslan Mohd Ridzuan from Faculty of Civil Engineering, Universiti Teknologi MARA, Malaysia.
- Asst. Prof. Chaleeda Borompichaichartkul, Department of Food Technology, Chulalongkorn University, Thailand.
- Haryanto, Ph.D, Department of Chemical Engineering, Universitas Muhammadiyah Purwokerto Indonesia
- Dr. Aqsha from Dept of Chemical Engineering, University Technology Petronas Malaysia.

There were 71 papers which were presented at the conference. The detailed information about the conference is available on the website http://inceas.ump.ac.id

IOP Conf. Series: Materials Science and Engineering **771** (2020) 011002 doi:10.1088/1757-899X/771/1/011002

EDITORIAL TEAM

IOP Conf. Series: Materials Science and Engineering **771** (2020) 011002 doi:10.1088/1757-899X/771/1/011002

THE 2nd INCEAS 2019 COMMITTEE

Advisory Committee

Dr. Anjar Nugroho, M.S.I., M.HI.

Dr. Jebul Suroso, S.Kep., Ns., M.Kep.

Ahmad Darmawan, S.E., M.Si.

Drs. H. Ikhsan Mujahid, M.Si.

Dr. Anwar Ma'ruf, S.T., M.T.

Steering Committee

M. Taufiq Tamam, S.T., M.T.

Haryanto, Ph.D.

Hindayati Mustafidah, S.Si., M.Kom.

Neni Damajanti, S.T., M.T.

Teguh Marhendi, S.T., M.T.

Harjono, S.T., M.Eng.

M. Agus Salim Alfatoni, S.T., M.T.

Official Committee

Wakhyu Dwiono, S.T., M.T.

Mukhlis Prasetyo Aji, S.T., M.Kom.

Besty Afriandini, S.T., M.Eng.

Arif Johar Taufiq, S.T., M.T.

Feri Wibowo, S.Kom., M.Cs.

Achmad Fauzan, S.Kom., M.Cs.

Dian Nova Kusuma Hardani, S.T., M.Eng.

Latiful Hayat, S.T., M.T.

IOP Conf. Series: Materials Science and Engineering **771** (2020) 011003 doi:10.1088/1757-899X/771/1/011003

2nd InCEAS 2019

IOP Conf. Series: Materials Science and Engineering **771** (2020) 011003 doi:10.1088/1757-899X/771/1/011003

2nd InCEAS 2019

Peer review statement

All papers published in this volume of *IOP Conference Series: Materials Science and Engineering* have been peer reviewed through processes administered by the proceedings Editors. Reviews were conducted by expert referees to the professional and scientific standards expected of a proceedings journal published by IOP Publishing.

IOP Conf. Series: Materials Science and Engineering **771** (2020) 012001 doi:10.1088/1757-899X/771/1/012001

Sound Recognition of Four Stroke Manual Transmission Motorcycle Engine's Damage

H Santoso[1,2]

[1] Universitas Atma Jaya Yogyakarta, Yogyakarta Indonesia
[2] Dili Institute of Technology, Dili East Timor

hariyantosantoso89@gmail.com

Abstract. Sound recognition that learns about the source of a sound, which is then used as a reference in learning. Sound recognition referred to in this study is the sound produced from the sound of a motorcycle engine. From the sound of the engine, the sound is then studied more and the results of the recorded sound are then used as objects of research. The management of motorcycle engine sound is carried out using Matlab and assisted by using the HMM (Hidden Markov Model) method with GMM (Gaussian Mixture Model) where HMM is used as chain pattern formation and GMM is used as a sound matching pattern. In this study indeed really learns from every sound of a motorcycle engine that will be used as a sample in the laboratory. From the sound of a motorcycle engine that is heard by the human ear can already indicate the conditions and eligibility limits of the motorcycle. Therefore, from this study trying to find answers to each condition and type of damage from a motorcycle through the sound of a motorcycle engine. With this research, it can help the vehicle owner, because the results of this study warn the vehicle owner to immediately conduct a scale check at the nearest official workshop and from this study also provide benefits and benefits for researchers and motorcycle vehicle owners to find out the condition of the damage early vehicles and help prevent them so as not to cause severe damage which can add air pollution in the city of Dili, East Timor. And the results of existing research are very helpful both for researchers, mechanics and motorcycle vehicle owners.

1. Introduction

In the case of learning being faced that is how to tell or explain to users of motorcycle vehicles to find out the condition of their vehicles. In the automotive world, especially in the current mode of transportation which is becoming popular and much favoured by the people in the City of Dili, East Timor, especially motorcycle vehicles, because of their simple and elegant models and shapes. The motorcycle itself has a classification of engine capacity starting from 50cc (Centimeter Cubic) and there are even more than 1000cc (Centimeter Cubic) with the type of engine there are some that use Two-Stroke and there are Four-Stroke [1], with manual and automatic transmission types that are often found along the city highway of Dili, East Timor [2] - [4]. Why is it necessary to recognize the type of damage to a motorcycle because of the sound of a motorcycle engine we can find out the conditions of eligibility and the type of damage. Because every sound of a motorcycle engine can show the presence status of each engine component or engine part. Therefore, the purpose of this study is to create a system that can recognize signals from the sound of machines that are processed using Matlab and supported by using the Hidden Markov Model (HMM) [5], [6] with the Gaussian Mixture Model (GMM) [7] - [9] and how to analyze system performance in providing accuracy to engine damage according to the symptoms caused by the engine sound signal.

IOP Conf. Series: Materials Science and Engineering **771** (2020) 012001 doi:10.1088/1757-899X/771/1/012001

2. Four-Stroke

Sound recognition engine damage motorcycles are learning to know and identify the eligibility conditions of the motorcycle, which learned through the sound of the engine noise when turned on. Before the machine can be switched on a motorbike motorcycle engine has a bunch of components or parts that are formed and arranged in order to switch on the machine and run it. In the structure of the component or part-own motorcycle, engines have different roles and functions [10] – [12]. The condition of the sound of a motorcycle engine noise, there are two conditions at the time of normal or standard and condition at the time of damage to components or parts. Conditions at the time, the damaged component or part is divided into four as to the condition of the engines suffered damage at the components of the Cylinder Head which consists of Camshaft and Valve, engine noise conditions suffered damage on the components of Cam Chain, engine noise conditions suffered damage at the components of the Clutch and engine noise conditions suffered damage at the components of the Crank Shaft/Piston.

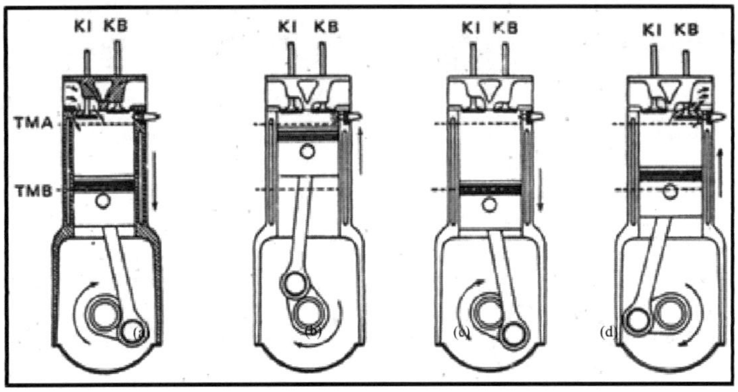

(a) (b) (c) (d)

Figure 1. Four Stroke Process [11]

In Figure 1. Above shows the workings of the crankshaft and piston in the position up and down.

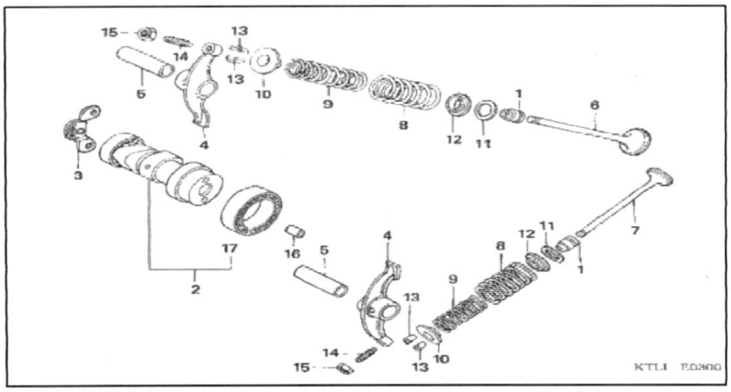

Figure 2. Camshaft and Valve Part [10]

In Figure 2. Above is the engine camshaft from the group with the valve and the arrangement of the parts of the engine group.

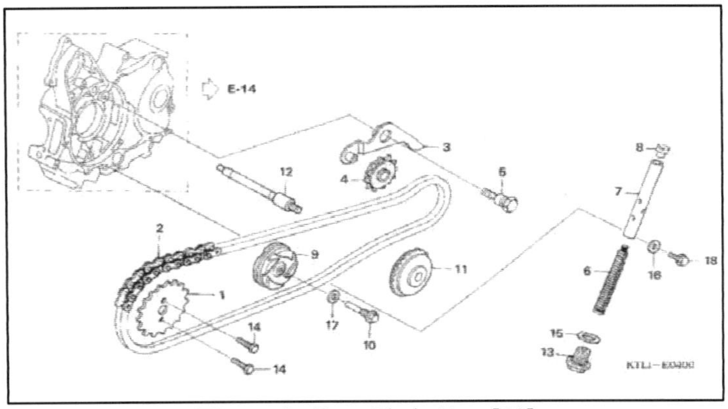

Figure 3. Cam Chain Part [10]

In Figure 3. Above is the engine cam chain part of the group and the arrangement of the parts of the engine group.

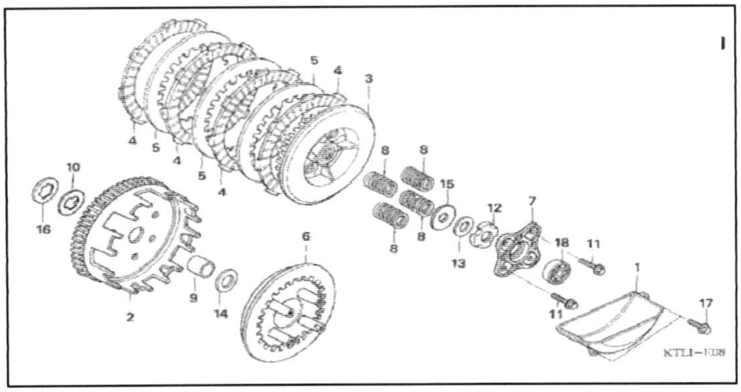

Figure 4. Clutch Part [10]

In Figure 4. Above is a group from the engine clutch and the arrangement of the parts of the engine group.

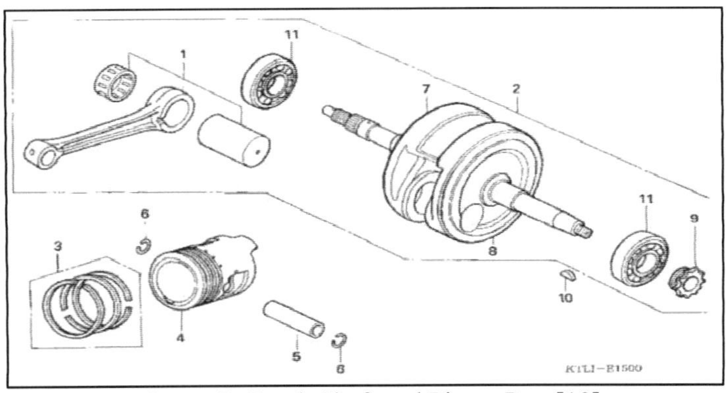

Figure 5. Crank Shaft and Piston Part [10]

In Figure 5. Above are the engine crankshaft of the group (piston) and the arrangement of the parts of the engine group. Sound recognition is a sound signal or longitudinal waves caused or resulting vibration from of an object in the form of solid, liquid and gas as an intermediary. The voice that can be heard by the human ear is limited, and humans can only hear the sound with a frequency of 20 Hz to 20 kHz or less [13].

IOP Conf. Series: Materials Science and Engineering **771** (2020) 012001 doi:10.1088/1757-899X/771/1/012001

On any conditions, the engine noise that has been collected starting from normal conditions until the turn of the spare parts can be view as shown in Table 1.

Table 1. Engine Group Information

Engine Group	Information	Conditions (%)	Solution for Item Repair
Cylinder Head	Smooth	50-70	Normal Part
	Noisy	71-80	Setting Valve (Intake and Exhaust)
	Very Noisy	80-90	Valve IN and EX Guide Replacement Includes Seal, Valve Stem
Cam Chain	Smooth	50-70	Normal Part
	Noisy	71-80	Setting Cam Chain
	Very Noisy	80-90	Cam Chain Replacement Include Tensioner(1set)
Clutch	Smooth	50-70	Normal Part
	Noisy	71-80	Disk, Clutch Friction Replacement Include Outer Clutch
	Very Noisy	80-90	Bearing, Radial Ball Replacement
Crank Shaft /Piston	Smooth	50-60	Normal Part
	Noisy	61-70	Ring Set Piston Replacement
	Noisy and Smoky	71-80	Piston Replacement Include Ring Set Piston and Cylinder
	Very Noisy	81-90	Rod Kit Connecting Replacement include Bearing Crankshaft

The sound that is used in processing and introduction is the sound of a motorcycle engine the sound with decent conditions of use and the condition of the voice can be recognized, which covers the mechanical condition of the spare part Cylinder Head, Valve, Cam Chain Tensioner, and Crank Shafts Piston [10].

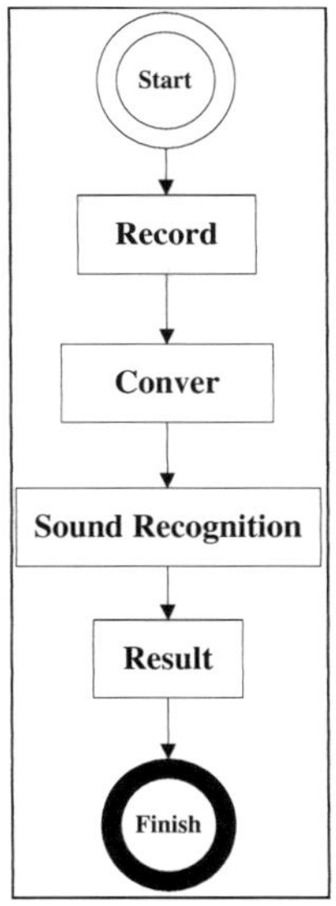

Figure 6. System Workflow

In Figure 6. above is a workflow system developed originally in the form of a sound record of damage to motorcycle engines after recording results in conversions and conversion results in a pattern and use it as a result of the pattern formed is collected data matching used as the information is a result of the program.

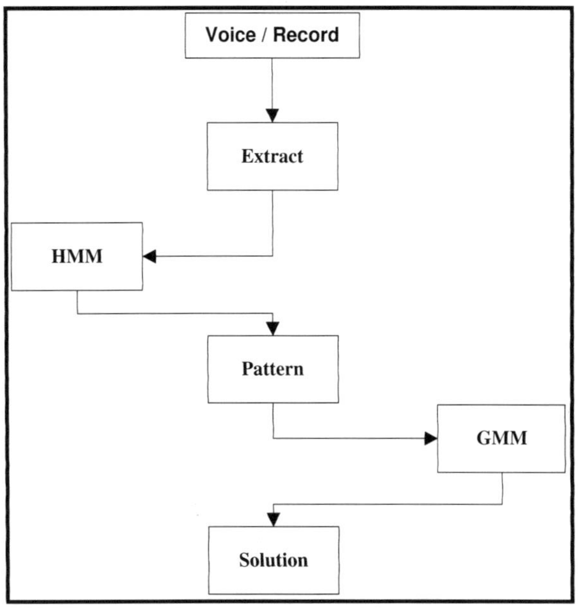

Figure 7. Pattern Recognition Process

In Figure 7. Above is the process of pattern recognition results from the sound extract needed by the HMM, if the recorded sound results that have been patterned by the HMM are then grouped by GMM to determine the results of information that determine the solution of the recorded sound condition of the engine sound that has been processed.

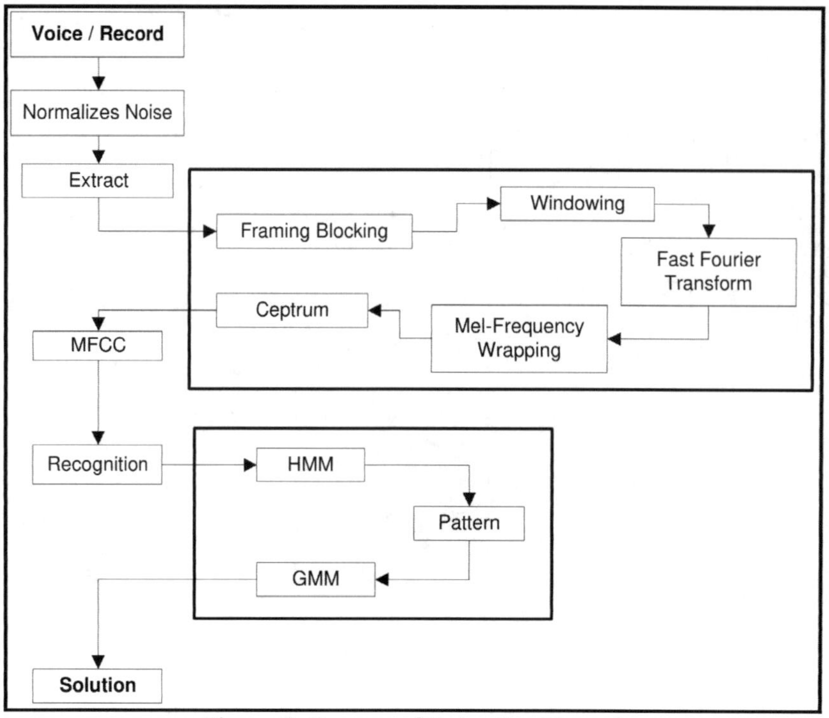

Figure 8. Process of Extracting Record

In Figure 8. Above is the process of extracting recorded sound from recorded machine sound so that it can be processed as needed from step by step here the role of MFCC [14] is needed because MFCC is used to extract recorded sound that has normalized noise so that it is easy to use to form the HMM chain to find the patterns needed by GMM.

3. Hidden Markov Model (HMM)

HMM is the model class statistics are frequently used in voice recognition [14], [15]. HMM, that helps model the occurrence is hidden and cannot be observed, but the result can be seen [16], [17]. Although the order of the State cannot be observed directly can only be estimated from the observations resulting from the sequence of the temptation of possibilities. HMM should build a probabilistic model to be able to explain the sequence of observations, in HMM for the probability to use parameters:

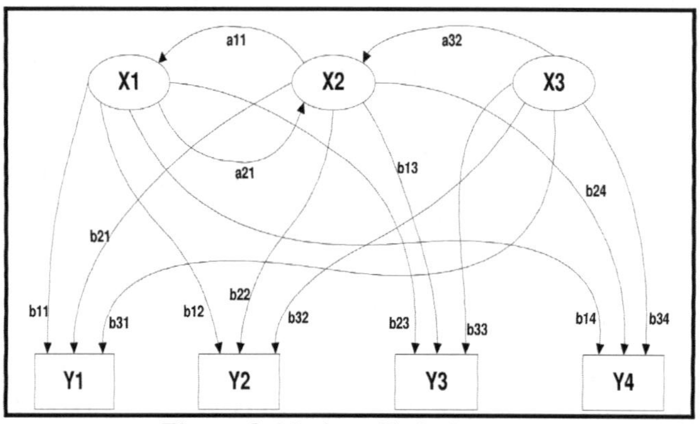

Figure 9. Markov Chain Process

In Figure 9. Above are parameters of the HMM in existing presenting State probability in accordance with condition X, which causes the transition state of the possibility of a_{ij}, with the possible results of the b_j, so the observation might happen Y.

From the HMM parameter with the flow of each Markov chain addressing the transition state and the possibility that will occur in a way that is if the condition [15]:

$$P[q_t = X_j | q_{t-1} = X_j, q_{t-1} = X_k, \dots]$$

$$= P[q_t = X_j | q_{t-1} = X_j] \quad (1)$$

$$a_{ij} = P[q_t = X_j | q_{t-1} = X_j], 1 \le i, j \le X \quad (2)$$

$$a_{ij} \ge 0 \quad (3)$$

$$\sum_{j=1}^{N} a_{ij} = 1 \quad (4)$$

The value of a_{ij}, is the transition probability that does not depend on the circumstances of the time t as the q_t, and a_{ij}, values amount ≥ 0. If it is assumed according to Prior and Likelihood probability estimation with hidden state is the value X, with a condition where X = B, X = R and probability for the observation of the circumstances is the value of Y, for the observations of the State of the components of Y = H, Y = T, Y = C, Y = P ... (5) the probability of the possibility:

$$P(X|Y) \quad (5)$$

$$P(X = B|Y = H) \quad (6)$$

$$P(X = R|Y = H) \quad (7)$$

Where the value of X = B is condition part good, X = R is the condition of the damaged part, Y = H for the observation part of the Cylinder Head, Y = T to observation part Cam Chain Tensioner/, Y = C for observation part Clutch, Y = P for Crank Shaft part observation/Piston.

4. Gaussian Mixture Model (GMM)

GMM is a model of the algorithm classifies the data from a dataset to form groups of data not previously defined [9] in a Gaussian distribution with Mean parameter is the centre point Gaussian distribution (μ), and Variance is a measure of the spread of the values of the data set (σ^2) :

$$P(X|\theta) = \sum_z \rho(x, z|\theta) \quad (8)$$

$$P(X|\mu, \sigma^2) = \frac{1}{\sqrt{2\pi\sigma^2}} e^{-\frac{(z-\mu)^2}{2\sigma^2}} \quad (9)$$

On a standard normal distribution when μ = 0 and σ = 1, with probability density functions such as:

$$\varphi(x) = \frac{1}{\sqrt{2\pi}} e^{-\frac{1}{2}x^2} \quad (10)$$

5. Experiments Result

Simulation for testing the results of voting already as expected and indicates the alignment of a sound physical condition with damage to the motorcycle engines such as in the following image:

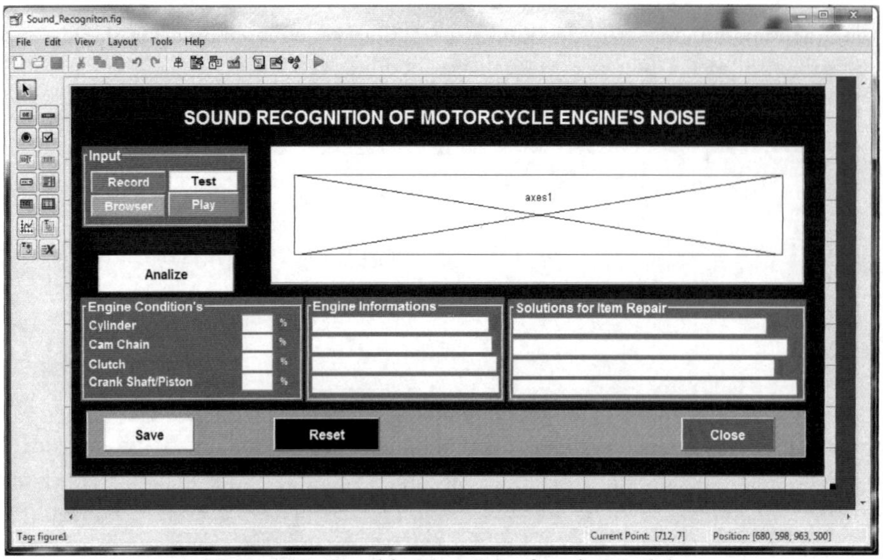

Figure 10. Graphic User Interface in Matlab

In Figure 10. Above displays, the main screen display from application development of this research was used as sound recognition performed on motorcycles.

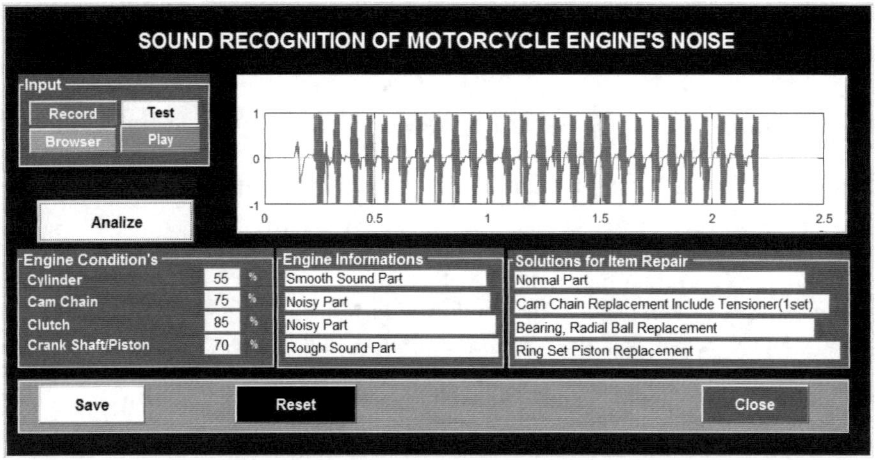

Figure 11. The Sound Recognition Result

In Figure 11. Above displays, the results of the recordings made by using Matlab GUI, then processed by a series of HMM from each transition probability already in assume according to Prior and Likelihood probability with estimation of the State of the hidden which is the value of X against the probability of observing the State of the value of Y to find out the percentage of probability the chances of that happening on the machine and the recorded sound will then be stored for use as sample data on further testing and the sound recognition results were done using Matlab GUI, indicate that collaboration between HMM with GMM can display the result of the vote has been recorded so that owners of vehicles, mechanical and researchers can know of any damage to the motorcycle's engine sound be heard.

The results of the motorcycle engine sound recorded can be recognized in accordance with the pattern that has been created. With the results of testing and recognition of the level of accuracy, the damage system is in accordance with the damage conditions of the motorcycle engine used as in Table 2.

Table 2. Recognition Level of Accuracy

Record Sound	Level of Accuracy (%)	
	Human Expert (Mechanic)	Program (HMM and GMM)
Record 1	85	84
Record 2	75	73
Record 3	80	78
Record 4	90	89
Record 5	70	69
Record 6	85	84
Record 7	86	85
Record 8	80	76
Record 9	87	86
Record 10	65	60

From the results of the comparison of the accuracy, level shows that the pattern that is built can recognize the sound produced from the recorded sound of a motorcycle engine with an average error value of 3% and at worst 5%, probably due to the not yet maximum noise normalization. But the results show that the system being built works well.

6. Conclusion

From the results of research that we do and can recognize the sound of a motorcycle engine in accordance with the conditions of any motorcycle engines are used. From the results of pattern recognition give 5%, a recording that has not been recognized by the system being developed, however, 95% of the recordings can be easily recognized and known conditions damage. But from our research did not cover the possibility that the program we develop can be worked perfectly on a type of motorcycle engines are different, therefore needed further research can apply other methods are more competent about voice recognition and can be developed further in the platform

Acknowledgement
At the end of this writing, the author would like to say many thanks to the University of Atma Jaya Yogyakarta, Indonesia and Dili Institute of Technology, East Timor which is very supportive of the guidance process in making this paper well and can be presented with good results in the participants of the conference.

References

[1] A. Pujanarso and D. Nursuhub, *Mesin Konvensi Energi*, III. Yogyakarta: CV. ANDI OFFSET YOGYAKARTA, 2013.
[2] "Timor Leste," *Wikipedia*, 2019. [Online]. Available: https://id.wikipedia.org/wiki/Timor_Leste.
[3] "Timor-Leste Population," *Worldometers*, 2018. [Online]. Available: http://www.worldometers.info/world population/timor-leste-population/.
[4] J. Macaulay, "Timor Leste : Newest and Poorest of Asian Nations," vol. 88, no. 1, pp. 40–46, 2018.
[5] P. Paramonov and N. Sutula, "Simplified scoring methods in HMM based speech recognition," *Proc. - 2014 Int. Conf. Soft Comput. Mach. Intell. ISCMI 2014*, pp. 154–156, 2014.
[6] T. Koumura and K. Okanoya, "Automatic recognition of element classes and boundaries in the birdsong with variable sequences," *PLoS One*, vol. 11, no. 7, 2016.
[7] R. Gajšek, F. Mihelič, and S. Dobrišek, "Speaker state recognition using an HMM-based feature extraction method," *Comput. Speech Lang.*, vol. 27, no. 1, pp. 135–150, 2013.
[8] L. Wang, N. Kitaoka, and S. Nakagawa, "Robust distant speaker recognition based on position-dependent CMN by combining speaker-specific GMM with speaker-adapted HMM," *Speech Commun.*, vol. 49, no. 6, pp. 501–513, 2007.

[9] K. Phapatanaburi, L. Wang, R. Sakagami, Z. Zhang, X. Li, and M. Iwahashi, "Distant-talking accent recognition by combining GMM and DNN," *Multimed. Tools Appl.*, vol. 75, no. 9, pp. 5109–5124, 2016.

[10] P. Astra Honda Motor, "Astra Honda Motor Indonesia," 2006. [Online]. Available: www.astra-honda.com.

[11] Philip Edward Irving, *Philip Edward Irving-Tuning for speed_ How to increase the performance of any standard motorcycle engine for racing and competition work (1967).pdf.* London: Temple Press Books, 1967.

[12] S. B. Hidayat, Tonny; Sikin, "Data Manajemen Dan Teknologi Informasi," *J. Ilm. DASI*, vol. 16, no. 4, pp. 5–9, 2015.

[13] "Sound," *Wikipedia*, 2019. [Online]. Available: https://id.wikipedia.org/wiki/Sound. [Accessed: 12-May-2019].

[14] S. Furui, *Digital Speech Processing, Synthesis, and Recognition*, Second Edi. New York: Marcel Dekker, Inc., 2001.

[15] L. R. R. Rabiner, "Tutorial on Hidden Markov Models and Selected Applications in Speech Recognition," *Ieee*, vol. 77, no. 2, pp. p257-286, 1989.

[16] R. Abdulah, Dedy; Ramadhan, "Implementasi Algoritma Hidden Markov Model Sebagai Pengenalan Perintah Suara Pada Aplikasi Winamp," 2016. [Online]. Available: https://yudiagusta.wordpress.com/2008/06/24/markov-chain/.

[17] M. D. Ahied, "Teknik Pengenalan Sinyal Wicara Dengan Metoda Hidden Markov Model - Neural Network (Syaraf Tiruan)," vol. 3, no. 1, pp. 10–20, 2016.

IOP Conf. Series: Materials Science and Engineering **771** (2020) 012002 doi:10.1088/1757-899X/771/1/012002

Design a Prototype of The Application System of Classification and Calculating Motor Vehicles on Highway

M T Tamam[1,*], W Dwiono[1], and R J P Safian[1]

[1]Electrical Engineering Depatement of Universitas Muhammadiyah Purwokerto
Jl. Raya Dukuhwaluh Kembaran, Banyumas, Jawa Tengah, Indonesia 53182

[*]e-mail: tamam@ump.ac.id

Abstract. The growth rate of the number of motor vehicles has a tendency that is always increasing. The increase in the number of motorised vehicles which is not followed by the addition of roads is a cause of traffic jams on the highway. This study aims to design an application system that is used to determine the level of traffic density on certain road sections by using the background subtraction method to classify and count the number of vehicles passing on that road section. From the results of software testing, three types of vehicles can be classified, namely motorcycles, cars, and trucks/buses with an accuracy of 96%, 83%, and 87.15%, respectively.

1. Introduction

Traffic density data is an indispensable component for planning a traffic condition. Until now to get traffic density data done manually, namely by assigning several people to be in the field, counting each vehicle that passes and then divided by a certain period. The background subtraction method is widely used for object detection applications, for example, traffic monitoring, human motion, and video surveillance. The object image is separated from the background image. The separated object image is then converted to a grayscale image to increase object segmentation [1]. The real condition of the traffic is recorded to determine the image processing techniques and simulation analysis data. The simulation method is then used to find out traffic problems. Then the traffic conditions are initiated and the actual conditions are modelled. Some changes are needed to make the model. Comparison of analysis between the conditions that are initiated and changes in the model will be performed [2]. Background registration and segmentation techniques use morphological operators. The system will detect and count objects on the highway. Computational complexity is proportional to the video size and the number of vehicles detected [3]. The first thing in making a video surveillance application is to detect and track the movement of objects. Detection and tracking of moving objects based on vision systems using a different image algorithm. With this method distance, frame per time and velocity can be determined [4]. Visual surveillance is widely used for security systems, law enforcement and military applications. Video surveillance applications can be made using background subtraction methods. A morphological method is used for further processing to eliminate noise [5]. Detection and segmentation of moving objects from two consecutive stereo frames by utilizing uncertainties in estimating camera movements and in calculating disparities. The relative camera movement and uncertainty are calculated by tracking and matching rare features in four images. The probability of motion at each pixel is estimated by calculating the uncertainty and ego-motion gap in the calculation procedure [6]. Many methods can be used for moving object detection. For example, temporal differences, background subtraction, optical flow. Each has advantages and disadvantages. The essence of this algorithm is a simple background subtraction method, it is easy and is used in real-time applications and does not need special hardware [7].

IOP Conf. Series: Materials Science and Engineering **771** (2020) 012002 doi:10.1088/1757-899X/771/1/012002

2. Materials and Methods

2.1 Material

Vidoe camera recorder, Microsoft visual studio 2015 software, visual basic software, OpenCV software.

2.2Methods

Placement of the video camera recorder for video shooting of traffic conditions on the highway on the flyover, then by analysing the work characteristics and suitability of each component supporting the design idea obtained a steady and optimal initial design. Overall the system built is a system used to classifications and calculate the number of motorised vehicles.

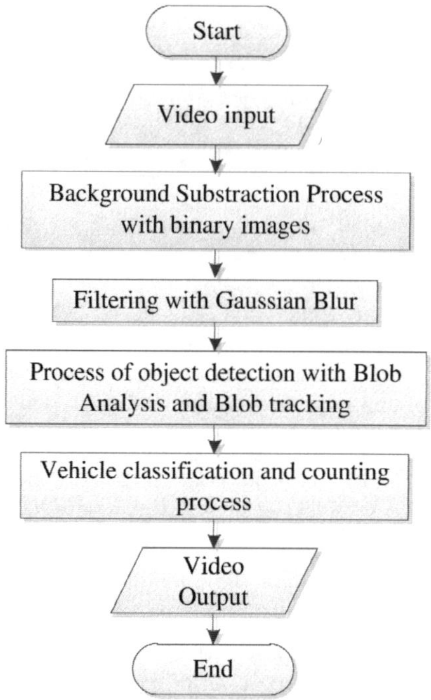

Figure 1. Flow chart system

The system built is software that is able to insert videos, then detect motorised vehicles, classify them and then count the number of each motorized vehicle. The process of inserting video as an image to be processed is an image preprocessing. Next will be the Background Subtraction process so that the object (foreground) will be separated from the background image (background). When an object in the form of a foreground begins to be separated from its background, it is then converted from an RGB image to a Grayscale image and filtered using the Gaussian Blur method, after which an absolute image difference is carried out, namely the process of reducing the background with the current frame. Absolute different functions are used to calculate the absolute value of the difference between two arrays or images. Absolute is used to avoid negative values. After that, the threshold stage is carried out for segmentation or separation of an object with others. The classification used is the blob detection method. This method is said to be successful if every passing vehicle is detected by the program as an object. After one or more objects have been detected, a vehicle classification mark such as a motorcycle, car and truck/bus is placed. After each object is classified, each object is then made into a Blob for object movement (tracking). When the previously made Blob meets the value of a y coordinate, the counting process will increase by one.

3. Results and Discussions

This system is a form of applications of classifying and calculating motor vehicle on the highway. This system consists of a video camera recorder that will capture traffic conditions on a highway section. Next will be the Background Subtraction process so that the object (foreground) will be separated from the background image (background). The separated objects are then converted from an RGB image to a

Grayscale image and filtered using the Gaussian Blur method. The classification process uses the blob detection method. Objects that have been classified are then made into Blob for object movement (tracking).

Figure 2. Condition of Pemuda street Purwokerto

Figure 2 shows the condition of Pemuda street Purwokerto during the day taken with a 1-minute video with a size of 1280 x 740 pixels.

Figure 3. The output system of Pemuda street Purwokerto condition

Figure 3 shows the system output for detecting traffic conditions on Pemuda street Purwokerto during the day, as shown in Figure 2.

Figure 4. Condition of Veteran street Purwokerto

Figure 4 shows the condition of Veterans Street Purwokerto in the afternoon taken with a 1-minute video with a size of 1280 x 740 pixels.

Figure 5. Output system condition of Veterans street Purwokerto

Figure 5 shows the system output for detecting traffic conditions on Veteran street Purwokerto in the afternoon, as shown in Figure 4.

Tabel 1. System accuracy test results

Location	Motorcycle	Car	Truck/bus
Pemuda street	92%	88%	96%
Veteran street	100%	78%	78,3%
Average	96%	83%	87,15%

Table 1 shows the results of testing the accuracy of the system in classifying and counting motorised vehicles. The level of accuracy to classify and count the number of motorbikes is 96%, cars are 83% and trucks/buses are 87.15%.

4. Summary

The application system created in this study has been able to work following the plan, which can classify and count the number of motorised vehicles on the highway. From the test results, it can be concluded that the level of accuracy to classify and count the number of motorbikes is 96%, cars 83% and trucks/buses 87.15%.

References

[1] Mahamuni. Patil, R. P. And Thakar, H.S. (2014). *Moving Object Detection Using Background Subtraction Algorithm Using Simulink.* IJRET: International Journal of Research in Engineering and Technology. Volume: 03 Issue: 06. eISSN: 2319-1163. pISSN: 2321-7308.

[2] Matsuhashi, N. Hyodo, T. And Takahashi, Y. (2005). *Image Processing Analysis Of Motorcycle Oriented Mixed Traffic Flow In Vietnam*. Proceedings of the Eastern Asia Society for Transportation Studies. Vol. 5. pp. 929 – 944.

[3] Daigavane, P. M. and Bajaj, P. R. (2010). *Real-Time Vehicle Detection and Counting Method for Unsupervised Traffic Video on Highways*. IJCSNS International Journal of Computer Science and Network Security. Vol.10. No.8.

[4] Jadav, K. R. Lokhandwala, M. A. and Gharge, A. P. (2011). *Vision-Based Moving Object Detection and Tracking*. National Conference on Recent Trends in Engineering & Technology. B.V.M. Engineering College, V.V.Nagar, Gujarat, India.

[5] Pawaskar, M.C. Narkhede, N. S. and Athalye1, S. S. (2014). *Detection Of Moving Object-Based On Background Subtraction*. International Journal of Emerging Trends & Technology in Computer Science (IJETTCS). Volume 3, Issue 3.

[6] Zhoua, D. Fr´emonta, V. Quosta, B. Daib, Y. Li, H. (2017). *Moving Object Detection and Segmentation in Urban Environments from a Moving Platform*. Image and Vision Computing, Elsevier, 68, pp.76-87. 10.1016/j.imavis.2017.07.006. hal-01609038.

[7] Amandeep. and Goyal, E. M. (2015). *Review: Moving Object Detection Techniques*. International Journal of Computer Science and Mobile Computing. Vol. 4. Issue. 9. pg. 345-349.

Design, Construction and Testing of Portable Systems for Temperature, Humidity and Ammonia Monitoring of Chicken Coop

H Supriyono [1,*], **U Bimantoro** [1], **K Harismah** [2]
[1] Department of Electrical Engineering, Universitas Muhammadiyah Surakarta, Surakarta, Indonesia
[2] Department of Chemical Engineering, Universitas Muhammadiyah Surakarta, Surakarta, Indonesia

[*]e-mail: Heru.Supriyono@ums.ac.id

Abstract. The air and environment quality have big impact for chicken health. The most dominant component polluting environment of chicken coop is ammonia which is produced by chicken manure. Other environment components that affect chicken health are including temperature and humidity of the air. The problem faced by chicken farmer is that the decreasing of chicken health which leads to the death caused by the bad air and environment condition. These environment variables were usually predicted by the farmer manually. The objective of the publication is to monitor the ammonia, temperature and humidity electronically by using computer-based systems for a chicken coop. The developed systems used DHT 22 sensor as a temperature detector and MQ-135 sensor as ammonia gas detector. The systems also able to control the fan and lamps to maintain temperature and humidity of the chicken coop stable. Data from the sensors then were processed by NodeMCU ESP8266 microcontroller and then was displayed on 20x4 LCD. Afterward, the data was sent to the Android smartphone by using social networking platform Telegram for distance notification systems. Testing results showed that the developed system works well.

1. Introduction

Newly hatched chicken seedlings (one day old), known as day old chick (DOC), in broiler farms, require good treatment so they will be able to grow healthy to produce high egg production or optimal weight and good carcass.

Because of the importance of temperature conditions, the level of temperature in the chicken coop must be known continuously. Temperature data on the chicken coop is used as a basis for consideration of farmer to take action so that the temperature of the chicken coop can be kept constantly at a comfortable range for chicken. In general, the temperature in a coop can be measured manually by using a mercury-based or electronic-based thermometer. Temperature data on the chicken coop may also need to be recorded for the purpose of further analysis related to the management of the coop. Measuring temperature in this way will take quite a long time to read and record the data.

Besides the temperature of the coop, another factor that greatly influences the growth of chicken is the air quality of the coop. The fermentation process of chicken manure mixed with a litter base produces ammonia gas (NH_3). The more manure produced by the chicken the more ammonia gas produced. The formation of ammonia gas will be affected by some factors including temperature, moisture, PH and nitrogen content of the litter or manure. The ammonia concentration combined with its exposure time will affect the health of both chicken and the people around especially the worker

[1]. Based on the results of research conducted by [2], [3] it can be noted that the high levels of ammonia gas in the coop affect the resistance of chicken to diseases that can lead to high mortality rates. Also, the high concentration of ammonia in the air of chicken coop reduce feed intake and then slowdown the growth rate, lowering egg production and decrease respiratory function [4].

Microcontroller-based electronic systems can be implemented to detect and record the presence of certain gases in nature automatically to replace the detection and recording manually by people such as for the detection of combustible gases. The results showed that the electronic system can works with sufficient precision and record it in a storage device in the form of a file faster than the manual process [5], [6]. Another research results reported in [7] showed that electronics system was employed for monitoring of air quality. Moreover, electronics based system also was reported to be used for detecting ammonia gas leakage with acceptable performance [8].

Based on the description above, taking into account that temperature, and ammonia gas levels are very important in chicken growth, it is very necessary to always be monitored automatically by using an electronic based system. Monitoring the main parameters of the chicken coop is very important to reduce the mortality rate of chickens, especially during the critical period when the chicks are aged between one day and three weeks due to the unfavorable condition of the coop. By reducing the mortality rate of chicks, it will reduce the losses suffered by farmers while increasing the profit of farmer.

The attempt for monitoring chicken coop environment variable especially temperature, humidity and hazardous gasses electronically has been published in [9] where computer with local network were used for monitoring environment and variables as well as for controlling devices like heating devices, moistturizing devices and ventilation. The research result showed that architecture, user would only be able to monitor the environment variable of chicken coop using designated computer statically, not mobile.

The objective of this publication is to monitor the main parameters of the chicken coop and display them electronically as well as for mobile device (smartphone) using Telegram social networking application so that it can be used as a basis for decision making. The use of mobile phone for displaying data and wifi for data communication in this published work is the main difference of this publication with research result in [9].

2. Method

2.1. Design process

The design process of the monitoring system consists of four main stages, namely architecture design, hardware design, software design, construction design and testing.

2.2. Tools and materials

The main hardware and software tools used in the design process were including:: (1) NodeMCU ESP8266 V3 microcontroller, (2) DHT22 sensor, (3) MQ-135 sensor, (4) Relay, (5) fan, (6) lamp, (7) real time clock (RTC), (8) Arduino IDE and (9) social networking Telegram.

2.3. Systems architecture

The block diagram showing the architecture design of the system from sensors to the display of mobile device can be seen in figure 1.

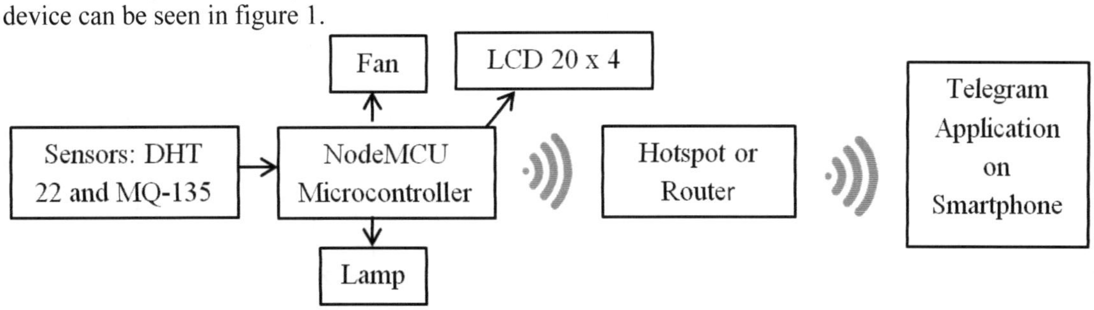

Figure 1. Diagram blok representing systems architecture.

The NodeMCU microcontroller is used as the main processor or brain of monitoring system for temperature, humidity, NH3 levels and remote control of fans and lights. Components for detecting temperature and humidity is DHT 22 while for NH3 gas content is MQ-135. Then, the data of the sensors reading are processed by using NodeMCU Microcontroller. Next, resulted data are displayed on the LCD 20x4 (20 columns and 4 rows) as well as on Android smartphone by using Telegram social networking application for remote monitoring. Besides for displaying the resulted data, Telegram application is also used as remote control of fan and lights in order for stabilization when the temperature and humidity of the chicken coop are not appropriate. Less humid of the chicken coop will prohibite the formation of ammonia. The fan, when it is turned on, functions like a ventillation that provide fresh air for the chicken coop. The fresh air will decrease the concentration of ammonia in the atmosphere of chicken coop [10]. Less ammonia in the chicken coop will decrease the death rates and produce healthier chicken so that increase the profit [11]. Figure 2 shows a wiring diagram of hardware components implementing system architecture depicted in figure 1.

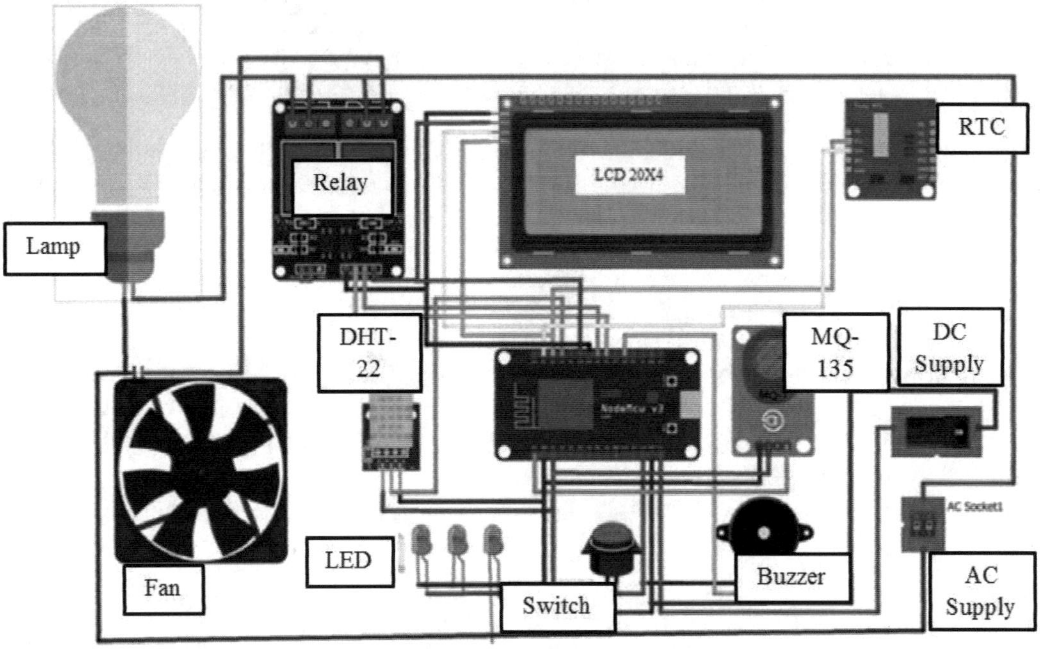

Figure 2. Wiring of hardware components.

2.4. Software design

The microcontroller handles all computation process by using software programmed in it. The general computational process of software in the NodeMCU microcontroller is described in figure 3.

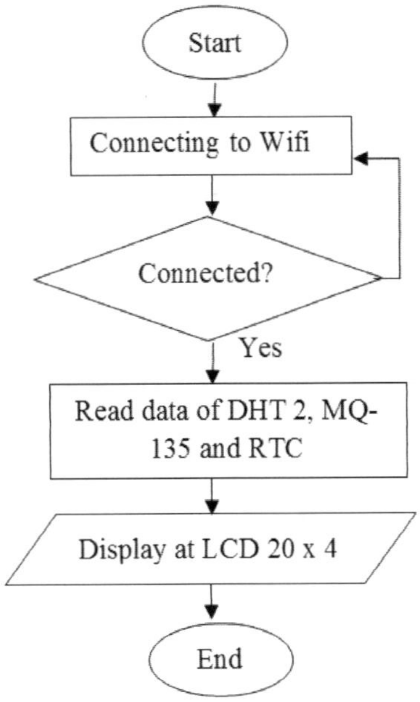

Figure 3. Flowchart showing general computational process of microcontroller program

The initial step in the computing process is to connect NodeMCU with a wireless internet network (wifi). After it is connected, NodeMCU will read the output data generated by the DHT 2 sensor, MQ-135 sensor and RTC. Then these data will be processed into final data in the form of temperature, humidity and ammonia gas levels which are then displayed on a 20 x 4 LCD viewer. Figure 4 shows the computation flow diagram of the remote ignition mechanism for fans and lights using a smartphone and then displays reports on temperature, humidity, ammonia gas concentrations, the status of whether fan and lights are on or not using the Telegram social networking application.

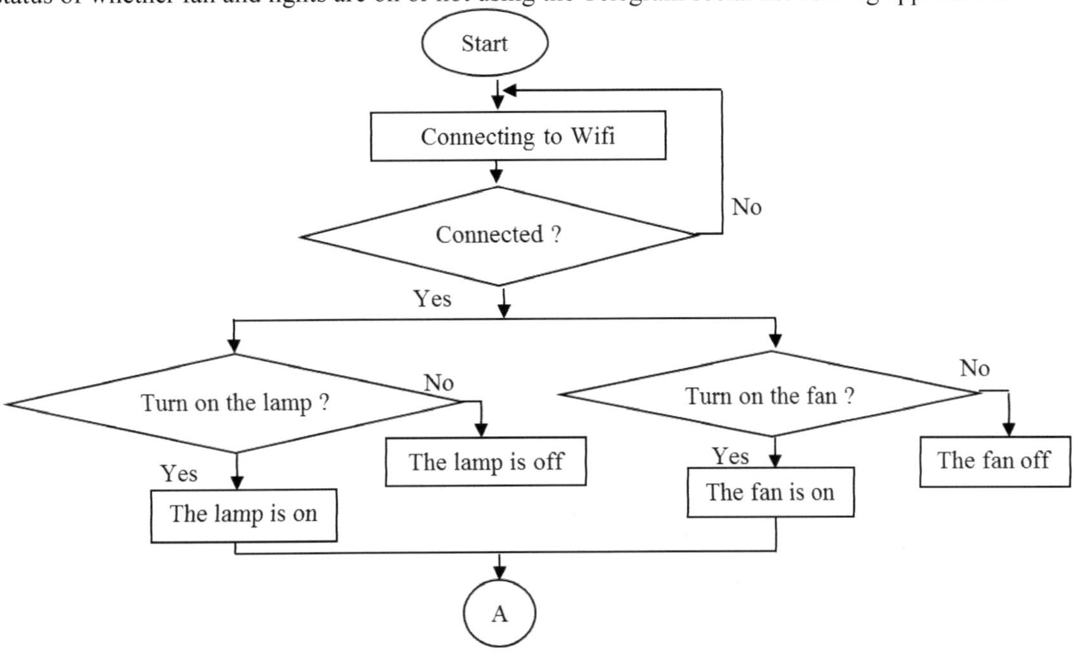

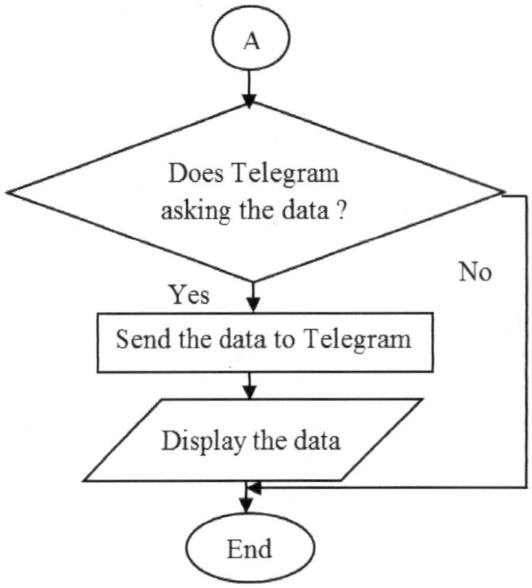

Figure 4. Flowchart of program for controlling lamp and fan remotely and display the data and status by using Telegram.

After successfully connected to wifi, the microcontroller will wait whether the user will command to turn on the light and the fan. After that, microcontroller will wait whether the Telegram application asking to send the data or not, if yes then microcontroller will send the data to Telegram to display, otherwise the program is finished.

2.5. Construction design

The hardware circuits involving NodeMCU microcontroller, RTC, relay, regulators (7805 and 7812 as source of 5 and 12 DCV power source respectively) and transformer CT 12 V. All of these components are then placed in an acrylic box with size of 17 centimetres in length x 12 centimetres in width x 8 centimetres in height. The RTC is used for displaying the time and date of the device. The box is equipped with AC socket to enter the source from main power line, the switch to turn off and turn on the appliance, the outlet to enter from the fan and the lamp after the relay.

3. Results and discussion

3.1. The resulted systems

Figure 5 shows actual outer view of the resulted systems in an acrylic box. Placing the hardware set in an acrylic box will protect the hardware set from water, dust and other external interference.

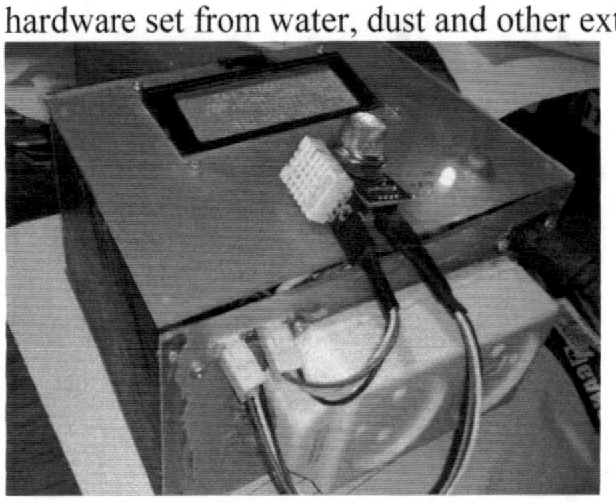

Figure 5. Actual view of resulted systems.

3.2. Testing scenario

The resulted system was then tested to monitor environmental conditions in laying chicks which have dimensions of 16 meters in length (P) x 9 meters in width (L) x 3.5 meters in height (T) which can be seen in figure 6. The coop has a capacity of 2000 chicks made of the gable roof because it is suitable for use in areas with high humidity levels.

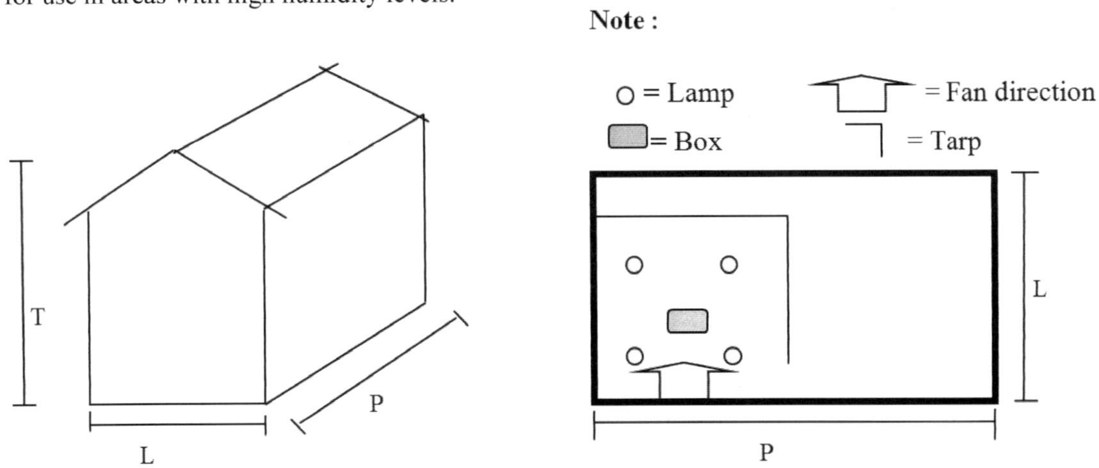

Figure 6. The dimension of the chicken coop for performance testing and the configuration of the systems inside the chicken coop.

Laying DOCs age from one day to 21 days initially are placed in a coop with a size of 5x7 meters and the entire roof wall is covered with a tarp so that the chicks are warm while the floor is sown in rice husk and above it is covered by paper.At the time of testing, the resulted system was placed in the middle of the coop (figure 6) while the lamps were placed hung of the roof spready throughout the enclosed area.

3.3. Testing of temperature and humidity testing

The temperature and humidity sensor used in this system is the DHT 22 sensor. In order to gain its validity, it is necessary to do a comparison between the measurement results using DHT 22 sensor and using the commercial available Thermo Hygrometter which is sold on the market. Testing of temperature and humidity measurements were done with a time span of five minutes. The test results showed that for the measurement of humidity the resulted system has an error of 0.337% while for the measurement of the temperature the tool has an error of 4.32% when compared with the results of measurements using a Thermo Hygrometer. Then the resulted system was tested for 14 days with the temperature and humidity data collection was carried out three times a day with the following time: (1) in the morning: between 06.00 am and 07.00 am, (2) afternoon: between 12.00 at noon and 13.00 pm, and (3) night: between 19.00 pm and 20.00 pm. The testing was done by looking the measurement result directly at the LCD and at Telegram display. The test results showed that the tool provides linear measurement results when compared to Thermo Hygrometer.

3.4. Testing of ammonia measurement

The emergence of respiratory disease cases in many chicken farms is caused by high levels of ammonia gas in chicken coop. Ammonia levels and their characteristics that can be recognized directly are as follows: (1) 20 ppm odor starts to smell, (2) 25-30 ppm odor is getting clearer, (3) 40-60 ppm sharp odor, 70-100 ppm can cause nasal irritation, (5) 200 ppm can cause eye irritation. The developed systems was tested in a stable environment for 14 days. Data is recorded three times a day: morning, afternoon and night. The results of observing NH3 ammonia gas in the coop can be seen in Table 1.

Table 1. Results of ammonia concentration measurement using developed systems for 14 days.

Day (i-th)	Ammonia concentration (ppm)		
	Morning	Noon	Night
1	-	101	99

2	98	102	107
3	100	105	103
4	98	109	110
5	100	104	111
6	110	116	120
7	111	119	115
8	120	125	122
9	122	117	119
10	105	104	109
11	117	119	100
12	90	100	105
13	95	110	116
14	112	120	116

3.5. Testing of telemonitoring and telecontrolling

The NodeMCU microcontroller used in this published research is equipped with an ESP 12E module that can be used for internet connection. To be able to connect to the internet, NodeMCU must be connected to a hotspot or router. Therefore the distance between the hotspot or router and the resulted system is very influential on the sucess of data communication since the bandwidth will decrease in accordance with the longer distance. The test results showed that the maximum distance of the hotspot with the device to stay connected is 20 meters with poor connection quality while the maximum distance for good connections is 10 meters.

In fact, in controlling fan and lights through the Telegram application on Android smartphone there is a delay between the the time commands of the Telegram application was send and the actual time of lamp and fan turned on. The time delay occurs because the transmission depends on the quality of the signal from the hotspot / router to the resulted systems. In general, time delay also occur because the distance from the hotspot / router with the device is getting further, the longer the distance the longer time delay will be longer. The test results for 5 trials with the distance less than 10 meters showed that the time delay has an average of 34.6 seconds.

4. Conclusion

Based on the test data have been done, it can be concluded that environment monitoring system and control of chicken coop for temperature, humidity, ammonia (NH_3) gas level have been successfully developed and connected to Android smartphone. The test results showed that for the measurement of humidity the resulted system has an error of 0.337% while for the measurement of the temperature the tool has an error of 4.32% when compared with the results of measurements using a Thermo Hygrometer. Test of wireless data communication suggested that hotspots / routers must be a maximum distance of 20 meters from the resuled system so that data transmission and control can run smoothly. Monitoring and control using Telegram application was successfully implemented with the time delay of 34.6 seconds in average.

Acknowledgments

Authors would like to thank to Research and Community Services of Ministry of Research Technology and Higher Education / (DRPM) *Kemenristekdikti* of Republic of Indonesia for providing research grant under PDUPT scheme year 2019.

References

[1] Ritz CW, Fairchild BD and Lacy MP 2004 *J. Appl. Poult. Res.* **13**:684
[2] Miles DM, Branton SL and Lott BD 2004 *J. Poul. Sci.* **10** 1650.
[3] Maliselo PS and Nkonde GK 2015 *Int. J. of Sci. & Tech. Res.* **4** 141
[4] Kristensen HH, and Wathes CM 2000 *W. Poul. Sci. J.* **56** 235
[5] Supriyono H and Hadi AN 2017 *Proc. of the Sec. Int. Conf. on Inf. and Com.* (Jayapura) p. 1.
[6] Supriyono H, Febriyanto ED and Harismah K 2019 *AIP Conf. Proc. 2114* **040014 (2019)**

IOP Conf. Series: Materials Science and Engineering **771** (2020) 012003 doi:10.1088/1757-899X/771/1/012003

(Surakarta) p. 040014-1.

[7] Kaur N, Mahajan R and Bagai D 2016 *Int. J. of Inn. Res. in Sci. Eng. and Tech.* **5** 9635

[8] Maddila TG, Santhossi S, and Muthukrishnan A 2017 *Int. J. of Eng. and Adv. Tech. (IJEAT)* **6** 168

[9] Xu L, Ren X, You Z, and Wu S 2014 *iJOE* **10** 37

[10] Alloui N, Alloui MN, Bennoune O and Bouhentala S 2013 *J. W. Poult. Res.* **3** 54

[11] Terzich M, Quarles C, Goodwin MA and Brown J 1998 *Av. Path.* **27** 566

IOP Conf. Series: Materials Science and Engineering **771** (2020) 012004 doi:10.1088/1757-899X/771/1/012004

Automatic Gate System Based on Water Flow Within the Intake of a Micro-Hydro Power Plant

P Pangaribuan[1], B S Aprillia[1*], A S Wibowo[1]

[1]Telkom University, School of Electrical Engineering, Bandung, Indonesia

**Corresponding author e-mail*: bandiyah@telkomuniversity.ac.id

Abstract. The micro-hydro power plant is an alternative to provide electricity especially in regions that are far from the PLN network. The micro-hydro power plant uses water flow to generate electricity. The micro-hydro power plant uses a gate that is utilised to distribute water to the turbine. However, the watergate is commonly controlled manually which results in drawbacks such as unfavourable stabilization and the need for direct monitoring that requires lots of constant expense as well as labour. The objective of this research is to make equipment that can work automatically in substituting a labourer's role in opening and closing the watergate according to the required water flow. A water flow sensor is used to measure the water flow, the result of the sensor reading is carried on to microcontroller (set up as Fuzzy Logic Control) and Controler sends the control signal to the actuator (DC motor). The Fuzzy Logic method as the watergate controller is used as the research's prototype system. Based on the data acquired by the research that has been done, the system is oscillated towards the change of the water flow. However, it's capable to approach the set value point. It can be seen from the water flow trial with a set point of 3 L/min the steady-state response of system undergoes at range of 2,92 L/min–2,95 L/min and with a set point of 3,5 L/min, the steady-state response of system undergoes at range of 3,44 L/min - 3,47 L/min.

1. Introduction

Indonesia has many potentials for water that are not utilized optimally, with around 75.000 MW and only 10,1% or as small as 7,572 MW being used. In 2014, the potential for mini/micro-hydro is as big as 1,2 GW for 240 micro-hydro power plant units with each generating 5 MW [1]. Micro-hydro is known as white resources because the installation of the power plant wields the resources given by nature and is environmentally friendly. The term micro-hydro is not standardized yet, but it is certain that micro-hydro power plant uses water as its main source of energy [2, 3]. The micro-hydro power plant is an electric generator system that can change water potential with a certain altitude and water flow to become electricity with the use of turbines and generators [4, 5].

Alternative energy can be divided into two types namely renewable and non-renewable energy. Micro-hydro is one of the renewable energy sources and an increasingly developed energy [6, 7]. For the use of an induction generator, there is the effect of the capacitor on the generator output voltage under load connected conditions. The greater the value of the load connected to the induction generator results in a large voltage shrinkage as well. Poor voltage regulation is one of the disadvantages of induction generators [8-10].

The micro-hydro power plant has a water gate in the inlet that is used to flow the water to the turbine. Watergates used commonly at the current moment are controlled manually and have some weaknesses namely unfavourable stabilization and the need for direct monitoring that requires lots of constant expense as well as labour [11].

The automatic device of a control system functions to minimize human error, while water is the most needed resources in daily life [12]. Almost all aspects of human have experienced rapid development [13]. This development is supported by advances in electronics and information technology. The use of tools that work manually has now been replaced with many automated systems and has been helping people do things.

The research is done to help solve the problem in the automatization of renewable energy in the form of micro-hydro power plant. Automatic controlling in the research is done to control the opening/closing of the watergate according to the established schedule using the Fuzzy Logic Controller.

2. System Design

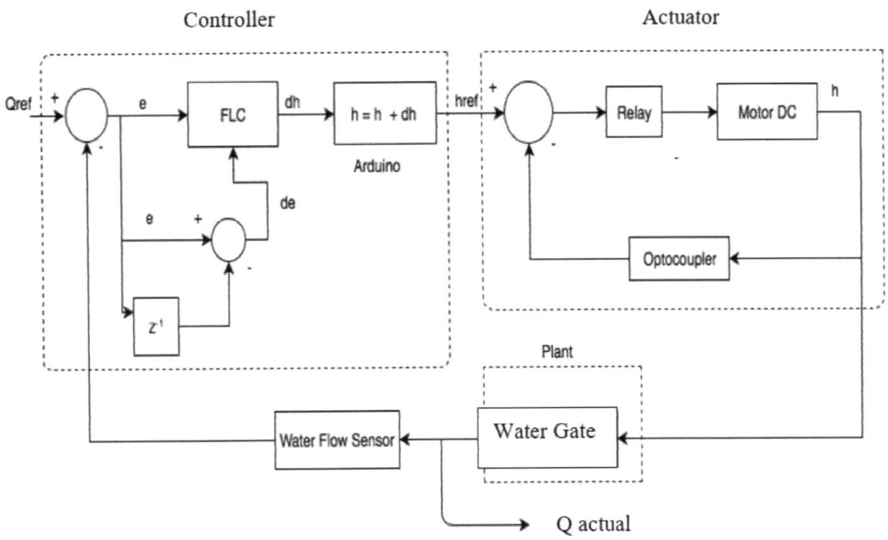

Figure 1. Overall system design.

Figure 1 explains the overall system design. The water flow sensor results will be processed by Arduino Uno and will then be using the fuzzy logic method. The value from the fuzzy logic method determines the elevation for the opening and closing of the watergate using a DC motor, as well as a velocity optocoupler sensor and a rotary decoder disc which controls the elevation position of the watergate whether it is according or not. The velocity optocoupler sensor outcome will be feedbacks, so Arduino Uno can control the direction of the DC motor, which rotates clockwise as well as counter-clockwise (gate opening or closing).

2.1. Software Design

The overall system flow diagram begins when turning on the Arduino Uno then initialisation of the pin and used variables, after that the water flow reading process using a water flow sensor. The results after reading the water flow sensor will be an input to the fuzzy logic control process which only process if the time is delayed > fuzzy interval. Delay time is a difference of the starting time of activating the Arduino Uno and the last time of doing the calculation for fuzzy, whereas the fuzzy interval is the time determined by the fuzzy to start the calculation. The outcome from the fuzzy logic is delta h that will be accumulated by h (the width of the door gap for now), therefore earning a new h value, the new h value is then used to determine the motor position so it can move according to the expected way. The diagram flow can be seen in Figure 2.

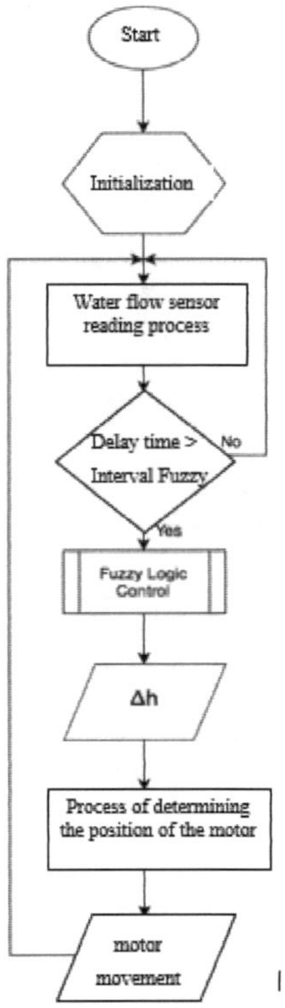

Figure 2. Software design

2.2. Fuzzy Logic Method

The main method used in this system is the fuzzy logic method. There are two input from the water flow data in this system called the error value and the delta error. The output from the fuzzy logic method is in the form of elevation of the watergate. A system using the fuzzy logic method for its main method could have several procedures that are linked.

Figure 3 shows the flow diagram on the fuzzy logic which starts with the input error value and delta error from the water flow sensor reading. The input value consists of the crisp input therefore the error value and delta error need to be changed into fuzzy value, after that the crisp value will be changed into fuzzy input value. Thereafter, the fuzzy input value will be converted into a fuzzy inference system process using the rules of "if-then" to a fuzzy output. Due to needing crisp value so it can be processed by the microprocessor so then the fuzzy output value is converted to a crisp output with the defuzzification so the crisp output becomes delta *h*.

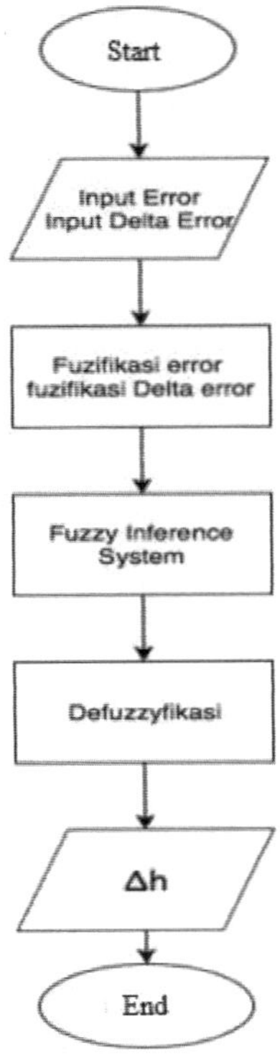

Figure 3. Flowchart of Fuzzy Logic

3. Results and Discussion

This subheading will discuss the testing and analysing of the device realisation. The testing is divided into 2 steps. The first step is to test the accuracy of the subsystem which are the sensor system and the actuator system on the plant. The second step is to test the integrated overall system. The testing is done to learn the reliability of the system that is made.

3.1. Water flow sensor calibration

The testing is done to learn about the accuracy of the water flow sensor. By comparing the water flow output manually and using the water flow sensor, these testing procedures are used for each measurement methods:

- Manual measurement by transferring the water in the plant within 10 seconds and collecting the water that passed the plant onto the measuring vial.
- Using water flow sensor by turning the Arduino Uno on and transferring the water in the plant so the propeller on the water flow sensor rotates. The output from the rotating propeller is converted into the velocity of the water flow.

Figure 4 shows the statistic descriptive analysis in the form of boxplot from the water flow measurement using a water flow sensor and the manual method. In general, the length in between different parts of the boxplot shows the spread and slope in the data and showing an outlier. Moreover, from bottom to top, each line represents the minimum value, lower quartile, upper quartile, median, and maximum value.

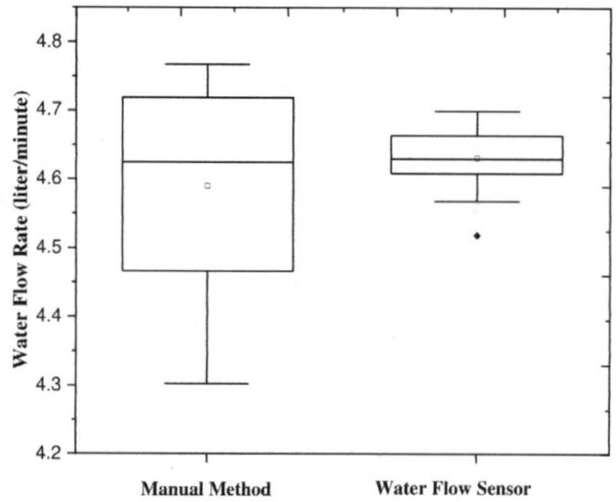

Figure 4. Water Flow Sensor Calibration

Figure 4 shows the output from the water flow measurement using the water flow sensor 30 times in 10 seconds. The average water flow acquired is 4,6316 litres/minute. Whereas manually for 30 times in 10 seconds using a stopwatch to fill up a measuring cup acquired a water flow of 4,7417 litres/minute. The long side from Figure 4 the water flow value measurement with the sensor is lower than water flow value from the manual measurements which means that the standard deviation of the water flow value sensor is lower than the manual measurement. The small value of the water flow sensor standard deviation shows the sensor's accuracy that is higher than the manual method.

3.2. Testing of watergate actuators

The testing purpose is to learn if the motor functions well in moving the watergate.

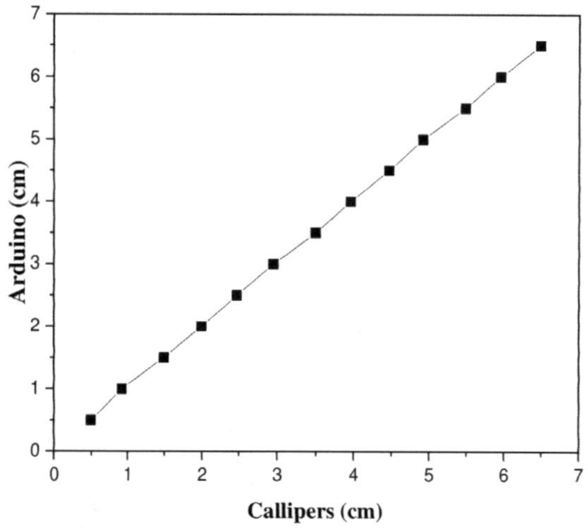

Figure 5. Opening the watergate actuators test.

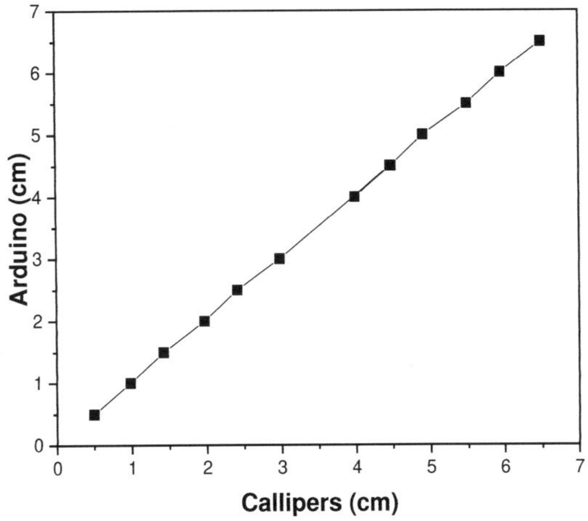

Figure 6. Closing watergate actuators test.

Figure 5 and 6 consecutively show the output of the watergate elevation testing manually using a calliper and automatically using the Arduino. By giving commands to the motor to move according to the elevation intended the testing is done, then the output of the elevation by the Arduino is compared to the watergate elevation by the calliper. The algorithm used in the actuator system calculation is stated below:

$$\text{watergate height output} = FK \times P \tag{1}$$

Whereas:

FK = Correction factor (335°/mm)
P = Watergate height orders

The correction factor is obtained by trial and error of the DC motor movements and the width of the watergate opening gap, therefore obtaining a certain degree value, below are the values acquired from the trial and error experimental results:

$$FK = \frac{D}{L} = \frac{19190°}{54\,mm} = 355°/mm \tag{2}$$

Whereas:

D = Degree value acquired from the gate movement results (19190°)
L = the width of the watergate gap after the DC motor moves (54 mm)

Figure 5 and 6 shows the test output from the calibrated actuator system, acquiring an average value from the percent error of 1,32% with the biggest error value of 0,089. The percent error value is obtained from (Error Value/measurement device) *100%. Therefore, it can be concluded as the accuracy value is 98,67% (100 – percent error average).

3.3. Testing of the overall of control system

Testing with the fuzzy logic method is done by determining the water elevation according to the desired setpoint, then the motor moves to open or close according to the already made fuzzy logic system output. The testing is done with 2 set points which are 3 litres/minute and 3,5 litres/minute.

3.3.1. First Testing at Setpoint 3.5 liter/minute

IOP Conf. Series: Materials Science and Engineering **771** (2020) 012004 doi:10.1088/1757-899X/771/1/012004

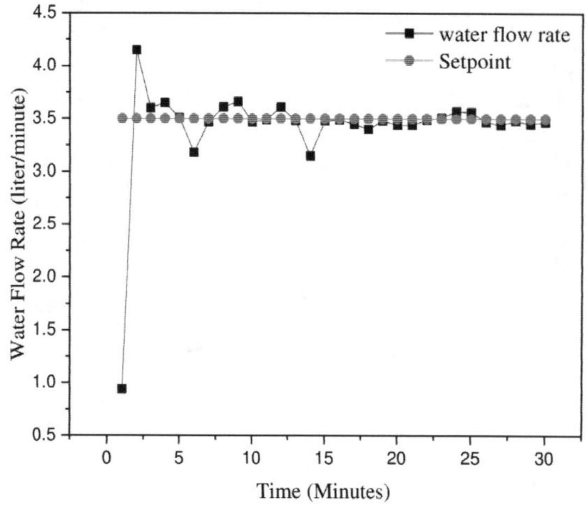

Figure 7. Water flow rate at setpoint 3.5 litres/minute.

Based on Figure 7, it can be seen that the characteristic of this system undergoes oscillation, this happens because of the changing water flow output. With those changes, the maximum overshoot can be seen happening in the second minute with 20,28%. On the second minute, the water flow achieved 4,15 L/min and starts to be in a steady-state after 26 minutes with a ranging water flow of 3,44 L/min – 3,47 L/min when the response comes in ±2% from the steady-state condition. The maximum percent overshoot value is acquired from the equation below:

$$\textit{Maximum percent overshoot} = \frac{c_{(tp)} - c_{(\sim)}}{c_{(\sim)}} \times 100\% \quad (3)$$

Whereas:

C_{tp} = The highest water flow value
$C\sim$ = The water flow value in its steady-state

3.3.2. Second Testing at Setpoint 3 liter/minute

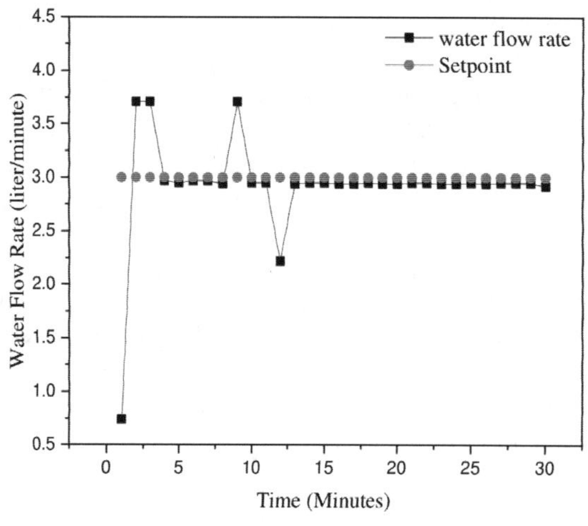

Figure 8. Water flow rate at setpoint 3 litres/minute.

Figure 8 shows that in minute-2 the maximum percent overshoot of 25,76% happening when the water flow becomes 3,71 l/m then the system experience steady state after 26 minutes with a range of flow water of 2,92 l/ m – 2,95 l/m when the responding system came in ±2% in its steady-state condition.

4. Conclusion

The application of the fuzzy logic in the system is done fairly well, this can be seen in the output of the flow water testing on 3,5 L/min and 3 L/min set points with a maximum percent overshoot not more than 25,76% and the respond system is able to approach the set points when in steady-state with flow water range of 3,44 L/min – 3,47 L/min on the testing of 3,5 L/min and flow water range of 2,92 L/min – 2,95 L/min on the testing of 3L/min when the respond system come in at ±2% from the steady-state condition.

References

[1] Kumara, N.S., *Telaah Terhadap Program Percepatan Pembangunan Listrik Melalui Pembangunan PLTU Batubara 10.000 MW.* Majalah Ilmiah Teknologi Elektro, 2009. **8**(1).
[2] Fraenkel, P., et al., *Micro-hydro Power*, in *Micro-hydro Power: A guide for development workers.* 1991, Practical Action Publishing. p. 1-127.
[3] Monition, L., M. Le Nir, and J. Roux, *Micro hydro-electric power stations.* 1984.
[4] Smith, N., *Motors as generators for micro-hydro power.* 1994: Intermediate Technology Publications London.
[5] Arismunandar, A. and S. Kuwahara, *Buku pegangan teknik tenaga listrik.* 1974: Association for International Technical Promotion.
[6] Akella, A., R. Saini, and M.P. Sharma, *Social, economical and environmental impacts of renewable energy systems.* Renewable Energy, 2009. **34**(2): p. 390-396.
[7] Hafez, O. and K. Bhattacharya, *Optimal planning and design of a renewable energy based supply system for microgrids.* Renewable Energy, 2012. **45**: p. 7-15.
[8] Bhadauria, P. and M. Ilyas, *Voltage and Frequency Control of Asynchronous Generator for an Isolated Wind Energy Conversion System.*
[9] Tischer, C.B., et al., *Proportional-resonant control applied on voltage regulation of standalone SEIG for micro-hydro power generation.* IET Renewable Power Generation, 2017. **11**(5): p. 593-602.
[10] Scherer, L.G., R.V. Tambara, and R.F. de Camargo, *Voltage and frequency regulation of standalone self-excited induction generator for micro-hydro power generation using discrete-time adaptive control.* IET Renewable Power Generation, 2016. **10**(4): p. 531-540.
[11] Firmansyah, E. and M. Isnaeni, *Pengendalian Beban Generator Secara Otomatis Dengan Algoritma Pid Pada Pltmh Berbasis Plc.* Jurnal Teknologi, 2012. **5**.
[12] Jeranyama, P., *Irrigation water management.* Cranberry 2009 chart book, management guide for Massachusetts, 2009: p. 53-54.
[13] Pramudita, D. and A. Ulinuha, *Protoype Sistem Buka Tutup Pintu Air Otomatis pada Persawahan Berbasis Arduino Uno.* 2017, Universitas Muhammadiyah Surakarta.

Design and Implementation of Battery Management System for Portable Solar Panel with Coulomb Counting Method

A R Afif [1], B S Aprillia[1,*], W Priharti[1]

[1]Telkom University, School of Electrical Engineering, Bandung, Indonesia

[*]e-mail: bandiyah@telkomuniversity.ac.id

Abstract. Secondary batteries are commonly used as the storage of energy produced by solar panels. However, the utilization of a battery without proper management can cause damage due to overcharging and over-discharging. This study aims to design a battery management system (BMS) on a Valve Regulated Lead-Acid (VRLA) battery. The method used was the battery State of Charge (SOC) estimation using Coulomb Counting (CC) method. The results showed that the BMS was successfully designed and implemented to automatically cut-off the current when the SOC value is 100% (charging limit) and 20% (discharging limit).

1. Introduction

The energy production of a solar panel can be optimized using an energy storage system such as a battery [1, 2]. A battery itself can be classified into two different types, i.e. primary and secondary. An example of a secondary battery that is commonly used is the Valve-Regulated Lead-Acid (VRLA) type. However, this type of battery has several drawbacks such as overcharging and over-discharging. In a long time of usage, these drawbacks could lead to the degradation of the battery lifetime and working quality.

A Battery Management System (BMS) is required to avoid battery damage due to battery incorrect usage. Several methods of BMS are available, one of them is by estimating the value of battery State of Charge (SOC). In this study, the Coulomb Counting (CC) method was used as the SOC estimation method. The basic principle of the CC method is to accumulate the electrical charge that is going in and out of the battery [3, 4]. The temperature and electric current could also affect the accuracy of the SOC [5, 6].

According to the problem stated above, this study aims to design a battery management system on Valve Regulated Lead-Acid (VRLA) battery using State of Charge (SOC) estimation with Coulomb Counting (CC) method. Several indicators were displayed on the battery monitoring system, i.e. the battery voltage, current and the percentage of SOC value. The CC method was selected due to its accuracy in determining the SOC. The accurate SOC is a crucial parameter required in battery monitoring and management system [7-9].

2. System Design

2.1. Hardware Design

In this study, a BMS for the portable solar panel by estimating the value of battery SOC using the CC method was designed. This system monitored the battery condition by displaying the percentage of SOC value calculated from the data obtained from the measurement. The measurement was divided into two-phase: charging and discharging. A 340 W solar panel was used as the power supply in the charging phase and a 5 W AC lamp and an inverter were used as the load in the discharging phase. The circuit model of the BMS for charging and discharging phase are respectively shown in Figure 1 and 2.

IOP Conf. Series: Materials Science and Engineering **771** (2020) 012005 doi:10.1088/1757-899X/771/1/012005

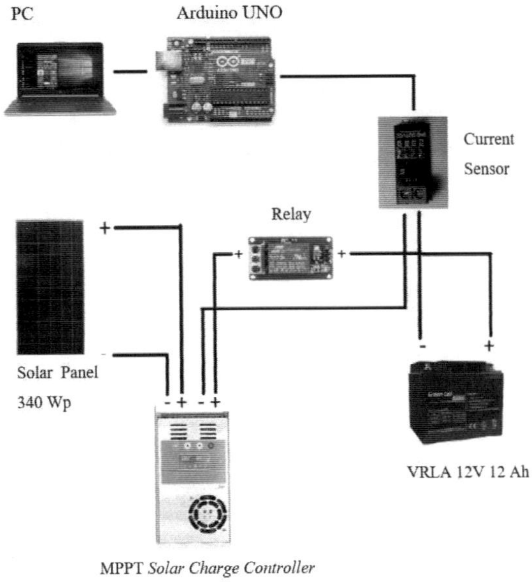

Figure 1. Design Hardware Charging System

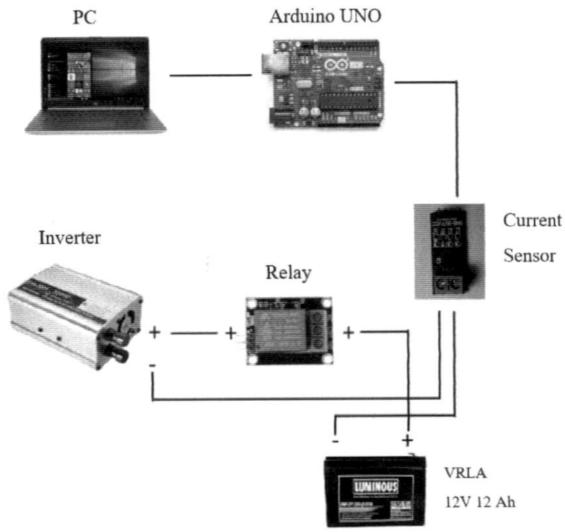

Figure 2. Design Hardware Discharging System

2.2. Coulomb Counting

Coulomb counting (CC) was the method used to estimate the SOC in this study. The basic principle of the CC method is to accumulate the electrical charge that are going in and out of the battery. The formula to calculate the operating charge, Q_{out} of a battery is

$$Q_{out} = I \ x \ t \tag{1}$$

where:

Q_{out} = operating charge

I = operating current

t = measurement time

IOP Conf. Series: Materials Science and Engineering **771** (2020) 012005 doi:10.1088/1757-899X/771/1/012005

Eq 1 was used to calculate the amount of charge at a certain condition. However, to estimate the battery SOC, the value of the total charge, Q_{total} is required. The formula is

$$Q_{total} = Q_{max} - Q_{min} \tag{2}$$

Hence, the estimation of SOC can be calculated as

$$SOC(t) = SoC(t_0) - \left(\frac{Q_{out}}{Q_{max}}\right) x\ 100\% \tag{3}$$

Where:
$SoC(t_0)$ = initial SOC before charging/discharging phase
Q_{out} = operating charge of the battery
Q_{max} = maximum charge of the battery

2.3. Software Design

Software design was required to integrate the system with the hardware. Arduino Uno was used as the microcontroller or the brain of the system. Similar to the hardware, the software also divided into two-phase, i.e. charging and discharging. The flowchart of the charging and discharging phase are shown in Figure 3 and 4, respectively.

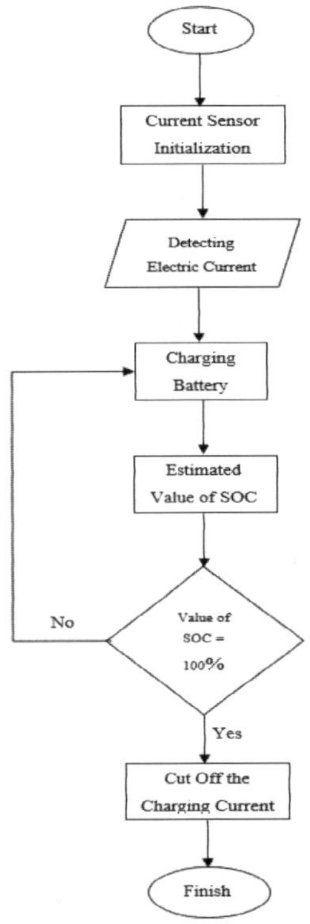

Figure 3. Flow Chart Charging System

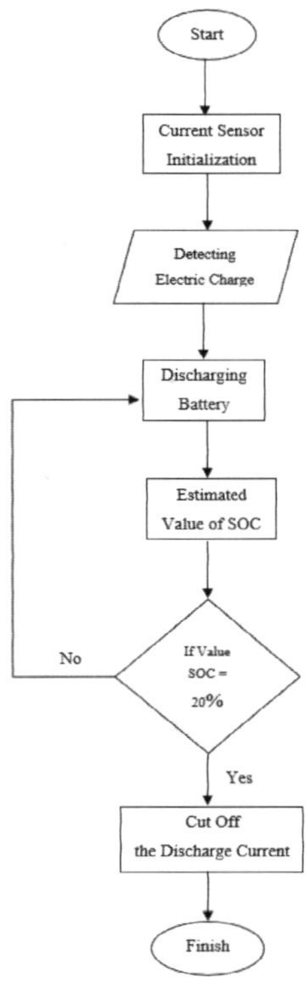

Figure 4. Flow Chart Discharging System

3. Results and Discussions

3.1. Current sensor calibration

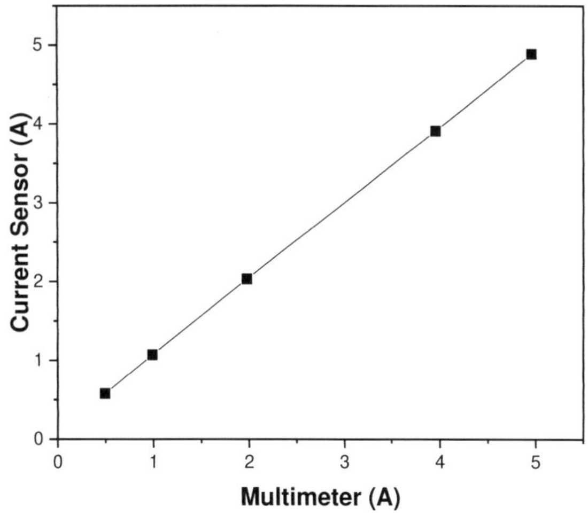

Figure 5. Current sensor calibration

Figure 5 shows the calibration result of ACS current sensor. The calibration was carried out using a current of 1 – 5 Ampere, with 30 repetitions for each current. The maximum current of 5 Ampere was used due to the maximum output current of the power supply. Based on these results, it can be concluded that the current sensor is accurate and in good condition.

3.2. Voltage sensor calibration

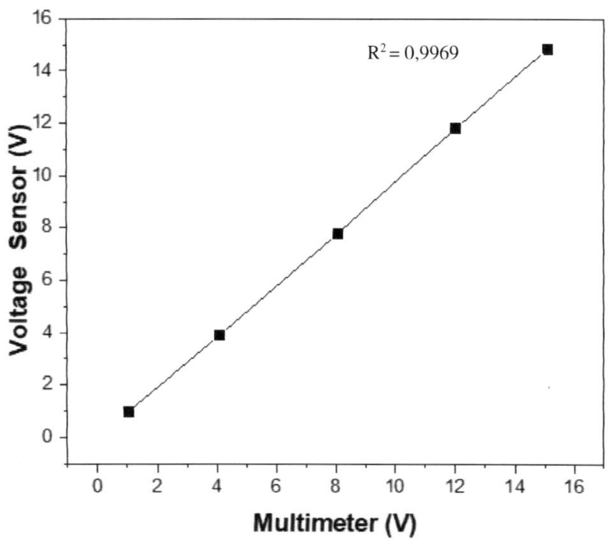

Figure 6. Voltage sensor calibration

The calibration of the voltage sensor was carried out using 5 different voltages with 30 repetitions for each voltage. The calibration obtained a value of $R^2 = 0{,}996$ as shown in Figure 6. It means that the voltage reading from the sensor matched the voltage reading from the standard instrument (multimeter) with an accuracy of 99,6%. It can be concluded that the voltage sensor is accurate and well function.

3.3. Battery Charging Phase

The solar panel of 340 W was used as the power supply in the battery charging phase. The time required to fully charged the battery was 1,07 hours with the initial SOC value 20% and maximum SOC value 100% (when the battery is fully charged).

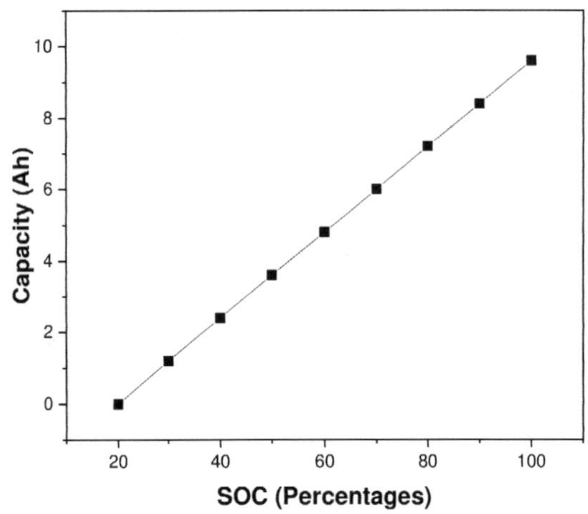

Figure 7. The graphical representation of battery charging capacity versus SOC

Figure 7 illustrates the results of battery charging phase when the value of the initial SOC is 20% and progressively increases with the increase in battery capacity. In the battery charging phase, the battery is fully charged when the SOC value reached 100%.

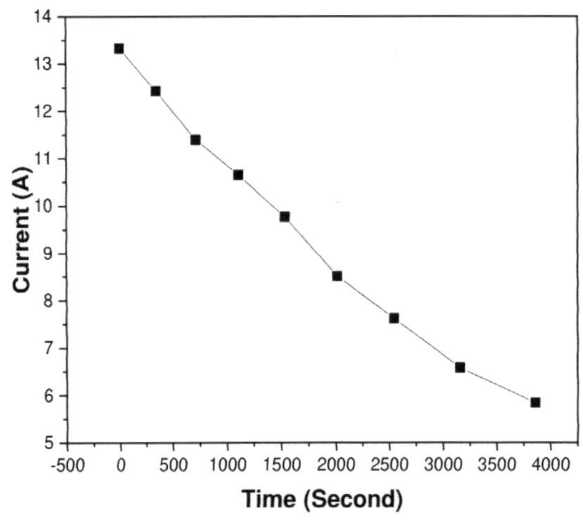

Figure 8. The graphical representation of current versus time during charging

In this system, the Maximum Power Point Tracking (MPPT) charge controller was utilized to extract the maximum available power (output current and voltage) from the solar panel [5]. In the charging phase, output current value decreased with time as shown in Figure 8. It occurred due to the reduction of charging current by MPPT when the battery is fully charged. This condition is required to minimize overcurrent during battery charging. Hence in this system, the charging current was automatically cut-off when the battery is fully charged or the value of SOC reached 100% [6].

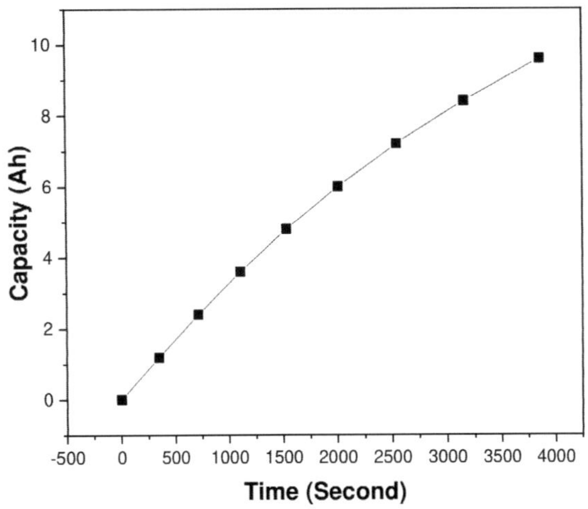

Figure 9. The graphical representation of battery charging capacity versus time

Figure 9 illustrates the battery capacity that was increased with time during charging. The rate of battery charging capacity is directly proportional with the current. The higher light intensity gave greater current [10] resulted in a faster rate of battery charging capacity. The maximum capacity of 9,6 Ah was charged in the charging phase as the output of 9,6 Ah was used during the discharging phase.

3.4. Battery Discharging Phase

A 5 W AC lamp and an inverter was used as a load during the battery discharging phase. The graphical representation of battery discharging capacity versus SOC is shown in Figure 10.

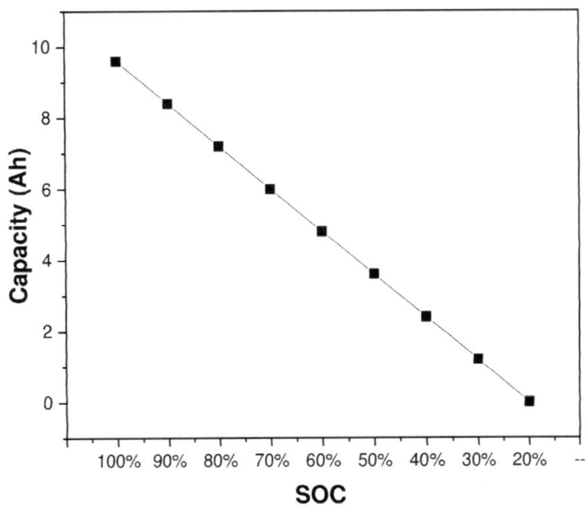

Figure 10. The graphical representation of battery discharging capacity versus SOC

Figure 10 shows that the SOC value was decreased with the reduction of battery capacity. The current was cut-off when the SOC value reached 20% or the amount of discharged battery capacity was 9,6 Ah.

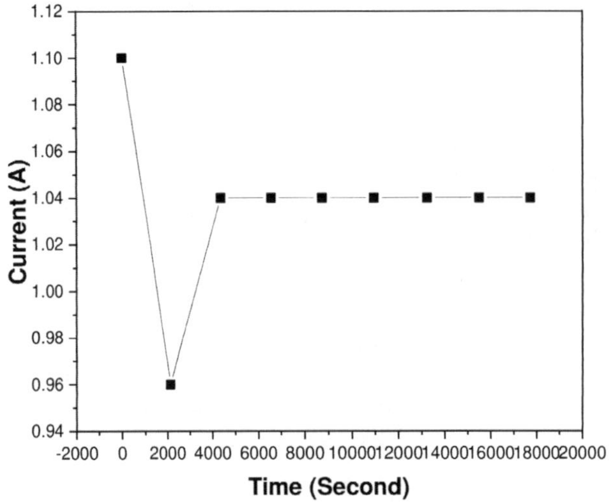

Figure 11. The graphical representation of current versus time during discharging

The reading of discharging current was constant at 1,04 A. The selection of load is important in the BMS design since it will affect the duration of discharge. The greater load gave a faster rate of battery discharging capacity, resulted in a shorter duration of discharging.

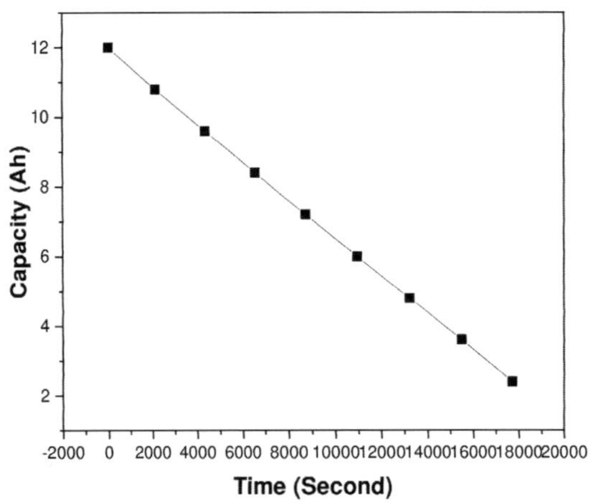

Figure 12. The graphical representation of battery discharging capacity versus time

The battery capacity decreased with time when connected to a load as shown in Figure 12. The time required to discharged 9,6 Ah of battery capacity was 17.712 seconds or 4,92 hours.

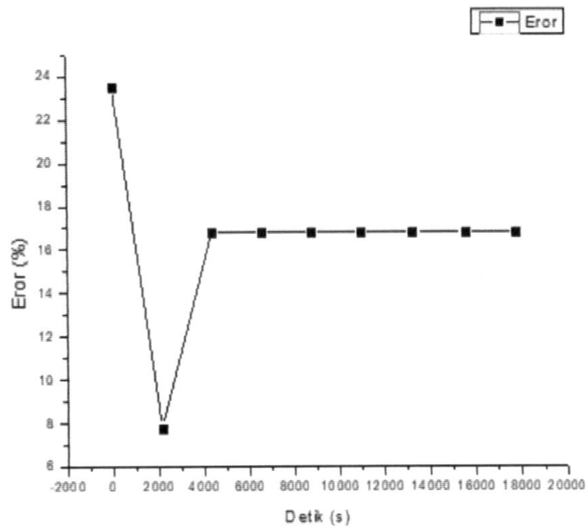

Figure 13. The graphical representation of error versus time

Using an operating current of 0,89 A during the discharging phase, SOC reading using CC method obtained quite significant value of error i.e. 23,5% at the first reading. To minimize this error and to increase the accuracy of SOC, the CC method can be combined with the I-observer method. The value of error using the combination of these two methods can be reduced to <3% [1].

4. Conclusions

A maximum battery capacity of 9,6 Ah was reached during charging phase with the required time of 1,07 hours using 340 W solar panel as the power supply. A battery capacity of 9,6 Ah was used during the discharging phase with the required time of 4,92 hours using 5 W AC lamp and an inverter as the load. The minimum and the maximum error of 7,8% and 24% respectively were obtained during discharging with an operating current of 0,89 A using CC method. The battery management system in this study was successfully designed and implemented to automatically cut-off the current when the SOC value is 100% (charging limit) and 20% (discharging limit).

References

[1] Aprillia, B.S., D.K. Silalahi, and M.A.F. Rigoursyah, *Desain Sistem On-Grid Energi Terbarukan Skala Rumah Tangga Menggunakan Perangkat Lunak HOMER.* JTIM: Jurnal Teknologi Informasi dan Multimedia, 2019. **1**(3): p. 174-180.
[2] Nurfaidah, Y., et al. *Analysis of smart house power savings with on-grid photovoltaic power system.* in *Journal of Physics: Conference Series*. 2019. IOP Publishing.
[3] Prajapati, V., et al. *A literature review of state of-charge estimation techniques applicable to lithium poly-carbon monoflouride (LI/CFx) battery.* in *India International Conference on Power Electronics 2010 (IICPE2010)*. 2011. IEEE.
[4] Samadani, S.E., R.A. Fraser, and M. Fowler, *A review study of methods for lithium-ion battery health monitoring and remaining life estimation in hybrid electric vehicles*. 2012, SAE Technical Paper.
[5] Duryea, S., S. Islam, and W. Lawrance. *A battery management system for stand alone photovoltaic energy systems.* in *Conference Record of the 1999 IEEE Industry Applications Conference. Thirty-Forth IAS Annual Meeting (Cat. No. 99CH36370)*. 1999. IEEE.
[6] Feng, F., R. Lu, and C. Zhu, *A combined state of charge estimation method for lithium-ion batteries used in a wide ambient temperature range.* Energies, 2014. **7**(5): p. 3004-3032.

IOP Conf. Series: Materials Science and Engineering **771** (2020) 012005 doi:10.1088/1757-899X/771/1/012005

[7] Partovibakhsh, M. and G. Liu, *An adaptive unscented Kalman filtering approach for online estimation of model parameters and state-of-charge of lithium-ion batteries for autonomous mobile robots.* IEEE Transactions on Control Systems Technology, 2014. **23**(1): p. 357-363.
[8] Nugroho, A., et al. *Battery state of charge estimation by using a combination of coulomb counting and dynamic model with adjusted gain.* in *2015 International Conference on Sustainable Energy Engineering and Application (ICSEEA)*. 2015. IEEE.
[9] Ferdausi, M., *designing smart charge controller for the solar battery charging station (SBCS)*. 2012, BRAC University.
[10] Aprillia, B.S., M.R. Zulfahmi, and A. Rizal, *Investigasi Efek Partial Shading Terhadap Daya Keluaran Sel Surya.* Jurnal Elektro dan Mesin Terapan, 2019. **5**(2): p. 9-17.

Design of Automatic Switch System of Residential Load From Solar Cell and Power Plant Resources using Neural Network

D K Silalahi[1,*], B S Aprilia[1], W Priharti[1], K Kumillayly[1], S Saidah[2]

[1]Electrical Engineering, Faculty of Electrical Engineering, Telkom University
[2]Telecommunication Engineering, Faculty of Electrical Engineering, Telkom University

*desrikristina@telkomuniversity.ac.id

Abstract. Residential loads are electronic equipment that is often used at home. Residential loads are supplied from two sources, they are solar panel and the State Power plant. Selecting the supply load source, so an automatic switch system is needed from the load. Optimizing residential load resources and avoiding overloading, load balancing techniques are needed. The switch system uses a relay, determined by the solar panel power output. This research discusses the design of automatic load switch systems for residential loads from solar panels and State Power plant using Artificial Neural Networks (ANN). ANN control system that is arranged using ANN Backpropagation consists of 4 inputs, four hidden layers, each consisting of 4 neurons and one neuron in the output layer. In this research, to determine the network that has been formed to provide changes of load. The results of testing that the parameters used to get the smallest error rate in the process of setting an automatic power load switch is best to use a number of repetitions of 2000 times with an error percentage 5.3%. Artificial Neural Networks can experience convergent failure or not close to output because the initial guess is not good. Initial experiments with actual solar panel data with as many as 98 data produced non-convergent output values. The solution to improving the initial experiment is to simplify learning data 40 data produces convergent output.

1. Introduction

The requirement for electrical energy is a major need for all people in the world. Fossil energy fuels are getting thinner so that continuous electricity is needed. Therefore, many efforts were made to create alternative power plants to support the increasing need for electricity supply. Indonesia is a tropical country that has high solar energy strength with radiation per day with an average of 4.8 kWh /m2 /day [1]. This power can be used as an environmentally friendly renewable energy source, one of which is by utilizing solar panel energy [2]. Solar panels are semiconductor materials that convert solar energy into electricity[3].

Solar panels used as an alternative energy source can be applied to household loads. Household expenses are electronic equipment that is often used at home [4]. Reduce dependence on electricity sources from the power plant of electricity supply continuous. One way to burden residential load can choose between two sources of electricity namely from solar panels and power plant. The selection of these two sources required a system of automatic residential load switches [5]. This system as a regulator of the electrical load power supply, so that the switch of electrical energy supply to the load can work automatically by prioritizing solar panel energy. Increasing the use of household loads and optimizing resources and avoiding overloading, the solution to these load problems is to load balancing techniques. With a switch system using a relay [6], determined by the solar panel power output.

Research that has been done there is a load power switch with a relay using a PLC [7]. The design of this research is based on the implementation of household load switches from the solar panel and power plant sources that will produce alternative energy optimally. Based on this research, this study designed a load power automatic switch equipped with a relay as a household load automatic switch from a solar panel source and power plant using an Artificial Neural Network.

Automatic switches function to control the two-source electricity system between solar panel electricity and power plant [8]. The load is supplied from one source at a time. The automatic switch works automatically by detecting the availability of solar panel sources, if the electrical energy of the solar panel can supply the load, the power plant switch will be off and the load is only supplied by the solar panel electrical energy. The opposite condition, if the solar panel electricity cannot supply the household load, the power plant switch will be on and the load will only be supplied by power plant electricity.

Artificial Neural Networks are simulations of the human brain that have the ability to learn [9]. In the learning process consists of an input layer and the output layer is inserted into the artificial neural network, then the network will be studied to provide acceptable answers. Artificial Neural Networks have the main characteristic that is the ability to learn. Learning in artificial neural networks can be interpreted as an adjustment process according to the interconnect weighting parameters possessed by the cell [10]. The learning process will stop if the error value is considered to be small enough for all pairs of learning data. The network that is doing the learning process is called in the training phase[11].

2. Experimental Design

The design of this system is divided into three parts, namely the system block diagram design, hardware design and software design. Block diagrams are used to design systems in general, hardware design includes the design of electronic systems, while software design uses Python programming as a programming language for Arduino Uno microcontrollers for reading switches to displaying data. The input current and load voltage to do an automatic switch process with the output of an electric energy source changing AC current. The working principle of this system is a residential load with different power capacities connected to an automatic switch to regulate the electrical load to determine the source of electricity used by first determining the source of electricity for solar panels and power plant in accordance with the required characteristics and specifications.

In this hardware design explains the automatic switch system using technology with Artificial Neural Network Algorithm needed hardware to perform system tasks easily such as solar panels, batteries, inverters, microcontrollers, relays, current and voltage sensors and panel boxes. The device used to store electricity from solar panels is a battery. The device for charging batteries is to use the solar charge controller. The device to convert DC current to AC is an inverter [12]. The voltage sensor is used to detect voltage while the current sensor is used to detect current. The microcontroller used is Arduino Uno. The load used is a household load or AC [13].

In general, the duration of solar radiation for 6 hours is 70%. Then the solar panel power received is 469 Wh. Total power consumption a day needs to be added by 20% to be used for three-load devices namely fans, rice cookers and lamps. Other than solar panels, namely inverters as converters of DC to AC currents because in general household appliances use AC currents, and charge controllers to cover current to the battery if the voltage is over the battery and stop taking current from the battery if the battery is empty.

This software architecture and interface describe the software to be realized. The software used is python software for learning the Artificial Neural Network algorithm before it is converted into Language C. Arduino Uno Programming is used to be realized into automatic switches. First, the system will read the voltage and current on the solar panel, then calculate the total power available on the solar panel. Also carried out measurements of the current at the load to calculate the load power where the power is the current times the voltage 220V. Based on the results of the calculation of the power on the solar panels and the load then read the relay to turn on or not on each load and determine which load should be turned off. The load will adjust the power supply of the solar panel. When the voltage of the solar panel drops, the source switches to the power plant. If the panel has increased current and voltage is sufficient, the load will be returned to the solar panel supply. Figure 1 is the design of the system block diagram.

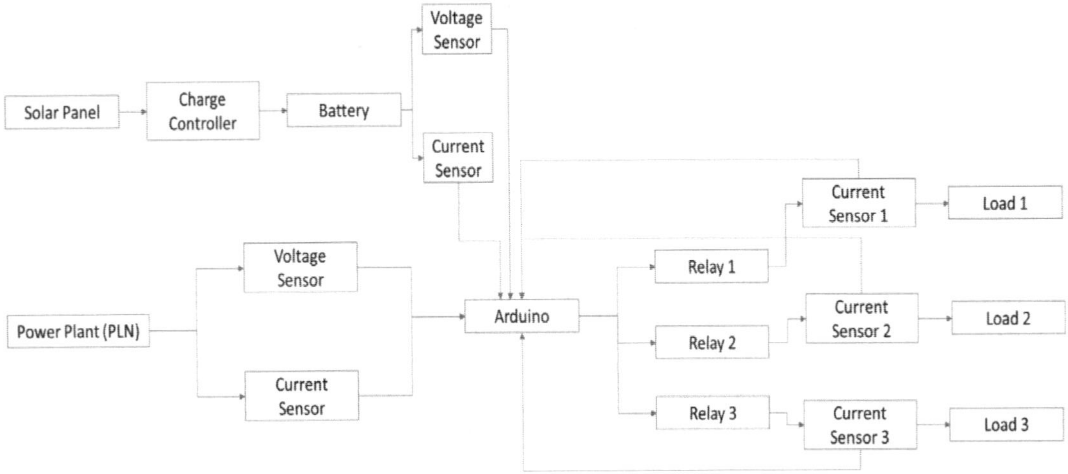

Figure 1. System Block Diagram

3. Result and Discussion

Current sensor calibration, the purpose of calibration of the current sensor is to see the accuracy of the output of the sensor so that when using an error does not occur. At the calibration of the current sensor is done comparing the output current of the power supply with the current sensor reading. This research uses three voltage sensors ACS712 current sensor separately. The current sensor calibration step has 3 main pins namely Vcc, GND and Vout. Vcc and GND pins are connected to Vcc and GND pins on Arduino Uni. While the Vout pin from the sensor is connected to pin A0 on Arduino Uno. The results of calibration of current sensor 1 have an average error of 1.38% while current sensor 2 has an average error of 1.25% and for current sensor 3 has an average error of 1.32%. The calibration results can be concluded that the current sensor has a high accuracy so that it can be used in testing.

Voltage sensor calibration, the purpose of the voltage sensor calibration is to see the accuracy of the output of the sensor so that when using an error does not occur. The voltage sensor has an average error of 2.0%. The data contained a large coma difference in the value of the reading of the voltage sensor with the calculated value. This happens because the ZMPT101B voltage sensor has a sensitivity of 100mV / A so the sensor cannot detect voltages below 1V properly. The results of the calibration, the voltage sensor has an average error of 01.00%. because the sensor has an error rate below 5%, it is suitable for use in testing.

Testing Phase, artificial neural network algorithm have a problem that can experience convergent failure or not close to the output. As a result, the number of iterations will produce less than expected or far-reaching output. Other things, the amount of data is too much, not more detailed. So, if there is too much data but the details lack artificial neural networks confuse because each calculation per line produces a difference that is always large. weight correction does not produce output close to the expected output. A large amount of data makes the experiment more complex, while a large amount of details helps the accuracy process.

At this stage, the data input matrix is sized 10 x 4, where the row is the amount of data while the column consists of solar panel power, load 1, load 2, load 3. So the total learning data is 10 x 4 = 40 pieces of data as training data and the output is the switch closure for each load. This data is then read by the python program to do the training process where at this stage a test is made which is the number of repetitions of training 2000 times. The results of this training are compared with test data. In solar panel power, the voltage ranges from 10-14V with a current of 35A so the solar panel power ranges from 350-460 watts.

Table 1. Simplification of the Amount of Data

Solar Panel Power (Watt)	Load 1	Load 2	Load 3	Load Index
350	350	40	25	1
385	350	40	25	1
420	350	40	25	0

455	350	40	25	0
350	40	350	25	2
385	40	350	25	2
420	40	350	25	0
455	350	40	25	0
350	25	40	350	3
350	25	40	350	3

So, be closer to the convergent results. First, the load and power of the solar panels are made into an index group, the panel power has an index of 1 or 2, meaning that 1 solar panel power is weak and 2 strong solar panel power. Load power has an index of 1.2 or 3 meaning 1 small load power, 2 medium load power and 3 high load power. Output index ranges from 0,1,2 and 3, when the learning index is normalized to 0, 0.333, 0.6667, 1. It means that if 0, then nothing is turned off, if 0.33 means load 1 dies, if 0.6667 means load 2 is dead, and if 1 means load 3 is dead.

To anticipate the range that experiences differences, if the output of the neural network algorithm is below 0.1 then the load index = 0, meaning that all load relays are on. If the index is between 0.1 and 0.3 then load 1 is dead, if the index is between 0.3 and 0.7 then load 2 is dead and if the index is above 0.7 then load 3 is dead. The results of this training show that the final synaptic range at 1000 repetitions is at an average of 0.406 while at 2000 iterations at an average of 0.404 and at a repetition of 3000 times at an average of 0.405. by multiplying the normalization value the accuracy results are as follows:

Table 2. Results of Error and Accuracy Percentage

Iteration	Error (%)	Accuracy (%)
Iteration 1000 times	7	93
Iteration 2000 times	5.3	94.7
Iteration 3000 times	6.5	93.5

In Table 2. Accuracy results obtained with 2000 repetitions of data are good accuracy. At iteration 2000 times get accuracy is 94.7%. This can be caused by the nature of the artificial neural network process which if more and more repetitions will lead to many error values while on repetitions that are too small will result in more errors. So, artificial neural networks algorithm at iteration 2000 times is the best regulation automatic power load switches.

4. Conclusion

The parameter used to get the smallest error rate in the process of regulating automatic power load switches is best to use 2000 repetitions with an accuracy of 5.3%. Artificial Neural Networks can experience convergent failure or not close to the output because the initial guess is not good. Initial trial results with many but complex data artificial neural networks with solar panel power data and the actual load with a total of 98 data result in non-convergent output values. Results of experiments with solutions: Artificial neural network with solar panel power data which is simplified into an index with a total of 40 data learning data, producing convergent output.

References

[1] A. Setiawan and E. A. Setiawan, "Optimization of a Photovoltaic Power Plant In Indonesia With Proper Tilt Angle and Photovoltaic Type Using a System Advisor Model," *Int. J. Technol.*, vol. 8, no. 3, pp. 539–548, 2017.

[2] O. Inganäs and V. Sundström, "Solar energy for electricity and fuels," *Ambio*, vol. 45, no. 1, pp. 15–23, 2016.

[3] C. J. Cleveland and C. Morris, *Handbook of Energy. Volume II_ Chronologies, Top Ten Lists, and Word Clouds*. 2014.

[4] A. J. Collin, G. Tsagarakis, A. E. Kiprakis, and S. McLaughlin, "Development of Low-Voltage Load Models for The Residential Load Sector," *IEEE Trans. Power Syst.*, vol. 29, no. 5, pp. 2180–2188, 2014.

[5] S. Dudhate, A. Attargekar, D. Desai, and A. Patil, "Power Supply Control from Different Sources," *Int. J. Sci. Eng. Res.*, vol. 7, no. 1, pp. 484–487, 2016.

[6] J. G. Kolo, "Design and Construction of an Automatic Power Changeover Switch," *AU J. Tech. Rep.*, vol. 11, no. 2, pp. 1–6, 2007.

[7] G. Seshagiri Rao, S. Raghu, and N. Rajasekaran, "Design of Feedback Controller for Boost Converter Using Optimization Technique," *Int. J. Power Electron. Drive Syst.*, vol. 3, no. 1, pp. 117–128, 2013.

[8] B. Onipede, S. Joseph, and O. Odiba, "Developing an Automatic Switch for Home or Industrial Power Supply Changeover," *J. Appl. Sci. Technol.*, vol. 21, no. 4, pp. 1–7, 2017.

[9] S. Shanmuganathan, "Artificial Neural Network Modelling," *Stud. Comput. Intell.*, 2016.

[10] R. Rojas and R. Rojas, "The Backpropagation Algorithm," in *Neural Networks*, 1996.

[11] K. Mehrotra, C. Mohan, and S. Ranka, *Elements of Artificial Neural Networks*. 2019.

[12] O. Akinyele *et al.*, "Construction and Evaluation of a Power Inverter," *Int. J. Latest Technol. Eng. Manag. Appl. Sci.*, vol. 4, no. 7, pp. 1–5, 2015.

[13] R. K. Chauhan, B. S. Rajpurohit, R. E. Hebner, S. N. Singh, and F. M. Gonzalez-Longatt, "Voltage Standardization of DC Distribution System for Residential Buildings," *J. Clean Energy Technol.*, 2015.

DC Power Supply using Digital Potentiometer on Switching Regulator Module Monitored and Controlled via NI LabView

F Nugraha[1,*], R A Priramadhi[1], M Ramdhani[1]

[1]Faculty of Electrical Engineering, Telkom University. Jl. Telekomunikasi Jl. Terusan Buah Batu, Sukapura, Kec. Dayeuhkolot, Bandung, Jawa Barat 40257

*fachryo97@gmail.com, rizkia@telkomuniversity.ac.id, mohamadramdhani01@gmail.com

Abstract. Smart Controlled Adjustable Power Supply is an automated power supply device in which the Voltage output from the device can be controlled using Digital Potentiometer as a voltage feedback manipulator and can be monitored using PC. Voltage and output current can be controlled in accordance with the specifications of the desired voltage 0-30V DC with 0.01 V resolution. This Study will show how Digital POT can change the output voltage of the LM2576 Module and show the correlation between the BIT number of POT and the Voltage output. Afterward, the device will be controlled and monitored via NI LabView so it can easily be accessed by everyone and can compete in industry 4.0 due to its accessibility. There will be some test runs to prove the designed device is working properly and match the desired spesifications.

1. Introduction

Nowadays Voltage DC power supply was used in many fields and power electronics are used to control the energy flow process by giving voltage and current in an optimal form and suitable for loads [1]. Traditional Power supply whose voltage regulating method was manually mechanical operation. In order to be able to adapt to industrial automation, the voltage regulating method must be upgraded by changing the mechanical operation onto automatic operation. This paper will show the design of a power supply DC that can be controlled. The keys of this method were the LM2576 module and digital potentiometer AD5292. LM2576 module is the power supply module that can produce output 1.25V – 30 V DC voltage with an input of 30DC constant voltage. AD5292 is connected to the module for controlling the output voltage of the module and communicate with the Arduino serial. So the main purpose of this paper was to design a new method for regulating dc voltage output of a power supply. The controlled adjustable power supply will be controlled and monitored via NI LabView. The voltage output of the device will be shown in graph voltage vs time, also the desired output will be controlled using a graphical interface that available in NI LabView.

2. Design of the Power Supply

2.1. Power Supply Specification

The power supply specifications were determined in the beginning so the author could choose what kind of voltage regulator module that can be used. Specs were determined near to PXI programmable power supply from National Instruments.

Table 1. Design Specifications

Num.	Parameter	Units
1	Input Voltage	30 V DC
2	DC output range	0 – 30 V DC
3	Max current output	3 A
4	Ratio level voltage	XX,X / 0.1 V per Level

2.2 Designing System

Design is made to suit all of the determined specifications. The switching regulator module picked due to high efficiency and has an adjusted voltage feature [2][3]. Another component to design was the voltage sensor and the current sensor.

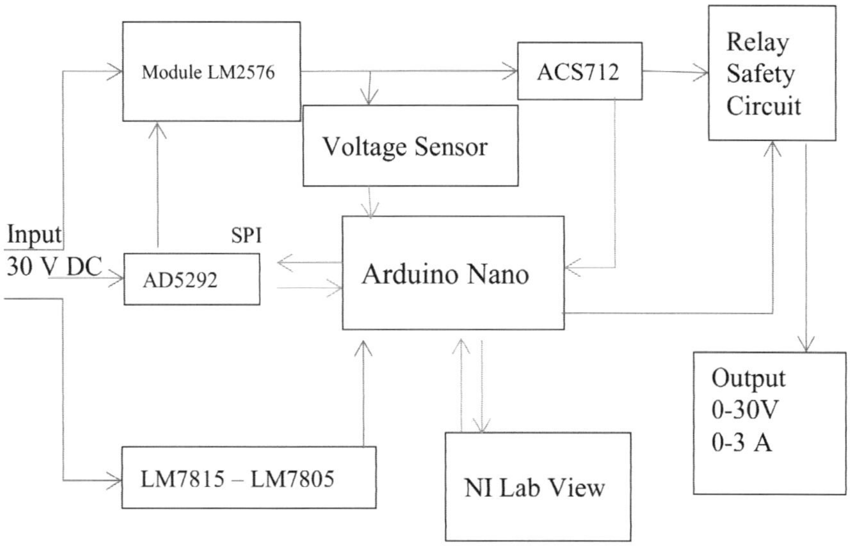

Figure 1. Circuit Block Diagram of the System.

In Figure 1 shows all of the system circuits. Voltage input of 30 V DC goes into module LM2576 which is the chosen switching regulator, also goes into LM7815 and LM7805 for powering Arduino Nano, and the last goes into potentiometer digital, AD5292 [4]. Current sensor ACS712 placed after the switching regulator. The output of this system is the voltage variance of 0-30V DC with 3 A max.

2.3 Control Algorithm

The program is planted in Arduino nano for doing all the activities for adjusting the desired voltage output and communicate in serial with NI LabView for exchanging data. Figure 2 shows how the program in Arduino works. That flow chart is the process of input voltage turning into an adjusted output voltage.

IOP Conf. Series: Materials Science and Engineering **771** (2020) 012007 doi:10.1088/1757-899X/771/1/012007

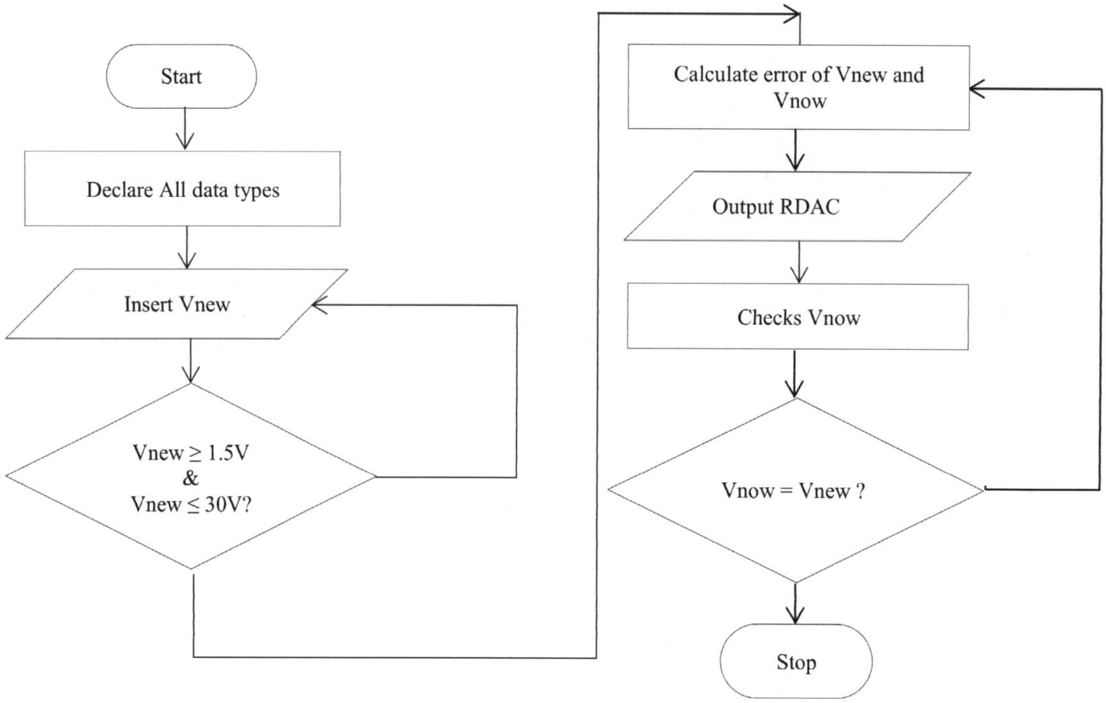

Figure 2. Flow Chart of the program.

2.4 NI LabView Design

LabView is used for monitoring and controlling the device that has been made. Block diagram is designed to make Arduino and LabView can communicate interchangeably and all of that made from sets of VIs. VIs were chosen to make a program so Arduino and LabView can communicate. Arduino passes voltage output data and then the data put into graphs. In Figure 3 shows a list of used VIs to communicate and manipulate the data.

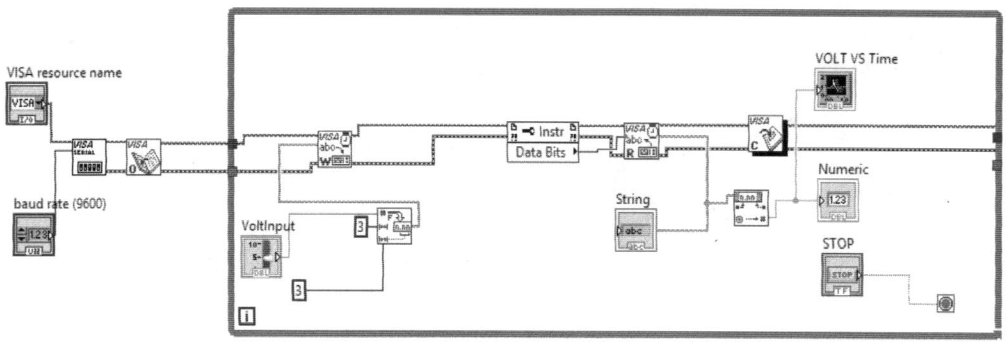

Figure 3. Design VIs in LabView

After made all of the block diagrams, there was also a front panel that works for the user interface. User interface made with the graphical metered cursor and a graph of voltage vs time. In Figure 4 are shown the display of a front panel in LabView. The red bar is the slider for voltage input, and there is the numeric input on the right side of the slider. Numeric input allows us to type input using the keyboard.

Figure 4. Front panel of NI LabView

2.5 Circuit Schematics and PCB Layout

The designing of the hardware circuit was done by Eagle PCB design software. In Figure 5 is shown the layout of the printed circuit board. Every modules and all the electronic components are implemented in one board. In Figure 6 is shown the schematic of the designed circuit. In the schematic also shown that there are pins that designed for Arduino Shield and ACS712 Shield.

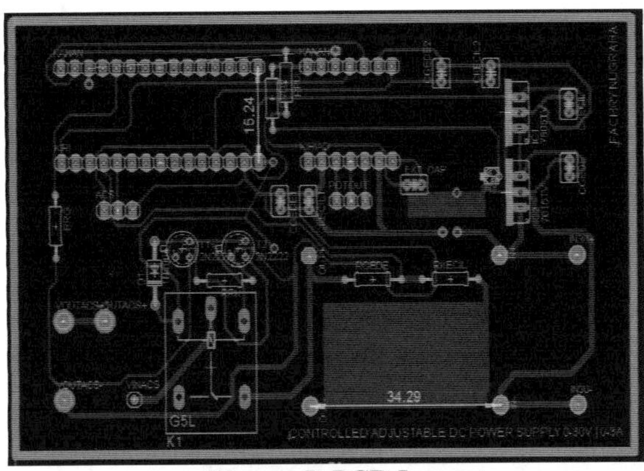

Figure 5. PCB Layout ;

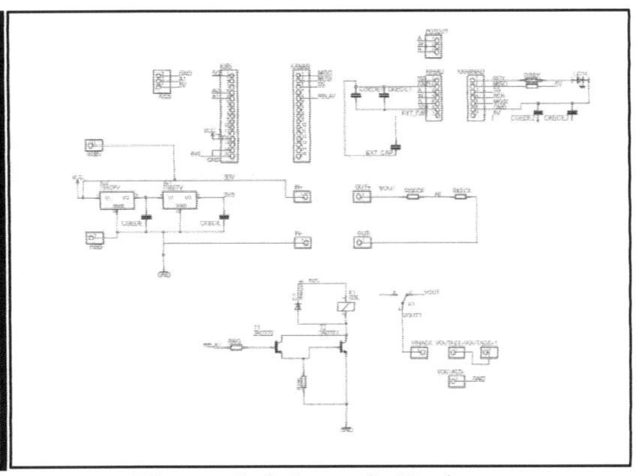

Figure 6. Schematic Design

3. Tests Runs and Analysis

This chapter will discuss the results of the tests that have been carried out along with the results of the analysis. There are 3 tests to be performed, including testing the output voltage that can be issued by the device by looking at the range of voltage that can be issued, then testing the resolution that can be issued by the device and finally testing the maximum current that can be given from the output by the device.

In Figure 7 shown the test of every bit position of the wiper of potentiometer digital. As the graph shows, the rise of the voltage is linear meaning its suitable to use the AD5292 for controlling voltage output. The resolution also covered with this device, it makes up to 0.08V, which means it covers the target spec at 0.1V resolution..

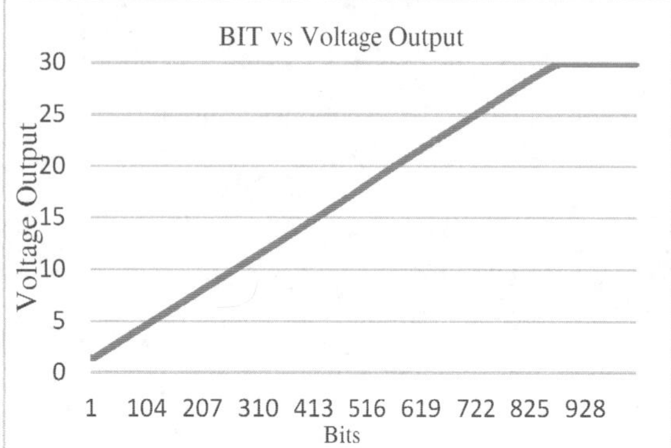

Figure 7. Graph AD5259 Bit wiper Position vs Voltage Output

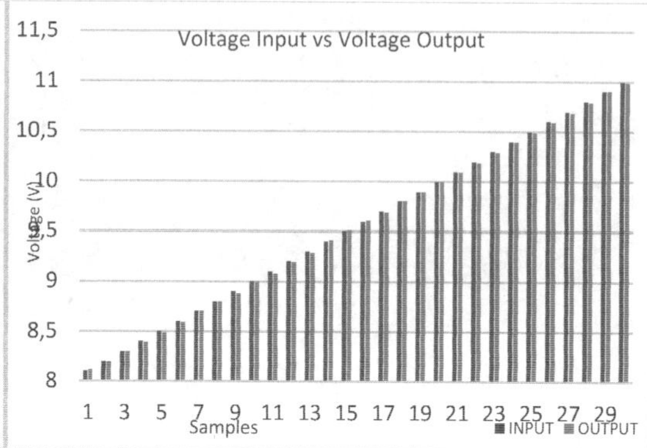

Figure 8. Voltage Input vs Voltage Output Controlled via LabView

AD5292 is compatible with our power supply specification, then move on to the other test. The next test is to measure the accuracy of voltage input vs voltage output. The device is controlled using LabView Via PC. Voltage inputs are compared with voltage outputs to see how the device can control the voltage outputs precisely. In Figure 8 are shown 30 samples of voltage inputs and outputs. The difference between the those approximately only 0.02V

The next test is to test the device by giving the load to it. Resistor given is at 1 Ω. This test measures how the current output flow if it's attached to a load. Voltage input is adjusted at constant 1.5V. In Figure 9 shown that the current that flows through the device and the load are fluctuates. It happened because of the current sensor ACS712 is not suitable for this kind of test condition. ACS712 suits in AC voltage measurement.

The last test was measuring how fast the voltage output change from 0-30V. The voltage input gradually increases from 0-30V then the output voltage is measured in time. In Figure 10 shown voltage output change in time. It needed the 90s to make a 0-30V change. It also shows the steady state time and settling time needed for the device to change its voltage output form 0-30V. All the tests have been done proving that the design is quite stable and it reaches the desired specification.

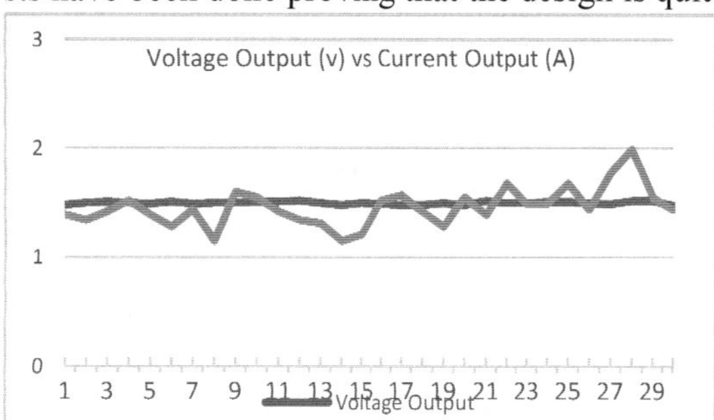

Figure 9. Current output vs Voltage output

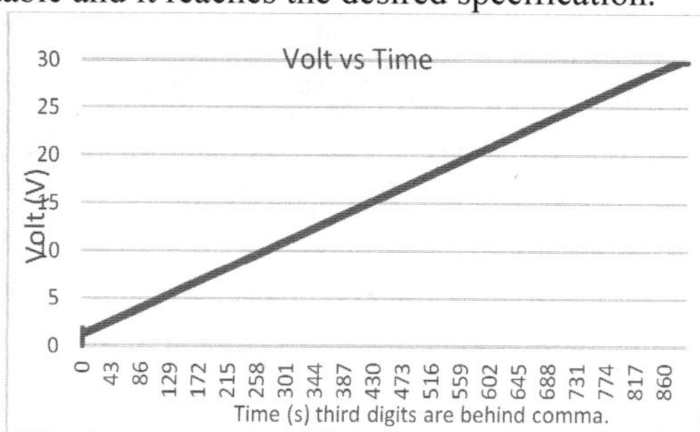

Figure 10. Voltage output vs Time

IOP Conf. Series: Materials Science and Engineering **771** (2020) 012007 doi:10.1088/1757-899X/771/1/012007

4. Conclusion

This study aims to design and implement a power supply whose output voltage can be set from the range of 0-30 V DC with a resolution of 0.1 V per level with a maximum current of 3 A and the most important thing, to be controlled and monitored by NI LabView. After doing this research, the results met all the required parameters. There are 3 parameter specifications: range volt, volt resolution, and maximum voltage output at 30V. The result of the required range of volts meets the desired specifications. For volt resolution, it can be adjusted to a precise 0.1V per level. The maximum current at the output can meet the target of 3A, it's just that the application is not enough because there will be the decrease in current due to automatic safety on ICLM2576. The current reading also not quite right because of the used unsuitable current sensor (ACS712), for further research, the current sensor should be replaced with more suitable current reading. On the other hand, controlling and monitoring can be done using NI LabView. The interface shows the exact data that was expected, and it is easy to be accessed so this device could be useful in the industry 4.0 environment.

References

[1] Ned Mohan, Tore M. Undeland, William P. Robbins, "*Power Electronics: Converters, Applications, and Design*",3rd Edition, Wiley

[2] Steve Roberts M.Sc.B.Sc., "DC/DC Book of Knowledge : Practical Tips for user", 3rd Edition,Austria

[3] Semiconductor Components Industries "3,0 A, 15 V, Step-Down Switching Regulator," LM2576 datasheet, Jan. 2006 – Rev. 8

[4] Analog Devices, Inc. "256-/1024-Position, Digital Potentiometers with Maximum ±1% R-Tolerance Error and 20-TP Memory" AD5291/AD5291 datasheet, 2014 Rev.E

IOP Conf. Series: Materials Science and Engineering **771** (2020) 012008 doi:10.1088/1757-899X/771/1/012008

Design of Automatic Switch System of Solar Panel and Power Plant for Residential Load using Artificial Neural Network

D Rahmawati[1,*], B S Aprillia[1], W Priharti[1], D K Silalahi[1], K Kumillayla[1]

[1]Electrical Engineering, Faculty of Electrical Engineering, Telkom University

*dienrahmawati@telkomuniversity.ac.id

Abstract. Electricity needs in Indonesia are getting bigger, following the increasing population. Today, most of Indonesia's electricity needs providing by PLN. However, most PLN electricity sources are non-renewable energy sources. Indonesia is a tropical country that receives sunlight almost all year long, so it is possible to use solar as an alternative energy. We designed an automatic switch system that could switch the electrical energy source from solar panel to the power plant from PLN if the power from the solar panel were insufficient. The automatic switch controller in this system uses backpropagation neural networks. The artificial neural network (ANN) model in this study consisted of two inputs, two hidden layers, and one output. The residential loads used are lamps, rice cooker, and fan. From the research results, the optimum epoch value is 2000, with an automatic switching accuracy value reaching 98.6%.

1. Introduction

Electricity consumption in Indonesia is increasing every year. Based on data from the Ministry of Energy and Mineral Resources (ESDM), electricity consumption per capita in 2018 reaches 1064 kWh [1]. The use of electricity is not only consumed by residential, namely electronic equipment used at home [2], but also business and industry. The State Electricity Company (PLN) primarily meets the fulfilment of power for the needs of the Indonesian population, which deals with all aspects of electricity in Indonesia. The electricity distributed by PLN mostly comes from hydroelectric power plants, steam power plants, diesel power plants, and gas power plants. Keep in mind that the power plant still uses a lot of fuels such as coal, diesel, fuel oil, and MFO. As a tropical country crossed by the equator, Indonesia is almost illuminated by the sun throughout the year. Therefore, the concept of renewable energy using solar power is suitable for use as alternative energy. This is following Ministerial Regulation (Permen) Number 49 the Year 2018 regarding the Use of Solar Roof Power Generation Systems by consumers of PT PLN. The Ministerial Regulation aims to reduce the cost of electricity and to increase the use of renewable energy.

Solar power generation is a power plant that converts solar energy into electricity. One of the ways to convert solar energy into electrical energy is to use photovoltaic, which for the change process uses photoelectric effects. Because there are not many solar power plants in Indonesia, if the household makes a solar power system, it can be hybridized by using electricity from PLN. This hybrid system is needed to maintain the stability of electricity availability and the maximum utilization of solar panel energy. Changing from one generator to another on a hybrid system requires a switching system.

The switch controller is designed to automatically detect the availability of generating sources that supply the load. The controller unit will check the battery voltage on the solar panel if the voltage from the solar panel starts to fall and is not able to meet the needs, then other sources will replace it [3]. In previous research, there is a source controller using the buck converter often decreases when there is loading [4]. The design is based on the implementation of solar panels as a power supply that will produce maximum utilization of alternative energy. Other studies include switching from solar power plants to power plants from PLN with its power supply regulator using fuzzy logic control [5], manual and automatic switches [6], and programmable logic control [7]. In this paper, we present an automatic switching system of solar panel and power plant for residential loads using artificial neural networks. Where the primary goal of developing this system to design an automatic switching system of solar cells and power plant that matches residential loads.

2. Proposed Automatic Switching System of Solar Panel and Power Plant

We developed an automatic switching system of solar panel and power plant for residential loads. The system that we developed to consist of hardware and software. The main function of the hardware in our system converts solar energy to electricity and saves it to the battery, measure voltage and current, and execute switching. The software in our system serves to determine the power supply to be used. Figure 1 presents the block diagram of the automatic switching system of solar panel and a power plant.

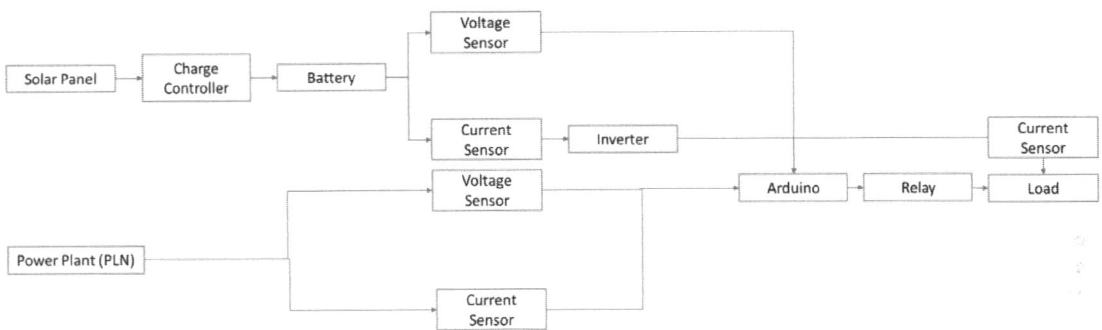

Figure 1. Diagram block of the proposed automatic switching system of solar panel and power plant

2.1. Hardware Design

In the proposed automatic switching system, a monocrystalline type solar panel 100Wp used to convert solar energy to electrical energy. The device for storing energy produced by the solar panel is a battery 12V 35Ah. We used solar charge controller 20A to charge the battery [8]. The tool for converting DC into AC using an inverter 1000Watt. Current sensor ACS 712 used to know the current read on the power supply (solar panel and PLN) and residential loads (lamps, rice cooker, and fan). Voltage sensor ZMPT101b used to know the voltage read on the power supply and residential loads. Arduino is a microcontroller from this system. We embed artificial neural network in this microcontroller. And then, relay two channels serve to turn on one source of electricity, solar panel or power plant.

The design of this study is a solar panel with changing sunlight intensity connected to a battery with a voltage that is considered constant. But the characteristics and specifications are determined in advance by considering the load data. Between the power plant and the solar panel is a switch with artificial neural network control with input voltage on the battery. With this input, it will be processed with an artificial neural network algorithm with the weight value of the results of the neuron experiment.

2.2. Software Design

We used Python to design the artificial neural network system before it is applied to Arduino Uno. An artificial neural network with a multilayer architecture consisting of three layers, namely, the input layer, is composed of several neurons whose numbers are adjusted by the input pattern. Furthermore, a hidden layer with several neurons whose name is done by trial and error, where chosen, is the one that produces convergence with the fewest iteration. The output layer consists of several neurons depending on the desired output pattern.

The input pattern used in this system is voltage data issued by the solar panel and PLN voltage for 24 hours. And then, the output pattern is the prediction of on/off switch pattern for 6 hours. The training data used in this system is the error value and duty cycle data from the solar panel. While the data for testing uses data transpose from the original data. The activation function applied is a binary sigmoid function that has a range of values from 0 to 1, so it is necessary to normalize the data using equation

$$\mathbf{X}' = \frac{\boldsymbol{X} - \boldsymbol{b}}{\boldsymbol{a} - \boldsymbol{b}}$$

X' is normalization data, X is preliminary data, a is the maximum value, and b is the minimum value.

Figure 2 shows the flowchart of automatic switch systems between the solar panel and PLN. The developed system is an automatic power source regulation system. The main function of this system works by utilizing the solar panel first. If the solar panel does not meet the need, then immediately switches to the power plant source. The flow of the system starts with reading the voltage and current of the solar panel. After that, the system will calculate the power generated by the solar panel. Then the system calculates the current and the power at the load. When the solar panel power is less than the load power, the power plant switch is "ON". Conversely, if the solar panel power is higher than the load power, the power plant switch is "OFF", and the program has finished.

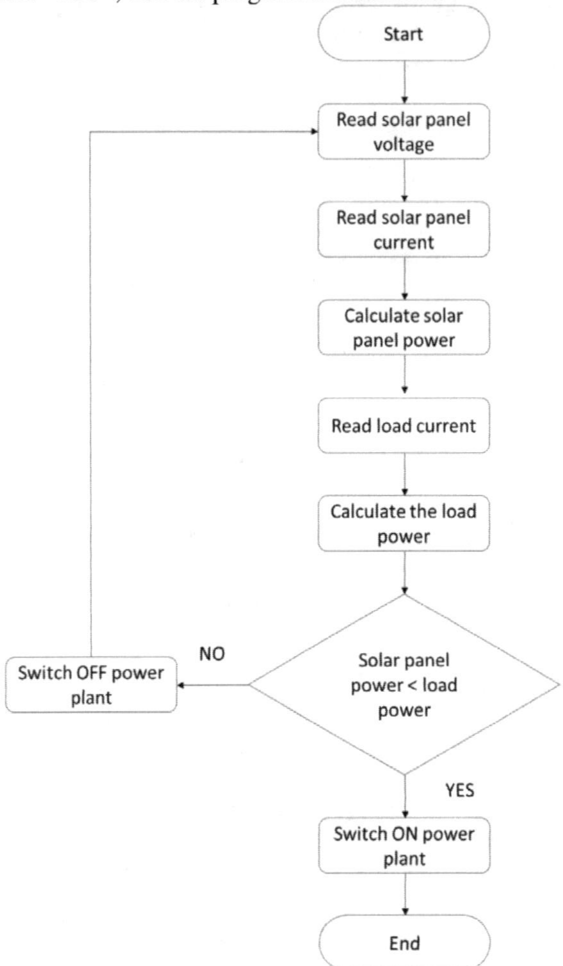

Figure 2. Flowchart of the system

3. Experiment and Analysis

There are four tests carried out, namely current sensor calibration, voltage sensor calibration, battery limit testing, simulation testing. We used residential loads such as six lamps, one rice cooker, and one fan.

3.1. Current Sensor Calibration

The purpose of the current sensor calibration is to find out whether the current sensor can work as expected or not. Current sensor testing carried out using a power supply source. The alternating current sensor calibration step is to connect the sensor pins to the Arduino UNO. The current sensor has three main pins, namely, VCC, GND, and VOUT pins. VCC and GND pins connected to VCC and GND pins on Arduino UNO. And then, the VOUT pin from the sensor is connected to pin A0 on Arduino UNO. Measurement is done by measuring the current with a power supply source. In the next step, the power supply output current is compared with the sensor reading. The results of the current sensor calibration show that the current sensor has pretty good accuracy, with an average percentage error of 1.57%. The current sensor has an error rate below 3%, so the current sensor is reasonable and can be used in the experiment.

3.2. Voltage Sensor Calibration

The purpose of the voltage sensor calibration is to find out whether the voltage sensor can work as expected or not. Calibration is done by comparing the measurement results of the voltage sensor readings with the calculation. The voltage sensor calibration step is to use one-way communication to exchange data between the sensor and Arduino UNO. Pin 1 of the line is connected to pin A1 on Arduino UNO devices. Pin VCC and GND are connected to pin VCC and GND on Arduino UNO. The voltage sensor calibration results show that the voltage sensor has pretty good accuracy, with an average percentage error of 2.0%. In the data, there is a quite significant difference in the number behind the comma, between the value of the voltage sensor reading and the voltage value on the kWh meter. This is caused, the voltage sensor used is ZMPT101B which has a sensitivity of 100 mV/A, so the sensor cannot detect voltage values below 1 volt accurately. Another factor that causes a large enough difference is the effect of loading on the kWh meter.

Calibration is done by comparing the measurement results of the voltage sensor reading with the voltage input from the power supply. The results show that the voltage sensor has an accuracy of 1.00%. In the data, there is a quite significant difference in the number behind the comma, between the value of the voltage sensor reading and the power supply output voltage. The cause of this is the voltage sensor used is a voltage sensor, which has a resolution of 0.00489V, so the sensor cannot detect voltage values below 1V accurately. Besides that, it is caused by the loading effect, which causes the value to be quite significant.

3.3. Battery Limit Testing

Battery limit testing is carried out to determine the voltage limits for the switching regulator. The test is carried out on batteries that have been charged using solar panels for 6 hours. This test was conducted to determine the use of batteries in residential loads such as lamps (42W), fan (12W), and magic com (350W). Table 1 shows the results of testing battery usage.

Table 1. Battery usage at residential load

Time	Load	Voltage	Information
15.54	Six lamps, rice cooker, and fan	13.50V	All load on
16.33	Six lamps, rice cooker, and fan	11.84V	All load on
17.02	Six lamps, rice cooker, and fan	11.55V	All load on
17.32	Six lamps, rice cooker, and fan	10.64V	Six lamps off

It is found that the switch limit is determined by the voltage on the battery. At a voltage of 13.50V, all load conditions are on, but, at 10.64V six lamps off. Because at 10.64 V there is a load off condition, 10.64V is used as a setpoint for the supply of the load. The setpoint means at 10.64V system switch from the solar panel source to the power plant source.

3.4. Simulation Testing

Simulation testing and analysis of artificial neural network responses in the automatic switch system obtained some values as shown in Table 2.

Table 2. Simulation result

No	Duty cycle	Error	Learning			Hidden layer
			1000	2000	3000	
1	0.985	0.005	0.064	0.015	0.045	2
2	0.998	0.005	0.055	0.023	0.039	2
3	0.998	0.005	0.054	0.024	0.040	2
4	0.999	0.993	0.944	0.980	0.953	2

5	0.978	0.993	0.945	0.957	0.964	2
6	0.999	0.993	0.944	0.980	0.953	2
7	0.999	0.993	0.944	0.980	0.953	2
8	0.999	0.993	0.944	0.980	0.953	2
9	0.998	0.993	0.944	0.980	0.954	2
10	0.999	0.999	0.944	0.980	0.953	2
11	0.999	0.999	0.944	0.980	0.953	2
12	0.998	0.999	0.944	0.980	0.955	2
13	0.998	0.999	0.944	0.980	0.955	2
14	0.997	0.005	0.055	0.022	0.039	2
15	0.999	0.005	0.054	0.024	0.040	2
16	0.999	0.005	0.054	0.024	0.040	2
17	0.916	0.005	0.156	0.025	0.133	2
18	1.001	0.999	0.944	0.980	0.951	2
19	1.001	0.999	0.944	0.980	0.951	2
20	0.999	0.999	0.944	0.980	0.953	2

In Table 2, the results obtained from this training show that the average synaptic values at 1000 repetitions are 0.638, at 2000 repetitions is 0.643, and at 3000 repetitions is 0.639. Table 3 shows the results of the error percentage of each epoch obtained from the average error value for each epoch multiplied by its normalization.

Table 3. Error Percentage

Iteration	Error (%)
Iteration 1000 times	9.063
Iteration 2000 times	1.349
Iteration 3000 times	8.287

In Table 3, the results of this simulation show that synaptic data at 2000 repetitions are the best accuracy. The accuracy achieved in the 2000 iteration is 98.6%. This can be caused by the nature of the artificial neural network process, which, if more and more repetitions will lead to many error values, while on repetitions that are too small will result in more errors.

4. Conclusion

We have presented an automatic switch system of solar panel and a power plant for residential load using an artificial neural network. Based on the results of the experiment, it can be concluded that the parameter used to get the lowest error rate in the process of regulating the best power supply is to use 2000 repetitions with an accuracy of 98.6%. Based on battery testing, the battery will be discharging at ± 10.6V and charging at 13.5V. An artificial neural network can experience convergent failure or not close to the output because the initial guess is not good.

References

[1] Anonym. PLN: Konsumsi listrik makin besar, negara semakin maju. https://industri.kontan.co.id/news/pln-konsumsi-listrik-makin-besar-negara-semakin-maju. March 23, 2019.

[2] A. J. Collin, G. Tsagarakis, A. E. Kiprakis, and S. McLaughlin, "Development of Low-Voltage Load Models for The Residential Load Sector," *IEEE Trans. Power Syst.*, vol. 29, no. 5, pp. 2180–2188, 2014.

[3] R. Rojas and R. Rojas, "The Backpropagation Algorithm," in *Neural Networks*, 1996.

[4] S. M. Shariatmadar, A. A. Jazi, and S. M. A. Esfahani, "Designing a Buck Converter with a New Switching Technique at Zero Current," *Int. J. Eng. Res. Technol.*, vol. 8, no. 2014, pp. 84–89, 3AD.

[5] A. Saleh, "Rancangan Sistem Kontrol Hibrid Energi Surya Fotovoltaik (SESF) dengan Sumber

Listrik PLN Menggunakan Fuzzy Logic Controller," Seminar Nasional dan Expo Teknik, 2015.
[6] L.E. Bien, I. Kasim, W. Wibowo, "Perancangan Sistem Hibrid Pembangkit Listrik Tenaga Surya Dengan Jala-Jala PLN Untuk Rumah Perkotaan", Universitas Trisakti. 2008.
[7] P. Raharjo, B. Sujanarko, T. Hardianto, "Perancangan Sistem Hibrid Solar Cell-Baterai–Pln Menggunakan Programmable Logic Controllers (Design Of Hybrid System Solar Cell-Baterry-Pln Using Programmable Logic Controllers)", BERKALA SAINSTEK, 2015 Dec 1;3(1):1-5.
[8] B. Onipede, S. Joseph, and O. Odiba, "Developing an Automatic Switch for Home or Industrial Power Supply Changeover," *J. Appl. Sci. Technol.*, vol. 21, no. 4, pp. 1–7, 2017.

IOP Conf. Series: Materials Science and Engineering **771** (2020) 012009 doi:10.1088/1757-899X/771/1/012009

LoRa Gateway as Internet of Things (IoT) Infrastructure Components on Undip Vocational School

D Y Tadeus[1,*], Yuniarto[1], F Mangkusasmito[1]

[1]Electrical Engineering, Department of Industrial Technology, Vocational School, Diponegoro University, Semarang, Indonesia

[*]e-mail: distayoel@live.undip.ac.id

Abstract. Internet of Things (IoT) is a concept that aims to broaden the benefits of continuously connected internet connectivity. The IoT concept infrastructure can be built using existing digital communication standard protocols such as Wifi, Zigbee, Bluetooth, but these three standards have limitations in terms of distance and power consumption. The LoRa communication protocol is a special communication standard that emphasises the aspects of long-distance, low power, low data transfer at relatively low prices. It utilises a gateway to forward messages from nodes. This research focuses on the implementation of open and low-cost LoRa gateway, which will be installed in the area of the Diponegoro University Vocational School Semarang. Communication functionality is tested using The Things Network server. The coverage range of gateway was tested by recording RSSI and SNR values received by the gateway. Test results show RSSI readings above -100 dBm at a radius below 100 m from the gateway. RSSI and SNR values appear to drop dramatically in low elevation conditions, especially if the position of the node-gateway is almost obstructed by the ground surface. The surface obstacle in the form of dense trees and thick buildings also contribute to reducing the quality of signal reception.

1. Introduction

Indonesia government has launched Roadmap Making Indonesia 4.0 for readiness to enter the 4th industrial revolution era. Internet of Things (IoT) is one of the mainstays of the road map that was initiated by the Ministry of Industry. The big aspiration in Making Indonesia 4.0 is to promise Indonesia to be among the ten countries that have the strongest economy in the world by 2030. IoT refers to a network of physical devices, vehicles, household appliances, and other items planted with electronic devices, software, sensors, actuators, and connectivity that make it possible to connect to the internet and collect and exchange data.

There are standard wireless network-based protocols such as Wifi, Bluetooth, Zigbee can be utilised as IoT infrastructure, but all of them have limitations in terms of low range coverage and relatively high power consumption and are not suitable when applied to distributed applications with limited resources. LoRa can be the main solution for this problem because it provides long-distance wireless connectivity capability, low power consumption, relatively low price, the limitation is the low data transfer capacity [1]. Technically, it's modulation performance has been evaluated, discussed and considered to be a good candidate in addressing the IoT challenges and low-cost remote sensing applications [2]–[4]. Figure 1 shows typical LoRa network. It is designed to allow low-powered devices to communicate with Internet-connected wide area networks applications over long-range wireless connections [4]. The LoRa hardware component consists of nodes and gateway, both a type of radio wave transceiver with the ability to communicate over long distances with low energy.

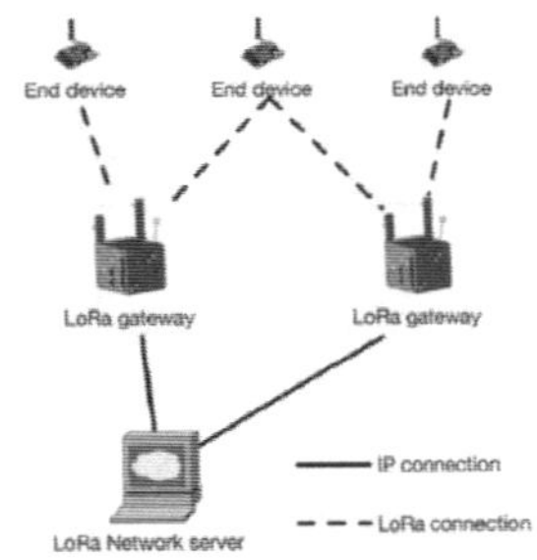

Figure 1. LoRa network

Performance and capability analysis of a currently available LoRa transceiver has been conducted, achieving reliability of 80%, through experimental demonstration covering 1.5 ha in a built-up environment [5]. The LoRa application for IoT based tactical troop tracking system in Thailand uses a Raspberry Pi-based gateway and a node containing Global Positioning Position (GPS), temperature, humidity and water sensors for tracking geological and physical conditions of troops in areas with a radius of 0.5 km [6]. LoRa is also used in the implementation of the concept of precision agriculture as a wireless and integrated data collector utilising node and gateway on farmland and tested on small scale melon farming [7]. Technical experiments have been carried out on LoRa's transceiver capability in scheduling transmission frames with less than 3 microseconds for smart metering, smart building, and process industry applications [8]. Furthermore, LoRa was also applied to a water level monitoring system where the LoRa node was placed as a medium between the sensor hub and the gateway and successfully tested its functionality using 50 node simulations [9].

Those examples describe LoRa with its gateway and nodes has been widely applied in various fields and applications, making it very potential to support the development of IoT. The existence of LoRa gateway is mandatory, so this research focuses on building open and low-cost LoRa gateway to emphasise its "low price" feature as the initiation of IoT infrastructure components in the area of Diponegoro University Vocational School, which in its current condition does not available. The gateway hardware is built using ESP8266 microcontroller and HopeRF RFM96 LoRa transceiver module. It has a task to forward messages or data payload from nodes to LoRa network server, The Things Network (TTN) via internet access. The availability of the proposed gateway will open the continues development of LoRa-based IoT applications on campus, specifically on Undip Vocational School.

2. Research Method

LoRa uses license-free sub-gigahertz radio frequency bands like 433 MHz, 868 MHz (Europe) and 915 MHz (North America). Indonesia adopts the US902-928MHz ISM Band as specified on LoRaWAN 1.1 Regional Parameters Document [10]. The Lora gateway hardware section consists of the NodeMCU ESP8266 controller module, HopeRF RFM96 915 MHz Lora transceiver module with a spiral antenna, and the power supply module. The HopeRF RFM9x families are LoRa long range modem that provides ultra-long range spread spectrum communication and high interference immunity while minimising current consumption. It has several key features [11] such as:

- 100 mW constant RF output
- +14 dBm high-efficiency PA
- High sensitivity: down to -148 dBm
- FSK, GFSK, MSK, GMSK, LoRa and OOK modulation
- 127 dB Dynamic Range RSSI
- Small form factor, 16*16mm

The controller is connected to the transceiver via the Serial Peripheral Interface (SPI) communication protocol. Figure 2 shows the Lora gateway block diagram.

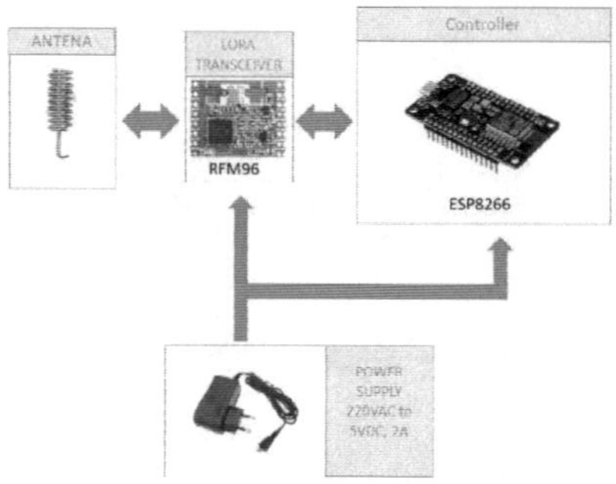

Figure 2. Lora hardware gateway block diagram.

The Lora gateway software section is built using an open-source library Arduino-lmic by Matthijs Kooijman [12] that is ported to the Lua programming environment by Jaap Braam [13]. This library is intended for constructing single-channel Lora gateway listening to all spreading factors on one specific channel. It has features that allow online setting for gateway parameter configurations via the preset command list as shown by Table 1.

With the help of Java-based software, ESPlorer, source code from the library can be opened, edited, compiled and downloaded to the controller. It is needed to set SSID and password of the internet-connected Wifi access point to the library source code so the connection between gateway and Lora network server can be established. This process is done by sending the following sequence command [9] in the ESPlorer command input terminal.

```
wifi.setmode(wifi.STATION)
station_cfg={}
station_cfg.ssid="SSID name"
station_cfg.pwd="password"
station_cfg.save=true
wifi.sta.config(station_cfg)
wifi.sta.autoconnect(1)
```

Gateway will remember it and automatically attempt to connect to Wifi access point. If it fail on the first attempt, the gateway will try again until it is successfully connected to the network marked by the appearance of a notification in the shell window of ESPlorer.

The frequency parameter configuration of the gateway needs to be matched with the frequency of LoRa transceiver module. So, GW_FREQ parameter is set to 902300000. This is the first channel of the US902-928MHz ISM Band specified by LoRa Alliance.

After the gateway is up and running, parameter configurations can be changed online (no need re-downloading) by sending the preset command from Table. in the ESPlorer command input terminal with this format: CONFIG.PARAMETER=VALUE. Figure 5 shows the example of changing gateway hostname with these commands: CONFIG.GW_HOSTNAME="LoRa 1Ch Gateway SV Undip", followed by CONFIG.save()to saves the current configuration and CONFIG.print()will show the current configuration as shown by figure 3.

The Things Network (TTN) is one of the Lora network servers that offers open-source, decentralised infrastructure for the Internet of Things, and supported by the large community around the globe. The next step is to register the gateway on TTN server so the gateway can be recognised and forward data payload from nodes to TTN server. This stage is the main indicator of success/failure on the gateway-Lora network server communication. After registration, it will receive gateway ID number from the server.

After the gateway registration process is complete and successfully connected, it is necessary to create an application on the TTN containing the node devices. The server will issue several identification numbers: Device EUI, Application EUI, Device Address, Network Session Key, and App Session Key specifically for each node that is registered.

Table 1. Default preset command for gateway parameter configurations.

Parameter	Description	Default
GW_HOSTNAME	A hostname for telnet server	"lorawangw"
GW_NTP_SERVER	DNS name of the NTP server to connect	"nl.pool.ntp.org"
GW_ROUTER	DNS name of the router to connect	"router.eu.thethings.network"
GW_PORT	Port number of the router to connect	1700
GW_FREQ	Frequency (in Hz) to listen to	868100000
GW_BW	BW to listen to	"BW125"
GW_SF	SF to listen to ("SF7".."SF12"|"ALL")	"ALL" (listen to all SF's)
GW_ALT	The altitude of your gateway location	0
GW_LAT	Latitude of your gateway location	"0.0"
GW_LON	Longitude of your gateway location	"0.0"
GW_NSS	NSS pin number	0
GW_DIO0	DIO0 pin number	1
GW_DIO1	DIO1 pin number	2

3. Result and Discussion

In the Gateway Overview menu on TTN console, the connection status of the gateway can be seen as shown in figure 4. According to its location, the up and running gateway will appear on Lora world map provided by TTN. It also displays information related to hardware of the gateway, especially the controller model.

The coverage area of the gateway is tested by placing nodes in several locations around the Undip Vocational School with a radius of approximately 100 m from the gateway. On each point, the RSSI and SNR value of the signal are recorded as seen from TTN's server. Google Map application is used for distance measurements of node-gateway. The position of the gateway is inside building D with elevation 2.5 m from the floor.

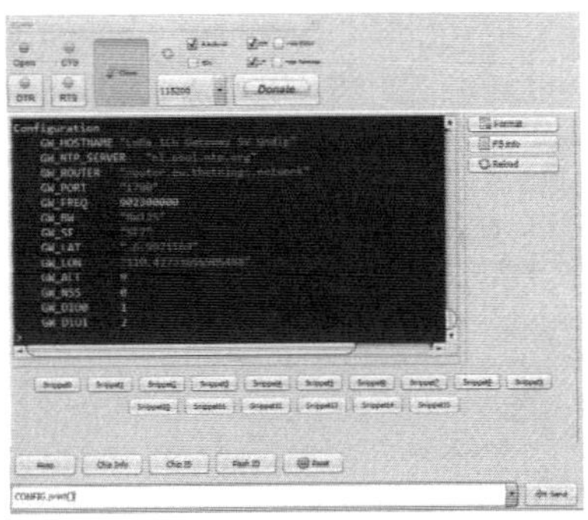

Figure 3. Example of changing gateway hostname parameter.

Location no.2 shows the test location in front of the microcontroller lab, with elevation of the node approximately 2 m above the floor of building D, there are obstacles in the form of buildings and trees, the distance of the node to the gateway is approximately 103.33 m, resulting in a reading of RSSI = -106 dBm, SNR = 9. Location no.3 shows the test location in the motorcycle parking lot, with elevation node about 7 m above the floor of building D, there are obstacles in the form of buildings and trees, the distance of the node to the gateway is approximately 49.62 m, resulting in a reading of RSSI = -104 dBm, SNR = 9.

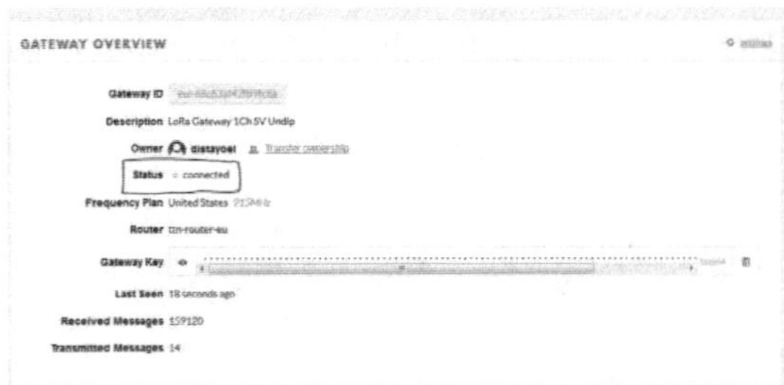

Figure 4. Gateway connection status on TTN server.

Table 2. Test results from several point locations

No	Location	RSSI (dBm)	SNR	Distance (m)
1	in front of the hall	-114	5	87.82
2	in front of the microcontroller lab	-106	9	103.33
3	motorcycle parking lot	-104	9	49.62
4	behind the mechanical technology lab	-117	-2	100.07
5	next to the Law Faculty	-119	-4	137.58
6	in front of the Law Faculty	-113	4	92.86

Location no. 4 shows the test location behind the mechanical technology lab, with the elevation of the node approximately 6 m above the floor of building D, at this point the ground surface almost obstructs the position of the node-gateway. There are obstacles in the form of buildings and trees, the distance of the node to the gateway is approximately 100.07 m, resulting in a reading of RSSI = -117 dBm, SNR = -2. Location no. 5 shows the test location next to the Law Faculty, with elevation of the node about 5 m below the floor of building D, there are obstacles in the form of buildings and trees, the distance of the node to the gateway is approximately 137.58 m, resulting in a reading of RSSI = -119 dBm, SNR = -4. Location no. 6 shows the test location in front of the Law Faculty, with node elevation of approximately 4 m below the floor of building D, there are obstacles in the form of buildings and trees, the distance of the node to the gateway is approximately 92.86 m, resulting in a reading of RSSI = -113 dBm, SNR = 4.

Test results show RSSI readings above -100 dBm at a radius below 100 m. RSSI and SNR values appear to drop dramatically in low elevation conditions, especially if the position of the node-gateway is almost obstructed by the ground surface. The surface obstacle in the form of dense trees and thick buildings contribute to reducing the quality of signal reception. The SNR even drops below the noise floor (negative value) at the location next to the law faculty and behind the mechanical technology laboratory. With this unique topographical land surface, the coverage service of LoRa gateway on Undip Vocational School may not be effective if the radius of nodes is more than 100 m.

4. Conclusion

The LoRa gateway has been successfully implemented on Undip Vocational School. Test results show the effective coverage range of gateway is approximately 100 meters, that is enough to cover LoRa based IoT applications on Undip Vocational School. The test result also proves the ability of LoRa to receive signal below the noise floor (negative values of SNR). Thick buildings, dense vegetations and trees, are major contributors limiting the range of LoRa communication. To extend its coverage service, the position of the gateway is recommended to be placed in a high location such as mounted on a dedicated tower or installing more gateway on another location. The success of test demonstrations also proved that LoRa can be realised using open and low-cost components. The up and running gateway is a part of the infrastructure for unlimited multidiscipline LoRa based IoT applications.

5. References

[1] R. S. Sinha, Y. Wei, and S. H. Hwang, "A survey on LPWA technology: LoRa and NB-IoT," *ICT Express*, vol. 3, issue 1, pp 14-21, March 2017.
[2] A. Lavric and V. Popa, "Internet of Things and LoRa™ Low-Power Wide-Area Networks: A

survey," in *ISSCS 2017 - International Symposium on Signals, Circuits and Systems*, 2017.

[3] A. J. Wixted, P. Kinnaird, H. Larijani, A. Tait, A. Ahmadinia, and N. Strachan, "Evaluation of LoRa and LoRaWAN for wireless sensor networks," in *Proceedings of IEEE Sensors*, 2017.

[4] A. Augustin, J. Yi, T. Clausen, and W. M. Townsley, "A study of Lora: Long range & low power networks for the internet of things," *Sensors (Switzerland)*, vol. 16, issue 9, pp 1466, 2016.

[5] M. C. Bor, J. Vidler, and U. Roedig, "LoRa for the Internet of Things," in *Proceedings of the 2016 International Conference on Embedded Wireless Systems and Networks*, 2016.

[6] W. San-Um, P. Lekbunyasin, M. Kodyoo, W. Wongsuwan, J. Makfak, and J. Kerdsri, "A long-range low-power wireless sensor network based on U-LoRa technology for tactical troops tracking systems," in *Proceedings - ACDT 2017: 3rd Asian Conference on Defence Technology: Advance Research Collaboration on Defence Technology*, 2017.

[7] R. Aquino-Santos, A. González-Potes, A. Edwards-Block, and R. A. Virgen-Ortiz, "Developing a new wireless sensor network platform and its application in precision agriculture," *Sensors*, vol. 11, issue 1, pp 1192-1211, 2011.

[8] M. Rizzi, P. Ferrari, A. Flammini, and E. Sisinni, "Evaluation of the IoT LoRaWAN Solution for Distributed Measurement Applications," *IEEE Trans. Instrum. Meas.*, vol. 66, issue 12, pp 3340-3349, 2017.

[9] Lukas, W. A. Tanumihardja, and E. Gunawan, "On the application of IoT: Monitoring of troughs water level using WSN," in *2015 IEEE Conference on Wireless Sensors, ICWiSE 2015*, 2016.

[10] LoRa Alliance, "LoRaWAN™ 1.1 Regional Parameters", LoRa Alliance Technical Committee Regional Parameters Workgroup

[11] HopeRF Electronic, "RFM95/96/97/98 Low Power Long Range Transceiver Module V1.0 General Description"

[12] Kooijman, Matthijs, "LoraWAN-in-C library, adapted to run under the Arduino environment", https://github.com/matthijskooijman/arduino-lmic

[13] Braam, Jaap, "A LoRaWan Gateway in LUA", https://github.com/JaapBraam/LoRaWanGateway

Acknowledgments

The authors would like to express heartfelt thanks to Electrical Engineering, Department of Industrial Technology, School of Vocation, Diponegoro University.

IOP Conf. Series: Materials Science and Engineering **771** (2020) 012010 doi:10.1088/1757-899X/771/1/012010

Solar Cell Output Optimization using Light Convergence Method

Bandiyah Sri Aprillia[1,*], Cahyantari Ekaputri[1], Muhammad Rafiqy Zul Fahmi[1]

[1]Telkom University, School of Electrical Engineering, Bandung, Indonesia

[*]e-mail: bandiyah@telkomuniversity.ac.id

Abstract. Light intensity is one of the many factors that can influence the power output of a solar cell. In this research, light convergence method is performed by using a Fresnel lens to increase the light intensity that are passes through the solar cells. 2 solar cells are used for this research. One of them is paired with a Fresnel lens positioned perpendicularly at a distance of 5cm. The output from both solar cells are then compared. The solar cell paired with the Fresnel lens contains a higher power output by up to 40.29% at 13.10 o'clock in the afternoon compared to a normal solar cell. However therein lies a decrease in power output at 10.00 – 10.30 o'clock and 13.30 – 14.00 o'clock by up to -30.65% because shadows from the Fresnel lens relative to the sun's position affects the light intensity a solar cell receives. Results from 3 days' worth of tests shows that at 10.00 – 14.00 o'clock, the solar cell paired with a Fresnel lens retain a higher average power output of around 6.08% compared to a solar cell without one.

1. Introduction

Solar energy is the ultimate source of nearly all the different types of energy on Earth. It is unique in that it can be directly converted to electric power upon contact with certain materials commonly referred to as solar cells. There are a few factors that affect the power output of a solar cell namely temperature, materials, light intensity, and etc [1]. Because of these factors, solar cells' have a low efficiency that ranges only from 7.1% to 46%.

The purpose of this research is to optimize the power output of solar cells based on one of the affecting factors, specifically, light intensity. A higher amount of light intensity a solar cell receives results in a higher power output [3-4]. The method used in this research is the light convergence method. A method that concentrates light to a point in a solar cell using mirrors or lenses [5]. The resulting lights will be concentrated on the solar cell, increasing the light intensity that passes through the solar cells, which in turn improves the power output.

In other researches, the media that are used for light convergence includes reflectors and convex lenses. The usage of a reflector for converging light can increase the power output of solar panels as high as 25% when installed perpendicularly to the solar cell [6]. Whilst the use of 30 convex lenses with a distance of 15cm from the solar cells can also increase the output of a solar cell as high as 26.67% [7].

In this research, Fresnel lens will be used as the media for light convergence because a Fresnel lens is relatively smaller in surface area compared to a reflector or a convex lens and can be purchased for a lower price.

2. System Design

2.1. Hardware Design

This research makes use of 2 solar cells with the same specifications. One of the solar cells will be paired with a Fresnel lens that can converge light into the solar cell.

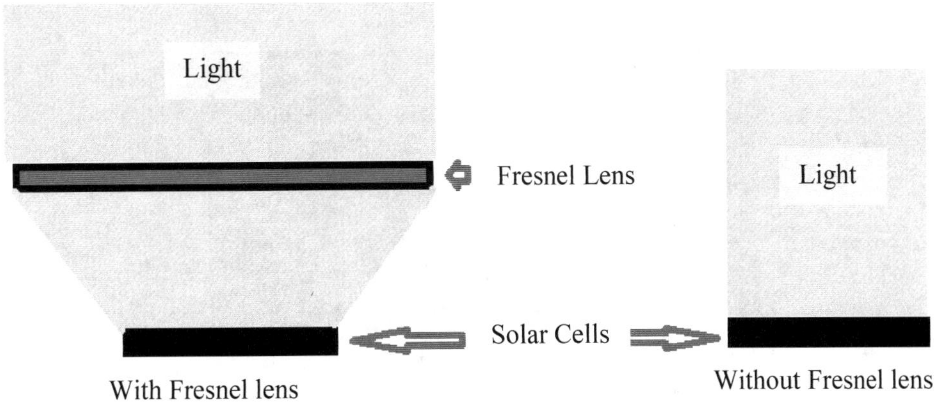

Figure 1. Illustration of Fresnel Lens Converging Light into Solar Cell

The use of a Fresnel lens that has a larger surface area than the solar cell ensures the amount of light that the solar cell receives will also increase in accordance with the surface area of the Fresnel lens. The more substantial the light intensity will result in higher power output from the solar cell in accordance with the following equation.

$$Voc = \frac{nkT}{q}\ln\left(\frac{XIsc}{Io}\right) \quad (1)$$

Where
Voc = Open Circuit Voltage (Solar cell voltage when I = 0)
nkT/q = Thermal voltage (Equation of voltage towards temperature (0.02586V when 26.85 Celsius))
X = Light Intensity (1kW/m^2)
Isc = Short Circuit Current (Solar cell current when V = 0)
Io = Solar Cell Current in Dark Conditions

2.2. Method of Finding Light Convergence Point

Because the distance between the Fresnel lens and solar cell directly impacts the light convergence. A test is required to find the best light converging point. The pursue of the convergence point is done by adjusting the distance between the Fresnel lens and solar cells at the time when the sun is at its peak, then specify which distance the Fresnel lens converges light optimally so that the resulting output of the solar cell is also maximized.

3. Results and Discussions

3.1. Test Results of Finding Light Converging Point

The testing of convergence is done to find the best distance between the Fresnel lens and solar cells, in which to maximize the value of the solar cell's power output.

Testing is done at 11:44 – 12:04 o'clock because those are the times when the position of the sun is at its peak. The data for timing is obtained from the website *https://www.suncalc.org*, a website that calculates the position of the sun.

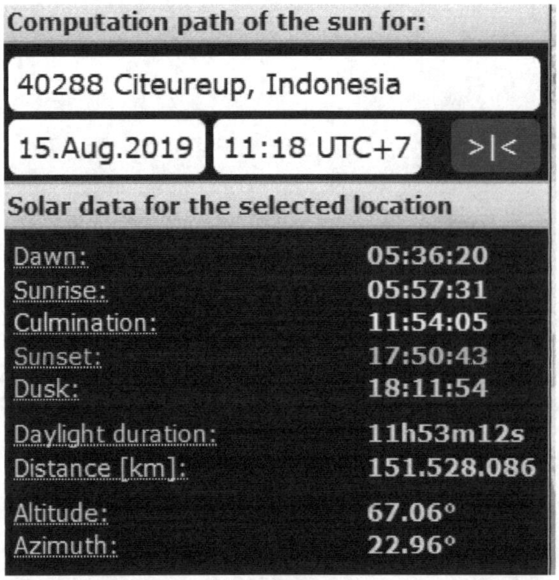

Figure 2. Sun's Positional Data from Date of Testing

Testing is done by adjusting the distance between the Fresnel lens and the solar cell, from 1cm until 10cm. The voltage and current output are calculated for every centimetre of difference. Results from the tests can be seen from the following Figure 3.

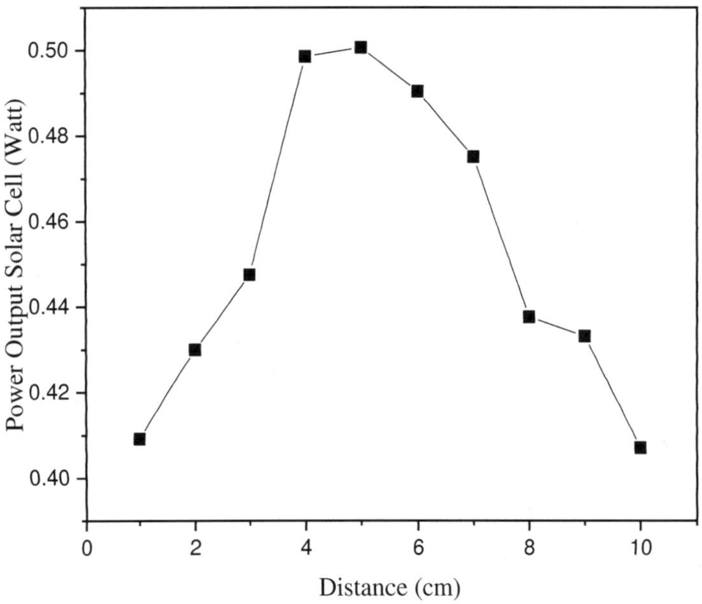

Figure 3. Data Showing Results of Light Convergence Point Testing

Table 1. Results of Light Convergence Point Testing

Distance Between Fresnel Lens and Solar Cells (cm)	Voltage (Voltage)	Current (Ampere)	Electrical Power (Watt)
1cm	6.2	0.066	0.4092
2cm	5.891	0.073	0.43004
3cm	5.665	0.079	0.44754

4cm	5.602	0.089	0.49858
5cm	5.442	0.092	0.50066
6cm	5.389	0.091	0.4904
7cm	5.338	0.089	0.47508
8cm	5.273	0.083	0.43766
9cm	5.219	0.083	0.43318
10cm	5.153	0.079	0.40709

From the result of Table 1, it can be seen that the highest output is 0.5006 watt with a distance of 5 cm between the Fresnel lens and the solar cell. It can be concluded from the data above that the distance of 5cm is the best possible distance for the light converging method because the power output is at its' highest compared to other distances.

3.2. Results of Tests

Tests are done by using 2 solar cells with the following specifications:

Table 2. Solar Cell Specifications

Dimension	Voltage	Current	Open-Circuit Voltage (Voc)	Maximum Load Voltage
70x55x3mm	5.5V	100mA	8.2V	6.4V

One of the solar cells is paired with a Fresnel lens separated at a distance of 5cm to converge light onto the solar cell. Temperature sensors connected to the microcontroller are implanted on both solar cells to calculate their temperature. Then, the two solar cells are placed where they will be exposed directly to the sunlight from 10.00 – 14.00 (UTC+07:00). Afterwards, calculation of Open-Circuit Voltage (Voc) and Short-Circuit Current (Isc) for the output of the solar cells are done every 5 minutes using a multi-meter. The tests operate at 10.00 – 14.00 o'clock (UTC+07:00).

3.2.1. Results of Solar Cell Power Output

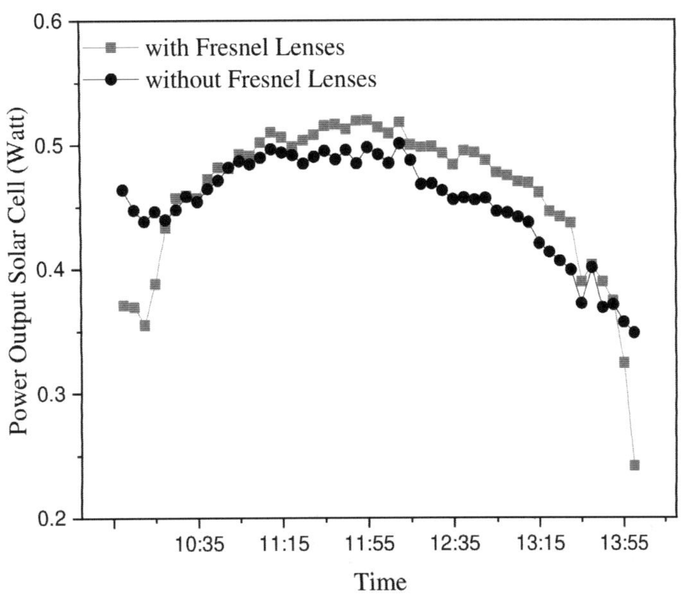

Figure 4. Graphics of Solar Cell Output

From Figure 4 we can see the difference in power output from the solar cell that is paired with the Fresnel lens and the solar cell without it from 10.00 to 14.00 o'clock. Results from the entirety of the tests show an improvement in the solar cell's power output, with an average improvement of around 1.61% or 0.008135 Watt.

The biggest improvement occurs at 13.15 o'clock where the increase of power is 9.75% or 0.041 Watt. However, a decrease in output also happens as far as -30.66% or -0.10697 Watt at 10.00 – 10.20 o'clock and 13.55 – 14.00 o'clock, which may happen because of the position of the sun leaning in the East or West creating the possibility of partial shading towards the solar cell that can cause a decrease in power output.

3.2.2. Results of Solar Cell Temperature Tests

Temperature measurements of solar cells are also done to see if the pairing of Fresnel lens would also impact the temperature of the solar cell and if it would impact the output of the said solar cell.

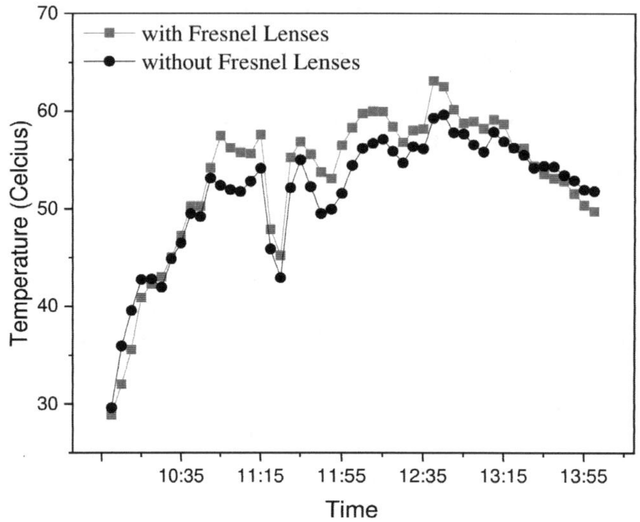

Figure 5. Graphics of Solar Cells' Temperature

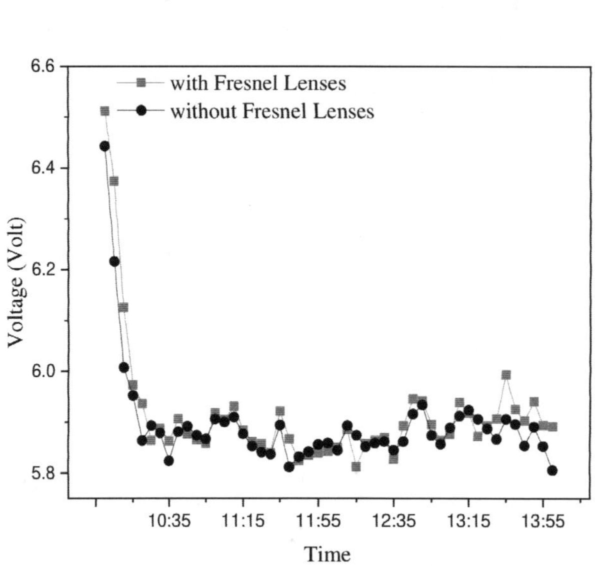

Figure 6. Graphics of Solar Cell's *Open-Circuit Voltage*

From Figure 5 it can be seen that the solar cell that is paired with a Fresnel lens possess a higher temperature for the majority of the time compared to the solar cells without it. This happens because the Fresnel lens concentrates light towards the solar cell which causes more light to hit the solar cell resulting in an increased temperature.

It can also be seen that at 10.00 – 10.20 o'clock when the difference in temperature is significantly higher, the value of the open-circuit voltage that is measured decreases. This is consistent with the theory where the higher the temperature of the solar cell, the lower the voltage of the solar cell.

4. Conclusion

The conclusion from this research is, that the pairing of a Fresnel lens for a solar cell can increase the power output of a solar cell by 9.75%, with an average increase of 1.61% when testing is done at 10.00 – 14.00 o'clock. The Fresnel lens that is used must have a surface area that is bigger than the surface area of the solar cell to ensure that it can pass through a higher amount of light intensity. If the position of the sun is in the east or the west such as at 10.00 – 10.30 o'clock or 13.30 – 14.00 o'clock then partial-shading will occur to the solar cells paired with the Fresnel lens causing the power output of the solar cell to decrease by up to -30.66%. The placement of the Fresnel lens can also impact the temperature of the solar cells. This happens because of the higher intensity of light being accepted by the solar cells, which also causes higher temperature. Consistent with equation (1), the higher the temperature of the solar cell, then the lower the open-circuit voltage from the affected solar cell.

References

[1] Sorensen, Bent. 2010. Renewable Energy Fourth Edition. Burlington : Academic Press.

[2] Green, M. A., Emery, K., Hishikawa, Y., Warta, W., & Dunlop, E. D. (2014). Solar cell efficiency tables (Version 45). Progress in Photovoltaics: Research and Applications, 23(1), 1–9. doi:10.1002/pip.2573

[3] Arshad, R., Tariq, S., Niaz, M.U, Jamil, M. 2014. Improvement in Solar Panel Efficiency Using Solar Concentration by Simple Mirrors and by Cooling. 2014 International Conference on Robotics and Emerging Allied Technologies in Engineering (iCREATE), Islamabad, Pakistan, April 22-24, 2014.

[4] Aprillia, B. S., Zulfahmi, M. R., & Rizal, A. (2019). Investigasi Efek Partial Shading Terhadap Daya Keluaran Sel Surya. *Jurnal Elektro dan Mesin Terapan*, *5*(2), 9-17.

[5] Shalaev, V. M. (2008). PHYSICS: Transforming Light. Science, 322(5900), 384–386. doi:10.1126/science.1166079.

[6] Mulyono,Dwi. 2003. Pengaruh Penambahan Reflektor Terhadap Karakteristik Arus-Tegangan Dan Efisiensi Sel Surya. Universitas Sebelas Maret, 2003.

[7] Priyanto,Budhi. 2013. Peningkatan Daya Keluaran Sel Surya Dengan Penambahan Intensitas Berkas Cahaya Matahari. Jurnal Neutrino Vol.5, No.2 April 2013

IOP Conf. Series: Materials Science and Engineering **771** (2020) 012011 doi:10.1088/1757-899X/771/1/012011

Design On-Grid Solar Power System for 450 VA Conventional Housing using HOMER Software

Bandiyah Sri Aprillia[1] and Muhammad Agung Foury Rigoursyah[1]

[1]Telkom University, School of Electrical Engineering, Bandung, Indonesia

[*]e-mail: bandiyah@telkomuniversity.ac.id

Abstract. The on-grid solar power system is a solution that can fulfil the ever-increasing demands for electricity. The important role of a solar grid system is to save and reduce the cost of electricity, particularly those with high energy usage in the daytime. This research aims to design a configuration for placing a solar grid system in the housings for the city of Bandung, Indonesia. The research method used is to collect solar radiation data, tool specifications, and other data needed, and then proceed to optimise on these bases. On-grid solar power system design requires the software HOMER. HOMER is used to find the correct system configuration, payback period, and the best NPC technically. Based on the result of optimization, the system configuration can fill the needs of electricity by about 42 % for daily consumption with a total NPC cost of around thirty eighty million one hundred thousand with four years payback period.

1. Introduction

Indonesia is a country that consists of 17,504 islands and have a total population of 267 million which continue to grow every year. The increasing total population of the country means that the need for electricity will also increase. According to the Central Bureau of Statistics (BPS), the amount of electricity consumers has increased to 62,068,283 people. Compared to 2016, this number has risen as much as 3,785,790 consumers, or 5,9%. When looking at the total number of consumers, residential sectors compared to other sectors consists of the most users comprising of 62,543,434 or 91.88% of consumers [1].

Most of the energy infrastructures in countries around the world, whether developed or in the process of development, use fossil fuels as their foundation [2]. Based on other researches, the use of fossil fuel results in problems which have been analyzed and overcame in many researches [3]. Problems such as environmental impacts, scarcity, supply risks, and instability of market and price place fossil fuels in the midst of a carbon-less economic shift [4].

Other than health and environmental issues, fossil fuels cannot be distributed equally causing unrest in the long-term security of electrical energy. This is a result of fossil fuel being the main player in the energy production infrastructure at this moment [5]. Furthermore, fossil fuel is a non-renewable resource which creates the problem in the availability for the next generation of humanity.

The most efficient renewable energy for Indonesia seems to be solar radiation. This is because Indonesia is a tropical country and is positioned in the equator line. Solar radiation can be changed into electrical energy using photovoltaics. Photovoltaics are sets of related components that have been arranged, balanced and loaded [6] which can change solar energy into electrical energy using the photovoltaic effect [7].

Designs of renewable power plants in its development can be classified as either single systems or hybrid systems [8-10]. Hybrid systems in here consist of the merging of two power suppliers which are the power network of PLN (State Power Plant) and solar energy. Installation of suitable solar panels for the household's needs are required to mesh the energy source from PLN with the installed solar panels. The ratio of power resulted from the hybrid system is 50:50. This installation is hopefully cheap whilst also reducing the amount of power from PLN source.

Energy from the sun can be used by solar panels. Solar panels can convert solar radiation directly into electrical energy so that it can be used to fulfil the needs of residents [8-10]. Design of solar power plants for residential-scale needs accurate calculation so that it can be economic, effective, and efficient. Therein lies the need for tools which can help us simulate and optimize. The tool that can be used in this instance is a software named HOMER (Hybrid Optimization Model for Electric Renewable) [9-10].

In this research, HOMER is used to evaluate the renewable on-grid system design on a scale of 450 VA per household in Bandung. Installations consist of three Trina solar 340 Wp and one inverter 1 kW capacity, with the assumption that the household needs a total power of 2,045 kWh/day.

2. System Design

2.1. Calculation of Electrical Load on a Household Scale

In Table 1. The electrical load is earned from a survey based on the power ratings.

Table 1. Electrical Load per Household

Types of Electrical Load	Section 1.01 Total	Active Hours (hours)	Load (W)	Total load (W/day)
Rice cooker	***Section 1.02 1***	24	250 *(cook)* *50 (heat)*	1400
Light	***Section 1.03 3***	12	12	432
Light	***Section 1.04 1***	13	24	312
TV	***Section 1.05 1***	16	106	1696
Fan	***Section 1.06 1***	14	46	644
Fan	***Section 1.07 1***	10	40	400
Dispenser	***Section 1.08 1***	1	70	70
Charger	***Section 1.09 2***	2	135	540
Charger	***Section 1.10 3***	2	10	60
Charger	***Section 1.11 1***	1	8	8
Charger	***Section 1.12 1***	1	4	4
Iron	***Section 1.13 1***	0.25	350	87
Total	***Section 1.14***			5653

2.2. Design of Solar Panels Circuitry Installation using Helioscope

Design of solar panel circuitry installation can be done by simulation using Helioscope. Helioscope is an online application that can simulate circuits with inputs such as the area of surface, solar panel brands, inverter, and etc. Data that are earned from Helioscope can become an input for HOMER as a custom input. Figure 1 is the simulation result from *Helioscope*, circuits in the simulation is located in Latitude of -7.5 and longitude of 107.5.

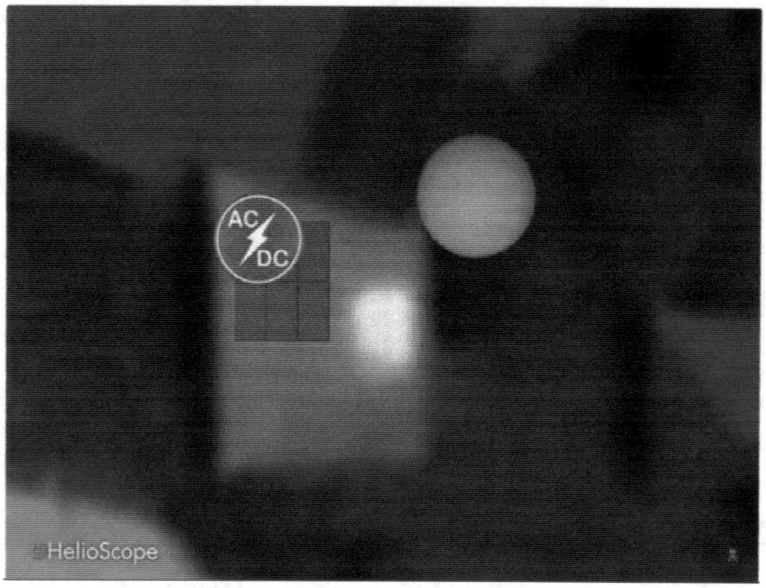

Figure 1. Installed Solar Panel Design in Helioscope

2.3. On-Grid System Design in HOMER

Generally, the data input in HOMER consists of electrical load per day (kWh/day), installation location, technical specifications of installed components, and cost of installed components. In this research, the specifications of solar panels in Helioscope are inputted with the custom mode. Figure 2 is an on-grid system design which is created using HOMER.

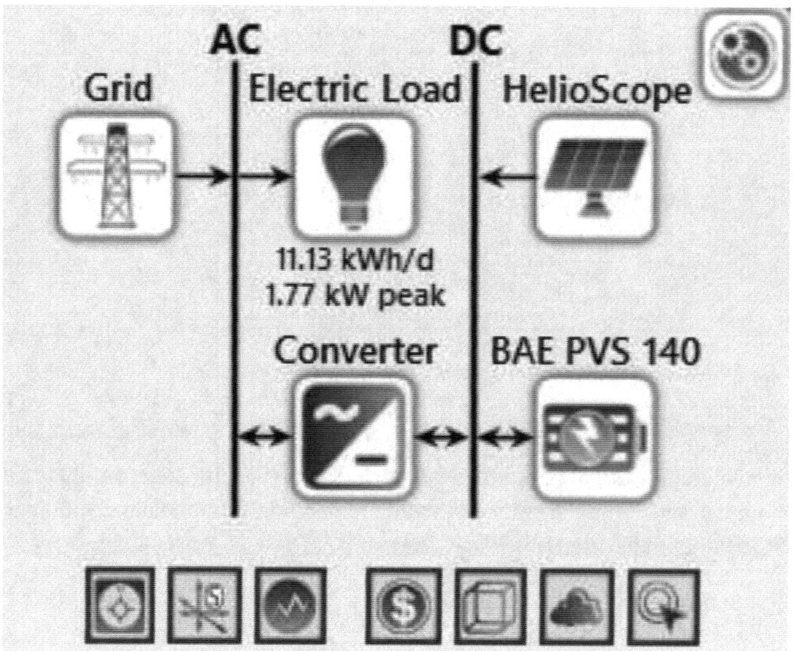

Figure 2. On-Grid Solar Power Design in HOMER

In Figure 2, solar panels function as the DC input which is directly connected to the battery. The battery's voltage is then changed into AC by the use of an inverter which then flows into an electrical load. If the energy from the battery is not enough to fulfil the needed load, then the electrical load will be powered by PLN.

3. Results and Discussion

Results from the evaluated system designed by HOMER gives several configuration options in table 2. Every configuration that are plausible are accompanied by the NPC value from the lowest to the highest. Optimal results will depend on certain factors such as solar radiation and the fraction of renewable energy. The main purpose of optimization using HOMER is to calculate the cost of power per kWh (COE), NPC, and greenhouse gas emissions (GHG).

The daily average solar radiation for the city of Bandung is 4,89 kWh / m^2 / day which is comparable with the global solar radiation of 4.9 kWh / m^2/ day and a minimum fraction of renewable energy is 50. Table 2, column 1 to 3 shows the existence of the gridline, PV cells, and PV cell capacity for each case. Column 4 to 9 shows the size of the converter in kW, size of gridline in kW, initial capital, cost of operation, notional principal contract (NPC) and cost of equity (COE) each on a scale of Indonesia Dollar Rupiah (IDR) per kWh. Column 10 shows the minimum fraction of renewable energy ranging at 42.4.

Table 2. On-Grid Solar Power System Configuration by HOMER

Component	PV (unit)	PLN	Inverter (KW)	Dispatch	NPC (IDR)	COE (IDR)	Operating Cost (IDR/year)	Initial Capital (IDR)	Ren fraction (%)
PV + PLN/ Grid + Inverter	3	~	1	LF	38.1M	673.4	1.62M	6.10M	42.4
PV +PLN/Grid + Inverter	3	~	1	CC	38.1M	673.4	1.62M	6.10M	42.4
PLN/Grid	0	~	0	LF	59.8M	1.467	3.03M	0M	0
PLN/Grid	0	~	0	CC	59.8M	1.467	3.03M	0M	0

On-Grid system for household scales sees 450 VA as the most optimal setting with a configuration for the PV of 2040 Wp and one kW inverter. This system configuration can fulfil electrical load needs as high as 42 % for daily consumption of 5.56 kWh/day. Costs of components can reach up to six million one hundred thousand. The cost is relatively expensive, but with the use of on-grid systems, those costs can be closed with the profit of selling renewable energy (PV) into the PLN or grid network. Because the rates of selling electrical energy from consumers who use renewable energy are much higher compared to the sales of electricity from the grid. The total cost that needs to be paid during the project's period (30 years) to configure all this is thirty eighty million one hundred thousand. Other than that, with the use of solar power system can lower the operational cost and lower the NPC cost compared to the use of the grid, because the use of solar power system can be sold to the grid. The higher the amount of electrical energy sold to PLN, then the higher the profit earned. This transaction process can be impactful to the amount of operational cost and NPC.

Table 3. Comparison of Cost per Year Between PLN and On-Grid Solar Power System.

System	COE	Total kWh/yr.	Total cost in IDR/yr.	Save in IDR./yr.
PLN/ Grid	1467	2062	3.024.954	1.636.404
Installed Solar Panels	673.4	2062	1.388.550	

IOP Conf. Series: Materials Science and Engineering **771** (2020) 012011 doi:10.1088/1757-899X/771/1/012011

Table 3 shows the amount of money that can be saved per year when using the on-grid system that is designed like that in figure 2. The money saved every year amounts to one million six hundred thirty-six thousand.

Payback period means that the amount of year to restore the amount of investment and analysis cost benefits from the system [11-12]. This parameter is influenced by the rates per Kwh/ hour, cost of fuel and electricity that increases from day today. In Table 4 it can be concluded that for the on-grid system that has been designed it will take four years to restore the investment's cost.

Table 4. Result of Calculation from the Payback Period in HOMER

Metric	Value
Present worth (Rp)	21.670.840
Annual worth (Rp/year)	1.097.240
Return on Investation (%)	19.7
Internal rate of return (%)	23
Simple payback (year)	4.33
Discounted payback (year)	4.71

4. Conclusion

The result of the on-grid system design for conventional housing of 450 VA using HOMER software recommends a configuration for the three PV of 2040Wp and one inverter 1 kW. The system configuration can fulfil the needs of the electrical load as high as 42 % from a daily load consumption with the total cost NPC as high as thirty eighty million one hundred thousand. On-Grid System that has been designed can reach a total savings of around one million six hundred thirty-six thousand compared to the use of conventional electricity (PLN). Moreover, based on HOMER analysis investment costs will be paid back after four years.

References

[1] Statistik Ketenagalistrikan 2017. Direktorat Jendral Ketenagalistrikan. Kementerian Energi dan Sumber Daya Mineral. Oktober 2018

[2] Y. Sugiawan and S. Managi, "New evidence of energi-growth nexus from inclusive wealth," *Renewable and Sustainable Energi Reviews*. pp. 40–48, 2019.

[3] B. Pillot, M. Muselli, P. Poggi, and J. B. Dias, "Historical trends in global energi policy and renewable power system issues in Sub-Saharan Africa: The case of solar PV," *Energi Policy*, pp. 113–124, 2019.

[4] M. M. Khan *et al.*, "Triangular relationship among energi consumption, air pollution and water resources in Pakistan," *J. Clean. Prod.*, pp. 1375–1385, 2016.

[5] K. Narula, "Global Energi System and Sustainable Energi Security," *Marit. Dimens. Sustain. Energi Secur.*, vol. 68, pp. 23–49, 2019.

[6] M. R. Patel, *Wind and Solar Power Systems*. 2005.

[7] Aprillia, B. S., Zulfahmi, M. R., & Rizal, A. (2019). Investigasi Efek Partial Shading Terhadap Daya Keluaran Sel Surya. *Jurnal Elektro dan Mesin Terapan*, *5*(2), 9-17.

[8] J. Ahmed and Z. Salam, "A Modified P & O Maximum Power Point Tracking Method with Reduced Steady-State Oscillation and Improved Tracking Efficiency," *IEEE Trans. Sustain. Energi*, vol. 7, no. 4, pp. 1506–1515, 2016.

[9] Bandiyah Sri Aprillia, D K Silalahi *et al.*, "Desain Sistem Panel Surya On-Grid Untuk Skala Rumah Tangga Menggunakan Perangkat Lunak HOMER," Jurnal Teknologi Informasi dan Multimedia., vol. 1, no 3, pp. 174 – 180, 2019.

[10] Nurfaidah, Y., Wibawa, I. P. D., Aprilia, B. N. D., Ekaputri, C., & Reza, M. (2019, November). Analysis of smart house power savings with on-grid photovoltaic power system. In *Journal of Physics: Conference Series* (Vol. 1367, No. 1, p. 012047). IOP Publishing.

[11] M. A. Matin, Bhuiyan, A. Deb, and A. Nasir, "Optimum Planning of Hybrid Energi System Using Homer for Rural Electrification," *Int. J. Comput. Appl.*, vol. 66, no. 13, pp. 45–52, 2013.

[12] W. M. Amutha and V. Rajini, "Cost benefit and technical analysis of rural electrification alternatives in southern India using HOMER," *Renewable and Sustainable Energi Reviews*. pp. 236–246, 2016.

IOP Conf. Series: Materials Science and Engineering **771** (2020) 012012 doi:10.1088/1757-899X/771/1/012012

IoT Based Logistics Vehicle Security Monitoring System

W Priharti[1,*], S Sumaryo[1] and T Saraswati[1] and M R Nurfadilah[1]

[1] School of Electrical Engineering, Telkom University, 40257 Bandung, Indonesia

[*]Corresponding email address: wpriharti@telkomuniversity.ac.id

Abstract. Speed, security and accuracy of a goods transportation play an important role in the logistics industry. This study proposes a logistics vehicle monitoring system that are consists of automated locking and tracking system to ensure the security and on-time arrival of the transported goods. Global Positioning System (GPS) combined with Global System for Mobile Communication/General Packet Radio Service (GSM/GPRS) internet connection method was utilized to provide real-time monitoring using an Android application display. The automatic locking command was successfully executed within 30.50 s delay. The logistics vehicle location accurate to 16.7 m was pinpointed within 28.72 s delay. The warning was successfully given when the logistic vehicle was moving away at minimum 500 m from the destination point to ensure the delivery route was exactly followed.

1. Introduction

The widespread use of the internet nowadays has increased public interest in electronic commerce activities or e-commerce. To support the e-commerce success, speed, security and accuracy in the delivery of goods play an important role. To achieve this, reliable and fast transportation, as well as efficient logistics services are required. This requirement can only be fulfilled by good and highly credible logistic companies.

Most logistics companies in Indonesia cover such a wide area throughout the nation. They offer a range of services, from simple transport services, through service forwarding, warehousing, palletizing, packing, packaging, to full service of supply chains [1]. The large numbers of logistics companies in Indonesia offer the customers the freedom to choose the companies that suit their need. However, the good performance and credibility of the logistics companies will decide whether the company will exist in the market at all [2].

The basic of a logistic company is not only sending parcels, documents and heavy freight, but also information about them. Hence, a well-developed operating system which consists of working people and infrastructure is required. To ensure a fast and correct flow of information and also to increase the efficiency, the logistics companies need to use the appropriate information technologies. Another objective of the application of information technology by logistics companies is to avoid problems and fulfil the expectations of potential and existing customers [3].

Many problems that arise in the logistics companies include the foul driver, delay of arrival, lost or broken goods, even the thievery of goods during travel [4]. These terrible things can occur due to lack of supervision, which can result in excessive costs for the company. Therefore, the logistics vehicle security monitoring system is required to avoid these problems.

Due to the security reason, the location monitoring of a vehicle or a vehicle tracking system has been commonly used in recent years. The utilization of Global Positioning System (GPS) combined with the latest technology of internet of things (IoT) has been widely implemented in the private cars, school buses and transport trucks [5]–[7]. The diverse connection methods were used such as GSM/GPRS (Global System for Mobile Communication/General Packet Radio Service) or Wi-Fi (Wireless Fidelity)

[8]–[13]. However, a GSM/GPRS method was convinced to be more suitable for the usage of a moving vehicle [14].

In this study, the logistics vehicle security monitoring system consists of the automated locking system and tracking system was designed. The locking system is required to ensure the safety of the goods inside the logistics vehicle. Whereas the tracking system is required to locate the position of the logistics vehicle during goods transportation. Its objective is to ensure that the logistics vehicle always follow the pre-determined route so it will not delay the arrival of goods transportation. GPS technology combined with IoT with GSM/GPRS connection methods was used along with Thinkspeak as the IoT platform.

2. Methods

The safety monitoring system for logistics vehicle designed in this study consists of two parts,i.e. the automated locking system and tracking system. To give the user the convenience and simplicity, both systems can be accessed via an android application using IoT. The diagram block of the overall system is illustrated in Figure 1.

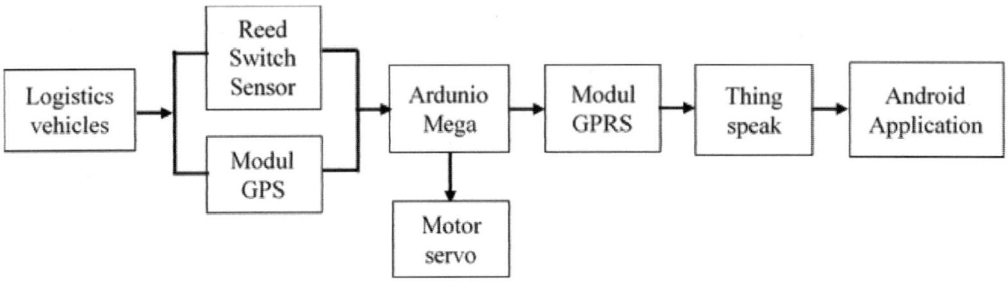

Figure 1. Block diagram of a logistics vehicle safety monitoring system using IoT.

2.1. Automated Locking System

A real logistic vehicle box with a dimension of 261×170×150 cm^3 was simulated using a ratio of 1:7 prototype. The dimension of the prototype is 37.2×24.2×21.4 cm^3 as illustrated in Figure 2. The locking system designed in this study is a logistic vehicle box door security system using a servo motor as an IoT-based key slot drive. This system can lock and unlock the box door and can monitor the state of the door remotely to provide a sense of security for the user. The controlling and monitoring of the system can be carried out via an android application using IoT. The reed switch sensor installed on the box door was used to send the data to the android so that the door state can be monitored in real-time. If the box door is opened by force, the reed switch sensor will send a notification to the android application.

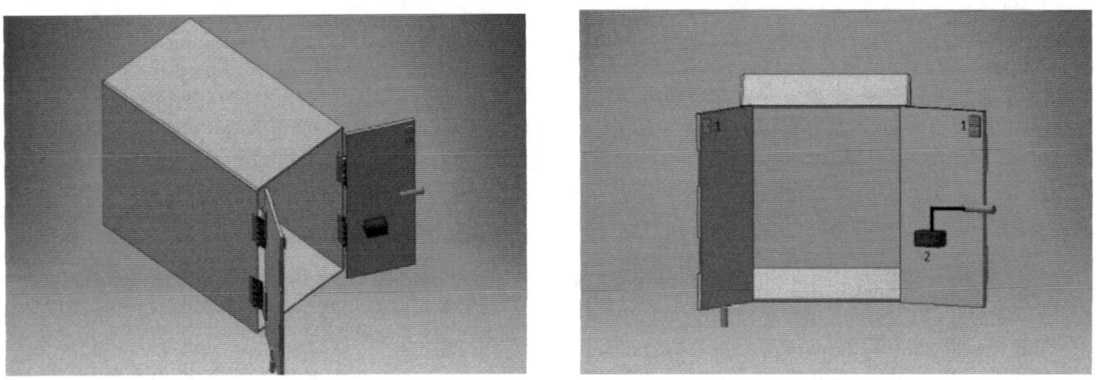

(a) (b)

Figure 2. The logistic vehicle box prototype: (a) top view, (b) front view

Figure 2(b) shows the installed position of the reed switch sensor (1) and servo motor (2). The reed switch sensor is an electrical switch operated by an applied magnetic field. The working principle of this sensor is quite simple based on 0 and 1 logic. If the magnet and the switch are closed, the sensor logic is 1 or High (the door is closed), otherwise, if the magnet and the switch are apart, the sensor logic is 0 or Low (the door is opened). In this prototype, the magnet and switch are placed adjacently as illustrated by Figure 2(b). Then, the servo motor was connected to the box door key. The microcontroller used in this system was Arduino ATMega2560 with a 12V Yuasa accumulator as the power supply.

2.2. Tracking System

The tracking system designed in this study utilized a GPS module based on an Android application. This system can locate the longitude and latitude position of a logistic vehicle and display it on the maps of android application. GPS Modul Ublox Neo-7M was connected to Arduino Mega ATMega2560 microcontroller to get the location of the logistic vehicle. That location was then sent to the webserver using GPRS (General Packet Radio Service) Modul SIM800C and displayed on the android application in real-time.

The objective of this system is to monitor the current position, the route needs to be taken, and the distance needs to be travelled by a logistic vehicle. If the vehicle is not moving, or if it travels outside the determined route, the system will give a warning to the user (administrator), so the user can contact the vehicle driver.

2.3. Data Transfer and Display

Thinkspeak was used in this study as an open-source of IoT cloud platform application. This application can retrieves and stores the data from the reed switch sensor and GPS through the internet that uses hypertext transfer protocol (HTTP). The data from the sensor and GPS were uploaded from the Arduino board connected to a GPRS module to the cloud. It updates all the data logs received from the sensor and giving the status application to the registered users. With internet-based monitoring, the data would be easily accessible through an online interface (computer or smartphone). An Android application was designed to facilitate users. The aims of this application is to give the users the convenience in the monitoring of logistics vehicle state of door and location to ensure its safety. The application was made using MIT App Inventor with a display shown in Figure 3.

Figure 3. Logistics vehicles safety monitoring system application designed using MIT App Inventor

3. Results and Discussions

3.1. Reed Switch Sensor Calibration

The calibration was carried out to determine the sensitivity of the sensor. The method was to measure the farthest distance between the switch and the magnet when the door was opened. Illustration of the calibration is shown in Figure 4.

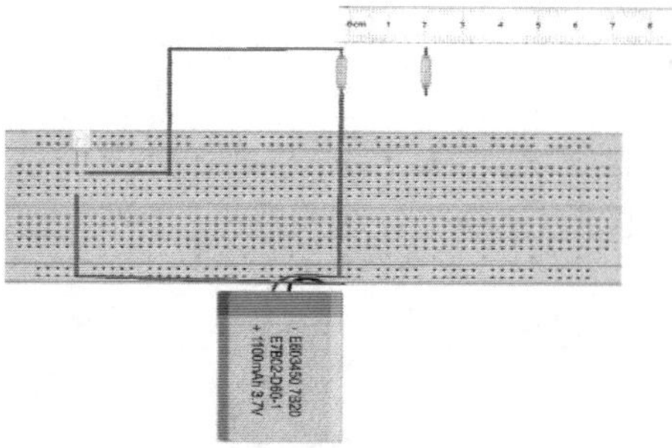

Figure 4. The illustration of reed switch sensor calibration

The distance covered in the calibration was in the range of 0 – 6 cm. The data were taken 30 times with a 0.2 cm distance for each data. The results show that the reed switch sensor was only active at a distance below 2.2 cm. Above that, the sensor was inactive or could not give any response.

3.2. GPS Ublox NEO-7M Module Calibration

The purpose of this calibration was to find out the reading of the coordinates (latitude and longitude) by the GPS Ublox NEO-7M GPS module. The readings were then compared to the GPS Garmin 62s as the standard measurement instrument. The results from 30 readings show that the mean distance discrepancy between GPS Ublox NEO-7M and GPS Garmin 62s is 16.97 m.

3.3. Data Transfer Test

This test was carried out to measured the duration of data transfer from: (a) microcontroller (MC) to Thinkspeak and (b) Thinkspeak to Android application (APP). Each measurement was carried out 30 times in two conditions: (i) open space (simulating the open-free road) and (ii) closed spaced (simulating the road surround by high buildings). The results of these measures are given in Table 1.

Table 1. The mean duration of data transfer for logistic vehicle locking and tracking system

Data transfer	Locking Mean Duration (seconds)		Tracking Mean Duration (seconds)	
	Open space	Closed space	Open space	Closed space
MC to Thinkspeak	21.90	32.80	14.35	28.30
Thinkspeak to APP	8.60	14.60	14.37	N/A
Total	30.50	47.40	28.72	–

3.4. Android Warning Test

This test aims to determine the capability of Android application warning system if the logistic vehicle moving further away from the intended route with a distance of more than 500 meters. This test was carried out by first routing the Android application and determined the destination point. Then the Android application will calculate the distance that will be travelled from the initial location to the destination location. After routing, the vehicle travels to the point of destination by taking a route that was located away from the destination location. The results are given in Table 2.

Table 2. The results of Android warning test

No	Distance away from destination point (meter)	Distance counter in Android Application	Warning System
1	500	*ON*	*ON*
2	1000	*ON*	*ON*
3	1500	*ON*	*ON*
4	2000	*OFF*	*OFF*
5	2500	*OFF*	*OFF*

The results show that the Android application could only give a warning if the logistic vehicle was located inside 1500 meters away from the destination point. Beyond that point, the warning was not triggered or could not give any response. It may be caused by the faulty distance measurement of the Android application so the warning system became inactive. To overcome this problem, the distance measurement of the Android application should be checked regularly to ensure an accurate reading.

4. Conclusions

The logistics vehicles security monitoring system consists of an automated locking and tracking system was proposed in this study. GPS system combined with GSM/GPR internet connection method was utilized to provide real-time monitoring using an Android application display. The application was successfully able to execute the automatic locking command within 30.50 s delay. The logistics vehicles location accurate to 16.7 m was pinpointed within 28.72 s delay. To ensure the delivery route was exactly followed, the warning was successfully given when the logistic vehicle was moving away at minimum 500 m from the destination point. Based on these results, the proposed system was considered promising to ensure the safety of the goods inside the logistics vehicle and to locate the position of the logistics vehicle during goods transportation.

Acknowledgements

Authors wishing to acknowledge the financial support from Direktorat Penelitian dan Pengabdian Masyarakat (PPM) Telkom University to attend this conference.

References

[1] A. Kawa, "SMART logistics chain," no. March 2012, 2015.
[2] M. F. Alrifaie, N. Harum, M. O. I. Fairuz, I. Roslan, and M. A. Shyaa, "Vehicle Detection and Tracking System IoT based : A Review," *Int. Res. J. Eng. Technol.*, pp. 1237–1241, 2018.
[3] S. Lakshmi, "The Role of Transportation in Logistic," *Int. J. Res. Appl. Sci. Eng. Technol.*, vol. V, no. XI, pp. 1267–1270, 2017.
[4] M. A. Romansyah, "Aplikasi sistem pelacakan kinerja pengiriman pada truk pengangkut barang berbasis android," Universitas Muhammadiyah Surakarta, 2015.
[5] A. Ahmed, E. Nada, and W. Al-Mutiri, "University Buses Routing and Tracking System," *Int. J. Comput. Sci. Inf. Technol.*, vol. 9, no. 1, pp. 95–104, 2017.
[6] R. Abhilash, R. Mahima, S. Monisha, and R. Nagashri, "Smart Tracking System for School Buses," *Int. Res. J. Eng. Technol.*, vol. 4, no. 4, 2017.
[7] V. Harish, "Logistics System," *Int. J. Bus. Policy Strateg. Manag.*, vol. 4, no. 2, pp. 17–24, 2017.
[8] T. Ning, "Vehicular GPS monitor system based on GPRS network," *IEEE 2007 Int. Symp. Microwave, Antenna, Propag. EMC Technol. Wirel. Commun. MAPE*, pp. 277–280, 2006.
[9] P. D. Hardo, Suprapto, and R. Pulungan, "Perancangan Sistem Pengawasan Pengiriman Barang Menggunakan GPRS, GPS, Google Maps, Android, dan RFID pada Intelligent Warehouse Management System," *Proc. First Symp. Ind. Technol.*, no. November, pp. C58–C66, 2012.
[10] M. Tegar and Y. Muslim, "Rancang Bangun Sistem Pemantauan Kendaraan Pengirim Barang Menggunakan GPS Berbasis Webgis," 2013.
[11] S. Kannimuthu, C. K. Somesh, P. D. Mahendhiran, D. Bhanu, and K. S. Bhuvaneshwari, "Certain investigation on significance of Internet of Things (IoT) and Big Data in vehicle tracking system," *Indian J. Sci. Technol.*, vol. 9, no. 39, 2016.
[12] S. K. R. Mallidi, "Iot Based Smart Vehicle Monitoring System," *Int. J. Adv. Res. Comput. Sci.*, vol. 9, no. 2, pp. 738–741, 2018.
[13] R. Somya, "Sistem Monitoring Kendaraan Secara Real Time Berbasis Android menggunakan Teknologi CouchDB di PT. Pura Barutama," *J. Nas. Teknol. dan Sist. Inf.*, vol. 4, no. 2, pp. 53–60, 2018.
[14] L. C. M. Varandas, J. J. P. C. Rodrigues, and B. Vaidya, "mTracker: A mobile tracking application for pervasive environment," *24th IEEE Int. Conf. Adv. Inf. Netw. Appl. Work. WAINA 2010*, pp. 962–967, 2010.

IOP Conf. Series: Materials Science and Engineering **771** (2020) 012013 doi:10.1088/1757-899X/771/1/012013

Root-Mean Adjustment Threshold for the Improvement of Anomaly Detection Algorithm in Hyperspectral Images

M Zaini[1] and M Rizkinia[1,*]

[1]Department of Electrical Engineering, Universitas Indonesia, Depok 16424, Indonesia.

[*]mia@ui.ac.id

Abstract. Hyperspectral anomaly detection aims at identifying unique objects with different spatial and spectral appearances from its surrounding pixels. One of the widely used conventional methods is Collaborative-Representation-Based Detector (CRD). CRD approximates each pixel on the background that can be represented by neighboring pixels, while the anomaly pixels cannot be. The detection image generated from this method is quite satisfying since it can detect anomaly pixels quite accurately. However, the resulting detection images still contain a lot of pixels which are not anomaly even with tiny intensity values. In this study, we apply the Root-Mean (RM) adjustment threshold to filter the false-positive pixels in the output images of CRD, so that it gives more accurate results. In the experiment, six hyperspectral data with the corresponding ground truth images are used. The results show that after applying the RM threshold, the accuracy increases with the decrease of the six RMSEs in the range of 31.51 – 63.58%.

1. Introduction

Hyperspectral imagery (HSI) is a type of satellite imagery that contains hundreds of narrow and contiguous spectral bands making it possible to distinguish different objects based on their spectral signatures. Various applications of HSI have been provided and still drawn researchers' attention. Over the past two decades, hyperspectral anomaly detection has indeed become a very interesting topic for researchers to develop under its high dimensionality and redundant information. Various algorithms in anomaly detection are very useful especially in civil and military affairs [1] - [4]. One algorithm that has been introduced is the Collaborative Representation-Based Detector (CRD). CRD has yielded satisfactory results so far, but there are still many unwanted pixels in the resulting image as false-positive pixels. Therefore, this study aims to create an appropriate threshold value to filter out these pixels.

In this study, the Root-Mean (RM) adjustment threshold will be applied to the CRD results to improve the accuracy by filtering the false-positive pixels in the CRD output images. Then, the RM threshold adjustment performance will be evaluated by comparing the results to those of CRD with a threshold taken from the ROC curve analysis.

2. Anomaly Detection in Hyperspectral Images

Hyperspectral images that are rich in spectral information are unique to each other, facilitating various applications in remote sensing technology, which one of them is target detection. The target can be a recognized object or an anomaly contained in certain pixels in the image. Anomaly detection is conducted by distinguishing certain objects that do have significant differences with other pixels, which then these other pixels will be referred to as background.

Anomaly is defined by referring to a background modeling in which the background modeling uses reference data both from local data at certain pixels and overall data from the image. The pixels in the background can be represented by certain mathematical modeling. There is a relationship between one pixel and another in the form of a linear equation. The pixels that are not related to the equation can be the pixels of the intended anomaly. Therefore, if the background is successfully identified, at the same time the anomaly can also be identified.

The purpose of anomaly detection is to detect pixels in an image that has a spectral different from its surroundings [5]. Based on an approach taken into the development of a method, there are five categories of anomaly detection methods, i.e., those that based on statistical modeling, Kernel, projection, segmentation, and representation [6]. The most popular statistical modeling-based algorithm is the one with the application of the Gaussian and RX distribution formulas, which is commonly used as a comparison in creating other algorithms. RX Global or also called Mahalanobis distance detector [7], models the background of all parts of the image with a Gaussian distribution formula. Similar to global, but the local difference is that the process occurs not all at once in the image as in global, but on a particular local bounded by a window. As for the widely used representation-based algorithm is Collaborative-Representation-Based Detector (CRD), which is based on the sparse representation of the anomaly pixels.

2.1. Collaborative-Representation-Based Detector

CRD is one method of target detection, in this case, an anomaly that also belongs to the unsupervised category, where the target object is unknown. The concept carried by CRD is to represent each pixel (normal) in the background, which is estimated from the surrounding spatial pixels. Representation is displayed with a linear combination of these pixels, in this case, pixels that cannot be represented will be considered as an anomaly or desired target. Therefore, once the background estimation can be represented, the desired anomaly pixel will also be obtained.

CRD applies two local windows to local pixels (dual-window strategy). The term window refers to the width of the local pixel matrix in the image. There is an inner window (w_{in}) and an outer window (w_{out}) whose size can be adjusted and measured which produces the most optimal anomaly pixels.

There are some uniqueness as well as advantages of the CRD method [8]. It is the method that was first applied to the detection of anomalies without estimating the covariant matrix in the background [11]. CRD applies a regularization matrix that allows pixels representing the background to be adaptive to conditions where there are also anomalous pixels on background pixels.

Given a three-dimensional (3D) hyperspectral image in the form $\mathbf{H}= \{\boldsymbol{h}\}_{i=1}^{n}$ in size of $\mathbf{R}^d$, where d is the number of spectral bands of the image, n is the number of samples that will be used. Next, there is $\boldsymbol{y}$, a pixel size $d \times 1$, which is the pixel to be tested. This pixel as a central pixel, located in the middle of a collection of data that forms a matrix [8].

The matrix is a two-dimensional matrix $\boldsymbol{H}_s = \{\boldsymbol{h}_i\}_{i=1}^{s}$ the result of the transformation of the three-dimensional matrix, s is the number of samples whose size depends on the size of two windows (hereinafter referred to as the window), i.e. the inner window (w_{in}) and outer window (w_{out}). In this case, $s = w_{out} \times w_{out} - w_{in} \times w_{in}$ and matrix $\boldsymbol{H}_s$ is contained in each pixel $\boldsymbol{y}$ (local) as illustrated in figure 1.

Refer to [9], the CRD algorithm is intended to find the weight vector $\boldsymbol{\lambda}$ obtained from minimizing the equation $||\boldsymbol{y} - \boldsymbol{H}_s\,\boldsymbol{\alpha}||_2^2$ and $||\,\boldsymbol{\lambda}||_2^2$. In the second term, a matrix regularization called distance-weighted Thikonov [12] is added which was previously applied to [9] and [10] shown in equation (1).

$$\Gamma_y = \begin{bmatrix} ||\boldsymbol{y} - \boldsymbol{h}_1||_2 & & \\ & \ddots & \\ & & ||\boldsymbol{y} - \boldsymbol{h}_s||_2 \end{bmatrix} \quad (1)$$

wherex$\boldsymbol{h}_1, \boldsymbol{h}_2, \ldots, \boldsymbol{h}_s$is the columns of matrix $\boldsymbol{H}_s$. This regularization will calculate the euclidean distance between the middle pixels and each pixel $\boldsymbol{H}_s$ whose role is to compare how similar the values of neighboring pixels are to the central pixel. This matrix will give a new equation to be

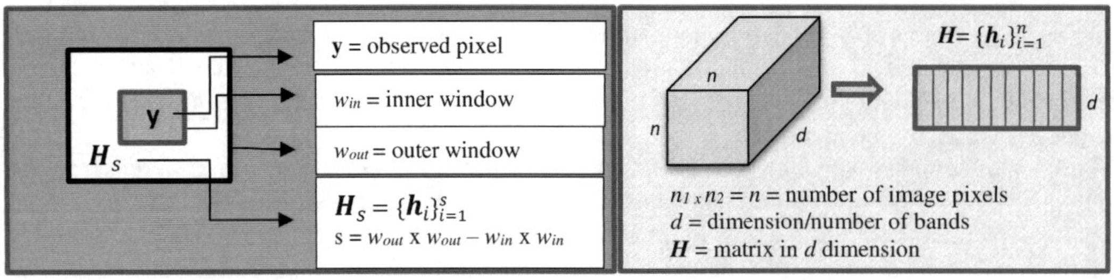

Figure 1. Visualization of observed pixel **y**

$$\arg\min_{\alpha} ||\boldsymbol{y} - \boldsymbol{H}_S \boldsymbol{\lambda}||_2^2 + \lambda|| \Gamma_y \boldsymbol{\lambda}||_2^2 \quad (2)$$

which implies that the penalty value for each element of $\boldsymbol{\lambda}$ depends on the similarity (shown by the distance) of these pixels to the central pixel. In other words, if the distance gets smaller on a certain pixel, it is almost the same as the central pixel (higher similarity), in this case, a relatively large coefficient value is allowed, and vice versa. Anomaly pixels can be obtained from the residual image, namely the reduction of the original image with a background representation as shown by the equation (3).

$$r = ||\boldsymbol{y} - \hat{\boldsymbol{y}}||_2 = ||\boldsymbol{y} - \boldsymbol{H}_s\boldsymbol{\lambda}||_2 \quad (3)$$

If r is greater than a particular threshold value, the pixel $\boldsymbol{y}$ can be claimed to be an anomaly pixel. Conversely, if a value smaller than the threshold is obtained, the pixel $\boldsymbol{y}$ is considered a normal pixel representing the background pixel. The final results of this detection process will be accommodated in a matrix $\mathbf{Y} = \{\boldsymbol{\lambda}_i\}_{i=1}^{n}$ where $\boldsymbol{\lambda}_i$ is equal to r for anomaly pixels and 0 for normal pixels, then the filter will proceed with the Root-Mean adjustment threshold.

2.2. T_{roc} Threshold

This T_{roc}-Threshold adjustment threshold hereinafter referred to as T_{roc}, is obtained by taking a threshold or the best value limit from the ROC process. The ROC graph is represented by a false alarm rate on the *x*-axis and probability of detection on the *y*-axis. The algorithm of T_{roc} is summarized in figure 2.

3. Root-Mean (RM) Threshold

Root-Mean Adjustment Threshold hereinafter referred to as RM is a value limit that will filter the pixels of the weight vector or output image from the CRD detector. The initial idea of this concept is that a boundary value that is sufficient to represent a value from group data is needed. In statistics, there are some centralized data values that can represent grouped data, one of which is Mean or average. Mean is believed to be more comprehensive than other data centralization because it involves and treats all data fairly so that one value is obtained as a representation of all the values that existed before. Because in this research the context is an anomaly, there will be several stages to get a fairly fair RM value. A group of data on the anomaly matrix consists of most of the zero pixels and pixels close to zero, while the anomaly pixels only amount to a few and as much as possible pixel value 1 or close to 1. Therefore, the Mean value of the anomaly matrix data is not directly used to filter the image CRD results but to filter pixels whose value close to 1 (pixels that are likely to be anomalous pixels).

The pixels obtained are then accommodated on a new matrix. This new matrix is whose mean value will be used to filter the CRD image. Because the mean value obtained is still relatively small compared to the anomaly pixel values which is relatively large, so it needs to be rooted to get better results. The value of RM is obtained from a mathematical equation which will be described below. Given a matrix $\mathbf{R} = \{t_i\}_{i=1}^{m}$ where t_i is an element of output matrix $\mathbf{Y}$ which is less than $\widehat{\mathrm{R}}$. $\widehat{\mathrm{R}}$ is the average of $\mathbf{Y}$.

$$\widehat{\mathrm{R}} = \frac{\sum Y}{n} = \frac{\sum_{i=1}^{n} \lambda_i}{n} \quad (4)$$

$$t_i = r_i > \widehat{\mathrm{R}} \quad (5)$$

while m is the number of elements from t_i, i.e. the number of elements of the output **Y** image that is greater than $\widehat{\mathbf{R}}$. Next, the RM value of the **R** matrix, which is the root of the mean of **R.**

$$\mathrm{RM} = \sqrt{\frac{\sum \boldsymbol{R}}{m} = \frac{\sum_{i=1}^{m} t_i}{m}} \tag{6}$$

The obtained RM value will be the limit in the filtering output image **Y**. Matrix **Y** will have a new value. To make it easier to distinguish the original **Y** value from the new one, the results of this filter will be accommodated into a new matrix that is $\widehat{\mathbf{Y}} = \{\lambda_j\}_{j=1}^{n}$, where λ_j is equal to 1 when $\lambda_j \geq RM$, otherwise 0. The results of this CRD-RM are predicted to provide better image quality than the results of the CRD detection without RM and also the CRD results with the best threshold taken from the ROC. Figure 2 explains the concept of applying the threshold RM compared to the T_{roc} threshold.

<u>T_{roc} Threshold Algorithm</u>

Input Matrix **Y**
For all pixels **do**
  For each value of threshold on ROC curve, count T_{roc} = TPR – FPR
End for
Find the maximum T_{roc}
For all pixel **do**
1. Compare each pixel in **Y** with the T_{roc} max.
2. If the pixel is greater than or equal to T_{roc}, the result will be 1 (white), otherwise 0 (black).
3. The pixels assemble matrix $\widehat{\boldsymbol{Y}}_{Troc}$

End for
Output: Image of anomaly detection result (matrix $\widehat{\mathbf{Y}}_{Troc}$)

<u>RM Threshold Algorithm</u>

Input Matrix **Y**
Count mean $\widehat{\mathrm{R}}$
For all pixels **do**
  For each pixel **Y**, compared with $\widehat{\mathrm{R}}$ respectively.
End for
For pixels **Y** > $\widehat{\mathrm{R}}$, saved in matrix **R**
Count mean of matrix **R** based on (6), resulting RM
For all pixels **do**
- Compare each pixel in **Y** with the RM.
- If the pixel is greater than or equal to RM, the result will be 1 (white), otherwise 0 (black).
- The pixels assemble matrix $\widehat{\mathbf{Y}}_{\mathrm{RM}}$

End for
Output: Image of anomaly detection result (matrix $\widehat{\mathbf{Y}}_{\mathrm{RM}}$)

Figure 2. Comparison of T_{roc} and RM thresholding algorithm

4. Result Evaluation

To seek the performance and accuracy of CRD-RM, an experiment was conducted to six data[1][13] by comparing CRD-T_{roc} and CRD-RM based on visual quality and RMSE. Table 1 shows the Root Mean Square Error (RMSE) values for the two methods. RMSE shows the quantitative differences between the constructed anomaly map and the ground truth anomaly map by taking the root of square different average for all pixels.

From the table, CRD-RM shows improvement with lower RMSE in all data as well as reduces error in the range 31.51 – 63.58%. Moreover, in terms of visual quality, CRD-RM yields satisfactory results for all data as can be shown in figure 3. From the figure, one can see that CRD-T_{roc} results in many false-positive pixels (background pixels are detected as anomaly). Meanwhile, CRD-RM encounters the problem but raises some false-negative pixels (anomaly pixels are detected as background). However, the false-negative pixels in CRD-RM are still fewer than the false-positive pixels in CRD-T_{roc} which can be seen from figure 3. The errors in CRD-RM are caused by the pixels that is rounded up to 1 even though there are only one or two pixels that are not desired.

The value of RM is determined by taking one of the centralized data values which in statistical theory is a mean. The mean can be said to be the best representation of a group of data. However, because in this case, in the context of an anomaly where there are only a few anomalies compared to thousands of data (pixels), the total value is zero, the mean value taken is the mean value of data greater than $\widehat{\mathrm{R}}$ as in

[1] http://xudongkang.weebly.com/data-sets.html

equation (5). Once again because of the anomaly context, the mean obtained will be squared so that it has a greater value and the undesirable pixels do not participate in the final output image.

Figure 3 shows the image detected in the two methods. It can be seen that the CRD-*Troc* column gives results with a lot of false-positive pixels caused by the small threshold value. Meanwhile, CRD-RM gives more satisfying results which reduce RMSE

Table 1. RMSE CRD-T_{roc} dan CRD-RM

Algorithm	Airport1	Airport3	Beach3	Beach4	Urban1	Urban2
CRD-T_{roc}	0.2317	0.1942	0.0316	0.1565	0.1667	0.1987
CRD-RM	**0.1166**	**0.1330**	**0.0200**	**0.0570**	**0.0663**	**0.1217**
Error reduction	49.68%	31.51%	36.71%	63.58%	60.23%	38.75%

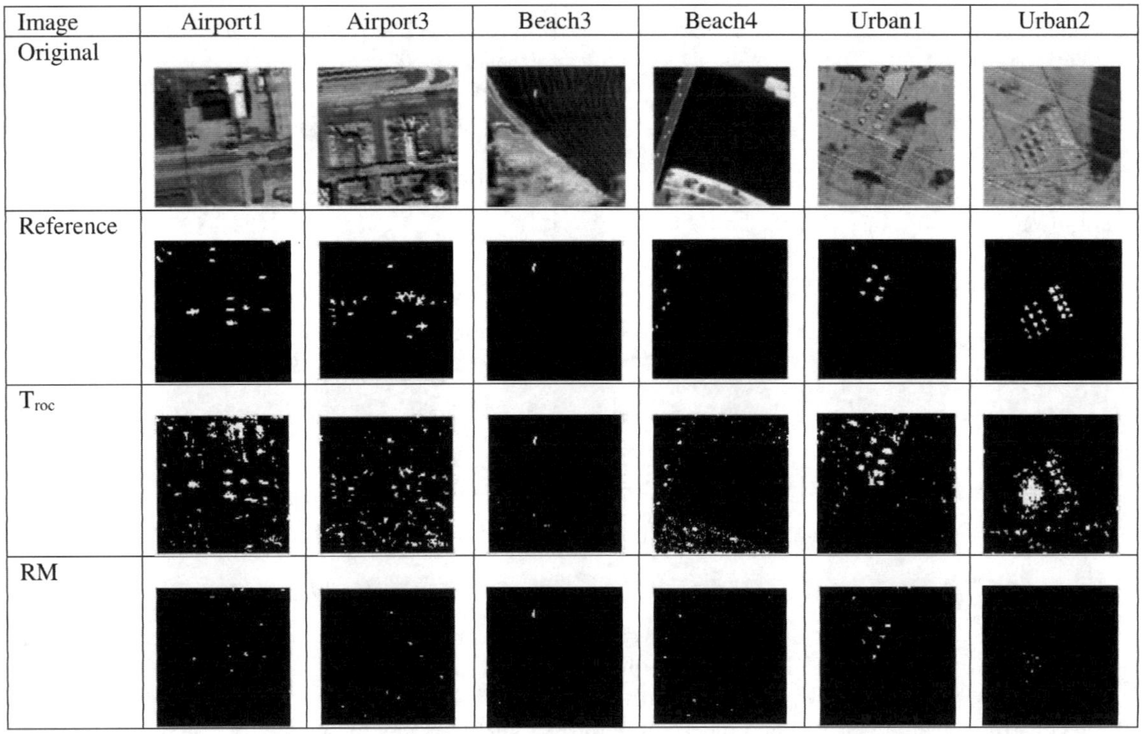

Figure 3. Comparison of anomaly images of CRD-T_{roc} dan CRD-RM algorithm.

5. Conclusions

Based on experiments conducted in this study, we can conclude that Root-Mean algorithm that is applied to the output images of CRD anomaly detection significantly improves the accuracy in terms of quantitative (RMSE) and qualitative for the visual appearance, compared to the conventional thresholding method, i.e. T_{roc}. It reduced error in the range 31.51 – 63.58% compared with the CRD-T_{roc}. Future work may lay on the development of anomaly detection algorithm itself.

6. References

[1] Manolakis D and Shaw G 2002 Detection algorithm for hyperspectral imaging applications *IEEE Signal Process. Mag.* vol 19 chapter 1 p 29–43

[2] Stein D W J *et al* 2002 Anomaly detection from hyperspectral imagery *IEEE Signal Process. Mag.* vol 19 chapter 1 p 58–69

IOP Conf. Series: Materials Science and Engineering **771** (2020) 012013 doi:10.1088/1757-899X/771/1/012013

[3] Schweizer S M and Moura J M F 2000 Hyperspectral imagery: Clutter adaption in anomaly detection *IEEE Trans. Inf. Theory* vol 46 chapter 5 p 1855–71

[4] Bajorski P 2012 Target detection under misspecified models in hyperspectral images *IEEE J. Sel. Topics Appl. Earth Observ. Remote Sens.* vol 5 chapter 2 p 470–7

[5] Kucuk S and Yuksel S E 2015 Comparison of RX-Based Anomaly Detectors on Synthetic and Real Hyperspectral Data *IEEE Whispers* (Turkey-Ankara: Hacettepe University, Department of Electrical and Electronics Engineering)

[6] Xie W, Jiang T, Li Y, Jia X and Lei J 2019 Structure Tensor and Guided Filtering-Based Algorithm for Hyperspectral Anomaly Detection *IEEE Trans. Geosci. Remote Sens* vol 57 chapter 7 p 4218-30

[7] Reed I S and Yu X 1990 Adaptive multiple-band CFAR detection of an optical pattern with unknown spectral distribution *IEEE Trans. Acoust., Speech, Signal Process.* vol 38 chapter 10 p 1760–70

[8] Li W and Du Q 2015 Collaborative Representation for Hyperspectral Anomaly Detection *IEEE Trans. Geosci. Remote Sens.* vol *53* chapter 3

[9] Zhang L, Yang , and Feng X 2011 Sparse representation or collaborative representation: Which helps face recognition?. *Proc. IEEE Int. Conf. Comput. Vis.* (Barcelona, Spain) p 471–8.

[10] Zhang L, Yang M, Feng X, Ma Y and Zhang D 2012 Collaborative Representation Based Classification for Face Recognition (arXiv: 1204.2358)

[11] Li W and Du D 2013 Unsupervised Nearest Regularized Subspace for Anomaly Detection in Hyperspectral Imagery *Proc. IEEE Int. Geosci. Remote Sens. Symp., Melbourne, Vic.* (Australia) p 1055–8

[12] Li W, Tramel E W, Prasad S, and Fowler J E 2014 Nearest Regularized Subspace for Hyperspectral Classification *IEEE Trans. Geosci. Remote Sens.* vol 52 chapter 1 p 477–89

[13] Kang X, Zhang X, Li S, Li K, Li J, Benediktsson J A 2017 Hyperspectral Anomaly Detection with Attribute and Edge-Preserving Filters *IEEE Trans. Geosci. Remote Sens.* vol 55 chapter 10 p 5600–11

Acknowledgments
Authors wishing to acknowledge support funding from Hibah Publikasi Internasional Terindeks Tugas Akhir (HIBAH PITTA B) Number NKB-0753/UN2.R3.1/HKP.05.00/2019 from Universitas Indonesia.

IOP Conf. Series: Materials Science and Engineering **771** (2020) 012014 doi:10.1088/1757-899X/771/1/012014

Coil Inductance Design for Low Power Hybrid Wireless Power Transfer

T A Kurniawan[1,*], R B Gumilang[1], G Wibisono[1]

[1]Department of Electrical Engineering, Faculty of Engineering, Universitas Indonesia

[*]email: taufiq.alif@ui.ac.id

Abstract. This paper presents a coil inductance design for low power hybrid wireless power transfer (WPT). The proposed coil inductance design of transmitter and receiver circuits of 7.1 μH are designed using handmade copper wire structure. Several coils with the variations of wire diameters, coil loop radius, and numbers of loops are proposed to find optimum structure in the low power hybrid WPT. The optimum coil structure is obtained at wire diameter of 0.5 mm, 2.25-inch coil loop radius and 1 MHz operation frequency. The measured transceiver coil design has exhibited peak power transfer efficiency (PTE) of 33.1% at 1 cm distance between Tx and Rx. Therefore, the proposed coil design is applicable to low power hybrid wireless power transfer circuits.

1. Introduction

Nowadays, the medical implant devices are very promising to conduct therapy in several conditions. Many therapies especially in neurological stimulation for sleep apnea, pain management, Parkinson's Disease, epilepsy, bladder control, gastrointestinal disorders, numerous autoimmune diseases and psychological disorders can be effectively conducted using implantable circuits [1]. In addition, the retinal prosthesis system using wireless power transfer in [2] for retinal degenerative patients is also interesting solution to reduce the cause of blindness in the world.

One of the successful keys to design wireless power transfer is the optimal wireless power links, especially on the accuracy of the model, i.e., self-inductance and the mutual inductance [3]. Since the two optimum parameters will determine the characteristic power transfer efficiency of the device, many researches have focused on the algorithm design to develop accurate models of wireless power links [4-11]. The most widely approach for designing coil inductors is by using copper wire with a resistivity of 1.68×10^{-8} Ωm [12] which is designed by optimizing the coil structure. In addition to using a mathematical model, this approach requires a coil model iteration that is measured separately from the whole WPT devices.

This paper is a preliminary research of the hybrid wireless power transfer (H-WPT) design for medical implant devices. The hybrid transceiver coil is designed to facilitate the circuit measurements. In our model, the transmitter block consists of a DC power source, oscillator and transmitter coil. While the receiver block consists of coil receiver, rectifier, and voltage regulator. The transceiver coils are designed in a circular shape with several iterations of copper wire diameter and coil loop radius. To evaluate the reliability of the proposed transceiver coils, we tested the proposed coil in the air with the distance variations between transmitter and receiver i.e., 0 cm, 1 cm, 2 cm and 3 cm.

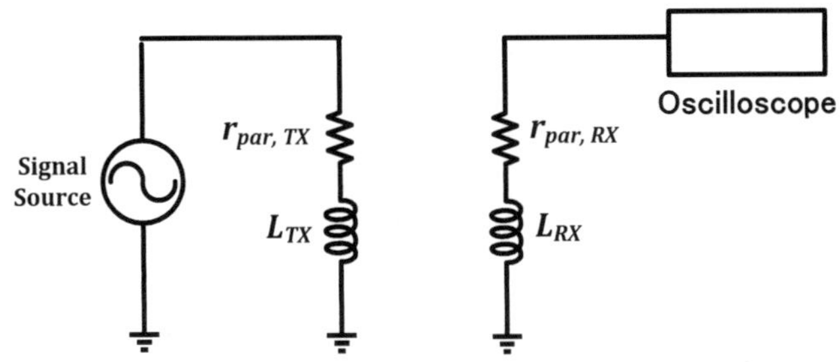

Figure 1. The circuit configurations

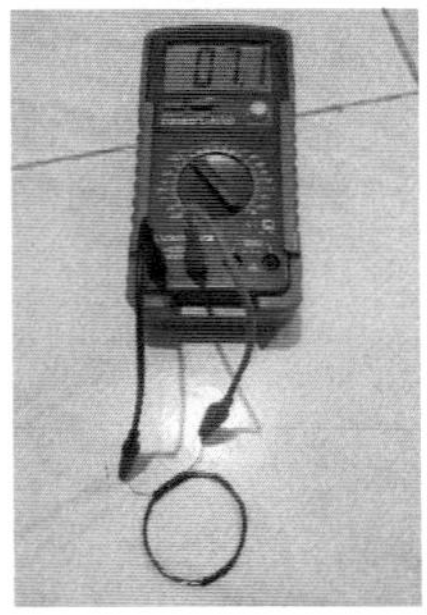

Figure 2. Coil design

2. The circuit configuration

The schematic circuit is shown in Figure 1 The signal source is connected directly to the transmission coil (L_{TX}), while the receiver coil is connected to the oscilloscope. $r_{par,\,TX}$ and $r_{par,\,RX}$ are the parasitic resistance in the coil inductor. Its value is determined by the Q-factor value of the coil, following the equation:

$$Q = \frac{\omega L}{r} \tag{1}$$

where,

Q = Q factor of transceiver coil
ω = $2\pi f$
r = parasitic resistance (Ω)

Because the required inductance coil values for the transmitter and the receiver circuits for the proposed WPT circuit are 7.1 µH, the transmitter and receiver coils are designed using copper e-mail wire. The structure design of the coil is shown in Figure 2.

Calculation of the coil inductance is performed by referring to equation [4]:

$$L = \mu R\left[ln\left(\frac{16R}{w}\right) - 2\right] \tag{2}$$

where,

μ = the permeability of the medium surrounding the coil (H/m)
w = wire diameter (m)
R = coil loop radius (m)

3. Measurement Results

Figure 3 illustrates the measurement setup of the proposed circuits. The signal source is Kenwood AG-203 wide Frequency Range Oscillator 10Hz-1MHz. The Oscilloscope GW-instek GDS-1102U 100 MHz is used for the output signal measurement. Several structure coils are designed by handmade using the copper wire of different wire diameters, the coil loop radius, and the numbers of loops. Figure 4 depicts the coil layout design for 7.1 µH modelled by using equation (2). In order to verify the self-inductance value, the LCR meter Dekko L-4070D is used.

Figure 5 shows the measurement of power transfer efficiency (PTE) at a wire diameter of 0.5 mm and the distance between Tx and Rx is 1 cm. It is illustrated that the highest PTE of 31.44% is obtained at the highest operating frequency and the largest coil loop radius. Increasing the coil loop radius and the frequency are directly proportional to the improvement of the PTE.

Figure 6 illustrates the PTE measurement on a 2.25-inch coil loop radius and 1 cm Tx-Rx distance. It is seen that the variations of the wire diameter influence the PTE, with the highest value at a wire diameter of 0.5 mm. Figure 7 shows the PTE measurement on a 2.25-inch coil loop radius and 1 MHz operation frequency. It shown that coil distance variations from 0 to 3 cm decrease the PTE from about 70 % to 10 %.

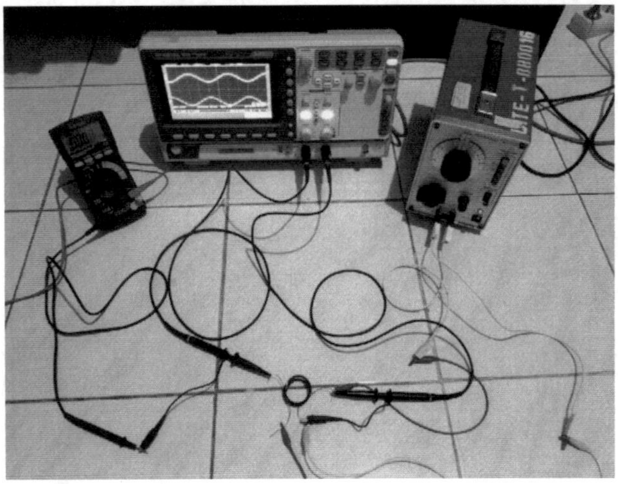

Figure 3. Measurement setup

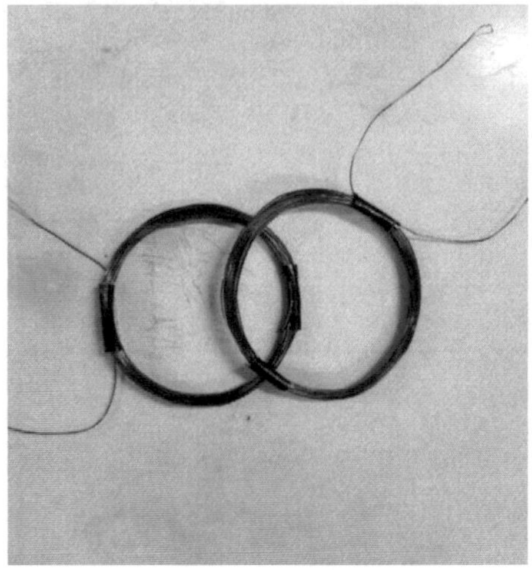

Figure 4. Coil layout design of 7.1 µH

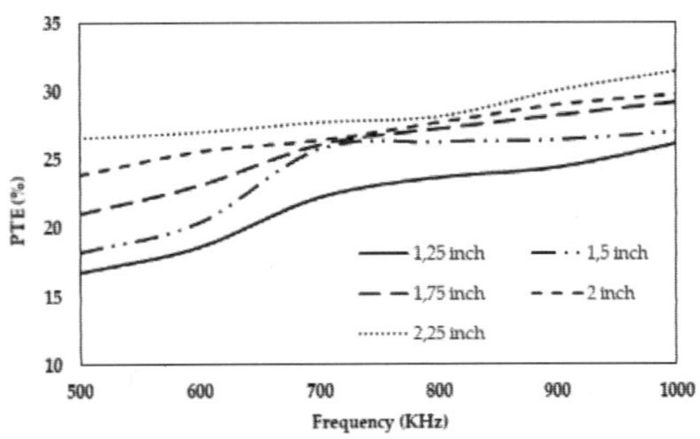

Figure 5. Measurement results of the power transfer efficiency (PTE) over 0.5 mm wire diameter and 1 cm distance between Tx to Rx

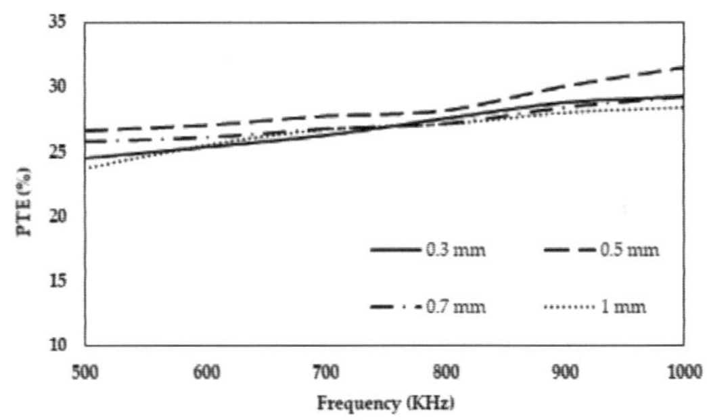

Figure 6. Measurement results of the power transfer efficiency (PTE) over 2.25-inch coil loop radius and 1 cm distance between Tx to Rx

4. Conclusion

A 7.1-µH coil inductance design has been demonstrated using handmade copper wire structure. The proposed transceiver coil inductance has 33.1% PTE at a 1 cm distance between transmitter and receiver. Based on the various wire diameters, the coil loop radius, and the numbers of loops, the optimum coil structure is obtained at a 0.5 mm wire diameter and a 2.25-inch coil loop radius. In order to improve the PTE, increasing operation frequency until several MHz is possible. However, the WPT applications in biomedical implant device is restricted about 20 MHz. Therefore, the proposed coil inductance is applicable to low power hybrid wireless power transfer systems.

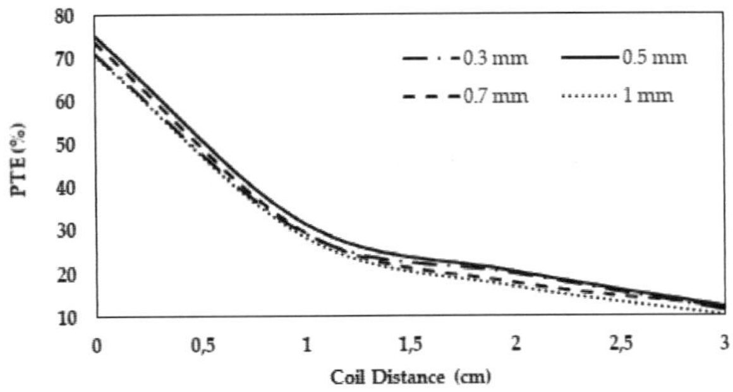

Figure 7. Measurement results of the PTE over 2.25-inch coil loop radius and 1 MHz frequency

Acknowledgment
The authors would like to thank Ministry of Research, technology and Higher Education Republic of Indonesia for funding through "Hibah Penelitian Terapan Kementerian Riset, Teknologi, dan Pendidikan Tinggi Republik Indonesia Tahun Anggaran 2019", under contract No. NKB-1802/UN2.R3.1/HKP.05.00/2019.

References
[1] I. A. Mashahadi, M. Pahlevani, S. Hor, H. Pahlevani, E. Adib, "A New Wireless Power Transfer Circuit for Retina Prosthesis", IEEE Transactions on Power Electronics, vol. 34, no. 7, pp. 6425-6439, July 2019.
[2] J. McDonald, "Integrated Circuits for Implantable Medical Devices", White paper on Freescale, Journal Title, electronic file available at: https://cache.freescale.com/files/32bit/doc/white_paper/ICIMDOVWP.pdf, August 2019.
[3] S. I. Babic and C. Akyel, "Calculating mutual inductance between circular coils with inclined axes in air," IEEE Trans. Magn., vol. 44, no. 7, pp. 1743–1750, Jul. 2008.
[4] Sadeque Reza Khan , Sumanth Kumar Pavuluri, and Marc P. Y. Desmulliez", Accurate Modeling of Coil Inductance for Near-Field Wireless Power Transfer", IEEE Transaction on Microwave Theory and Technique, vol. 66, no. 9, pp. 4158-4169, Sept 2019.
[5] A. K. RamRakhyani, S. Mirabbasi, and M. Chiao, "Design and optimization of resonance-based efficient wireless power delivery systems for biomedical implants," IEEE Trans. Biomed. Circuits Syst., vol. 5, no. 1, pp. 48–63, Feb. 2011.
[6] C. M. Zierhofer and E. S. Hochmair, "Geometric approach for coupling enhancement of magnetically coupled coils," IEEE Trans. Biomed. Eng., vol. 43, no. 7, pp. 708–714, Jul. 1996.
[7] S. Atluri and M. Ghovanloo, "Design of a wideband power-efficient inductive wireless link for implantable biomedical devices using multiple carriers," in Proc. 2nd Int. IEEE EMBS Conf. Neural Eng., Mar. 2005, pp. 533–537.
[8] X. Li et al., "A wireless magnetic resonance energy transfer system for micro implantable medical sensors," Sensor, vol. 12, no. 8, pp. 10292–10308, Aug. 2012.
[9] Y. Yi, U. Buttner, Y. Fan, and I. G. Foulds, "Design and optimization of a 3-coil resonance-based wireless power transfer system for biomedical implants," Int. J. Circuit Theory Appl., vol. 43, no. 10, pp. 1379–1390, 2015.
[10] T. P. Duong and J.-W. Lee, "A dynamically adaptable impedance matching system for midrange wireless power transfer with misalignment," Energies, vol. 8, no. 8, pp. 7593–7617, 2015.
[11] D.-H. Kim, J. Kim, and Y.-J. Park, "Optimization and design of small circular coils in a magnetically coupled wireless power transfer system in the megahertz frequency," IEEE Trans. Microw. Theory Techn., vol. 64, no. 8, pp. 2652–2663, Aug. 2016.
[12] 1. Giancoli, Douglas C., Physics, 4th Ed, Prentice Hall, (1995).

Analysis of Arduino Uno Application on Control System Based on Industrial Scale

A J Taufiq[1,*], I H Kurniawan[1], T A Y Nugraha[1]
[1]Electrical Engineering Department of Universitas Muhammadiyah Purwokerto
Jl. Raya Dukuhwaluh Kembaran, Banyumas, Jawa Tengah, Indonesia 53182

[*]arifjt@ump.ac.id

Abstract. This research analyses the use of Arduino Uno as a controller, replacing the industrial based controller on the oil and gas industry with the plant: one of the control loops in the Refinery Fuel Oil Complex I PT. Pertamina (Persero) Refinery Unit IV Cilacap, Indonesia. Research is done by making a simulation plant using industrial-scale instrumentation with the Arduino Uno as the controller, and compare the results with the controller for the actual plant. The result is Arduino Uno is able controlling industrial-scale plant-based instrumentation for one control loop that was tested and was able to eliminate the difference between Process Variables and Sets Points at different control modes are proportional to industrial class controllers by using the Proportional, Integral and Derivative (PID) algorithms together with the combination.

1. Introduction

At this time, the world of technology is developing rapidly in all fields. With the rapid advancement of science and technology today, much available equipment that can work automatically using a digital control system. Application of automatic control system technology is vast, ranging from household items such as room temperature control system, up to the level of process automation industries such as oil and gas processing.

Arduino Uno is a microcontroller board open-source platform that is widely used and widely recognised by students, especially for students studying electrical engineering material control system, both in lectures and lab work [1-2].

Previous studies have shown that an Arduino-based PID controller application can be used to control a process automatically [3-6]. Arduino Uno makes it easy for students to learn digital PID control in real-time, some PID control parameters can be changed easily, and responses can be observed directly [7].

Previous studies have shown that an Arduino-based PID controller application can be used to control a process automatically [8-9]. In this study, we will discuss the application of an Arduino-based PID controller to control vessels level automatically but use hardware in the form of industrial-scale instrumentation equipment to examine possibilities the use of Arduino as an alternative to the control system in industrial.

2. Materials and Methods

2.1 Material

Arduino Board, compressor, Current to Voltage Converter module, Voltage to Current Converter module, Level Transmitter Honeywell ST700 Series, PWM to Voltage Converter module, Smart Valve Positioner Azbil AVP301, Regulator lock, software Intouch Wonderware, Honeywell Experion Station and Software Arduino IDE.

2.2 Methods
Prepared a control simulation that reflects operational in the Industry. The plant is set to be controlled with an Arduino-based controller with PID control. To be compatible with industrial instrumentation, the control, actuator and adjust to the industry standard through the conversion of the required signal that is the conversion of voltage to current and current to voltage instrumentation current signals 4-20mA ISA-50.00.01-1975 (R2012). For testing the application of Arduino as a controller in the industry, it is tried in one control inserted through the control room to control in one of the control loops.

3. Results and Discussions
The working principle of this control system is to create a plant or process that is equipped with industrial-based instrumentation, with a fluid level control system as a control loop representing industrial processes. This control system can be controlled centrally through the HMI (Human Machine Interface), with the actual level, must be controlled automatically through the controller following the setpoint specified by the operator. After the control system is complete, calibration will be performed so that all parameters obtained are accurate following the actual conditions, then proceed with the tunning controller to get a stable system response according to the set point.

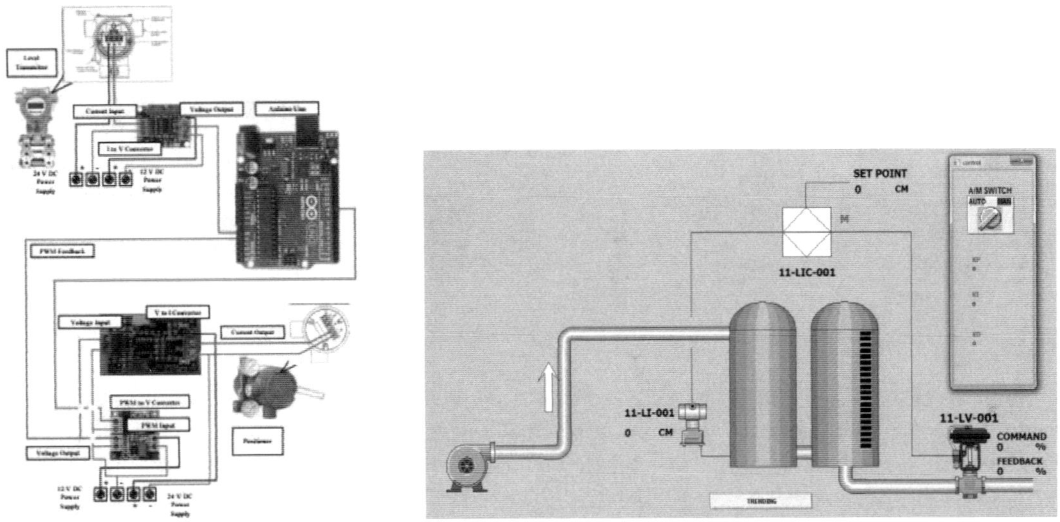

Figure 1. Arduino-based control system simulation and HMI Display

First of all, after the power on the system, The level transmitter will read the fluid level available in vessels. The level transmitter sends a 4-20mA signal (according to the standard signal used by industry) which is equivalent to 0-100% fluid level that will be accepted by current to voltage converter. Here the signal is 4-20mA converted to 0-5V DC which is connected directly to analogue input Arduino. Arduino that works as the controller will execute the system controlling according to parameters determined by the operator through HMI. Execute the controller here to control the same PV value with approaching the SP value (corresponding to tuning parameters).

Response from The controller is the output signal produced by Arduino in the form of a PWM pulse Comparable to 0-100% openings valve. Because Valve works using pneumatic power, the signal PWM is converted to 0-5V DC by PWM to Voltage Converter module. Then this voltage converted currents between 4-20mA to be accepted by the positioner as the final control element. This positioner will regulate valve openings pneumatically from 0-100% is proportional to the 4-20mA signal accepted by positioners. This control system works simultaneously and continuously in the loop closed.

The control loop is divided into two modes. Manual mode: operators are given access to set the output value in %. Comparable to 0-100% of the control valve opening. Auto mode: the controller will fully control output. Operators are given access to the Kp, Ki, and Kd parameters to set the controller response. This industrial-based Instrumentation control system simulation facility for the Arduino application analysis facility on the industrial-based instrumentation-based control system aims to answer two questions: How is the response based Arduino control system to controlling industrial-scale instrumentation equipment? Can the Arduino be used on industrial scale control systems? To answer the two questions above, testing is needed using the simulation facility that has been made, and then an analysis will be conducted to conclude. The testing stages to be carried out are: Manually testing the loop control simulation equipment to determine plant characteristics and tuning control loop to get the best response from the plant, conducting controlled trials in industrial processes.

Simulating changes in the parameters of the microcontroller tuning and analysing the response. The controller made with Arduino integrated with the original plant (according to the control valve specifications), i.e. laver 11V1 (with Tag No. control 11LIC010A) at the Refinery Fuel Oil Complex I, PT. Pertamina (Persero) Refinery Unit IV Cilacap (Figure 2).

Figure 2. Arduino integration with DCS

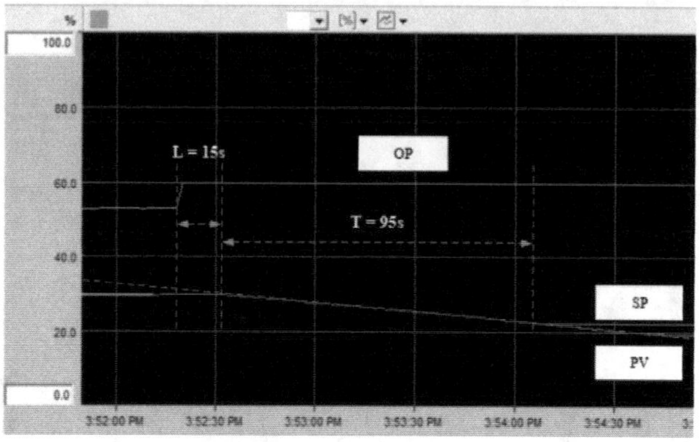

Figure 3. Transient response controller

With Modbus communication connected to the computer, access points can be sent through the controller to the DCS Honeywell Experion PKS station so that the controller can be tuned with the Ziegler-Nichols method or manually input through the station. After Arduino is installed into the real plant, the tuning is then performed using the trial and error method to determine the effect of changing the tuning parameters to the stability of the system. The process is still using the same plant, namely the control system level 11LIC-010A on 11V1 Refinery Fuel Oil Complex I PT. Pertamina (Persero) Refinery Unit IV Cilacap. The following is Arduino's response in handling industrial-scale control with different tuning parameters.

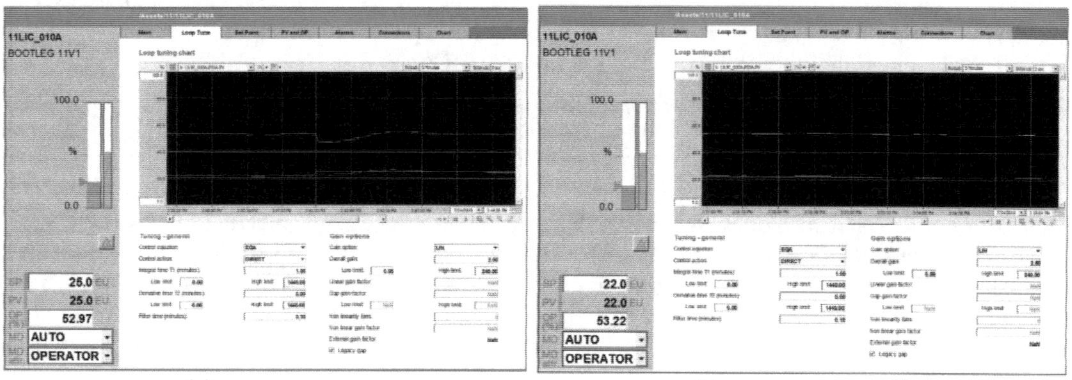

(a) (b)

Figure 4 Controller response to industrial plants a. (P = 2; I = 1; D = 0) b. (P = 2.5; I = 1; D = 0)

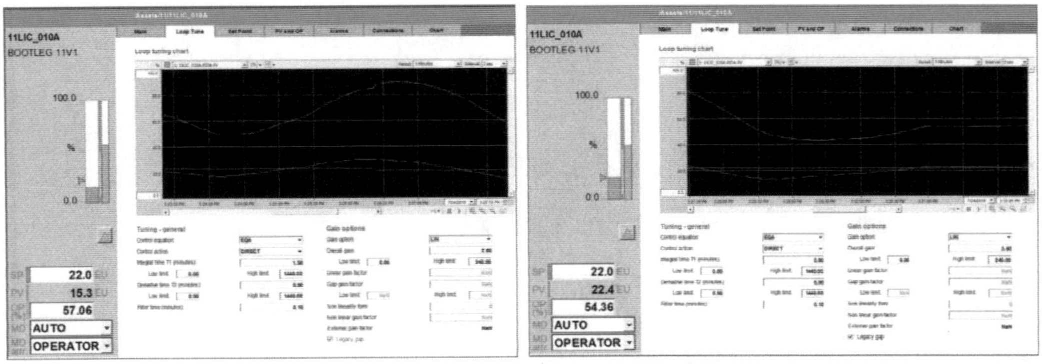

(a) (b)

Figure 5 Controller Response in Industrial Plants a. (P=2.5; I=1,5; D=0) b. (P=2.5; I=2; D=0)

4. Conclusion

The results of the study concluded as follows: Arduino-based control system can handle the process of industrial-scale control. Arduino technically can be used in industrial-scale control systems. The only constraint is the Arduino does not have the features Safety Instrumented Level (SIL) it should be added yourself. Arduino can be used to control the industrial process after going through the calibration process using industry-standard calibration equipment with high accuracy.

References

[1] Perenc, I., Jaworski, T., Duch, P. 2019 *Teaching programming using dedicated Arduino Educational Board. Computer Applications in Engineering Education*, [s. l.], v. 27, n. 4, p. 943–954.

[2] Jen-Hsing Li 2018 *Control System Laboratory with Arduino.* International Symposium on Computer, Consumer and Control (IS3C). ISBN: 978-1-5386-7037-8. IEEE.

[3] Allam, Thirupathi 2016 *Design of PID Controller for DC Motor Speed Control using Arduino Microcontroller.* International Research Journal of Engineering and Technology (IRJET).

[4] Theopaga, Adhi Ksatria 2014 *Design and Implementation of PID Control Based Baby Incubator.* Journal of Theoretical and Applied Information Technology (JATIT).

[5] Chen, X., & Li, H. 2017 *ArControl: An arduino-based comprehensive behavioral platform with real-time performance.* Frontiers in Behavioral Neuroscience.

[6] Bista, Dinesh 2016 *Understanding and Design of an Arduino-based PID Controller.* Virginia Commonwealth University. Virginia

[7] Z. Adel, A. A. Hamou and S. Abdellatif 2018 *Design of Real-time PID tracking controller using Arduino Mega 2560 for a permanent magnet DC motor under real disturbances.* International Conference on Electrical Sciences and Technologies in Maghreb (CISTEM), *Algier*s, pp. 1-5.

[8] SALIM, S. N. S. et al. 2019. *Enhanced self-regulation nonlinear PID for industrial pneumatic actuator.* International Journal of Electrical & Computer Engineering (2088-8708), [s. l.], v. 9, n. 4, Part 2, p. 3015–3024.

[9] Mir, Arsheen & Rajguru, Swarnalatha 2018 *Implementation of an industrial automation system model using an Arduino.* Journal of Engineering Science and Technology. 13. 4131-4144.

IOP Conf. Series: Materials Science and Engineering **771** (2020) 012016 doi:10.1088/1757-899X/771/1/012016

Simulation and Analysis of ANFIS (Adaptive Neuro-Fuzzy Inference System) for Music Genre

A H Pratama[1,*], W Dwiono[1] and M T Tamam[1]

[1]Department of Electrical Engineering, University Muhammadiyah Purwokerto Indonesia

*e-mail: anugrahhp.ws@gmail.com

Abstract. Music comes in many different genres and styles according to its content, which is easy for a human listener to distinguish but hard for a machine to do. This limitation encourages the creation of a system that can help a computer to classify music genre better. This paper proposed a simple method that can classify music genre on a classical, jazz, pop, and rock using frequency analysis feature extraction and ANFIS classification method. There are two types of ANFIS model, which is model A and model B. The most accurate model, is model A with an average accuracy of 53,33% across all genre. The best model for single genre classification is Model B with 80% accuracy for Classic type.

1. Introduction

Music comes in many different genres and styles according to its content. It is easy for a listener to distinguish them but hard for a machine to do. Many researchers used many algorithms to classify the music genre by computer utilisation. ANFIS algorithm has been used for the grouping step in a speaking/melody perception system [1], the neural network in [2-4]. A different approach has been made in [5]. The proposed methods, which are collections of a feature extraction phase tailed by a grouping technique, search both the differences of parameters used as input and the classifier design. Another method is utilising some audio feature [6] and unsupervised feature pre-training [7].

Brendan Petty[3], has created a model of music genre classification system using Back Propagation Artificial Neural Network. This system classifies across four genres; pop, classic, heavy and jazz. This model resulted in an overall accuracy of 66% across all type. This model proves that ANN can be used as a classification method for song genre, albeit the low accuracy.

Michael Haggblade, Yang Hing, and Kenny Kao[8] created model of song classification using four methods, they are DAG SVM (directed acyclic graph support vector machine), Neural Networks, k-means, and k-NN. This model is used to classify four genres; classical, jazz, metal, and pop, which resulted in 86%, 96%, 87%, and 87% accuracy respectively. This condition proves that song can be classified using a machine learning method.

Imam Ikhsan [2] created a model of song classification using hidden Markov model which utilised a probability model. This method was used to classify three genres which are pop, rock, dance and resulted in 80% overall accuracy.

Based on those, this paper proposes a method to classify four types of music genre by using ANFIS (Adaptive Neuro-Fuzzy Inference System). ANFIS has been chosen as a classification method because ANFIS is a combination of Artificial Neural Network (ANN) and Takagi-Sugeno fuzzy logic inference system. This property makes ANFIS has almost identical features as human brains in decision making which theoretically should work great as a classification method, especially in classifying analogue signal such as song. The accuracy of the created ANFIS model is expected to be around 80% across all genre which is, Classic, Jazz, Pop, and Rock.

2. Research Method

The model in this paper is constructed as a simulation using MATLAB, while ANFIS is used as a classification method[9][10]. The amount of data used is 220 songs where 160 songs are used for training (40 for each genre) and 60 songs for testing (15 for each style) The overall classification model could be simplified as below.

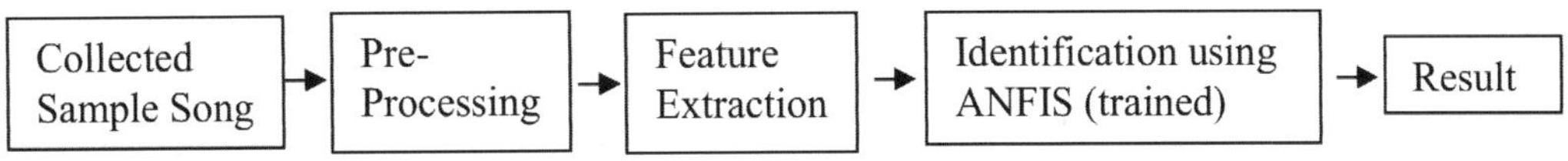

Figure 1. Simplified block diagram of music genre classification model using ANFIS

This model works by pre-processing the song first to remove an unnecessary audio element, and after that, the 'clean' audio file will have its feature extracted by feature extraction. These 'features' that have been extracted will be used by ANFIS to classify the song's genre.

2.1. Pre-processing

In pre-processing, the song will be polished so that the feature extraction can be done efficiently. First, the song will be filtered between 1 – 16384 Hz, and then it will be downsampled to 8000 Hz, and its channel will be reduced to mono. Lastly, a DC removal method will be applied to remove DC offset noise.

2.2. Feature Extraction

Feature extraction in this model was done by analyzing the content of each frequency range. The audio file will first be divided into four frequency category, which then for each frequency, the specified variable can be extracted. Fourteen variables can be extracted across all frequencies.

a. Low Frequency (0-200Hz)
   Variable extracted: bass frequency variation, spectral power - low
b. Low-mid frequency (200Hz-600Hz)
   Variable extracted: spectral power - low-mid
c. Mid Frequency (600Hz-1250Hz)
   Variable extracted: mid-frequency beat likelihood, mid-frequency beat offset, mid-frequency variation, Spectral power - mid
d. High Frequency (above 5000Hz)
   Variable extracted: high-frequency strength of half-beat, spectral power – high

Those variables are special to each frequency range, other variables that are not specified to certain frequency are stereo spread, tempo, the strength of half-beat, dynamic range, attack velocity - fast, and attack velocity – slow.

2.3. Constructing ANFIS (Training method)

In this stage, the ANFIS system will be constructed by training the system using a database which was obtained from the feature extraction stage. The database consists of 7 selected variables which have the most distinctive value difference. The training method is a hybrid method, which combines the least-squares estimator method on the forward training and gradient descent on the backward training

3. Result and Analysis

3.1. Feature Extraction

The result of feature extraction is represented in Table 1. All variables are in average value to make observation and analysis easier.

Table 1. The average result of all variable in feature extraction

	Classic	Jazz	Pop	Rock
dynamic range	1.4734856	1.163313774	0.844799464	0.840871379
fast peak strength	0.889252	1.20585283	0.877501786	0.591762069
high low weight to double	0.914536	1.243662264	0.933158929	1.087165517
high RMS	0.392766	0.338833962	0.343866071	0.425398276
low beat dev	1.320644	0.971219623	0.631375	0.72216069
low, mid RMS	0.181534	0.301966038	0.564892857	0.450426316
Low rms	0.041402	0.040498113	0.125091071	0.062781034
Mid beat dev	1.214354	0.995069811	0.601792857	0.722679483
mid beat likely hood	1.125552	1.032001887	1.021205357	1.030113621
mid beat *offset*	0.416104	0.44007717	0.417451786	0.425796552
Mid-RMS	0.323302	0.349067925	0.368128571	0.487885965
slow peak strength	1.374362	1.342766038	0.2569	0.319444828
stereo RMS	1.30038	0.682358491	0.631148214	0.977753448

The data in this table then converted to graph (Can be seen in Figure 2) to make it easier to observe which variable has the most distinctive value for each music genres.

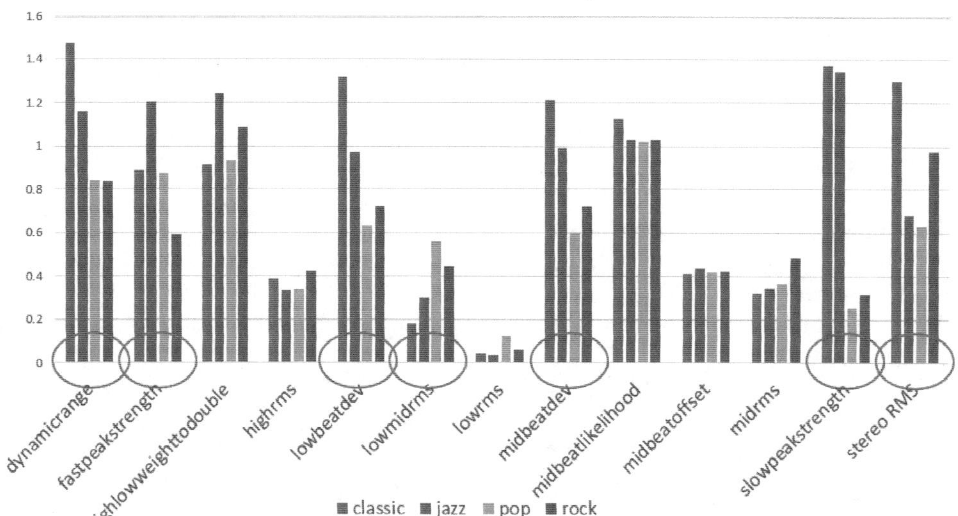

Figure 2. Graphical representation of feature extraction result for all variables in average

From Figure 2 it shows that from 14 variables, only seven variables can be used as ANFIS variable, they are the dynamic range, fast peak strength, low beat dev, low, mid-RMS, mid beat dev, slow peak strength, and stereo RMS. All of these variables have the most distinct value for every genre, which make them ideal for classifying song genre.

3.2. ANFIS Training Result

Using seven variables from the previous stage, the ANFIS model will be constructed by training the ANFIS system first. The training parameters are as follow:

a. 7 ANFIS input variables Gaussian type, with 2 Membership Function for every variable
b. ANFIS output variable using linear type Membership Function
c. 160 training data (40 for each genre)
d. hybrid training method

The training result can be observed in Table 2.

Table 2. ANFIS training result

No	Epoch Value	Training Error Value
1	20	0.063235
2	30	0.0533
3	40	0.046984
4	50	0.046984

From table 2 we can observe that the number of training value is stopped at 40 because from epoch 40 and above the value of training error is the same which is 0.046984 so it's pointless to continue training with higher epoch value.

3.3. *ANFIS Result*

After it was trained, the ANFIS model is completed, which means it can make a decision just like the human brain based on the knowledge it got from training. The data for this test is 15 songs for each genre. The purpose of this test is to measure the accuracy and performance of the created model, whether this model can be used to classify the music genre or not. The ANFIS testing result can be observed in Table 3.

Table 3. ANFIS testing result

No	Song Genre	Correct	Wrong	Accuracy
1	Classic	9	6	60%
2	Jazz	8	7	53,33%
3	Pop	7	8	46,67%
4	Rock	8	7	53,33%
	Total	**32**	**28**	**53,33%**

From table 3 it shows that the overall accuracy of the model is only 53,33%, which is below the expected value, which is 80%. Several factors cause the low accuracy of this model:

a. Variables used in ANFIS is too few
b. The number of membership function used is too few
c. Feature extraction system doesn't work as intended
d. Low sampling frequency at downsampling process in pre-processing

3.4. *Overfitting/Underfitting Analysis*

Overfitting is a condition in machine learning where the system matches (fit) too close with training data, which leads to error when the input has a slight difference with training data. This error can make the ANFIS system unable to identify the correct data, which leads to lower accuracy. While underfitting is an error in machine learning where the system is unable to locate the input because of the lack of training data. Both can lead to poor ANFIS performance.

One solution for overfitting is reducing the number of training data. In this analysis, a new ANFIS model (model B) will be constructed using less training data to prove whether overfitting present or not. The ANFIS parameter for ANFIS model B are as follow:

a. 5 ANFIS input variables Gaussian type, with 3 Membership Function for every variable
b. ANFIS output variable using linear type Membership Function
c. 80 training data (20 for each genre)
d. hybrid training method

ANFIS model B testing result can be observed in table 4, together with model B as a comparison.

Table 4. ANFIS model A & model B testing result

No	Song genre	Accuracy	
		model A	model B
1	Classic	60%	80%
2	Jazz	53,33%	33,33%
3	Pop	46,67%	26,67%
4	Rock	53,33%	0%
Total		53,33%	35%

From table 4 it shows that ANFIS model B, which has lower training data, also resulted in lower accuracy. This fact proves that overfitting does not occur on ANFIS model A. Instead, underfitting occurs in both the ANFIS model, which resulted in low accuracy. However, it can be noted that ANFIS model B has particularly high accuracy for classifying a single genre which is a classic genre with accuracy 80%. This performance shows that ANFIS model B can be used correctly for classifying classic type.

4. Conclusion

Both the ANFIS model constructed in this paper can be used for classifying music genre despite its low accuracy, where model B, in particular, has high accuracy for classifying classic type. For improve the system's accuracy, the number of training data can be increased to reduce underfitting or by using better feature extraction system.

References

[1] J.E. Muñoz-Expósito, S. García-Galán, N. Ruiz-Reyes, P. Vera-Candeas, 2007, Adaptive network-based fuzzy inference system vs. other classification algorithms for warped LPC-based speech/music discrimination, Engineering Applications of Artificial Intelligence, Volume 20, Issue 6, Pp 783-793.

[2] Ikhsan, Imam, 2013, Simulasi dan Klasifikasi Genre Musik berbasis Hidden Markov model, Theses Report, Institut Teknologi Telkom.

[3] Brendan Petty, 2010, Music Genre Classification Using a Back Propagation Neural Network. https://www.brendanpetty.com/projects/music-genre-classification project/musicgenreclassification-report.pdf (Accessed at September 23, 2017)

[4] Yandre M.G. Costa, Luiz S. Oliveira, Carlos N. Silla, 2017, An evaluation of Convolutional Neural Networks for music classification using spectrograms, Applied Soft Computing, Volume 52, Pages 28-38.

[5] Antonio Jose Homsi Goulart, Rodrigo Capobianco Guido, Carlos Dias Maciel, 2012,Exploring different approaches for music genre classification, Egyptian Informatics Journal, Volume 13, Issue 2, Pages 59-63.

[6] Aisha Gemala Jondya, Bambang Heru Iswanto, 2017, Indonesian's Traditional Music Clustering Based on Audio Features, Procedia Computer Science, Volume 116, Pages 174-181.

[7] Mariusz Kleć, Danijel Koržinek, 2014, Unsupervised Feature Pre-training of the Scattering Wavelet Transform for Musical Genre Recognition, Procedia Technology, Volume 18, Pages 133-139.

[8] Haggblade, Michael. Yang Hong, dan Kenny Kao. 2011. Music Genre Classification. https://www.semanticscholar.org (Accessed at September 24, 2017).

[9] Kusumadewi, Sri, Hartati, Sri, 2006, Neuro Fuzzy Integrasi Sistem Fuzzy dan Jaringan Syaraf, Graha Ilmu.

[10] Kumar, Satish, 2004, Neural Networks, A Classroom Approach. New Delhi: Tata McGraw-Hill Publishing Company Limited.

Feature Selection Using Rough Set Theory Algorithm for Breast Cancer Diagnosis

D N K Hardani[1,*] and H A Nugroho[2]

[1]Department of Electrical Engineering, Universitas Muhammadiyah Purwokerto
[2]Department of Electrical and Information Engineering, Universitas Gadjah Mada

[*]e-mail address: diannova.kh@ump.ac.id

Abstract. Feature selection is one of the pre-processing stages of classification carried out by selecting relevant features that affect the results of classification. The advantage of feature selection is that it increases the value of accuracy. Data mining in the medical world has excellent potential for knowing hidden patterns in medical data sets. However, medical data sets often have large dimensions and have irrelevant features that can decrease the performance of the algorithm. This study aims to analyse the performance of the rough set approach as an algorithm used for feature selection in breast cancer diagnosis cases. This study conducted a feature selection process on the Wisconsin Breast Cancer (Diagnostic) Data Set provided by the UCI machine learning repository. There are several steps taken in research to realize these goals, such as data pre-processing, feature selection, data randomization, classification and performance evaluation. The result shows that feature selection using the rough set of methods has proven to be effective in reducing a large number of features in the data set.

1. Introduction

Cancer is a condition which cells have lost control and standard mechanisms so that they experience abnormal growth, fast and uncontrolled. Cancer can begin to grow in the mammary glands, milk ducts, fat tissue, or connective tissue in the breast. The key to surviving breast cancer survivors is to detect breast cancer as early as possible before cancer has a chance to spread. The breast cancer diagnosis method used by a doctor depends on the doctor's knowledge, intuition, and experience. The results of the diagnosis can affect the recommendation to the patients.

As technology advances, especially in the area of artificial intelligence, machine learning techniques are introduced to help improve automatic detection capabilities. With the help of this system, the possibility of misdiagnosis made by experts can be avoided, and medical data can be examined in a short time and more detail. The diagnosis of using computer-aided utilizes many data mining methods. Data mining with intelligent algorithms can be used to overcome diagnostic problems with medical data sets that involve multiple data inputs [1]. Data mining in the medical world has excellent potential for knowing hidden patterns in medical data sets. However, medical data sets often have large dimensions and have irrelevant features that can decrease algorithm performance. Feature selection is a pre-processing stage of classification carried out by selecting relevant features to improve the effectiveness and efficiency of the performance of the classification algorithm [2]. The advantage of feature selection is that it increases the value of accuracy [3].

Research on feature selection has been carried out in various cases. Feature Selection research for breast cancer diagnosis uses a wrapper approach based on genetic algorithm (GA) and case-based reasoning (CBR). GA is used to find the problem space that is aimed to find all possible subsets of features, and CBR is used to estimate the evaluation of the results in each subset. The experimental results show that the proposed model can be compared with other models in the Wisconsin breast cancer data set (WDBC) [4]. Research using illustrative case studies in the Indian cosmetics industry illustrated the advantages of a rough set approach compared to conventional techniques for extracting decision rules from data sets, which can be useful in a variety of marketing applications. The rule generated through the methodology can perform as an expert, which may be referred to in future strategic decision-making. The approach gave similar results with the results obtained by statistical methods but without making any assumptions [5]. Ruhe [6] applied a rough set approach to analyse software engineering data that is resulted from a goal-oriented measurement. One method of feature selection that has been widely used and has a high degree of accuracy is the rough set theory approach. The attribute reduction for a decision system contains many decision attributes [7]. Other research work proposed a reasonable definition of parameterization reduction of soft sets and compared it with the attribute reduction in Rough Set Theory [8]. Zeng [9] conducted the development of a noise source classification system based on new methods for feature selection. The method used was a new rough set-based feature selection. It successfully reduced the number of features and improved classification accuracy.

This study aims to analyse the performance of the rough set approach as an algorithm used for feature selection in breast cancer diagnosis cases. The rough set approach was chosen because it can identify significant features and eliminate features that were not relevant to produce a good learning model and to reduce the dimensions of the data without the lack of information contained in the data set.

2. Methods

This study performed a feature selection process in the Wisconsin Breast Cancer (Diagnostic) Data Set provided by the UCI machine learning repository [10]. The data set contained of 32 attributes, 569 instances and two classes. The number of instances consisted of 357 benign and 212 malignant. These attributes were computed from a digitized image of a fine needle aspirate (FNA) of breast mass. The attributes information is shown in Table 1. Ten real-valued features are computed for each cell nucleus.

Table 1. WBCD Feature Information [10]

No. Attribute	Attribute/ Feature
#1	ID number
#2	Diagnosis
	Radius (mean of distances from center to points on the perimeter)
	Texture (standard deviation of gray-scale values)
#3	Perimeter
.	Area
.	Smoothness (local variation in radius lengths)
.	Compactness (perimeter^2 / area - 1.0)
#32	Concavity (severity of concave portions of the contour)
	Concave points (number of concave portions of the contour)
	Symmetry
	Fractal dimension ("coastline approximation" - 1)

There are several steps taken in this research work to realize these goals. Figure 1 shows an overview of the research flow.

2.1. Data Preprocessing

Pre-processing data is the first step in diagnosing breast cancer. Data pre-processing conducted in this research is numerical data discretization. Some classification algorithms can provide better performance results when the data has been discretized. This research used the Boolean reasoning algorithm as a method of discretization process [11][12] and Rosetta toolkit [13].

IOP Conf. Series: Materials Science and Engineering **771** (2020) 012017 doi:10.1088/1757-899X/771/1/012017

2.2. Feature Selection

The feature selection process carried out to reduce the number of features in the WBCD data set and select features that are relevant to the results of breast cancer diagnosis decisions. In this study, two types of feature selection methods were used, namely Rough Set Theory (RST)-based feature selection and Correlation Based Feature (CFS)-based feature selection.

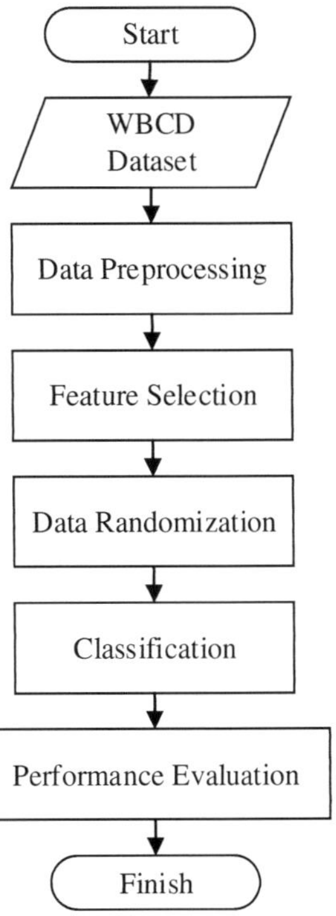

Figure 1. Research flow

RST provided a minimum set of features that can distinguish objects as well as the entire set of features used. However, the minimum feature set may not necessarily provide optimal performance when the classification process was carried out. In the rough set-based feature selection, the method used to reduce data is based on Johnson's algorithm so that a minimum reduction was obtained [14]. The feature selection process using the RST algorithm is shown in Figure 2.

CFS assumed that a subset of features containing of features that highly correlated with class and that each uncorrelated feature can improve classification performance. However, this subset of features was not necessarily a minimal set of features, such as reduction generated by RST. CFS was combined with the best first search strategy to select the correlation-based features. The feature selection process using the CFS algorithm is delineated in Figure 3. After the feature selection process was complete, a data set of some of the attributes selected in the feature selection process is obtained.

2.3. Data Randomization

The data randomization process in a data set can affect the performance of classification because the classification model formed during the training was different. Therefore, when the model tested on testing data, it would give different performance values. This data randomization process carried out using 10-fold cross-validation with ten iterations, any 100 data scrambling performance obtained for each classifier.

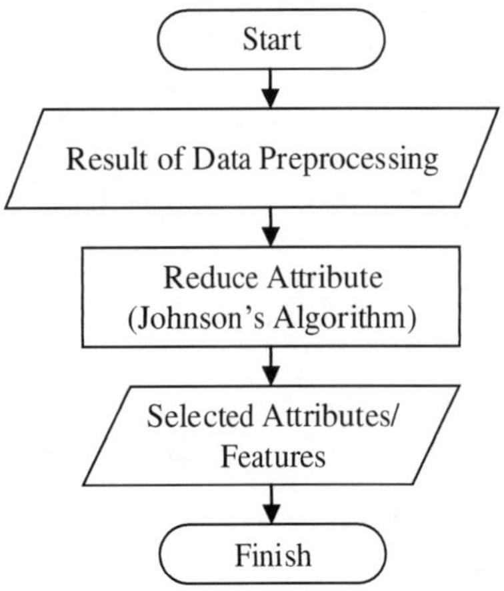

Figure 2. The feature selection process using RST method

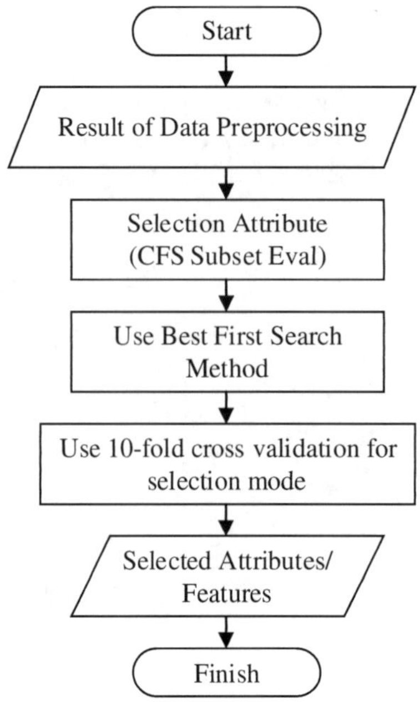

Figure 3. The feature selection process using CFS method

2.4. Classification
The computer-based breast cancer diagnosis was done by classifying the WBCD data set using a classification algorithm. Classification performance was presented by performing 10-fold cross-validation on the randomized data set. This method was a common method used in data mining research. 10-fold cross-validation divided the data set into ten parts (fold). In this study, the classifiers used to classify data sets were Naïve Bayes, Deep Learning, Multilayer Perceptron (MLP), Sequential Minimal Optimization (SMO), and J48. Every classifier used in this study followed the default settings and parameter provided by the Weka software [15].

2.5. Performance Evaluation
A comparison of classifier performance was presented by using three performance matrices, namely accuracy, sensitivity and specificity. Analysis of accuracy, sensitivity, and specificity can provide information about learning scheme trends that produce the best performance.

3. Result and Discussion
Based on the steps that have been taken for feature selection, the results obtained for each stage are described in this following section.

3.1. Data Discretization

Data discretization process was used to change the attributes of a numeric type data set to discrete. Some numeric data type attributes owned by the WBCD data set are converted to discrete type attributes using the Boolean reasoning algorithm discretization method. The results of the discretization are shown in Table 2.

Table 2. Discretization Result Value

id	texture_mean	texture_se	area_se	symmetry_se	perimeter_worst	smoothness_worst	Concavepoints_worst
[*, 87905)	[*, 16.68)	[*, 0.8941)	[*, 33.01)	[*, 0.01649)	[*, 105.95)	[*, 0.1256)	[*, 0.1358)
[87905, 892202)	[16.68, 19.66)	[0.8941, 1.3605)	[33.01, *)	[0.01649, *)	[105.95, *)	[0.1256, *)	[0.1358, *)
[892202, *)	[19.66, *)	[1.3605, *)					

Table 2 above shows the discretization process divided attributes at specific intervals. From 32 existing attributes, only 8 attributes discretized. The id attribute, texture_mean, and texture_se were discretized into 3 (three) intervals while the attributes area_se, symmetry_se, perimeter_worst, smoothness_worst, and concavepoints_worst divided into 2 (two) intervals. The '*' mark in the result of discretization, which states that the value is 'less than' if the position on the left. While the position on the right represents the value 'more than'. If there is no '*' sign that the discrete value is 'between'. For example, the texture_mean feature has a discrete value [*, 16.68) meaning that the texture_mean value < 33.01, [16.68, 19.66) means 16.68 ≤ texture_mean < 19.66, and the value [19.66, *) means texture_mean ≥ 19.66.

3.2. Feature Selection

Determination of training data in the Rough Set Theory (RST) based feature selection process using Johnson's Algorithm. This algorithm was very fast because it used a greedy search to find one reduction (or one reduction per object in the case of object-related reduction). Based on the attribute reduction process, the results show that 3 (three) attributes selected with 480 rules. Table 3 shows the results of the feature selection.

Table 3. The result of the Rough Set Theory (RST) based feature selection process

No	Attribute
1	Radius_mean
2	Perimeter_mean
3	Area_mean

In the feature selection process using the CFS algorithm combined with the best first search strategy was obtained 6 (six) features as shown in Table 4. The selection method used 10-fold cross-validation. This feature selection was processed based on correlation values. The features will be selected if have a high correlation of decision attributes and a low correlation between features.

Table 4. The results of feature selection using the CFS algorithm

Number of folds	%	Number of Attributes	Attribute
10	100	7	compactness_mean
10	100	21	fractal_dimension_se
10	100	26	smoothness_worst
10	100	27	compactness_worst
10	100	28	concavity_worst
10	100	30	symetry_worst

Table 3 shows that the RST-based feature selection method can choose fewer features than the CFS-based feature selection method in Table 4. If the computer-based feature selection method combined between RST and CFS algorithm will produce 9 (nine) features that can be used as input for the classification process to diagnose breast cancer.

3.3. Performance Evaluation

After conducting the randomization process by using 10-fold cross-validation with ten iterations, it produced a new data composition that was continued to classification process by using some classifiers such as Naïve Bayes, Deep Learning, MLP, SMO, and J48. The performance evaluation of each classifier is shown in Figures 4, 5 and 6 below.

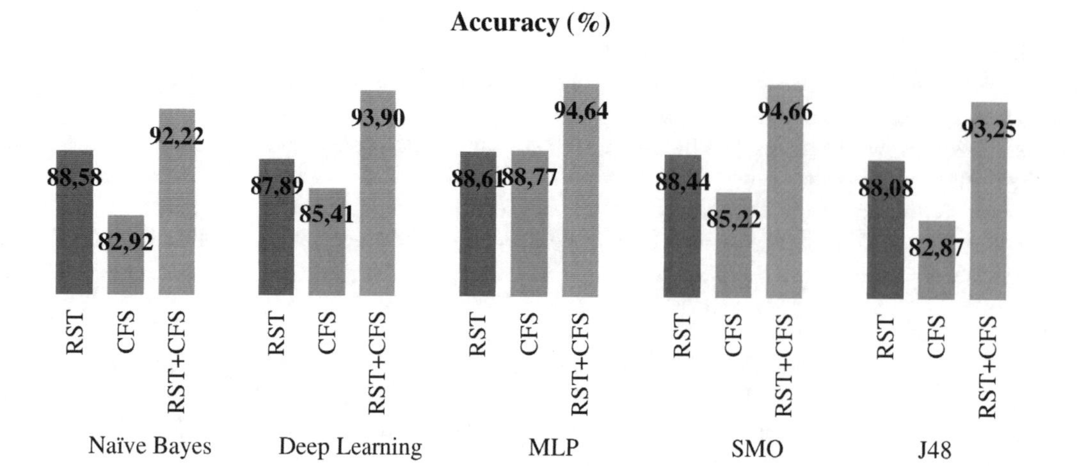

Figure 4. Accuracy of Classification

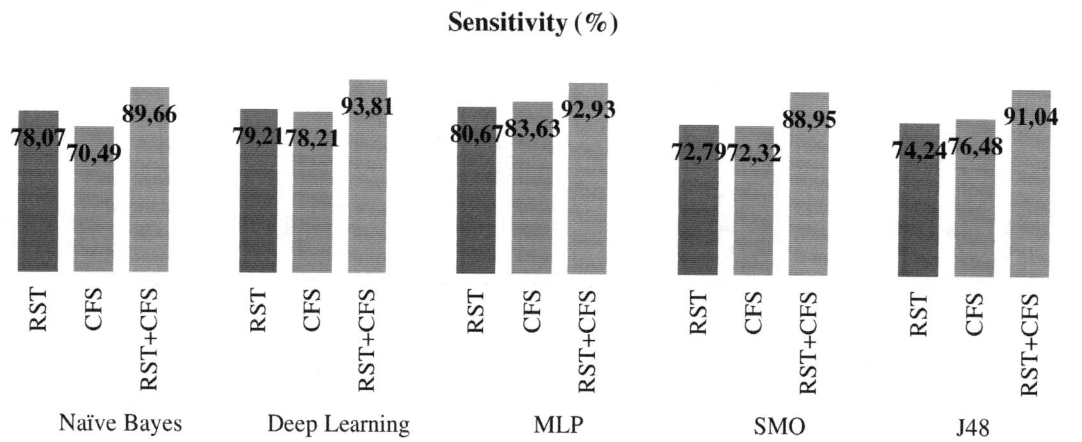

Figure 5. Sensitivity of Classification

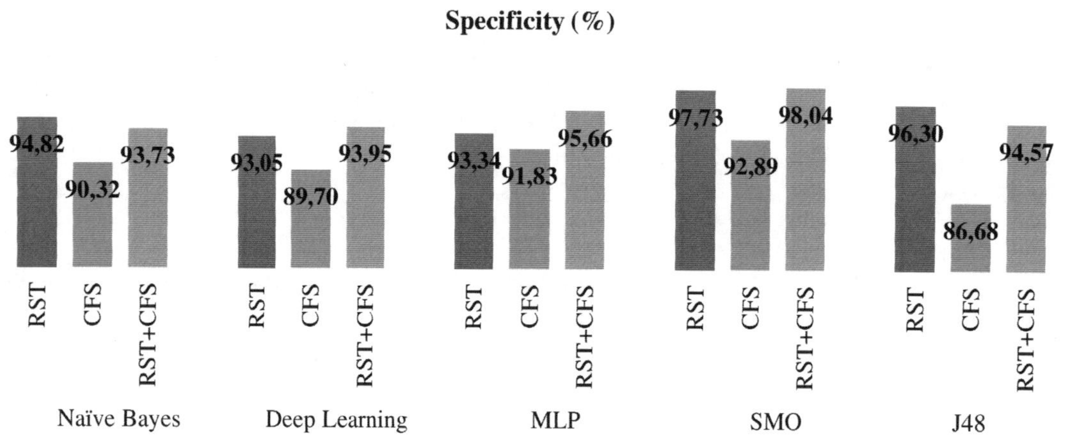

Figure 6. Specificity of Classification

Accuracy and sensitivity show the ability of the system to diagnose breast cancer. A classification system must be measured its performance so that the next step is to evaluate the performance of the classification method. Figures 4 and 5 show that the performance of the RST algorithm is more powerful than CFS algorithm. However, when the two algorithms were combined, the results show a significant improvement in classification performance. The ability of the system to predict conditions that are not true; in general, the RST algorithm has the most backward performance, as shown in Figure 6.

4. Conclusion
Feature selection using the rough set of methods has been proven to be effective in reducing a large number of features in the data set. It is beneficial for reducing computing time and improving the performance of the classification process. This system can improve the quality of life by predicting cancer in early development.

5. References
[1] M. F. Akay, "Support vector machines combined with feature selection for breast cancer diagnosis," *Expert Syst. Appl.*, vol. 36, no. 2, Part 2, pp. 3240–3247, 2009.

[2] A. Kustiyo, H. N. Firqiani, and E. P. Giri, "Seleksi Fitur Menggunakan Fast Correlation Based Filter pada Algoritma Voting Feature Intervals 5," *J. Ilm. Ilmu Komput.*, vol. 6, no. 2, 2008.

[3] K. Thangavel and A. Pethalakshmi, "Dimensionality reduction based on rough set theory: A review," *Appl. Soft Comput.*, vol. 9, no. 1, pp. 1–12, 2009.

[4] M. Darzi, A. Liaei, M. Hosseini, and H. Asghari, "Feature selection for breast cancer diagnosis: A case-based wrapper approach," *World Acad. Sci. Eng. Technol.*, vol. 53, pp. 1142–1145, May 2011.

[5] S. Mahapatra, Sreekumar, and S. S. Mahapatra, "Attribute selection in marketing: A rough set approach," *IIMB Manag. Rev.*, vol. 22, no. 1, pp. 16–24, 2010.

[6] G. Ruhe, "Rough set-based data analysis in goal-oriented software measurement," in *Proceedings of the 3rd International Software Metrics Symposium*, 1996, pp. 10–19.

[7] H. Li, X. Zhou, J. Zhao, and D. Liu, "Attribute Reduction in Decision-Theoretic Rough Set Model: A Further Investigation BT - Rough Sets and Knowledge Technology," 2011, pp. 466–475.

[8] D. Chen, E. C. C. Tsang, D. S. Yeung, and X. Wang, "The parameterization reduction of soft sets and its applications," *Comput. Math. with Appl.*, vol. 49, no. 5, pp. 757–763, 2005.

[9] X. Zeng and Y. Zhan, "Development of a noise sources classification system based on new method for feature selection," *Appl. Acoust.*, vol. 66, no. 10, pp. 1196–1205, 2005.

[10] UCI, "Breast Cancer Wisconsin (Diagnostic) Data Set," 1995. [Online]. Available: https://archive.ics.uci.edu/ml/datasets/Breast+Cancer+Wisconsin+%28Diagnostic%29. [Accessed: 21-Aug-2019].

[11] H. S. Nguyen, "Approximate Boolean Reasoning Approach to Rough Sets and Data Mining BT - Rough Sets, Fuzzy Sets, Data Mining, and Granular Computing," 2005, pp. 12–22.

[12] A. Øhrn, "Discernibility and Rough Sets in Medicine: Tools and Applications," *Nor. Univ. Sci. Technol.*, 2000.

[13] M. Kierczak, "ROSETTA A Rough Set Toolkit for Analysis of Data," *The Linnaeus Centre for Bioinformatics, Uppsala University, SWEDEN*, 2009. [Online]. Available: http://bioinf.icm.uu.se/rosetta/.

[14] D. S. Johnson, "Approximation algorithms for combinatorial problems," *J. Comput. Syst. Sci.*, vol. 9, no. 3, pp. 256–278, 1974.

[15] "WEKA The workbench for machine learning," *Department of Computer Science at the University of Waikato, New Zealand*. [Online]. Available: https://www.cs.waikato.ac.nz/ml/weka/.

IOP Conf. Series: Materials Science and Engineering **771** (2020) 012018 doi:10.1088/1757-899X/771/1/012018

Multilevel Thresholding Image Segmentation based on Autonomous Groups Particle Swarm Optimization

Murinto[1]

[1]Informatic of Engineering, Faculty of Information Technology, Universitas Ahmad Dahlan Yogyakarta, Indonesia

[1]murintokusno@tif.uad.ac.id

Abstract. Multilevel thresholding problems based on Otsu criteria are discussed in this paper. One weakness of the Otsu method is that computational time increases exponentially according to the number of thresholding dimensions. In this paper, a modified Particle Swarm Optimization (PSO) algorithm called Autonomous Groups Particles Swarm Optimization (AGPSO) is proposed to reduce two problems trapped in local minima and a slow convergence rate in solving high-dimensional problems. AGPSO is used for multilevel thresholding image segmentation. The performance of AGPSO is compared with standard PSO on three natural images. The parameters used to compare the performance of AGPSO and PSO are SSIM, PSNR, Computation Time, optimal threshold obtained from each algorithm. From the experimental results show that AGPSO is better when compared to PSO in image segmentation, from the resulting fitness value and higher SSIM and PNSR values.

1. Introduction

There are various methods of image segmentation that currently exist in the literature. One of the most commonly used segmentation methods is thresholding [1]. Among the several thresholding methods available, the Otsu method that deals with optimization issues has received much attention in the research that has been conducted [2]. Although the performance of the thresholding method is very good in bilevel thresholding, in multilevel thresholding this will require a lot of calculation time so it is not effective to overcome optimization problems [3]. On the other hand, there are various bioinspired algorithms that have advantages in their ability to learn to overcome optimization problems, some of which are genetic algorithm (GA) and particle swarm optimization (PSO). Both of these algorithms are widely used in optimization problems [4]. From previous studies in [5] [6] [7], PSO is more efficient and appropriate when used to find the optimal threshold of more than two. However, it is well known that one of the disadvantages of PSO is trapped in optimal local and premature convergence in handling complex optimization problems [8].

In this paper, a modified Particle Swarm Optimization (PSO) algorithm called Autonomous Groups Particles Swarm Optimization (AGPSO) is proposed to reduce two problems trapped in local minima and a slow convergence rate in solving high dimensional problems. The main idea of the AGPSO algorithm is inspired by the diversity of individuals in groups of birds or insects. Different functions with various slopes, arcs, and interception points are used to adjust the social and cognitive parameters of the PSO algorithm to provide different behavioural particles as in natural colonies. This paper is organized into several sections. Part 1 is an introduction section that contains background problems and proposed solutions. In Section 2, the conventional PSO algorithm and the proposed AGPSO algorithm are briefly explained. Section 3 provides a description of the multilevel thresholding problem based on Otsu criteria. Section 4 presents the results of the experiments to evaluate the performance of the proposed algorithm. Finally, conclusions are given in Section 5.

2. Autonomous Group Particle Swarm Optimization

The particle swarm optimization (PSO) technique was first introduced by Kennedy and Eberhart (1995) [9], is a stochastic optimization technique similar to the behaviour of bird flocks or sociological behaviour of a group of humans. The basic idea of PSO is to involve a scenario where a flock of birds in the search for food in an area. All birds do not know exactly where the food is, but with each iteration, they will know how far the food will be found. The best strategy will be followed by birds close to food and also from the previous best position achieved. PSO was built with the concept of optimization through a swarm particle. The PSO algorithm is a multiagent parallel search technique that maintains a particle swarm and each particle represents a potential solution in the swarm. All particles fly through the multi-dimensional search space by adjusting their position based on their own experience and from their neighbours.

The original PSO algorithm is written in the form of velocity updated and position updated (Kennedy and Eberhart, 1995) as shown in equations (1) and equation (2) respectively.

$$v_{id}^{t+1} = w.v_{id}^{t} + c_1 .r_1.(p_{id}^{t} - x_{id}^{t}) + c_2 .r_2(p_{gd} - x_{id}^{t}) \quad (1)$$
$$x_{ij}^{t+1} = x_{ij}^{t} + v_{ij}^{t+1} \quad (2)$$

where c_1 and c_2 are positive constants, called the acceleration coefficient. r_1 and r_2 are two random numbers with values within range [0,1]. w is the weight of inertia (inertia weight). A large value of inertia weight will facilitate a global exploration while a small inertia weight facilitates local exploitation. The i particle is represented as $X_i = (x_{i1}, x_{i2}, \dots, x_{iD})$. The best previous position of particle i is stored and represented as $P_i = (p_{i1}, p_{i2}, \dots, p_{iD})$. The position provides the best fitness value (the best fitness value). The index of the best particles among all the particles in the population is represented by the symbol g. The rate of change of position (velocity) for particle i is represented as $V_i = (v_{i1}, v_{i2}, \dots, v_{iD})$. During the update process, the speed of each dimension of a particle is limited to Vmax. D is the dimension of each search space.

The concept of autonomous groups is inspired by the diversity of individuals in a group of animals or a group of insects. In some cases, individuals are not quite similar in terms of intelligence and ability, but they all carry out their respective duties as members of the group. Each individual is able to work in any situation. In standard PSO, all particles have the same behaviour in terms of local and global search (local and global search), so the particles are considered as a group with the same strategy. However, using autonomous groups with a common goal in any population-based optimization algorithm theoretically can produce more randomized and directed searches simultaneously. In the autonomous mathematical model, groups consist of different strategies for updating $c_1$and c_2. The renewal strategy of autonomous groups can be applied with various continuous functions at intervals [0,1]. Some functions can be used to update cognitive factors (c_1) and social factors (c_2). These functions consist of ascending or descending linear and polynomial, as well as exponential and logarithmic functions. In Figure 1 the c_1 and c_2 updates are shown, where the blue curve is for c_1 update while the red curve is for c_1 update. From the picture, it can be seen that c_1 decreases according to the iteration, whereas c_2 increases. From this, it is clear that the particles tend to have higher local search capabilities when c_1 is higher than c_2. On the other hand, the search for the search space particle is more global when c_2 is higher than c_1. In autonomous groups how to find a balance between c_1 and c_2 as dynamic coefficients core of this algorithm [10].

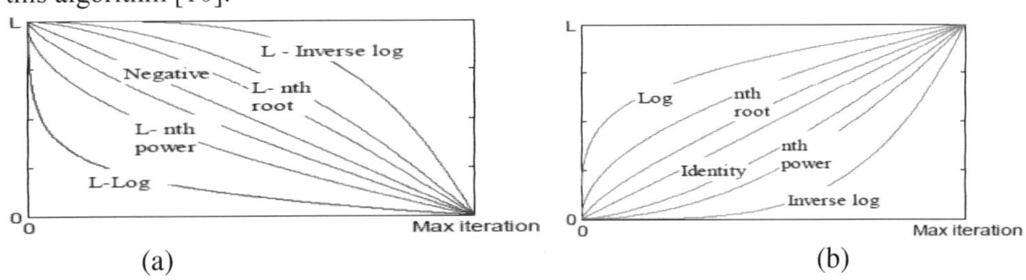

Figure.1. (a) Specific functions for updating c_1, (b) Specific functions for updatingc_2 when L is upper bounded for c_1 and c_2 [10]

3. Multilevel Thresholding Multispectral Image Segmentation

The Otsu method is a non-parametric method for segmentation that divides the entire image into classes so that the variance of the different classes is maximum. In his research, Otsu (1979) defines variance between classes as a sum of the Sigma functions of each class, which are written as equations:

$$g(t) = \sigma_0 + \sigma_1 \quad (3)$$
$$\sigma_0 = w_0 + (\mu_0 - \mu_A)^2 \quad (4)$$
$$\sigma_1 = w_1 + (\mu_1 - \mu_A)^2 \quad (5)$$

Where μ_T is intensity mean of origin image.

In bi-level thresholding, mean level each of levels (μ_i) can be written as:

$$\mu_0 = \sum_{i=0}^{T-1} \frac{ip_i}{w_0} \quad (6)$$
$$\mu_1 = \sum_{i=T}^{L-1} \frac{ip_i}{w_1} \quad (7)$$

The optimal threshold is obtained from the maximization function between class variances, which is written as an equation:

$$T^* = argmax\, g(t) \quad (8)$$

As for the multi thresholding problem, the Otsu method can be written as

$$\sigma_0 = w_0 + (\mu_0 - \mu_A)^2, \sigma_1 = w_1 + (\mu_1 - \mu_A)^2, \sigma_j = w_j + (\mu_j - \mu_A)^2,$$
$$\sigma_m = w_m + (\mu_m - \mu_A)^2 \quad (9)$$

In multilevel thresholding, the mean level for each of the classes (μ_i) can be written as:

$$\mu_0 = \sum_{i=0}^{T_1-1} \frac{ip_i}{w_0}, \mu_1 = \sum_{i=T_1}^{T_2-1} \frac{ip_i}{w_1}, \mu_j = \sum_{i=T_j}^{T_{j+1}-1} \frac{ip_i}{\mu_j}, \mu_m = \sum_{i=T_n}^{L-1} \frac{ip_i}{\mu_n} \quad (10)$$

The optimal multilevel threshold is obtained by maximizing the variance between classes of functions, which is written as an equation:

$$T^* = argmax(\textstyle\sum_{i=0}^{n} \sigma_i) \quad (11)$$

Threshold method is a method that is generally used in image segmentation to separate important targets from the background image. This is the basic principle for dividing image pixels into different regions through determining threshold features. A common threshold segmentation can be explained as follows: Assume there is a certain threshold and the image is divided into two groups of pixels with, one (target) with a threshold greater than the other (background) with a lower threshold value (T). There are two types of threshold segmentation methods: single and multi-threshold.

At the bi-level thresholding, the image pixels with grey levels below the T value will be assigned to one group K_0, while m pixels t with grey levels above T will then be included in another group, K_2. At the end of the thresholding process, an image I will be grouped into two groups. If an image I is assumed to be represented through the gray level L scale, then bilevel thresholding can be defined as an equation:

$$K_0 = \{f(x, y) \in I | 0 \le f(x, y) \le T - 1\}$$
$$K_1 = \{f(x, y) \in I | 0 \le f(x, y) \le T - 1\} \quad (12)$$

Multilevel thresholding uses more than one threshold value and produces an image with several groups.

$$K_0 = \{f(x, y) \in I | 0 \le f(x, y) \le T_1 - 1\}$$
$$K_1 = \{f(x, y) \in I | T_1 \le f(x, y) \le T_2 - 1\}$$
$$K_i = \{f(x, y) \in I | T_i \le f(x, y) \le T_{i+1} - 1\}$$
$$K_n = \{f(x, y) \in I | T_n \le f(x, y) \le L - 1\} \quad (13)$$

Where $T_i = 1,2,..,n$ are threshold values and n are number of thresholds.

4. Results and Discussion

In this research, the proposed strategy is applied to multilevel thresholding image segmentation which was tested on 3 test images, namely house image, mountain image and zebra image. Each test image has a size of 256 x 256 pixels, 8-bit gray level. Before the experiment, the parameters of AGPSO and PSO and other image segmentation parameters are determined first. The simulation is done using MATLAB R.2016b programming on an Intel Core i5-4210U 2.40 GHz processor with 4 GB RAM. The effectiveness and feasibility of the proposed approach for multilevel thresholding is carried out by maximizing between class variances (Otsu method), with the aim of selecting the best thresholding value and higher objective function values with rapid convergence. In Figure 2, three natural images and corresponding histogram images are shown. The AGPSO algorithm used in this study is compared with the standard PSO algorithm

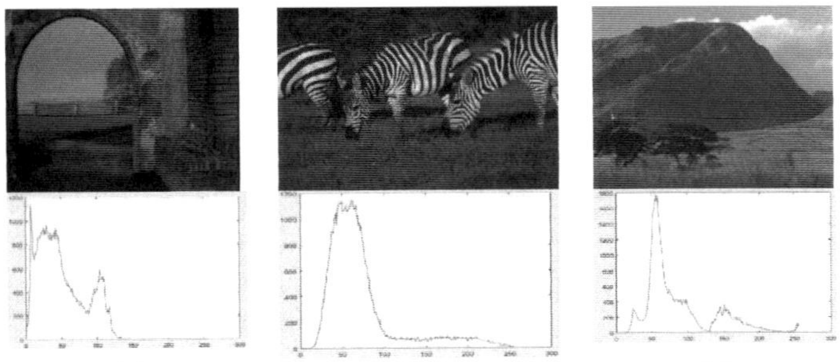

(a) (b) (c)

Figure.2 Natural Images and their histogram (a) house (b) zebra (c) mountain

Table 1 shows the results of the application of the PSO and AGPSO algorithms to the mountain multilevel thresholding image segmentation, where the level applied in this study is the level of m = 2,3,4,5.

Table 1 Image results and histograms at level m = 2,3,4,5 Multilevel Thresholding Using PSO and AGPSO on Mountain Image

Threshold (m)	Mountain			
	PSO method		*APSO method*	
	Origin Image	*Histogram*	*Result Image*	*Histogram*
m=2		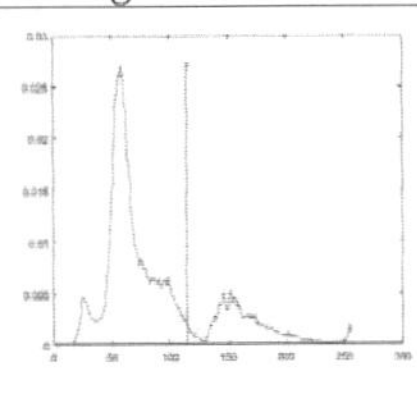		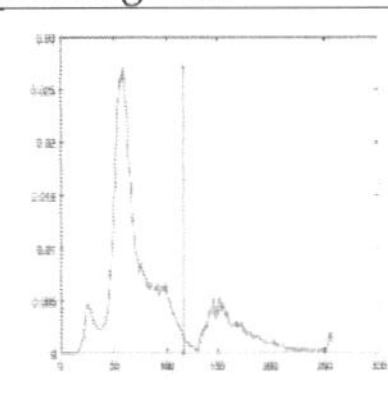
m=3		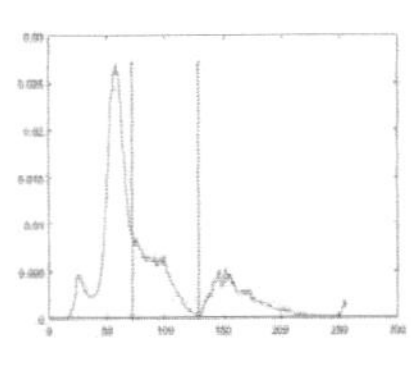	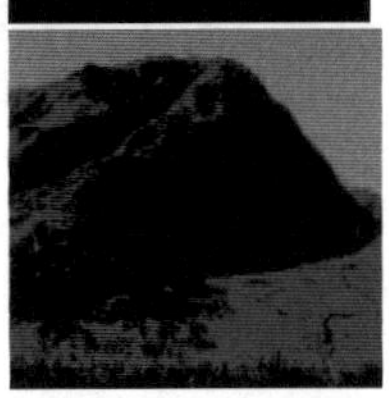	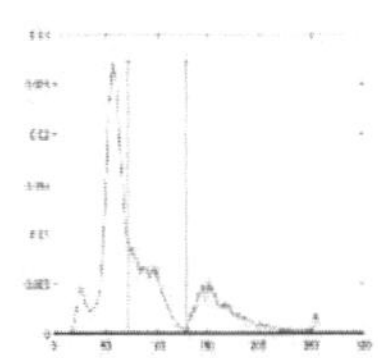

m=4

m=5

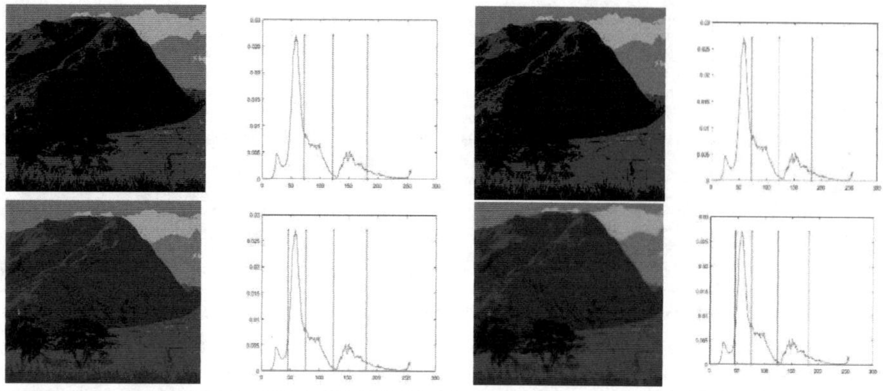

Optimal threshold values obtained using PSO and AGPSO for 1500 iterations, fitness values, CPU processing time, and PSNR, SSE, SSIM values are shown sequentially in Table 2, Table 3, Table 4 and Table 5.

Table 2. Comparison of optimal threshold values of the Otsu method using the PSO and AGPSO optimization algorithms

Image	m	PSO	AGPSO
House	2	61	61
	3	35 75	35 75
	4	28 54 86	28 54 86
	5	21 39 61 89	21 39 61 89
Zebra	2	117	118
	3	62 130	62 130
	4	58 105 170	58 105 170
	5	51 74 117 177	50 73 116 176
Mountain	2	115	116
	3	72 129	72 129
	4	72 122 182	72 122 182
	5	46 76 124 182	46 76 124 182

Table 3. Comparison of Fitness Otsu methods using PSO and AGPSO optimization algorithms

Citra	m	Fitness		Deviation Standard	
		PSO	*AGPSO*	*PSO*	*AGPSO*
House	2	810.6528	810.6528	0	0
	3	958.4464	958.4464	28.2843	28.2843
	4	1.0085e+03	1.0085e+03	29.0517	29.0517
	5	1.0292e+03	1.0292e+03	29.3201	29.3201
Zebra	2	1.4861e+03	1.4861e+03	0	0
	3	1.7017e+03	1.7017e+03	48.0833	48.0833
	4	1.7992e+03	1.7992e+03	56.2406	56.2406
	5	1.8450e+03	1.8449e+03	55.3918	55.3918
Mountain	2	1.7068e+03	1.7068e+03	0	0
	3	1.9285e+03	1.9285e+03	40.3051	40.3051
	4	2.0404e+03	2.0404e+03	55.0757	55.0757
	5	2.0907e+03	2.0907e+03	59.4306	59.4306

Table 4. Comparison of Processing Times (in seconds) the Otsu method uses the PSO and AGPSO optimization algorithms

Citra	m	PSO	AGPSO
House	2	4.134520	4.885279
	3	6.456734	6.746695
	4	8.387650	8.652968
	5	9.345620	10.582818
Zebra	2	4.097349	4.533585
	3	6.283603	6.608194
	4	7.649079	8.950233
	5	9.373399	10.953475
Mountain	2	4.074655	4.557945
	3	5.867928	6.518023
	4	7.639760	9.288086
	5	9.485907	10.171466

Table 5. Comparison of the Otsu method PSNR, SSE, SSIM using the PSO and AGPSO optimization algorithms

Citra	m	PSO			AGPSO		
		PSNR (dB)	*MSE*	*SSIM*	*PSNR (dB)*	*MSE*	*SSIM*
House	2	17.3799	1.1887e+03	0.2017	17.3562	1.1953e+03	0.2666
	3	21.1764	495.9554	0.5246	21.1764	485.9554	0.5426
	4	23.6710	279.2392	0.6469	23.6710	279.2392	0.6469
	5	25.7122	174.5272	0.7581	25.7122	174.5272	0.7581
Zebra	2	12.2313	3.8900e+03	0.1841	12.2315	3.8898e+03	0.1837
	3	15.8922	1.6744e+03	0.4402	15.8922	1.6744e+03	0.4402
	4	17.3867	1.1869e+03	0.5063	17.3867	1.1869e+03	0.5063
	5	19.2580	771.4082	0.6357	19.4757	733.69314	0.6480
Mountain	2	11.7879	4.3081e+03	0.1699	11.7884	4.3077e+03	0.1692
	3	14.5919	2.2589e+03	0.3309	14.5919	2.2589e+03	0.3309
	4	15.0430	2.0360e+03	0.3371	15.0430	2.0360e+03	0.3371
	5	20.8853	530.3386	0.7002	20.8853	530.3886	0.7002

5. Conclusion

In this paper, a modified particle swarm optimization algorithm, AGPSO, is proposed to overcome two problems that arise in solving high dimensions. Both problems are trapped in the local minimum and the slow rate of convergence. The results show that PSO with autonomous particle groups outperform conventionally and some recent modifications of PSO in terms of escaping from the local minimum and convergence speed. The performance of AGPSO is compared to the PSO standard on three natural images. The parameters used to compare the performance of AGPSO and PSO are SSIM, PSNR, Computation Time, optimal threshold obtained from each algorithm. From the experimental results show that AGPSO is better when compared to PSO in image segmentation, from the resulting fitness value and higher SSIM and PNSR values.

Acknowledgement

This research is funded by The Research Institution of Universitas Ahmad Dahlan (LPPM) through the Competitive Research Grant scheme in 2019. Thanks to LPPM.

References

[1] Sezgin, M., & Sankur, B. (2004). Survey over image thresholding techniques and quantitative performance evaluation. *Journal of Electronic imaging*, *13*(1), 146-166.Another reference

[2] Vala, H. J., & Baxi, A. (2013). A review on Otsu image segmentation algorithm. *International Journal of Advanced Research in Computer Engineering & Technology (IJARCET)*, *2*(2), 387-389.

[3] Otsu, N. (1979). A Threshold Selection Method for Grey Level Histograms. IEEE Transactions on System, Man and Cybernetics 9(1), 62–66.

[4] Paulinas, M., & Ušinskas, A. (2007). A survey of genetic algorithms applications for image enhancement and segmentation. *Information Technology and control*, *36*(3).

[5] Murinto, M., Astuti, N. R. D. P., & Mardhia, M. M. (2019). Multilevel thresholding hyperspectral image segmentation based on independent component analysis and swarm optimization methods. *International Journal of Advances in Intelligent Informatics*, *5*(1), 66-75.

[6] Sathya, P. D., & Kayalvizhi, R. (2010). PSO-based Tsallis thresholding selection procedure for image segmentation. *International Journal of Computer Applications*, *5*(4), 39-46.

[7] Maitra, M., & Chatterjee, A. (2008). A hybrid cooperative–comprehensive learning based PSO algorithm for image segmentation using multilevel thresholding. *Expert Systems with Applications*, *34*(2), 1341-1350.

[8] Gao, H., Xu, W., Sun, J., & Tang, Y. (2009). Multilevel thresholding for image segmentation through an improved quantum-behaved particle swarm algorithm. *IEEE Transactions on Instrumentation and Measurement*, *59*(4), 934-946.

[9] Mirjalili, S., Lewis, A., & Sadiq, A. S. (2014). Autonomous particles groups for particle swarm optimization. *Arabian Journal for Science and Engineering*, *39*(6), 4683-4697.

[10] Eberhart, R., & Kennedy, J. (1995, November). Particle swarm optimization. In *Proceedings of the IEEE international conference on neural networks* (Vol. 4, pp. 1942-1948).

IOP Conf. Series: Materials Science and Engineering **771** (2020) 012019 doi:10.1088/1757-899X/771/1/012019

Sudomy: Information Gathering Tools for Subdomain Enumeration and Analysis

Rizdqi Akbar Ramadhan[1], Redho Maland Aresta[2], Dedy Hariyadi[3]

[1]Islamic University of Riau, Indonesia.
[2]University of Amikom Yogyakarta, Indonesia.
[3]Universitas Jenderal Achmad Yani Yogyakarta, Indonesia.

[1]rizdqiramadhan@eng.uir.ac.id, [2]screetsec@gmail.com, [3]dedy@unjaya.ac.id

Abstract. In general, system security has become a crucial element in the digital aspect. Technically, evaluating the system there are aspects, one of which is by conducting a security assessment. Specifically, each system that will be evaluated is the essence of vulnerability search. The information security assessment is a form of awareness regarding cyberattacks that always increases from year to year. The assessment process can be carried out by an internal team and/or external auditor. The stages of evaluation by internal teams are certainly different from those of external parties. External auditors in conducting assessments need to learn or get as much information as possible related to the target, which in this case is in the form of a subdomain. Therefore, we need applications that support effective and efficient Information Gathering to assist in analysis and reporting. There are still many Information Gathering applications that do not include reporting and data validation systems. In this study, it is proposed to develop applications to support the Information Gathering stage which makes it easier for Cyber Security researchers/analysts.

1. Introduction

Based on the Indonesia Security Incident Response Team on Infrastructure / Coordination Center (ID- SIRTII / CC) that the level of internet security in Indonesia in 2018 has a poor ID-SIRTII Index value. When compared with Japan through the Mejiro Analysis has a higher risk, see Figure 1 even the total cyberattacks during 2018 totalling 232,447,974 attacks. The attack was monitored by ID-SIRTI / CC from January to December 2018 [1]. Bad predicate on internet security in Indonesia in 2018 needs to be introspected by network infrastructure managers that are connected to the internet. Therefore it is necessary to conduct security testing of information technology infrastructure on a regular basis to determine vulnerabilities in systems and information technology that can harm an organization [2]. To avoid cyber attacks on information technology infrastructure, the Ministry of Communication and Information has issued Ministerial Regulation No. 4 of 2016 which regulates the information security management system by electronic system administrators. The regulation stipulates that every electronic system operator must implement SNI ISO / IEC 27001 and/or Information Security Index guidelines [3].

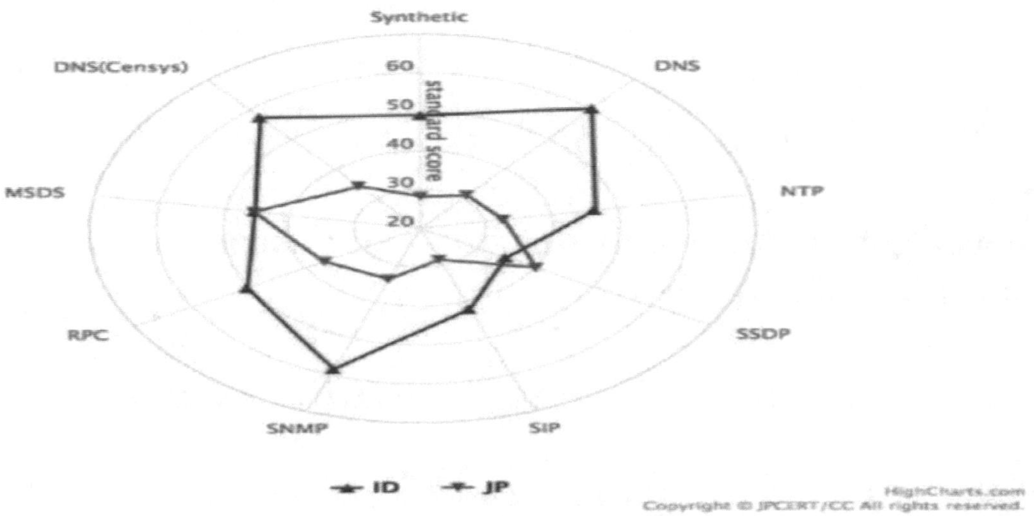

Figure 1. Mejiro Analysis

Testing of information technology infrastructure security systems can be carried out by experts in the field of information security (pentesters). The types of tests carried out by the pentester are 6 types [2]:

- Blind, the target party has prepared the system to be tested while the pentester when testing does not know the detailed information related to the target system to be tested.
- Double-Blind, the target party does not prepare the system to be tested while the pentester when testing does not know the detailed information related to the target system to be tested.
- Gray Box, the target party has prepared a system that will be tested while the pentester will conduct testing when given simple information by the target party related to the system to be tested.
- Double Gray Box, the target party has prepared the system to be tested while the pentester will conduct testing when given the scope of testing by the target party related to the system to be tested.
- Tandem, the target party and the pentester prepare together regarding the target system to be tested.
- Reversal, the pentester is fully aware of the operational and target work systems. The implementation of the test is unknown to the target party. Both the target and the pentester are from the same organization.

Penters with Blind and Double-Blind types did not realize the details of the system to be tested. Unlike the Gray Box type, Double Gray Box, Tandem, and Reversal already know the details of the system to be tested. Therefore the Blind and Double-Blind pentester types require tools to find out the basic or initial information of a system to be tested. To get the basic information needed by the pentester, in this study we built Sudomy, an information-gathering tool. Sudomy was built to complement the tools needed by the pentester following the rules of the National Institute of Standards and Technology (NIST) and/or the Information Systems Security Assessment Framework (ISSAF).

2. Related Works

2.1. Domain Name System

Domain Name System (DNS) is a system that functions to convert a domain name that is easy to remember in the form of an IP Address by requesting information to a system that has a hierarchy and spread. The existence of DNS makes it easier to connect computing resources both through the internet and internal networks [3]. The global implementation of the DNS system is the implementation of three roles, namely the Domain Name Registry Operator, Domain Name Registrar, and Service Providers and Customers. The overview of the Domain Name Registry can be seen in Figure 2 [4].

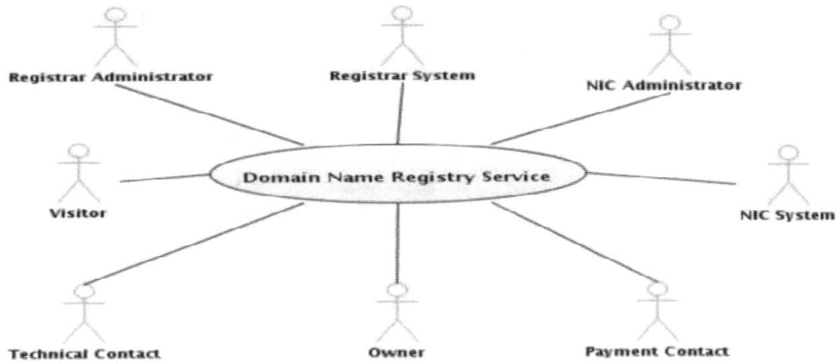

Figure 2. Overview of Domain Name Registry

2.2. National Institute of Standards and Technology SP 800-42

The US Department of Commerce published recommendations on Network Security Testing as set out at the National Institute of Standards and Technology Special Publication 800-42 (NIST SP 800-42). The basic methodology for penetration testing according to NIST SP 800-42 consists of four phases, namely Planning, Discovery, Attack, and Reporting, see Figure 3 [5]. In the initial phase of Discovery, the pentester can identify and gather the information that is potentially related to the target. Information gathering can be carried out various techniques including Domain Name System interrogation, InterNIC queries, Search of the target organization's web server (s) for information, Search of the organization's Lightweight Directory Access Protocol server (s) (LDAP) for information, Packet capture, NetBIOS enumeration, Network Information System, and Banner grabbing.

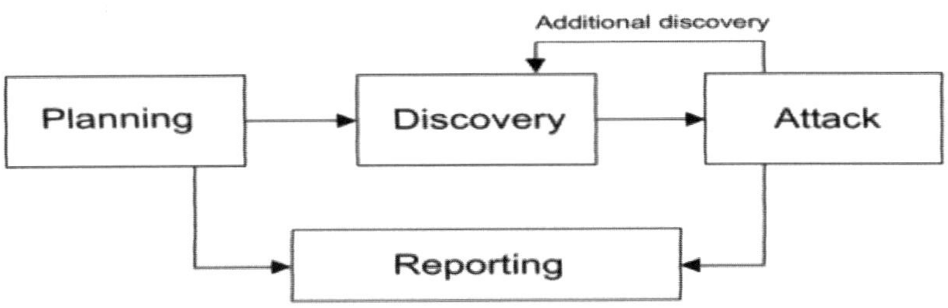

Figure 3. NIST: Penetration Testing Methodology

IOP Conf. Series: Materials Science and Engineering **771** (2020) 012019 doi:10.1088/1757-899X/771/1/012019

2.3. Information Systems Security Assessment Framework (ISSAF)

The Information Systems Security Assessment Framework (ISSAF) has three methodological phases, namely Planning and Preparation, Assessment, and Reporting, Clean-up, Destroy Artefacts. Whereas in the Assessment phase there are nine steps: Information Gathering, Network Mapping, Vulnerability Identification, Penetration, Gaining Access & Privilege Escalation, Enumerating Further, Compromise Remote Users / Sites, Maintaining Access, and Covering Tracks, see Figure 4 [6].

Figure 4. ISSAF: Penetration Testing Methodology

In the Information Gathering step, finding information from the target can be done technically, non-technically, or a combination of both. This aims to get information that is potentially related to the gap of the target. Information Gathering can be divided into two parts, namely passive and active. In the Passive Information Gathering section, the techniques used to obtain information are not directly related to the target. It could also be that the Passive Information Gathering section uses a third party to obtain information. Whereas the Active Information Gathering section in the process of getting information is directly related to the target. So this in some countries declared illegal [6].

2.4. Web Application Programming Interfaces

The advantage of developing using Web Application Programming Interfaces (Web APIs) is that it speeds up the development of an application. Therefore the implementation of Web APIs in the span of 2005- 2013 is always increasing, see 'figure' 5 [7]. Web APIs use the HTTP / HTTP protocol in communicating with major applications with Web APIs providers [8]. Passive Information Gathering application development utilizes Web APIs to speed up getting information from third parties, such as SecurityTrails, BinaryEdge, VirusTotal, Censys, and Shodan.

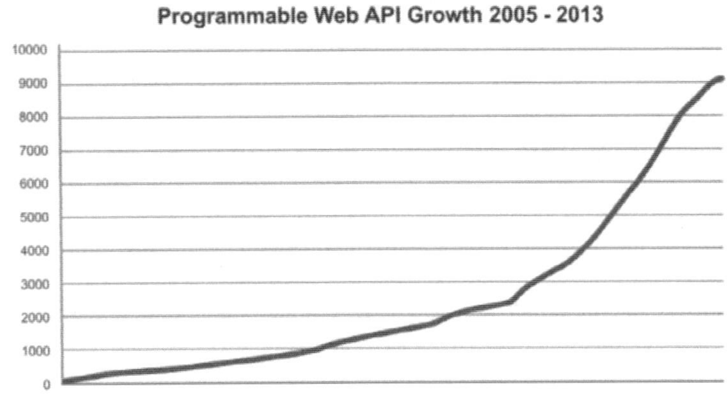

Figure 5. Programmable Web API Growth 2005-2013

2.5 Bash Script

Bash (Bourne-Again Shell) Script [9] can be chosen as a simple and easy-to-understand interface for users while choosing the Python programming language because it supports object-oriented programming

[10] and has library support that facilitates development. The use of Bash Script has three main components: lexical analysis and parsing, text expansion, and command execution see 'figure 6' [11]. Input can start from the interactive interface through the console or Bash Script file. The first component is the lexical analysis and parses each command into the data structure. Then the second component performs a series of expansions and replaces the variables of each command parsing by following the sophisticated rules. Each command execution is then interpreted.

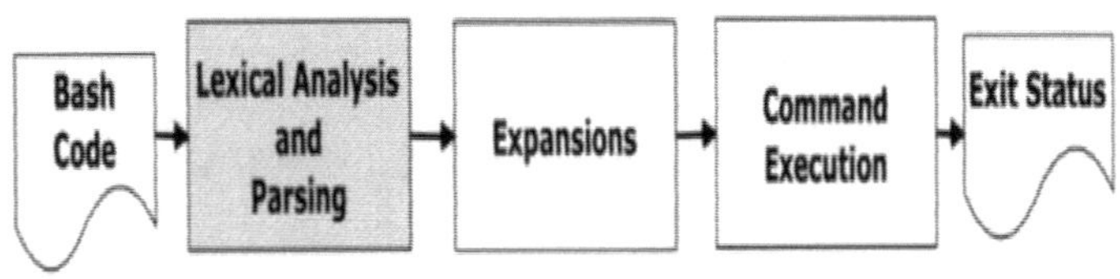

Figure 6. Architecture of Bash

3. Tools & Methods

3.1 System Design and Testing of Sudomy

Development of Information Gathering applications follows ISSAF rules [6] by applying two techniques, namely passive and active. Passive techniques obtain information through a number of ways by utilizing third-party resources such as using Web APIs, Information Gathering libraries or through OSINT Source

[12] with scraping processes. While the active technique uses applications that are installed with similar features, namely the Information Gathering function either by brute force, word lists or other new methods. The sudomy application diagram in conducting the Information Gathering stage can be seen in 'figure 7'. The Passive Information Gathering section uses resources from third parties using either the Web API, library or web scrapping. The web scrapping time testing process uses the regular expression (regex) technique on HTTP Body which uses three tests, namely the utilization of Native Apps that are called through Bash Script and libraries in two programming languages (Python and Go). The fastest time when the regex process uses Native Apps via Bash Script, which is Curl. The regex test results on the HTTP Body OSINT Source can be seen in Table 1.

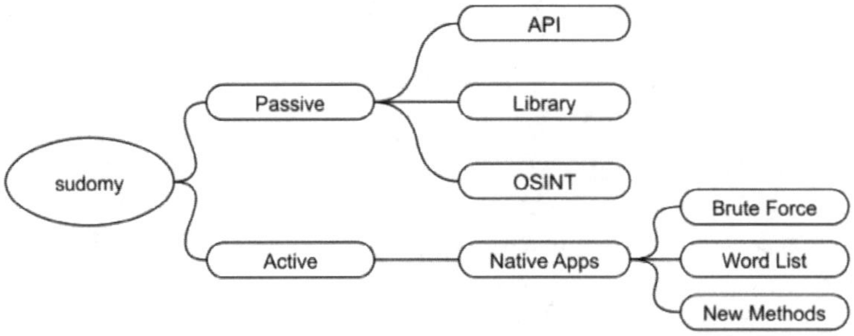

Figure 7. Overview of Sudomy Application

Table 1. RegEx HTTP Body Comparation

Command	Native/Library	Real	Users	System (CPU)
time ./curl.sh	curl (Native)	0m0.189s	0m0.032s	0m0.017s
time ./curl.py	import requests (Python)	0m0.693s	0m0.318s	0m0.110s
time ./curl.go	package req (Go)	0m0.308s	0m0.016s	0m0.010s

Whereas the Active Information Gathering section utilizes the Native Apps Gobuster. Gobuster application does not have a graphical interface, so it is easier to combine on sudomy which both have Command-line Interface (CLI) interfaces. The technique used in Active Information Gathering is the brute force on a target's directory, DNS, and Virtual Host.

3.2 Testing Materials

In testing Information Gathering tools using several supporting devices both hardware and software. The device used is a notebook with Intel® Core TM i7- 7700HQ @ 2.80Ghz CPU specifications and 8GB DDR4 RAM, with Virtual Box virtualization software running on the MS Windows 10 operating system. Virtual Box has a 2 core CPU configuration and 4GB RAM installed Kali Linux operating system. The Information Gathering process requires an internet connection, the provider used is PT Telekomunikasi Indonesia. The material details needed in the research are shown in Table 2.

Table 2. Materials

Materials	Description
Notebook	ASUS ROG GL553VD (CPU: Intel® Core TM i7- 7700HQ @ 2.80Ghz ; RAM: 8GB DDR 4)
Software Host OS	Virtual Box MS Windows 10
Guest OS	Kali Linux (RAM 4GB; No DE; Using Windows Manager)
Ping Connection	2 ms
Download/ Upload 30	
Mbps / 7 Mbps	
Provider	PT Telekomunikasi Indonesia Tbk (Telkom)

4. Result and Discussion

Information Gathering Method in sudomy application, there are two Passive and Active techniques. The passive technique uses third party resources such as DNSdumpster, WebArchive, Shodan, Total Virus, Certsh, BinaryEdge, SecurityTrails, Certspotter, Censys, Threatminer, Bufferover, Hackertarget, Entrust, ThereatCrowd, and Riddler. To improve the enumeration results sudomy application needs to add an API Key for Shodan, Censys, Total Virus, BinaryEdge, and SecurityTrails in the sudomy.API section. Whereas Active technique uses a combination of the Gobuster application with the wordlist provided by SecLists. SecLists has a collection of approximately three million wordlists. Testing is done by comparing the sudomy application with other applications such as sub finder and sublist3r with the target domain bugcrowd.com. The sub finder application uses 25 resources, sublist3r uses 11 resources, and sudomy uses 16 resources. The time needed to search for a subdomain from bugcrowd.com, the sub finder application takes 1 minute 38,621s, sublist3r takes 0 minutes 27,216s, and sudomy takes 0 minutes 6,946s. The subdomain results from bugcrowd.com found by the sub finder application are 25 subdomains, sublist3r are 23 subdomains, and sudomy is 49 subdomains. The results of enumeration looking for the bugcrowd.com subdomain using the sub finder, sublist3r, and sudomy applications can be seen in table 3. To further speed up the enumeration process and save CPU, RAM, and bandwidth, you can use third party resources as needed.

Table 3. Information Gathering bugcrowd.com Comparation

Tools	Time	Resources (Source)	Results (Subdomain)
subfinder	1m 38.621s	25	25
sublist3r	0m 27.216s	11	23
sudomy	0m 6.946s	16	49

The sudomy application has been equipped with a reporting system with HTML and CSV output format that makes it easy for Cyber Security researchers and / or analysts. The HTML formatted report displays subdomain enumeration results including Ping Sweep, HTTP status, IP Address, top ten open protocols, TakeOver Subdomain checking, screenshots, httprobe (utility for HTTP / HTTPS validation) and graphs resulting from enumeration, see Figure 8.

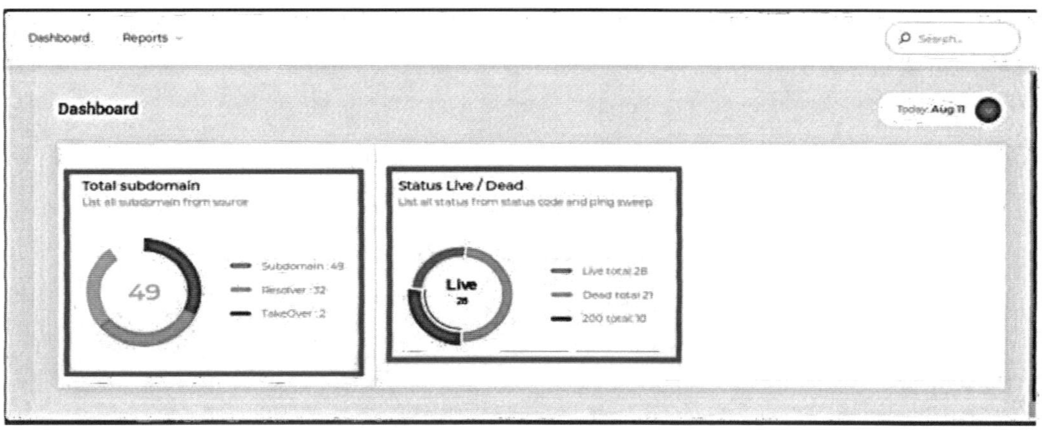

Figure 8. Dashboard Report Sudomy

5. Conclusion

Based on testing with similar tools, sudomy is still faster in the enumeration process of searching subdomains. Besides that, Sudomy also has features that really help Cyber Security researchers and / or analysts. The features that can be used include: Host checking on active subdomains, HTTP status code response information, conversion from subdomain list to IP Resolver, port scanning from IP Resolver, checking of TakeOver Subdomain attacks, screenshots through domain list, DNS Bruteforce Subdomain attack, and reporting. In addition sudomy application can be downloaded freely at https://github.com/Screetsec. Further development is expected to be able to integrate sudomy with other information security assessment applications.

Acknowledgement

Authors would like to say thank you very much to KEMENRISTEKDIKTI Indonesia and Universitas Islam Riau, University of Amikom Yogyakarta, Universitas Jenderal Achmad Yani Yogyakarta, Indonesia for funding this research project and support the facilities.

References

[1] Indonesia Security Incident Response Team on Infrastructure / Coordination Center 2019, *Indonesia Cyber Security Monitoring Report 2018* (Jakarta), 2018
[2] Institute for Security and Open Methodologies, *The Open Source Security Testing Methodology Manual,* 2010.
[3] P. V. Mockapetris, *RFC 1035: Domain Names - Implementation and Specification,* 1987
[4] A. A. Sinaci, O.T. Sehitoglu, M.T. Yondem, G. Fidan and I, Tatli, SEMbySEM in Action: Domain Name Registry Service Through a Semantic Middleware *eChallenges, 2010* 1–8, 2010.
[5] K, Scarfone, M, Souppaya, A, Cody and A, Orebaugh, *NIST SP 800-42: Guideline on Network Security Testing* vol 115 (Gaithersburg, MD), 2015.
[6] Open Information Systems Security Group, *Information Systems Security Assessment Framework (ISSAF),* 2006.
[7] C. Rudrakshi, A, Varshney, B, Yadla, R, Kanneganti and K, Somalwar, *API-fication,* 2014
[8] A. Mendoza, and G. Gu, Mobile Application Web API Reconnaissance: Web-to-Mobile Inconsistencies & Vulnerabilities *Proc. - IEEE Symp. Secur. Priv.* 2018-May 756–69, 2018.
[9] C. Ramey and B, Fox, *Bash Reference Manual* (Free Software Foundation, Inc.), 2019
[10] D. Hariyadi and Fazlurrahman, Membangun Telegram untuk Crawling Malware OSINT Menggunakan Raspberry Pi *Indones. J. Bus. Intell.* 2 18–24, 2019.
[11] I. J. Davis, M. Wexler, C. Zhang, R.C. Holt and T. Weber, Bash2py: A bash to Python translator *2015 IEEE 22nd Int. Conf. Softw. Anal. Evol. Reengineering, SANER 2015 - Proc.* 508–11, 2015.
[12] F.E. Nastiti, D. Hariyadi and Fazlurrahman, TelegramBot : Crawling Data Serangan Malware dengan Telegram *J. Comput. Eng. Syst. Sci.* 4, 2019.

IOP Conf. Series: Materials Science and Engineering **771** (2020) 012020 doi:10.1088/1757-899X/771/1/012020

Applying Data Mining to Analysis of Mismatch Between Education and Labor Market in Indonesia

Meilany Nonsi Tentua*[1,2], Edi Winarko[2]

[1]Teknik Informatika, Fakultas Teknik, Universitas PGRI Yogyakarta, Indonesia
[2]Department of Computer Science and Electronics, FMIPA UGM, Yogyakarta, Indonesia

*meilany@upy.ac.id, [2]ewinarko@ugm.ac.id

Abstract. One of the labor problems in Indonesia is the occurrence of mismatch (incompatibility between education and job). The incompatibility of education and job (labor market) can be categorized into several criteria such as vertical mismatch, horizontal mismatch, overqualification, under qualification, over skilling, under skilling, credentialism and economic skills. In several countries, analysis of mismatch between education and labor market can be used with several methods such as the Maximum Likelihood equation, the Mincer-type income equation, and the use of clustering algorithms in data mining. Indonesia can adopt a method of measuring the mismatch between education and labor market which suitable with the condition in Indonesia. The use of the K-Means algorithm or the hierarchical clustering algorithm will be able to see how much mismatch is occurring in Indonesia. Clustering is carried out referring to the Republic of Indonesia's Presidential Regulation No. 8/2012 and the Indonesian Classification of Work (KBJI).

1. Introduction

One of the labor problems in Indonesia is the occurrence of mismatch (incompatibility between education and job). The incompatibility of education and job (labor market) can be categorized into several criteria such as [1], vertical mismatch, horizontal mismatch, overqualification, under qualification, over skilling, under skilling, credentialism and economic skills.

This mismatch phenomenon is due to the transformation of the labor market in Indonesia. The first transformation occurred in the early 1990s which was marked by changes in the primary sector to the secondary sector in the labor market [2]. This condition is in line with the economic transformation from the agricultural sector to the industrial sector. There has been a decline in demand for labor from the agricultural sector and shifting to the industrial sector [3]. The second transformation occurred in the early 2000s. The rapid development of information technology and science has increased the absorption of labor in the service sector. Figure 1 shows a decrease in the demand for labor from the agricultural sector to the industrial sector in Indonesia from 2009 to 2016 [4].

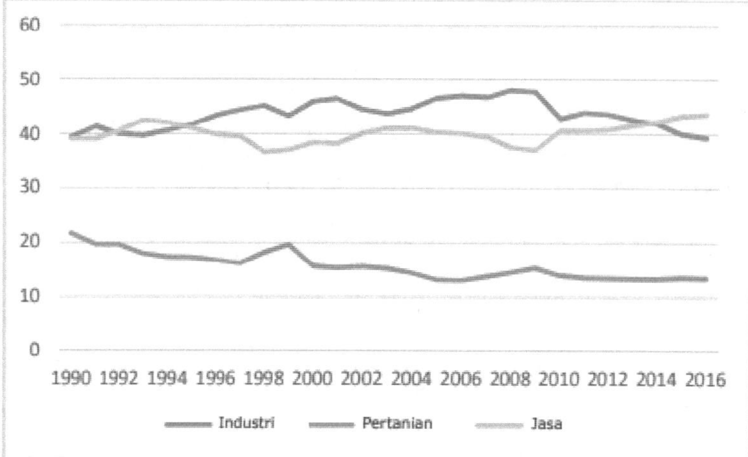

Figure 1. Workforce requirements per sector year in 1990-2016

On the supply side of labor market, there has been a large change with the growth of population and an increase in the level of education. Increasing the level of education in the workforce is an impact of the increasing access to education for the people of Indonesia. There is a gap between the highest level of education completed by staff with educational qualifications for certain types of work in the labor market. This is what causes overeducation and undereducation in the labor market. The phenomenon of overeducation and undereducation shows the existence of labor market imbalances in terms of supply and demand. In terms of the long term, this is a dilemma because it will lead to open unemployment, income problems, labor regulations, and education policies.

Statistical data shows that the labor force in Indonesia based on the ILO survey in 2018 was 125 million people or around 64.2% [5]. According to the 2015 national labor force survey processed by the University of Indonesia Demographic Institute, vertical mismatch or work mismatches with education and wages reached 53.33 percent. Meanwhile, horizontal mismatch or job qualification mismatch with educational background reaches 60.52 percent [6]. The youth unemployment rate in Indonesia is also much higher than the global average and when compared to various income groups [4].

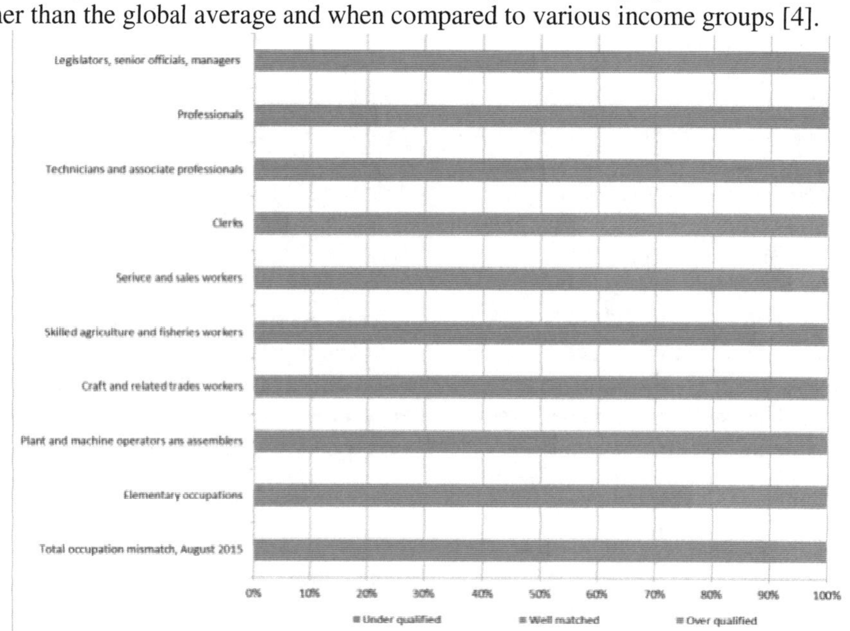

Source: BPS (2015) Labor force situation: August 2015, Badan Pusat Statistik, Jakarta. * ADB staff calculations based on revised population estimates. ** Excludes armed forces occupations.

Figure 2. Mismatch Between Education and Labor Market in Indonesia

The mismatch between the education system and labor market needs is a very serious problem for economic growth and development. The difference between supply and demand for labor ultimately results in a decrease in the relevance of labor and inadequate use of the most important manufacturing factors. Figure 2 seen that mismatch between education and labor market in Indonesia in 2015 [7].

Several approaches have been used to analyse the mismatch of education and job [8]. Some of the causes of job mismatches with education level are inadequate skills of job seekers. "Functional" skills that must be completed in the education and training system can be divided into three main categories: [9]

a) Self-assessment (direct approach) is an approach based on direct interviews with workers about the education needed for the current job position.
b) Job analysis (normative approach) is the use of a measurement of the relationship between work and education needed based on job analysis evaluation.
c) Realized match (statistical approach) by measuring the amount of education available to the job offered.

Some developed countries have made a discrepancy between the education system and the job that is very much considered by making a calculation model. This hospitality model will measure the level of discrepancy between education and employment. Several approaches are used in measuring to analysis mismatches between education and job.

In this article, we will review several models of measuring that have been carried out in several countries. Based on the review, a model that uses a data mining algorithm will be proposed to analyze the existing mismatch in Indonesia.

2. Literatur Review

Mismatch occurs when the level of education someone has is not suitable for the job. Vertical incompatibility can occur in two ways, excessive education or undereducation. Excess education is when someone is recruited for a job that requires a lower level of education than that of the individual. On the other hand, low education is where individuals have lower levels of education than expected for the job. The logical result of vertical mismatch is the presence of too many educated workers who bring skills to their jobs beyond the skills needed for the job, or the skills of less-educated workers are lower than those needed for that particular job [10]. Figure 2 shows the relationship between discrepancies that exist in education and labor market [11].

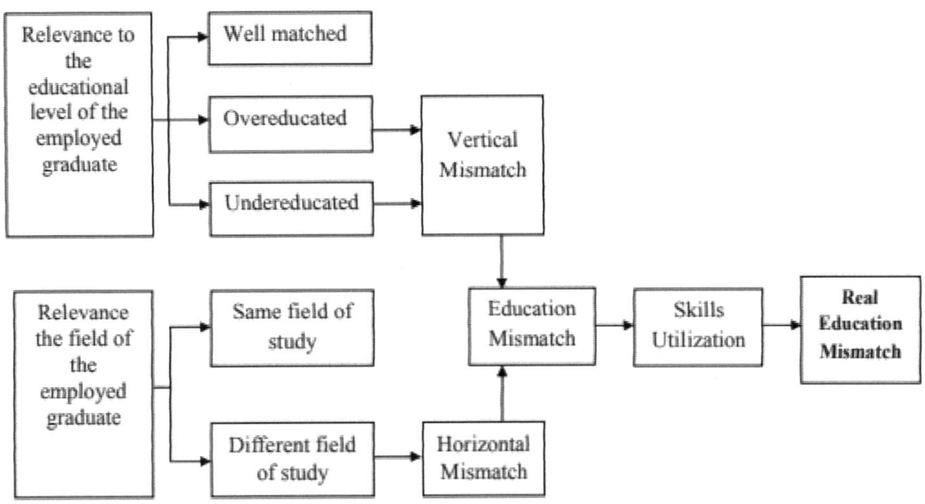

Figure 3. Relationship of mismatch that exist in Education and Job

a. Vertical mismatch
   A situation in which the level of education or skill is less or more than the required level of education or skill [12] [13]. There are several labor measurements for vertical mismatches such as subjective methods, empirical methods, job evaluation methods [14].
b. Horizontal mismatch
   A situation in which the level of education or skill matches job requirements, but the type of education or skill is inappropriate for the current job [15].
c. Overqualification
   A situation in which an individual has a higher qualification than the current job requires [16] [17] [18] [19].
d. Underqualification
   A situation in which an individual has a lower qualification than the current job requires [20] [21] [22].
e. Overskilling
   A situation in which an individual is not able to fully utilise his or her skill and abilities in the current job [23] [24].
f. Underskilling
   A situation in which an individual lacks the skill and abilities necessary to perform on the current job to acceptable standards [25] [26].
g. Credentialism
   A situation in which the level of education required to obtain the job exceeds the level of education required to perform the job adequately. It relies upon a belief of the employer that the possession of certificates and diploma implies higher productivity on the part of the individual [27] [28].
h. Economic skill obsolescence
   A situation in which skills previously utilised in a job is no longer required or have diminished in importance [22].

Several calculation models are conducted to see the amount of discrepancy between education and job in several countries. The results of this calculation are usually used to consider existing education policies. In Spain, especially in Barcelona, the calculation of inconsistencies between education and work is done by considering the availability of private transport to represent potential determinants. The equation model uses the Limited Information Maximum Likelihood equation, namely:[29]

$$mismatch_i = \propto + \beta' Xi + \gamma . pta_i + \delta . nca_i + \varepsilon_i \quad (1)$$

Equation (1) associates job-education mismatch with control variable vector (Xi) and with work accessibility with public transportation (PTA) and with private transportation (NCA). In the Netherlands, the calculation of nonconformity is done using the formulation of human capital H_{ct} available for company c in year t on the entrepreneur's perception of the mct match between skill demand and supply of skills in the company. The equation is written as:

$$m_{ct} = \alpha_0 + \alpha_1 H_{ct} + \alpha_2 X_{ct} + \eta_r + \eta_s + \varepsilon_{ct}, \quad (2)$$

where mct is equal to 1 if company c∈{1,2,..., N} reports mismatch in year t, and 0 otherwise; XctXctconstitutes a vector of company characteristics; ηrηr and ηsηs represent, respectively, region and sector fixed effects and ε_{ct} is the error term. [30]

More than a quarter of Australian graduates were found to be unsuitable. Graduates from fields of study and work were found. Mismatches are analyzed by looking at the income earned. The Ordinary Least Squares regression model is used in this calculation, which can be stated as:

$$\ln(sal) = X\beta + M + \varepsilon \quad (3)$$

where ln(sal) denotes annual salary, expressed in natural logarithmic format. The vector M are dummy variables that represent the three forms of mismatch, namely, i) vertically mismatched; ii) horizontally mismatched, and; iii) residual skills mismatched. In order to address the endogeneity of the mismatch variables in the earnings models, the two-stage residual inclusion approach of Terza et al [12].

2.1 Using Mincer-type income equation to analysis mismatch between education and job

The use of the variable labor income associated with a mismatch between employment and fields of study is usually estimated by the Mincer-type income equation in the country of Brazil. The equation used, i.e.

$$\ln(w_i) = \beta_0 + \beta_1 X_i + \beta_2 F_i + \beta_3 M_i + \beta_4 P_i + \beta_5 Dist_i + e_i. \quad (4)$$

where wi represents hourly labor income, Xi is a vector of demographic characteristics, Fi represents the field of study, Mi is the value of work not related to their field of study (0 or 1), Pi shows those who are partially suitable for employment, and ui is an error term. [31]

Almost the same as the country of Brazil, in New Zeland the measurement of the mismatch of education and labor market also uses income as a variable. The formula used is also almost the same shape but uses variables

$$\ln(E_i) = \beta_0 + \beta_1 S_i + \beta_2 Exp_i + \beta_2 Exp_i^2\ \beta_3 M_i + \alpha_4 X_i + u_i. \quad (5)$$

Ei represents individual income, Si is the level of education achieved, Expi is post-school labor market experience, Xi is a vector of variables that defines the individual (other than education or experience) and ui is a random error term. This specification reflects labor market outcomes based on employee information and is therefore entirely dependent on information about educational attainment from the supply side. [32]

Mismatch analysis in the United States also uses income as a benchmark in seeing how large the mismatch is between education and job.

$$\ln w_i = \delta_r S_i^r + \delta_0 S_i^0 + \delta_u S_i^u + x_i'\beta + \varepsilon_i \quad (6)$$

where wi denotes wages, Sr is the year of education needed for the job, S0 So the year of excessive education, and Su the year without education. xi is a vector of control variables including experience and the square of experience. δr, δo, and δu are returns for the required education years, excessive years of education, and fewer years of education. [33]

Demographic characteristics are represented by age, age squared, sex, race, and distance between work and residence. Some other countries such as Armenia, Georgia, Macedonia, Ukraine [8] Jepang [16] dan Afrika [34], also use Mincer-type income equality exchanges to analyze the mismatch of education and job.

2.2 Using data mining to analysis mismatch between education and job

Some mismatch analysis at this time has begun to use existing algorithms in data mining. As in Hungary [35], the analysis was carried out by utilizing an integrated database of the National Tax Administration, the National Health Insurance Fund, and higher education information systems. Several clustering methods including the BRIM algorithm [36] (bipartite, recursively induced modules) developed specifically for bipartite graphs, were chosen in analyzing the distribution of over education that is spread across several regions of the Hungarian country.

In the United Kingdom [37], education and work mismatch analysis are carried out using the clustering method in grouping work data at an individual level and looking at trends in overeducation that occur. The algorithm used in classifying is K-Median Clustering [38]. Machine Learning [39] is used in analyzing how graduates are employed in the country of Spain. Predictive models are built using machine learning algorithms. The data they used was taken from the Spanish Observatory for Employability and Employment (OEEU).

3. Method

We propose the K-Means algorithm or the hierarchical clustering algorithm approach in clustering mismatch between education and labor market. K-means is a method for dividing a set of observations into a user-defined number of subsets, based on how close the observations are to each other in the feature space[38]. There are k the number of subsets. The grouping is divided into five groups namely mismatch by level education, a field of study, job-specific level, level of basic skills and level of portable skills.

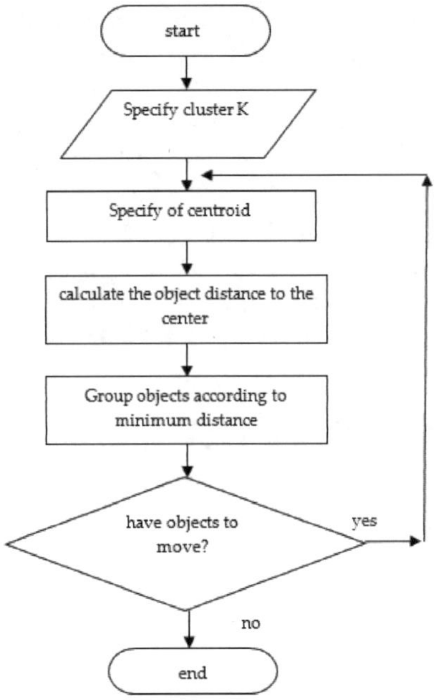

Figure 4. Flowchart of the K-means algorithm

The steps of the K-means algorithm are as follows, 1) Setting the initial midpoint value, 2) Calculating the distance of an object to a centroid using the Euclid distance formula, 3) Clustering an object by entering each object into a cluster (group) based on its minimum distance, 4) determine the centroid, based on the members of each group, 5) calculates the distance of the object to the new centroid until the grouping of each object has a minimum distance.

4. Discussion

Based on the 2016 Asian Development Bank report [7] job mismatches that occur in Indonesia tend to be related to the low level of education of workers who qualify for their jobs. Lack of qualifications is also a challenge in higher-level jobs. The low level of qualifications and the low level of over-qualifications lead to the problem of lack of skills. This is an important issue because high levels of skills and qualification mismatch are usually associated with lower levels of labor productivity.

With the importance of the mismatch problem between education and labor market, an analysis is needed that can measure the extent of the mismatch that occurs in Indonesia. Some methods that have been used to analyze in several countries are the Maximum Likelihood equation, the Mincer-type income equation, and the use of clustering algorithms in data mining.

Analysis of the measurement of education and work world mismatch that has been used in several countries pay attention to criteria related to education such as education level, experience, length of education and demographic factors. Demographic factors are represented by age, age squared, sex, race, and distance between workplace and residence. A mismatch analysis on education and types of work in Indonesia can be done by taking into account Presidential Regulation of the Republic of Indonesia Number 8 of 2012 [41] Concerning the Indonesian National Qualification Framework (KKNI) or based on the Indonesian Classification of Work Standard (KBJI) [42] made by BPS. The Indonesian National Qualification Framework (KKNI). KKNI provides nine qualification levels, starting from the 1st level qualification as the lowest qualification and the 9th level qualification as the highest qualification.

Table 1. Education, KKNI, and Level of Job

Education	KKNI	Level of Job
Basic Education (SD, SMP)	Level 1	Operator
Middle Education (SMA/MA/SMK)	Level 2	Operator
Diploma I	Level 3	Operator
Diploma II	Level 4	Technician/Analyst
Diploma III	Level 5	Technician/Analyst
Diploma IV/Bachelor	Level 6	Technician/Analyst
Profession	Level 7 or 8	Expert
S2	Level 8	Expert
Specialist	Level 8 or 9	Expert
S3	Level 9	Expert

Source: Presidential Regulation of the Republic of Indonesia Number 8 of 2012

Based on Table 1, it can be seen that there are 9 levels of KKNI according to Presidential Regulation of the Republic of Indonesia Number 8 of 2012. Basic, middle, and Diploma I have 1, 2, and 3 levels of KKNI with occupational positions as operators. Education DII, DIII, DIV / Bachelor have KKNI levels 4, 5, and 6 with a job position as a technician/analyst. As for professional education, S2, specialist, and S3 have KKNI levels 7, 8, and 9 with occupational positions as experts/professionals. The division into these three occupational positions is still considered to be less detailed.

The division of types of work in detail can be based on the Indonesian Classification of Occupational Classification (KBJI) issued by BPS. Based on KBJI 2000_1 there are 8 types of work ranging from professionals to operators and others. The types of work are in the order of types of work from the highest to with the lowest. The combination of KKNI levels with KBJI 2000_1 can be connected between the types of work and the level of education needed.

By observing the Republic of Indonesia's Presidential Regulation No. 8/2012 and the Indonesian Classification of Work (KBJI) Classification, classification can be made for workers in Indonesia. Analysis of mismatch measurements between education and labor market in Indonesia can see clustering using the K-Means algorithm or the hierarchical clustering algorithm. The use of clustering with the algorithm refers to Hungary, the United Kingdom and countries in Europe. Clustering can be done by grouping the types of mismatch that occur in Indonesia, as shown in Figure 4.

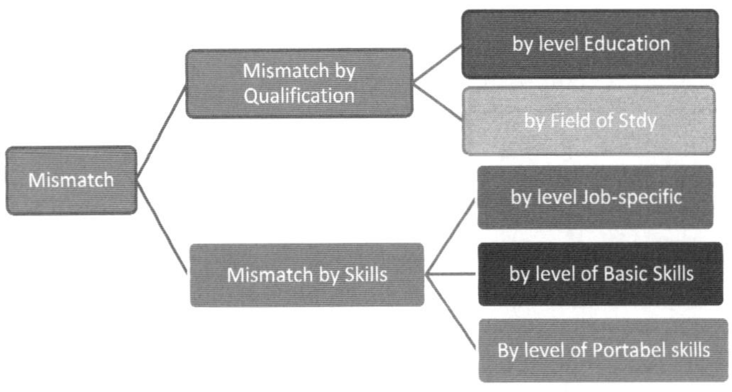

Figure 5. Clustering on the mismatch between education and labor market

The mismatch is grouped into parts namely mismatch based on qualifications seen based on the type of education and the field of study and based on the skills seen based on specific job level, basic skill level and level of portable skill. The data needed for classification and clustering are: personal data (date of birth, county of address, citizenship, gender)

a) occupational data (employment relationship, occupation, gross wage, etc.)
b) employer data (county of company headquarters, company size, company activity, etc.)
c) educational data (institute, faculty, program where the graduate graduated)
d) certification (skill)

5. Conclusion

Almost all countries have problems with the mismatch of education and labor market. In overcoming these problems they need analysis with several methods to measure how big the mismatch of education and labor market that occurs in the country. The method used is the Maximum Likelihood equation, the Mincer-type income equation, and the use of clustering algorithms in data mining.

Indonesia as a developing country is also inseparable from this problem. This mismatch phenomenon is due to the transformation of the labor market in Indonesia. The first transformation occurred in the early 1990s which was marked by changes in the primary sector to the secondary sector in the labor market.

Based on measurements that have been made in several countries, Indonesia can adopt a method of measuring the mismatch between education and labour market. The use of the K-Means algorithm or the hierarchical clustering algorithm will be able to see how much mismatch is occurring in Indonesia. Clustering is carried out referring to the Republic of Indonesia's Presidential Regulation No. 8/2012 and the Indonesian Classification of Work (KBJI).

References

[1] Montenegrin Employers Federation (MEF) and Prof Dr Milivoje Radovic and Dr Jovan Djuraskovic, "Report: Mismatch between Education System and Labour Market Needs – Enabling Environment for Sustainable, Enterprises in Montenegro" , Montenegrin Employers Federation (MEF), Podgorica, May 2016.

[2] C. Manning, "Labour Market Adjustment to Indonesia's Economic Crisis: Context, Trends and Implications," *Bull. Indones. Econ. Stud.*, vol. 36, no. 1, pp. 105–136, Apr. 2000.

[3] H. Hill, B. Resosudarmo, and Y. Vidyattama, "Indonesia's Changing Economic Geography," Department of Economics, Padjadjaran University, Working Papers in Economics and Development Studies (WoPEDS) 200713, Nov. 2007.

[4] Miyamoto, Michiko, "Laporan Ketenagakerjaan Indonesia 2017", International Labour Organization, 2017 .

[5] "Summary Report." [Online]. Available: https://www.ilo.org/ilostat/faces/oracle/webcenter/portalapp/pagehierarchy/Page3.jspx?MBI_ID=7&_afrLoop=135560966713369&_afrWindowMode=0&_afrWindowId=qz5pmvxii_1#!%40%40%3F_afrWindowId%3Dqz5pmvxii_1%26_afrLoop%3D135560966713369%26MBI_ID%3D7%26_afrWindowMode%3D0%26_adf.ctrl-state%3Dqz5pmvxii_129. [Accessed: 18-Jul-2019].

[6] "Ketidaksesuaian antara Pendidikan dan Kebutuhan Tenaga Kerja Masih Besar | Databoks." [Online]. Available: https://databoks.katadata.co.id/datapublish/2019/05/28/ketidaksesuaian-antara-pendidikan-dan-kebutuhan-tenaga-kerja-masih-besar. [Accessed: 18-Jul-2019].

[7] E. R. Allen, "Analysis of Trends and Challenges in the Indonesian Labor Market," Asian Development Bank, 2016 p. 38.

[8] O. Kupets, "Education in transition and job mismatch: Evidence from the skills survey in non-EU transition economies,", Departement of Economic, National University of "Kyiv-Mohyla Academy", p. 57, February 2015

[9] __, "Indonesia Skills Report Trends in Skills Demand, Gaps, and Supply in Indonesia" , Human Development Department East Asia and Pacific Region, 2010 .

[10] Senarath, Patabendige, "Job-Education Mismatch Among the Graduates: A Sri Lankan Perspective," *Ruhuna Journal of Management and Finance, ISSN 2235-9222, Volume 1 Number 2 - July 2014*].

[11] Cedefop, "Annual report 2010" , Publications Office of the European Union, ISBN Publications Office of the European Union, 2011 .

[12] I. W. Li, M. Harris, and P. J. Sloane, "Vertical, Horizontal and Residual Skills Mismatch in the Australian Graduate Labour Market," *Econ. Rec.*, vol. 94, no. 306, pp. 301–315, 2018.

[13] P. Pholphirul, "Educational mismatches and labor market outcomes: Evidence from both vertical and horizontal mismatches in Thailand," *Educ. Train.*, vol. 59, no. 5, pp. 534–546, Jun. 2017.

[14] R. Iryanti, "Education & Skill Mismatch di Indonesia:," Paparan, Agustus 2017, p. 28.

[15] J. Schweri, A. Eymann, and M. Aepli, "Horizontal Mismatch and Vocational Education," , Universitas Zurich, Februari 2019, p. 29.

[16] S. Flisi, V. Goglio, E. C. Meroni, M. Rodrigues, and E. Vera-Toscano, "Measuring Occupational Mismatch: Overeducation and Overskill in Europe—Evidence from PIAAC," *Soc. Indic. Res.*, vol. 131, no. 3, pp. 1211–1249, Apr. 2017.

[17] D. Erdsiek, "Overqualification of graduates: assessing the role of family background," *J. Labour Mark. Res.*, vol. 49, no. 3, pp. 253–268, Nov. 2016.

[18] S. LaRochelle-Côté and D. Hango, "Overqualification, skills and job satisfaction,", the Minister responsible for Statistics Canada , p. 16, 2016.

[19] V. Meliciani and V. Romania, "Overeducation and overskill in the Italian labour market: the role of fields of study," CIMR Research Working Paper Series, p. 32. 2016

[20] T. Sparreboom and A. Tarvid, "Imbalanced Job Polarization and Skills Mismatch in Europe," *J. Labour Mark. Res.*, vol. 49, no. 1, pp. 15–42, Jul. 2016.

[21] D. Rohrbach-Schmidt and M. Tiemann, "Educational (Mis)match and skill utilization in Germany: Assessing the role of worker and job characteristics," *J. Labour Mark. Res.*, vol. 49, no. 2, pp. 99–119, Oct. 2016.

[22] Adrian Nikolov; Desislava Nikolova, PhD; and Peter Ganev; Yavor Aleksiev, "Research to understand the extent, nature and impact of skills mismatches in the economy," *Inst. Mark. Econ. Eur. Econ. Soc. Comm. EESC*, p. 106, Jul. 2018.

[23] K. Mavromaras, S. McGuinness, S. Richardson, P. Sloane, and Z. Wei, "Over-skilling and job satisfaction," IZA Discussion Paper No. 5083, p. 58. July 2010

[24] A. Chłoń-Domińczak and A. Żurawski, "Measuring Skills Mismatches Revisited - Introducing Sectoral Approach," p. 46.

[25] Y. E. Pravitasmara Dewi and L. Suharti, "Understanding Skill Mismatch from Learning Organization Perspective: A Literature Review," *KnE Soc. Sci.*, vol. 3, no. 10, Nov. 2018.

[26] B. Manuela and M. Simona, "Social Integration Dynamics for Migrants: PIAAC to Measure Skill and Qualification Mismatch," *Univers. J. Educ. Res.*, vol. 6, no. 5, pp. 970–982, May 2018.

[27] T. Karmel, "Skills Deepening or Credentialism? Education Qualifications and Occupational Outcomes, 1996-201," vol. 18, no. 1, p. 23, 2015.

[28] A. Marin and S. Hayes, "The Occupational Context of Mismatch: Measuring the Link Between Occupations and Areas of Study to Better Understand Job- Worker Match," Canadian Journal of Sociology, p. 22. 2017

[29] A. Di Paolo, A. Matas, and J. L. Raymond, "Job Accessibility, Employment and Job-Education Mismatch in the Metropolitan Area of Barcelona," *SSRN Electron. J.*, 2014.

[30] S. J. Cabus and M. A. Somers, "Mismatch between education and the labour market in the Netherlands: is it a reality or a myth? The employers' perspective," *Stud. High. Educ.*, vol. 43, no. 11, pp. 1854–1867, Nov. 2018.

[31] M. Reis, "Measuring the mismatch between field of study and occupation using a task-based approach," *J. Labour Mark. Res.*, vol. 52, no. 1, 2018.

[32] J. Z. Yeo and S. A. Maani, "Educational Mismatches and Earnings in the New Zealand Labor Market," p. 31.

[33] E. Leuven and H. Oosterbeek, "Overeducation and Mismatch in the Labor Market," in *Handbook of the Economics of Education*, vol. 4, Elsevier, 2011, pp. 283–326.

[34] E. Grapsa, "How well matched are South African workers to their jobs?," p. 44.

[35] "Graph configuration model based evaluation of the education-occupation match." [Online]. Available: https://www.ncbi.nlm.nih.gov/pmc/articles/PMC5839539/. [Accessed: 09-Aug-2019].

[36] R. Tackx, F. Tarissan, and J.-L. Guillaume, "ComSim: A Bipartite Community Detection Algorithm Using Cycle and Node's Similarity," in *Complex Networks & Their Applications VI*, vol. 689, C. Cherifi, H. Cherifi, M. Karsai, and M. Musolesi, Eds. Cham: Springer International Publishing, 2018, pp. 278–289.

[37] "The changing graduate labour market: analysis using a new indicator of graduate jobs | SpringerLink." [Online]. Available: https://link.springer.com/article/10.1186/s40173-016-0070-0. [Accessed: 09-Aug-2019].

[38] D. Dohan, S. Karp, and B. Matejek, "K-median Algorithms: Theory in Practice," p. 21.

[39] S. Shalev-Shwartz and S. Ben-David, *Understanding Machine Learning: From Theory to Algorithms*. Cambridge: Cambridge University Press, 2014.
[40] "Analysis of the UK labour market - estimates of skills mismatch using measures of over and under education - Office for National Statistics." [Online]. Available: https://www.ons.gov.uk/employmentandlabourmarket/peopleinwork/employmentandemployeetypes/articles/analysisoftheuklabourmarketestimatesofskillsmismatchusingmeasuresofoverandundereducation/2015. [Accessed: 30-Jul-2019].
[41] Peraturan Presiden Republik Indonesia Nomor 8 Tahun 2012 Tentang Kerangka Kualifikasi Nasional Indonesia
[42] Klasifikasi Baku Jabatan Indonesia KBJI 2014, Kementrian Ketenagakerjaan dan Badan Pusat Statistik

IOP Conf. Series: Materials Science and Engineering **771** (2020) 012021 doi:10.1088/1757-899X/771/1/012021

Detection of Object Changes in Low Contrast Image Using Template Matching Method

I K Amri[1], Murinto[1,*]

[1]Department of Informatics Engineering, Universitas Ahmad Dahlan Yogyakarta, Indonesia

[1]indrakhairulamri@gmail.com, *murintokusno@tif.uad.ac.id.

Abstract. This research uses a contrast stretching method to improve the detected object image. This process is needed because of the lack of lighting in the room so it is difficult to identify properly. This research develops an image object change detection system to detect the movement of objects contained in a room through a Closed Circuit Television (CCTV) camera which is conducted simulations using a webcam camera using a template matching method with normalized cross-correlation techniques. The results of detecting changes in image objects are then made to compare between the reference image and image before there is an object change with the image after the object changes. System testing is carried out at a certain time, namely 07.00-17.00 WIB. The level of accuracy to identify the detection of changes in image objects using the template matching method with normalized cross-correlation technique is 85% for the category of no change in min (0.01025) and max (0.02088) and changes in min (0.01640) and max (-0.00513).

1. Introduction

Current technological advances, especially in the field of digital image processing technology (Digital Image Processing) cannot be separated from the role of humans, as happened with the detection of changes in image objects. Currently detecting changes and classification in image objects is still a large percentage of its success is still low [1] [2].

This is due to the lack of lighting in the room or lighting that is less stable, so the dimensions of the change in the object image captured in the program cannot be properly defined. Generally, low light conditions can be overcome with some technologies such as image processing. In low light, the video becomes blurred or experiences noise, causing a much greater video flow. The ability of Closed-Circuit Television (CCTV) to capture images more cleanly over long distances and capture changes in the object image that is moving current constraints [3]. With the condition of nighttime CCTV or lack of lighting can cause noise, the results are not clear, thus doing the wrong processing. Monitoring is still manual that is only monitoring using the eyes of a practicum assistant in the room. In the basic computing laboratory room, there have been various events with poor lighting conditions in the basic computing laboratory room, such as frequent unconsciousness about changes in objects that have occurred, changes in the object in question include the layout of computers, desks, chairs in each basic computing laboratory room and the location of the practitioners' seats in the room. With these constraints, we need a system that can improve the results of capturing changes in image objects so that the desired results can be obtained easily and simply so that users can define the desired object change in a more modern way without the need to use the naked eye to supervise practitioners.

2. Method

2.1. Contrast Stretching Method

Improvement of the digital image is a stage that is subjective, optional, and experimental because there is no standard algorithm in improving image quality. Operations in improving image quality include point operations, spatial operations, and transformation operations. The purpose of improving the image of an image, among others, is highlighting certain aspects of the display to be more easily understood or interpreted by human vision, as well as reducing or eliminating the display aspects of an image that is not needed such as noise.

This method serves to improve the quality of the image in contrast to the image that was previously still not good which is caused by poor lighting or image experiencing noise. In a contrasting image affected by lighting, errors in the settings of the camera during the process of image acquisition and the dynamics of the image sensor. The following is the formula for contrast stretching in equation (1).

$$O\,(i,j) = \frac{u(i,j)-c}{d-c}\,(L1) \qquad (1)$$

In equation (1) where $O\,(i, j)$ and $u(i,j)$ are pixels before and after being transformed to coordinates (i,j), c and d state that the maximum and minimum values obtained from the input image pixel, and L are maximum grayscale value. If the pixel value is less than 0 then the pixel value becomes 0 and if the pixel value is greater than (L-1) then the pixel value becomes (L-1). the grayscale color sequence where each pixel represents a gray degree with the lowest intensity value from 0 (black) to the highest that is 255 (white). In this grayscale color sequence, the pixels have a size of 8 bits or 1 byte [4] [5].

2.2. Template Matching Method

The template matching method is one of the techniques in digital image processing that functions to match each part of an image with an image that becomes a template (reference) [8]. An input image containing a particular template is compared to a reference template. The template is placed at the center of the image to be compared and counted how many points best fit the template. This step is repeated for the entire input image to be compared. The greatest point of conformity between the input image and the template image indicates that the template is the template image that best matches the input image.

2.3. Normalized Cross-Correlation Method

Normalized Cross-Correlation (NCC) is a technique in photogrammetry to measure the conjugation point. In the Normalized Cross-Correlation process, the first image is determined beforehand the reference image (template) that has pixel points inside it that the partner will look for in the second image. In the second image, it is determined in advance the search area (search window) which has a larger size than the reference image area (template). Therefore, the virtue of the Normalized Cross-Correlation technique is to find pairs of pixel titles between the first image and the image/pair with NCC value limits between -1 and +1. The size of the distance (Euclidean distance squared) is very influential in the use of cross-correlation for template matching.

$$d^2_{f,t}(u,v) = \textstyle\sum_{x,y}[f(x,y) - t(x-u,y-v)] \qquad (2)$$

Where in equation (2) f is the target image, t is a feature and d^2 is the Euclidean distance squared. Equation (3) is an extension of d^2.

$$d^2_{f,t}(u,v) = \textstyle\sum_{x,y}[f^2(x,y) - 2f(x,y)t(x-u,y-v) + t^2(x-u,y-v)] \qquad (3)$$

Equation (3) has the form $\sum t^2(x-u,y-v)$ constant. If $\sum f(x,y)$it is constant, then use equation (4) to measure the equation between the target image and the feature.

$$c(u,v) = \textstyle\sum_{x,y} f(x,y)t(x-u,y-v) \qquad (4)$$

If the image energy varies in position, equation (4) can fail, not invariant because it is caused by changes in lighting conditions at the image collapse when there is a change in image amplitude [8].

$$r(u,v)=\frac{\sum_{x,y}[f(x,y)-\bar{f}_{u,v}][t(x-u,y-v)-\bar{t}]}{\left\{\sum_{x,y}[f(x,y)-\bar{f}_{u,v}]^2\sum[t(x-u,y-v)-\bar{t}]^2\right\}^{0.5}} \quad (5)$$

In equation (5) is a correlation coefficient to overcome difficulties by normalizing image vectors and features or called NCC where $\bar{t}$ feature mean (ROI) and $\bar{f}_{u,v}$ average of image templates under the feature area (ROI)[6][7]. The flow of the normalized cross-correlation technique is shown in Figure (1).

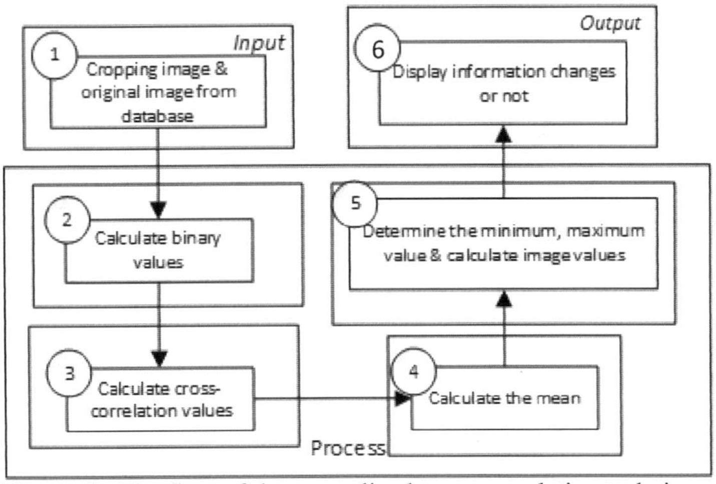

Figure.1. The flow of the normalized cross-correlation technique.

In Fig. 1 the flow of the normalized cross-correlation technique explains that in process 1 runs the input process of cropping the image and original image from the database, then in process 2 the binary values of the two images are calculated. In process 3 the process of calculating the value of cross-correlation is done, then at stage 4 calculate the mean. At stage 5 is determine the minimum, maximum value and calculate image values. And in the last process that is display information changes or not.

3. Results and Discussion

The original image before processing is a video image with **.mp4* extension with a size of 1080x1920 pixels which is then captured in the template, then the image enhancement method used in this study is tested on the captured image obtained from the video image that changes the image object with a pixel size of 1080x1920 with a **.png* image extension and converted to a grayscale image to facilitate processing. In the template matching method used to identify the object image that has been preprocessed in the form of image images with a size of 174x270 pixels compared to the original image in the database. Both images are identified in binary image format to facilitate processing. In the System Analysis Results, detection of changes in image objects is carried out by the system and Expert Judgment as in

Table 1 Results of Detection Changes in Keyboard Objects Using Matching Templates.

No	*Cropping* image	*Template*	NCC	*Expert*	System	conclusion
1			0.01025	Not changed	Not changed	correct
2			0.01033	Not changed	Not changed	correct
3			0.01143	Not changed	Not changed	correct
4			0.01094	Not changed	Not changed	correct

5				changed	changed	
6			0.02021	Not changed	Not changed	correct
7			0.01685	Not changed	Not changed	correct
8			0.01546	Not changed	Not changed	correct
9			0.01459	Not changed	Not changed	correct
10			0.01452	Not changed	Not changed	correct
11			0.00506	Not changed	Not changed	correct
12			0.00499	Not changed	Not changed	correct
13			0.01622	changed	Not changed	Not correct
14			0.00960	changed	changed	correct
15			0.01433	changed	Not changed	Not correct
16			0.01640	changed	Not changed	Not correct
17			0.00489	changed	changed	correct
18			-0.00166	changed	changed	correct
19			-0.00322	changed	changed	correct
20			-0.00513	changed	changed	correct

After the system and Expert Judgment analyzes the detection of changes in image objects, there are 3 images that do not match between the system and Expert Judgment. Where the value of N is the total number of images tested, that is 20 images consisting of 10 images that have not changed the image object and 10 images that have changed the image object. The following calculations to determine the level of accuracy: Known: N = 20, Matching results = 17, results not matching = 3 then obtained the level of accuracy to identify the detection of changes in image objects using the template matching method with normalized cross-correlation technique that is 85%.

4. Conclusion

The results of the identification of 20 test data using the Template Matching method with Normalized Cross-Correlation technique produce NCC values for the category of no change in min (0.01025) and max (0.02088) and changes in min (0.01640) and max (-0, 00513). Lighting is very influential at all on the value generated using this method, so the accuracy obtained is 85%. This method is not suitable for use in this research.

IOP Conf. Series: Materials Science and Engineering **771** (2020) 012021 doi:10.1088/1757-899X/771/1/012021

References

[1] Boyat A & Joshi B. (2015). Noise Models in Digital Image Processing. *An International Journal* (SIPIJ) (Vol.6, No.2, April 2015).

[2] Winiarti, S., Prahara, A., Murinto, Ismi, D.P. (2018). Pre-trained convolutional neural network for classification of tanning leather image. International Journal of Advanced Computer Science and Applications

[3] Moradi S & Zayed T. (2017). Real-Time Defect Detection in Sewer Closed Circuit Television Inspection Videos. ASCE (Pipelines 2017)

[4] Deka A & Mahanta L. (2018). Applications of Image Processing Techninques in Handwritten Signatures. (Vol.5, No.11, November 2018).

[5] Parikh M, Patel M & Bhatt D. (2013). Animal Detection Using Template Matching Algorithm. *Journal of Engineering and Emerging Technology*. (Vol.1, No.3 April 2013).

[6] Kaso A. (2017). Computation of The Normalized Cross-Correlation by Fast Fourier Transform. *Journal of Diagnostic Radiology and Nuclear Medicine.*

[7] Mohammed A, Farup I, Pedersen M, Hovde O & Yayilgan S. (2018). Stochastic Capsule Endoscopy Image Enhancement. *Journal of Imaging.*

[8] Hisham M, Yaakob S, Raof R, Nazren & Wafi. (2015). Template Matching Using Sum of Squared Difference and Normalized Cross Correlation. *IEEE Student Conference on Research and Development (SCOReD).*

IOP Conf. Series: Materials Science and Engineering **771** (2020) 012022 doi:10.1088/1757-899X/771/1/012022

Dragon Fruit Maturity Detection Based-HSV Space Color Using Naive Bayes Classifier Method

I S Khisanudin[1], Murinto[1,*]

[1]Department of Informatics Engineering, Universitas Ahmad Dahlan Yogyakarta, Indonesia

[1]irfansoudry@gmail.com, *murintokusno@tif.uad.ac.id.

Abstract. Color is one of the elements to determine the level of maturity of the fruit because each level of maturity of the fruit has a different color level. At present, the dragon fruit is only classified based on a visual analysis of the skin color of the human eye. This study aims to classify dragon fruit using image processing with the naïve Bayes method based on HSV color space. The color feature used in this study is RGB which has been converted to an HSV value, after getting an HSV value then classified using the naïve Bayes method. Image data used amounted to 120 training images and 30 test images. The results of the classification of dragon fruit using the naïve Bayes method based on HSV color space can show an accuracy rate of 86.6%.

1. Introduction

Red flesh dragon fruit (Hylocereus polyrhizus) is a type of cactus plant or in the family Cactaceae and subfamily hylocereanea that grows in many American regions but many are also cultivated in the Asian region. Dragon fruit, in general, contains water nutrients, carbohydrates, phosphorus, vitamin C, calcium, and specifically for red fleshy dragon fruit also contains beta carotene which functions as an antioxidant. In Indonesia, dragon fruit is planted on a small scale (household) or on a large scale (garden), dragon fruit can be said to be old enough to be harvested when it is ± 50 days after the flowers bloom perfectly. In fruit harvesting, it is necessary to consider the level of fruit maturity, because this is one of the factors in determining the quality of the dragon fruit. If the fruit is harvested when it is still not ripe even though the color has started to red, the result is that the fruit will have a less sweet taste. Likewise, when the fruit is harvested in conditions for too long it will produce a sweet taste but the fruit also has a short shelf life because it rots quickly.

Image processing is one technique that can be used in processing images by manipulating them into image data to obtain the desired information. Transforming the system from color space is one of the methods used in image processing. The basic basis of color composition is RGB or red, green, blue, which is also good for displaying color information from images but there are also some image processing applications that are not suitable. To overcome these deficiencies, one way to display image information is by converting the RGB value of the image into an HSV value.

The use of image processing technology is expected to help improve accuracy in the classification of fruit maturity levels. The condition of fruit maturity can be observed through the color of the image when taken against a background that contrasts with the object of study. This is what underlies the conduct of this research so that there are thoughts to build an image processing application to predict the maturity of dragon fruit based on the dragon fruit skin color features. The application of fruit maturity detection previously investigated include star fruit [1], banana [2], apple [3], manga [4], persimmon [5], palm oil [6], and there are also applications used in image processing for beef eligibility classification

IOP Conf. Series: Materials Science and Engineering **771** (2020) 012022 doi:10.1088/1757-899X/771/1/012022

[7], human skin recognition [8], and email spam classification based on text and images [9], classification tanning of leather [10].

2. Method

In this study the data obtained through the study of literature, observation and interviews. A literature study is one of the information gathering techniques related to the theme of image processing and detection of maturity based on fruit skin to support the theoretical basis of the research conducted by referring to research that has previously been done. An interview is a technique of collecting information directly from sources or experts that are useful for the design of the image processing system model. Furthermore, there are observations where we make observations directly on the object to get information and then it will be processed by the system.

Based on the image data of the red dragon fruit taken using a smartphone camera, we can obtain information values and the maturity classification of the dragon fruit based on their skin color. Dragon fruit which is still green means the dragon fruit is still unripe if it is reddish-green, it can be categorized as half-ripe, and dragon fruit can be said to be ripe when it is red. In Table I can be shown the level of maturity of the dragon fruit based on changes in color.

Table 1. Maturity Level of Dragon Fruit

Color Index	Fruit picture	Fruit Feature
1	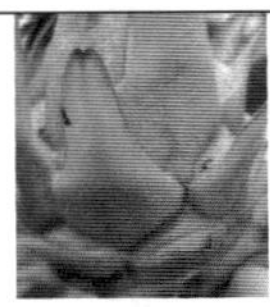	The entire surface of the skin is green and has a red color on the tip of the fruit scales
2	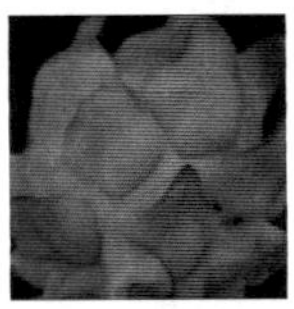	The skin is purple mixed with yellow and green, the fruit is a little hard
3	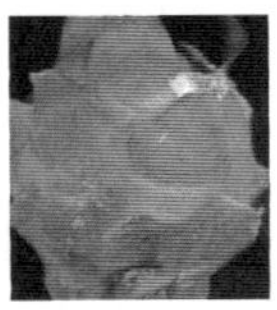	The surface of the skin is red and has a soft texture on the skin

The implementation of the dragon fruit maturity detection system based on skin color in the HSV color space with classification using the naïve Bayes method is shown in Figure 1.

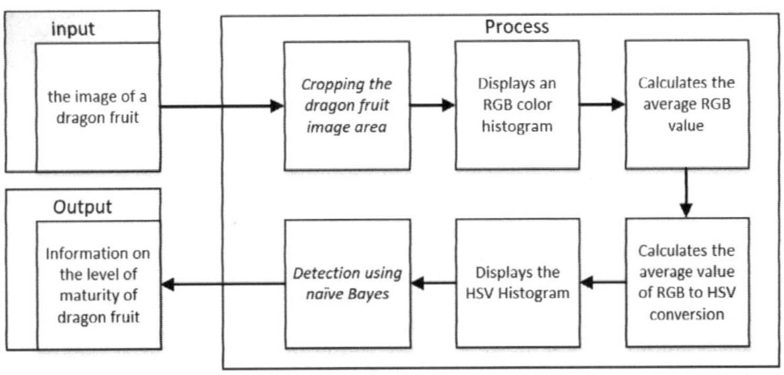

Figure 1. Flowchart of maturity determination program based on HSV color space

The steps taken in the program are:
1. Displays the original image taken from the file directory.
2. Propping the original image into the desired area to facilitate the classification process.
3. Calculate the average value of each RGB component of the cropped image
4. Transform the average RGB value into the HSV value by calculating the value of equation (1) first and then looking for the value of H by using equation (2), the value of S is calculated using equation (3), and the value of V is calculated using equation (4)[10].

$$r = \frac{R}{R+G+B}, g = \frac{G}{R+G+B}, b = \frac{B}{R+G+B} \quad (1)$$

$$H = \begin{cases} 0 & jika\ S = 0 \\ \frac{60+(g-b)}{S+V} & jika\ V = r \\ 60 * \left[2 + \frac{b-r}{s+v}\right], & jika\ V = g \\ 60 * \left[4 + \frac{r-g}{s+v}\right], & jika\ V = b \end{cases} \quad (2)$$

$$S = \begin{cases} \theta & jika\ V = 0 \\ 1 - \frac{\min(r,g,b)}{V}, & V > 0 \end{cases} \quad (3)$$

$$V = \frac{R+G+B}{3} \quad (4)$$

5. Calculate the maximum and minimum average values of H, S, and V from the results of trials of 10 samples of unripe dragon fruit, 10 samples of half-cooked dragon fruit, and 10 samples of ripe dragon fruit.
6. Classify the maturity level of the dragon fruit based on skin color using the naïve Bayes method using equation (5)[12].

$$\boldsymbol{f(x)} = \frac{\mathbf{1}}{\sqrt{(\mathbf{2}\pi)^{D}\,\mathbf{det}(\Sigma\ \ )}} * \mathbf{exp}\left(-\frac{1}{2}(\boldsymbol{x}-\boldsymbol{\mu})^{T}\,\Sigma^{-1}(\boldsymbol{x}-\boldsymbol{\mu})\right) \quad (5)$$

3. Result and Discussion

The system testing phase in this study was carried out by processing 30 images with *. Jpeg, *. Jpeg and * .bmp extensions with 10 images divided by 10 unripe images, 10 half-cooked images, and 10 ripe images. The image is cropped on the skin where you want to calculate the average RGB histogram value which is then transformed into the HSV color space. After obtaining the average value of HSV from the image, classification is done using the naïve Bayes method. Table II shows the values of H (hue), S (saturation) and V (value).

IOP Conf. Series: Materials Science and Engineering **771** (2020) 012022 doi:10.1088/1757-899X/771/1/012022

Table 2. Range of HSV values of unripe dragon fruit

Test Image Data	Hue	Saturation	Value
1	0.22425	0.85481	0.44024
2	0.21237	0.83432	0.63619
3	0.25027	0.32475	0.71868
4	0.20195	0.84625	0.66707
5	0.25828	0.51748	0.68915
6	0.20992	0.70125	0.42190
7	0.19355	0.72828	0.55938
8	0.19683	0.81568	0.67465
9	0.20561	0.72721	0.67178
10	0.20894	0.63579	0.70850
total	2.16197	6.98582	6.18754
Average	0.21620	0.69858	0.61875
Min	0.19355	0.32475	0.42190
Max	0.25828	0.85481	0.71868

Table 3. Range of half-ripe dragon fruit HSV values

Test Image Data	Hue	Saturation	Value
1	0.27505	0.48658	0.90462
2	0.55273	0.35087	0.60747
3	0.36432	0.31128	0.54466
4	0.25564	0.46469	0.84682
5	0.27481	0.54343	0.62916
6	0.31326	0.26961	0.69397
7	0.36014	0.40330	0.86763
8	0.29003	0.35765	0.78002
9	0.34182	0.47223	0.80405
10	0.16776	0.33256	0.61703
Total	3.19556	3.99220	7.29543
Average	0.31956	0.39922	0.72954
Min	0.16776	0.26961	0.54466
Max	0.55273	0.54343	0.90462

Table 4. Range of HSV values of ripe dragon fruit

Test Image Data	Hue	Saturation	Value
1	0.85200	0.56615	0.94806
2	0.91995	0.56648	0.96886
3	0.93848	0.58993	0.98491
4	0.89152	0.44855	0.97257
5	0.87888	0.53696	0.97311
6	0.86429	0.57267	0.93820
7	0.91878	0.70491	0.91073
8	0.82015	0.62398	0.96775
9	0.80875	0.54576	0.97071
10	0.83701	0.58775	0.96600
total	8.72981	5.74314	9.60090
Average	0.87298	0.57431	0.96009
Min	0.80875	0.44855	0.91073
Max	0.93848	0.70491	0.98491

Table 5. Range of minimum and maximum HSV values based on the level of maturity

	Hue	Saturation	Value	Note
Min	0.19355	0.51748	0.42190	Unripe
Max	0.27481	0.85481	0.70850	Unripe
Min	0.16776	0.26961	0.54466	Half Ripe
Max	0.82015	0.62398	0.96775	Half Ripe
Min	0.55273	0.35087	0.60747	Ripe
Max	0.93848	0.70491	0.98491	Ripe

Table 6. Result of classification maturity level red dragon fruit

No	Test Data	Expert Judgment	Average Value			System Test Results	Conclusion
			H	*S*	*V*		
1	Citra 1	Unripe	0.22425	0.85481	0.44024	Unripe	Passed
2	Citra 2	Unripe	0.21237	0.83432	0.63619	Unripe	Passed
3	Citra 3	Unripe	0.25027	0.32475	0.71868	Half ripe	Not passed
4	Citra 4	Unripe	0.20195	0.84625	0.66707	Unripe	Passed

5	Citra 5	Unripe	0.25828	0.51748	0.68915	Unripe	Passed
6	Citra 6	Unripe	0.20992	0.70125	0.42190	Unripe	Passed
7	Citra 7	Unripe	0.19355	0.72828	0.55938	Unripe	Passed
8	Citra 8	Unripe	0.19683	0.81568	0.67465	Unripe	Passed
9	Citra 9	Unripe	0.20561	0.72721	0.67178	Unripe	Passed
10	Citra 10	Unripe	0.20894	0.63579	0.70850	Unripe	Passed
11	Citra 11	Half ripe	0.27505	0.48658	0.90462	Half ripe	Passed
12	Citra 12	Half ripe	0.55273	0.35087	0.60747	Ripe	Not passes
13	Citra 13	Half ripe	0.36432	0.31128	0.54466	Half ripe	Passed
14	Citra 14	Half ripe	0.25564	0.46469	0.84682	Half ripe	Passed
15	Citra 15	Half ripe	0.27481	0.54343	0.62916	Unripe	Not passed
16	Citra 16	Half ripe	0.31326	0.26961	0.69397	Half ripe	Passed
17	Citra 17	Half ripe	0.36014	0.40330	0.86763	Half ripe	Passed
18	Citra 18	Half ripe	0.29003	0.35765	0.78002	Half ripe	Passed
19	Citra 19	Half ripe	0.34182	0.47223	0.80405	Half ripe	Passed
20	Citra 20	Half ripe	0.16776	0.33256	0.61703	Half ripe	Passed
21	Citra 21	Ripe	0.85200	0.56615	0.94806	Ripe	Passed
22	Citra 22	Ripe	0.91995	0.56648	0.96886	Ripe	Passed
23	Citra 23	Ripe	0.93848	0.58993	0.98491	Ripe	Passed
24	Citra 24	Ripe	0.89152	0.44855	0.97257	Ripe	Passed
25	Citra 25	Ripe	0.87888	0.53696	0.97311	Ripe	Passed
26	Citra 26	Ripe	0.86429	0.57267	0.93820	Ripe	Passed
27	Citra 27	Ripe	0.91878	0.70491	0.91073	Ripe	Passed
28	Citra 28	Ripe	0.82015	0.62398	0.96775	Half ripe	Not passed
29	Citra 29	Ripe	0.80875	0.54576	0.97071	Ripe	Passed
30	Citra 30	Ripe	0.83701	0.58775	0.96600	Ripe	Passed

Based on Table 2 it can be seen that the range of HSV values in every level of Nature for the Unripe category is minimum (0.19355; 0.51748; 0.42190) and maximal (0.27481; 0.85481; 0.70850), Half-ripe minimum (0.16776; 0.26961; 0.54466) and maximal (0.82015; 0.62398; 0.96775), as well as minimal Ripe (0.55273; 0.35087; 0.60747) and maximal (0.93848; 0.70491; 0.98491).

After analyzing the results of testing the level of maturity of the dragon fruit conducted by the system and also the Expert Judgment, the results show that the overall image of the dragon fruit has a match between the system and Expert Judgment. Where the value of N is the sum of all dragon fruit test images, 30 images consisting of 10 unripe images, 10 half-cooked images, and 10 ripe images

Table 7 Results of the classification of dragon fruit

		Prediction Class			Total
		Unripe	*Half Ripe*	*Ripe*	
Actual Class	*Unripe*	9	1	0	10
	Half Ripe	1	8	1	10
	Ripe	0	1	9	10
Total		10	10	10	30

Then the accuracy of the system can be calculated in a way

$$\text{Accuracy} = \frac{26}{30} x100\% = 86,6\% \quad (6)$$

4. Conclusion

Based on the results of research conducted it can be concluded that:

1. The level of maturity of the dragon fruit we can know besides the manual way can also automatically use a computer system
2. Detection results performed by the image processing system using the naïve Bayes method based on the HSV color space have an accuracy rate of 86.6% with a test data of 10 images in each category.

References

[1] M. Z. Abdullah., J. M. Saleh, A.S. Fathinul Syahir, B.M.N. Mohd. Azemi, Discrimination and classification of fresh-cut starfruits (Averrhoa carambola L.) using automated machine vision system, *Journal of Food Engineering* 76 (2006) 506–523, 2006.

[2] D. S. Prabha & J. S. Kumar, *Assessment of banana fruit maturity by image processing technique.* Association of Food Scientists & Technologists (India), 2013.

[3] R. Misigo and M. Evans, Classification Of Selected Apple Fruit Varieties Using Naive Bayes. *Indian Journal of Computer Science and Engineering (IJCSE)*, 0976-5166, 2016.

[4] Wanitchang, Padungsak, A. Terdwongworakul, J Wanitchang, N. Nakawajana, *Non-Destructive Maturity Classification of Mango Based On Physical, Mechanical and Optical Properties,* Journal of Food Engineering 105 (2011) 477–484, 2011.

[5] M. Vahid, K. Kheiralipourb. And M. G. VarnamkhastiA, Detecting Maturity of Persimmon Fruit Based on Image Processing technique, *Scientia Horticulturae* 184, 123–128, 2015

[6] Saeed, Osama Mohammed B. S. and team, Classification of oil palm fresh fruit bunches based on their maturity using portable four-band sensor system. *Computers and Electronics in Agriculture 82 (2012) 55–60*, 2015

[7] S. Kazuhiko and team, *Grading meat quality by image processing*. Pattern Recognition 33 (2000) 97}104, 2000.

[8] H. Rein-Lien, M. A. Motalleb, and A. K. Jain, Face Detection in Color Image. *IEEE Transactions on Pattern Analysis and Machine Intelegence*. (Vol. 24 No. 5), 2002

[9] A. Harisinghaney, A. Dixit, S. Gupta, and A. Arora, Text and image based spam email classification using KNN, Naïve Bayes and Reverse DBSCAN algorithm. *2014 International Conference on Reliability Optimization and Information Technology (ICROIT)*. (2014).

[10] S. Winiarti, A. Prahara, Murinto, D. P. Ismi, Pre-trained convolutional neural network for classification of tanning leather image. *International Journal of Advanced Computer Science and Applications*, 2018.

[11] V. Chernov, J. Alander, and V. Bochko, Integer-based accurate conversion between RGB and HSV color spaces. *Computers & Electrical Engineering, 46, 328–337.* 2015.

[12] J. Ren, S. D. Lee, X. Chen, B. Kao, R. Cheng, and D. Cheung, Naive Bayes Classification of Uncertain Data, *Ninth IEEE International Conference on Data Mining,* 2009.

Eliciting Requirements of Order Fulfilment in A Company

Johanes Fernandes Andry[1], Hendy Tannady[2], Filscha Nurprihatin[3]
[1]Department of Information System, Universitas Bunda Mulia, Jakarta, Indonesia.
[2]Department of Management, Universitas Pembangunan Jaya, Tangerang Selatan, Indonesia.
[3]Department of Industrial Engineering, Universitas Bunda Mulia, Jakarta, Indonesia.

e-mail:[1]jandry@bundamulia.ac.id,[2]hendy.tannady@upj.ac.id,
[3]fnurprihatin@bundamulia.ac.id

Abstract. The Order Fulfillment Process (OFP) is one of the core business processes in an organisation. OFP in a manufacturing company, the wide range of activities involved is carried out by people from different functional units. OFP is complicated and demands activity elicitation and coordination. Hence, it important for companies to re-engineer their OFPs to support their activities for the better. Oil distribution manufacturing is a company that acts as a manufacturer of Coolant Brake, and Radiator in Indonesia needs an application to fulfilled all its business process activities. In this research, the authors have elicited requirements of order fulfilment in this company used elicitation and collaboration methodology to gain knowledge about the user's requirement or need. From this research, an order fulfilment application that has been computerised has been produced following their needs to support organisation activities for the better.

1. Introduction

Current technological developments are the trigger for the company to immediately explore the company's potential to improve the company's performance further to generate revenue for the organisations. Order fulfilment application is one of the core business processes in the company [1], [2], [3] to assist in the implementation of daily business operations in overcoming business problems such as company needs, helping management in making decisions. One of the processes in order fulfilment is distribution.

Distribution is the mechanism through which products and services are transferred from producers to end consumers. In other words, the purpose of distribution is to connect producers of a product or service to end-users [4]. This case discusses companies that distribute oil. Oil distributors buy fuel from suppliers and sell it, by a margin, to customers. Besides, they can act as transporters; distribute oil to other people's customers at certain agreed costs [5]. An oil distribution company is a company that acts as a manufacturer of Cooling and Radiator Brakes in Indonesia. Manufacturing industry development more likely towards the developing country [6]. The company's goal is to increase the use of motorised vehicles, heavy equipment and machine tools to maintain the condition of their assets in a prime and resilient condition.

In the Order Fulfillment Process (OFP) in any form, various activities involved are carried out by people from different functional units (e.g. sales personnel to process customer orders, planning personnel to make plans and production schedules, and staff production staff to physical manufacture of products). As a result, OFP becomes very complicated and demands integration and coordination of activities in the company [1], [7], [8]. Hence, it underscores the importance for companies to re-engineer their OFPs to exploit opportunities to better integrate and cooperate with supply chain partners, thus contributing to SCM.

Companies must always be aware of things that can disrupt the smooth running of business processes such as the availability of inappropriate goods, the accumulation of debt and receivables that cause the operating cash flow system is not good, complaints from customers for dissatisfaction with the services provided by the company and decreased sales due to delays in decision making. It is hoped that the results of this study can be analysed for the current system, identify weaknesses in the current system, provide recommendations for improvement and development an information system need for order fulfilment that is in line with company needs so that organisational activities are better. In this case, only collect or eliciting requirement from stakeholder in an oil distribution company.

2. Theoretical Foundation

2.1. Eliciting Requirement

(1) A condition that is needed by the user to help him solve a problem or achieve his goals;
(2) The ability that must be possessed by an information system or application to meet its needs such as work contracts, quality standards, specifications, or other documents that are formally imposed;
(3) Documented condition and capability representation refer to (1) or (2).

In addition to the statement of requirements, requirements for information systems must contain other contextual information, such as [9], [10]:

- Background information that places requirements in context (eg the business needs of the company and system users, the current system or situation, and the reasons for using the new system).
- Domain knowledge is needed to understand the associated OFP system, including glossary and references to other documents needed.
- Definition of information system objectives and scope of requirements.
- A narration that links from one document to various parts of another document.
- Description of the stakeholders involved.
- Other project-specific topics those are important so that stakeholders can fully meet and understand specifications.

Requirement elicitation is the most important thing phase of the software development life cycle (SDLC) [11], [12]. Software development processes consist of many knowledge-intensive processes, among which requirement elicitation is perhaps the most critical for the success of the information system or software. The requirements elicitation process is intended to gain knowledge about the user's requirement or need [11], [13]. The purpose of requirement elicitation is to understand the problem that arises from the needs of stakeholders, including, but not limited to, customers and end-users and transform it into requirements to define and design the related solution. It covers the capture and discovery of stakeholder needs. Its aim is to identify information determining what features the software system should have [11], [14]. This stage is used to translate the needs and desires of software users who are not precise, incomplete, to complete, precise, and formal specifications as desired by management [11], [15].

2.2. Order fulfilment Process (OFP)

OFP starts with receiving a customer's order and ends by sending the final product. This includes activities such as order processing, stock checking, purchasing decision making, supplier selection, a purchase order (PO), planning component manufacturing, and final product assembly and delivery of goods. In different companies, OFP is run differently according to their unique business characteristics (for example operating modes: make-to-stock, make-to-buy or order, engineer-to-order by service, and assembly-to-order), they fulfil two requirements for a common goal, including:
(1) Delivery of products to meet customer expectations at the right time, the right place, the right amount, and the right price; and
(2) Acquisition of agility to deal with uncertainty from the internal and external environment of the company [16], [12].

Unbalanced and unstable integration of manufacturing and transportation systems which weakens the competitiveness of the supply chain. Especially in dynamic situations and environments, production and transportation systems must be well integrated so that efficiency, responsiveness, and flexibility can be achieved and maintained properly [13], [17].

3. Research Method

Research on Eliciting Requirements of Order Fulfillment uses quantitative and qualitative method approaches to answer research questions, conducting document reviews and surveys to oil distribution companies for collecting relevant data for the development of an application. Base on Figure 1 Steps of elicitation and collaboration methodology [18]. The methodology used in this research is to prepare for elicitation until manage stakeholder collaboration for developing information systems of OFP.

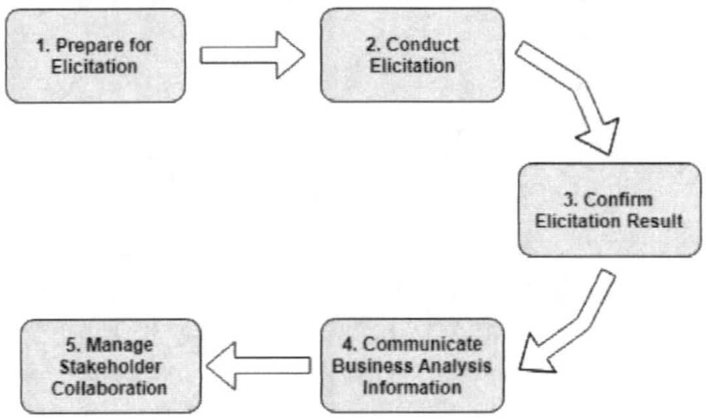

Figure 1. Steps of methodology elicitation and collaboration [18].

1. Prepare for elicitation.
Step one at this stage, ensuring that stakeholders have the information needed for analysis and ensure that they understand the nature of the activities they will undertake. It also establishes a set of objectives for the activity. At this area, also identify or prepare research resources to be used.
2. Conduct elicitation.
Step two at this stage, elicitation is carried out to understand the needs of stakeholders and find the best solution that can meet those needs. Excitation is done by direct interaction with stakeholders, conducting research, or conducting experiments.
3. Confirm elicitation results.
At this stage, ensuring that stakeholders have a shared understanding of the results of elicitation, the information obtained is properly recorded, and that business analysts get the information sought from elicitation activities.
4. Communicate business analysis information.
Provide information needed by stakeholders when they need it. Information is given in a format that gives those benefits and uses the right concepts.
5. Manage stakeholder collaboration.
And the final stage is collaborated with stakeholders in analysing business processes as a whole and ensuring the business analysis process gets the desired results.

4. Discussion and Analysis

The following is the analysis that we did for the eliciting requirements of OFP in a company using the methodology elicitation and collaboration. The stages of the method can be seen in Figure 1.

4.1. Prepare for elicitation.

Included are understand the scope of elicitation, select elicitation techniques, Set up logistics, secure supporting material and Prepare stakeholders. Output form this stage are scheduled all resource can be people and equipment from OFP process. Notify appropriate parties of the plan included in this phase, prepare questionnaire and distribute to stakeholders in oil Distribution Company.

Understand the scope of elicitation, the first step we did before elicitation is to determine and understand the scope of the elicitation activities. This aims to respond if the activity deviates from the intended scope. The scope of the elicitation development OFP consists of:

- The process of the purchase order and sales order transactions as the main data source.

- Important data such as user data, finished goods data, customer data, distributor and retail price data, raw material data, auxiliary material data, and supplier data.
- Financial reports such as profit or loss, balance sheet, journals, etc.

Select elicitation techniques, the next step after determining and understanding the scope of elicitation activities is to determine the elicitation technique. We used elicitation techniques by surveying, interviewing and studying data in the company that is related to scope. Choosing the right techniques and ensuring each technique is performed very important to the success of the elicitation activity.

Set up logistics, Logistics is planned before elicitation activities. The logistics we prepared for each elicitation activity are to identify the objectives of the activity, the participants and their roles, the scheduled resources, including people, rooms, tools, locations, communication channels, techniques, languages used by stakeholders (oral and written) and agenda-setting.

Secure supporting material, we identified the sources of information needed to carry out elicitation activities. Information needed for elicitation is people, systems, historical data, material, and documents. Prepare stakeholders, the final step before elicitation is that we educated stakeholders about how elicitation techniques work or what information is needed. We prepared stakeholders by asking them to review supporting materials before elicitation activities to make them as effective as possible.

4.2. Conduct elicitation consists of guide elicitation activity and captures elicitation outcomes.
Guide elicitation activity, after we finished preparing for elicitation, we began to do elicitation. During the elicitation activity, we continued to supervise and provide guidance to stakeholders to ensure that elicitation activities are focused on producing intended information at the desired level of detail so that the results of the elicitation are following the expected requirements.

Capture elicitation outcomes, In addition to supervising and providing guidance, we recorded and responded to the results of the elicitation obtained to ensure that the information produced during the elicitation activities is under the expected requirements. Results of conduct elicitation are including documentation appropriate from the technique and capture the information provided by management.

4.3. Confirm elicitation
Confirm elicitation results consist of compare elicitation results against source information and Compare elicitation results against other elicitation results; Confirmation is done by comparing information received with other information to look for inconsistencies or gaps.

Compare elicitation results against source information, after we did elicitation, information is needed to answer existing requirements. However, the information needs to be re-confirmed. We did this by comparing information received with other information to ensure accuracy.

Compare elicitation results against other elicitation results, in addition to comparing the results of elicitation with other information, we also compared the results of elicitation with other elicitation results to ensure that the information is consistent and accurate. Technique this stages are interview and observation concern all participant in these organisations.

4.4. Communicate business analysis information.
Communicate business analysis information consist of Determine the objectives and format of communication and Communicate business analysis package. Input for business analyst information is any kinds of information at any level of detail that the information system of development OFP.

Determine the objectives and format of communication; after we got accurate elicitation results, we processed the information to answer existing requirements. Then, the information was conveyed to stakeholders and in a format that can be used to continue the change activities.

Communicate business analysis package, the next step is that we communicated the information with stakeholders so that they can understand it. We provided stakeholders with the opportunity to review packages, ask questions about information, and raise concerns about whatever they may have.

4.5. Manage stakeholder collaboration

Manage stakeholder collaboration consist of gain agreement on commitments, and monitor stakeholder engagement and collaboration. Gain agreement on commitments, after communicating the information with stakeholders, we explained the importance of working together to involve them in the overall business analysis process to provide the results that are needed. Then, stakeholders and we identified and agreed on commitments about expectations and desired outcomes from shared commitments.

Monitor stakeholder engagement, after agreeing to a shared commitment, we monitored the participation and performance of stakeholders to ensure that experts and stakeholders participate effectively, the attitudes and interests of stakeholders remain constant or increase, the results of elicitation are confirmed promptly, agreements and commitments are maintained.

Collaboration, the last step we did is collaborating with stakeholders to encourage the flow of information, ideas, and innovation. In addition to making it easier to solve problems and achieve the desired results. Table 1. Summary Stakeholder Management Plan, to show members department of stakeholder in the company about elicitation requirement of development OFP.

Table 1. Summary of Stakeholder Management Plan

Stakeholder	Interest	Influence	Goal	Contribute	Best Way To Manage
IT	2	1	Systemized	High-level milestones	More specific
Purchasing	3	2	Priceless	Will delegate	Occasional in-person meetings
Marketing	1	1	Promotion	Very little, but want to be informed	Coordination with others
Finance	1	1	Budget	Compare data with manually	Not Spending Money
Logistic	1	1	Timeline	Daily lead	Careless

Interest: How much does project affect them? (1 = high, 2 = middle, 3 = low)
Influence: How much do they have? (1 = high, 2 = middle, 3 = low)

5. Conclusion

Based on the analysis of the order fulfilment application that has been running at the company, it can be concluded that every activity on the fulfilment order which exists, starting from purchase order and sales order transactions, recording user data, customer data, supplier data, price data, raw material data, data on supporting materials, finished goods data, making financial reports, and reporting on financial statements were still done manually. The company needs an application to fulfil all its business process activities. From this research, an order fulfilment application that has been computerized has been produced by their needs to support organisational activities for the better.

References

[1] L. L. Zhang, R. J. Jiao, and Q. Ma, "Accountability-based order fulfilment process reengineering towards supply chain management: A case study at a semiconductor equipment manufacturer," *J. Manuf. Technol. Manag.*, vol. 21, no. 2, pp. 287–305, 2010.

[2] D. Kritchanchai and B. L. MacCarthy, "Responsiveness of the order fulfilment process," *Int. J. Oper. Prod. Manag.*, vol. 19, no. 8, pp. 812–833, 1999.

[3] T. J. Turner, K. Mendibil, U. S. Bititci, P. Daisley, and T. H. J. Breen, "Improving the reliability of the customer order fulfilment process in a product identification company," *Int. J. Prod. Econ.*, vol. 78, no. 1, pp. 99–107, 2002.

[4] R. D. Soliani, "A Review of Distribution Logistics in the Brazilian Oil Industry," vol. 6, no. 2, 2016.

[5] G. Bobby, "Modelling the Cost Structure of UK Oil Distributors," *Int. J. Phys. Distrib. Mater. Manag.*, vol. 19, no. 3, pp. 24–29, 1989.

[6] H. Tannady and Y. Maimury, "Increasing the efficiency and productivity in the production of low

voltage switchboard using resource constrained project scheduling," *J. Ind. Eng. Manag.*, vol. 11, no. 1, pp. 1–33, 2018.

[7] Keely L. Croxton., "The Order Fulfillment Process," *Int. J. Logist. Manag.*, vol. 12, no. 1, pp. 19–32, 2003.

[8] Lin, Fu-Ren, "Reengineering the Order Fulfillment Process in Supply Chain Networks," *Int. J. Flex. Manuf. Syst.*, vol. 10, pp. 197–229, 1998.

[9] R. Vieira, D. Ferreira, J. Borbinha, and G. Gaspar, "A requirements engineering analysis of MoReq," *Rec. Manag. J.*, vol. 22, no. 3, pp. 212–228, 2012.

[10] Hull, E., Jackson, K. and Dick, J. in Requirements Engineering, 2nd ed., Springer, New York, NY, 2005.

[11] A. Saad and C. Dawson, "Requirement elicitation techniques for an improved case based lesson planning system," *J. Syst. Inf. Technol.*, vol. 20, no. 1, pp. 19–32, 2018.

[12] M. Christopher, *Logistics & Supply Chain Management*, vol. fourth Edi, no. 2. 2011.

[13] K. P. Steven L. Goldman, Roger N. Nagel, *Agile Competitors and Virtual Organizations, Strategies for Enriching the Customer*. 1995.

[14] D. Carrizo, O. Dieste, and N. Juristo, "Systematizing requirements elicitation technique selection," *Inf. Softw. Technol.*, vol. 56, no. 6, pp. 644–669, 2014.

[15] Chakraborty, A., Baowaly, M.K., Arefin, A. and Bahar, A.N. "The role of requirement engineering in software development life cycle", in Journal of Emerging Trends in Computing and Information Sciences, Vol. 3, No. 5, 2012.

[16] S. Tiwari, S. S. Rathore, and A. Gupta, "Selecting requirement elicitation techniques for software projects," *2012 CSI 6th Int. Conf. Softw. Eng. CONSEG 2012*, no. September, 2012.

[17] J. F. Andry, J. S. Suroso, and D. Y. Bernanda, "Improving quality of smes information system solution with ISO 9126," *J. Theor. Appl. Inf. Technol.*, vol. 96, no. 14, pp. 4610–4620, 2018.

[18] A Guide to the Business Analysis Body of Knowledge, 3rd Edition." in International Institute of Business Analysis, Toronto, Ontario, Canada, 2015, pp. 53-74.

Logical Acquisition in the Forensic Investigation Process of Android Smartphones based on Agent using Open Source Software

Mukhlis Prasetyo Aji[1], Dedy Hariyadi[2], Tri Rochmadi[3]

[1]Department of Informatics Engineering, Universitas Muhammadiyah Purwokerto
[2]Department of Informatics Engineering, Universitas Achmad Yani Yogyakarta
[3]Department of Information System, Universitas Alma Ata

[1]prasetyo-aji@ump.ac.id, [2]milisdad@gmail.com, [3]trirochmadi@almaata.ac.id

Abstract. Smartphones have an architecture with integrated components. Maintain the process of procuring digital goods on smartphones when asking for special assistance. According to the National Institute of Standards and Technology the process of procuring digital evidence on mobile devices There are 5 levels of acquisition: Manual Extraction, Logical Extraction, Hex Dumping / JTAG, Chip-Off and Micro Read. Which means buying digital evidence using logical acquisition methods or logical extraction still approved in SNI ISO 27037: 2014 also supports the process of procuring logistical permits in Special requirements, limited space. Dengen uses the live procurement method for logical acquisition on smartphones using AFLogical OSE can be obtained digital proof in the form of Contact List, Call Log, SMS and MMS. The use of digital proof Larons can be obtained through basic data from the application installed on the smartphone.

1. Introduction

Smartphones have an architecture with integrated components. For example, a smartphone with an Android operating system has components including: applications, application framework, libraries, runtime, and Linux kernel [1], see Figure 1. Based on this architecture, the process of acquiring digital evidence on smartphones generally requires special handling. According to David Ashfield, the acquisition process on smartphones is divided into four: logical acquisition, file system acquisition, physical acquisition, and manual acquisition [2].

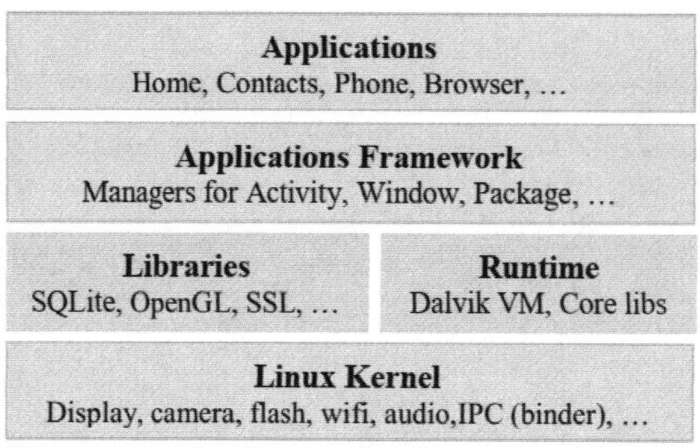

Figure 1. Android Operating System Architecture

According to the National Institute of Standards and Technology, U.S. The Department of Commerce for the acquisition of digital evidence on mobile devices has been translated back into 5 acquisition levels that are illustrated in the form of a pyramid [3]. The lowest level, **Level 1** is Manual Extraction, extracting digital evidence from electronic evidence by looking directly at it. **Level 2**, Logical extraction, extracts digital evidence from electronic evidence utilizing electronic evidence connectivity with computer forensic analysis using wired or wireless media. **Level 3**, Hex Dumping / JTAG, Hex Dumping / JTAG requires tools in the form of Flasher Boxes on the market to support digital evidence extraction by changing the boot loader. **Level 4**, Chip-Off, the acquisition process focuses on flash memory on the device. **Level 5**, Micro Read, conducts acquisitions on NAND or NOR gates that require an electron microscope, see Figure 2.

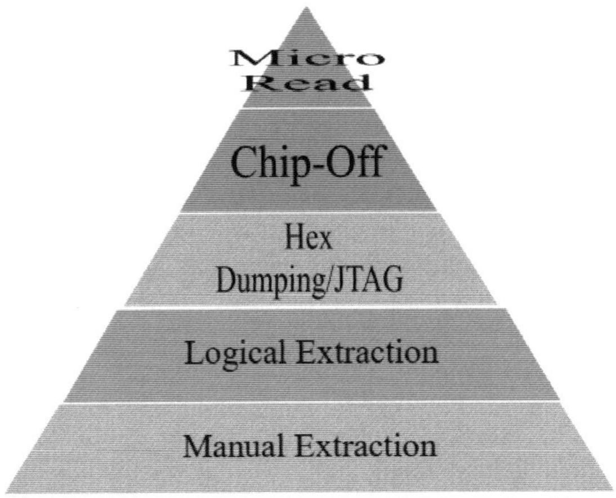

Figure 2. Pyramid Level of Acquisition of Evidence

This means that the acquisition of digital evidence using logical acquisition or logical extraction methods is still allowed. SNI ISO 27037:2014 also supports the acquisition process logically in certain conditions, such as limited scope [4]. The logical acquisition process on mobile devices can indeed save time, because the process of copying data that is relevant to the crime, for example, mobile numbers on contacts, Short Message Service (SMS), or Multimedia Messaging Service (MMS). This is as in Figure 3, about the standard diagram of the acquisition of electronic evidence in a lit condition.

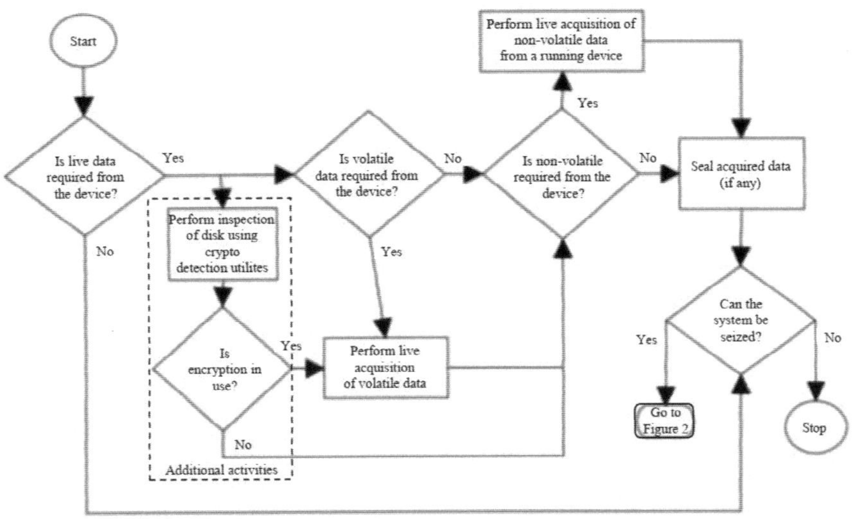

Figure 3. Procedure for Acquisition of Electronic Evidence in Lighted Condition

Background the logical acquisition process does not violate the rules of digital forensics, so in this study proposed the implementation of the logical acquisition on smartphones with agent-based

Android operating systems. This agent will be embedded or installed on a smartphone with an Android operating system. Test this logical acquisition-based agent, using an agent that has an open-source license.

2. Related Works

The handling of crimes involving digital devices following the principle should minimize interaction with electronic evidence and digital evidence. However, this is different from electronic evidence in the form of smartphones that sometimes require further interaction [5]. In this study, the acquisition process of digital evidence on smartphones with intelligent operating systems uses agent-based systems. The principle is an ecosystem that implements an agent-based system, adding special applications that are embedded in a target to execute certain commands [6]. So that the smartphone will be installed by an agent whose function is to collect digital evidence, while the additional applications are AFLogical OSE and Laron.

2.1. AFLogical OSE

AFLogical OSE[1] is a forensic application that is open-source licensed so that it can be used easily by non-law enforcement personnel [7]. The AFLogical OSE function is to collect or extract information on a smartphone with an Android operating system such as Contacts Phones, CallLog Calls, SMS, MMS messages, and MMSParts. AFLogical OSE there are two models available, the AFLogical OSE application is available in the form of an Android application and bundling with the Santoku Linux operating system in the form of a virtual machine sponsored by the company NowSecure[2] [8].

2.2. Laron

Laron is a forensic application that acquires logically digital evidence in the form of information installed on an Android system smartphone with an MIT license. The acquisition process using Laron must meet several requirements, namely: the cellphone is on, the cellphone is unlocked, Debugging mode is activated, the busy box is installed, and the cellphone is rooted. Laron development has been adjusted accordingly SNI ISO/IEC 27037:2014, see Figure 3 [9].

2.3. Android Debug Android (ADB)

Android Debug Bridge (ADB) is a tool or application part of Android Software Developers Kit (SDK) which connects between Android devices and computers so that it can be used to control and access files on Android-based smartphone devices [10]. The use of the Android Debug Bridge (ADB) can acquire digital evidence stored on smartphones but cannot return digital evidence that has been deleted. The requirements for implementing ADB on an Android-powered smartphone are the USB Debugging Mode must be active [11].

3. Acquisition Methods

The complexity of smart cellphone architecture requires adjustments in the process of acquiring digital evidence. So the approach to the acquisition process is not like the acquisition of digital evidence from a computer [12]. In this study, it was proposed to obtain digital evidence with various considerations that need to be installed on additional applications or agents on smartphones. Figure 4 shows the flow of agent-based acquisition processes on smartphones.

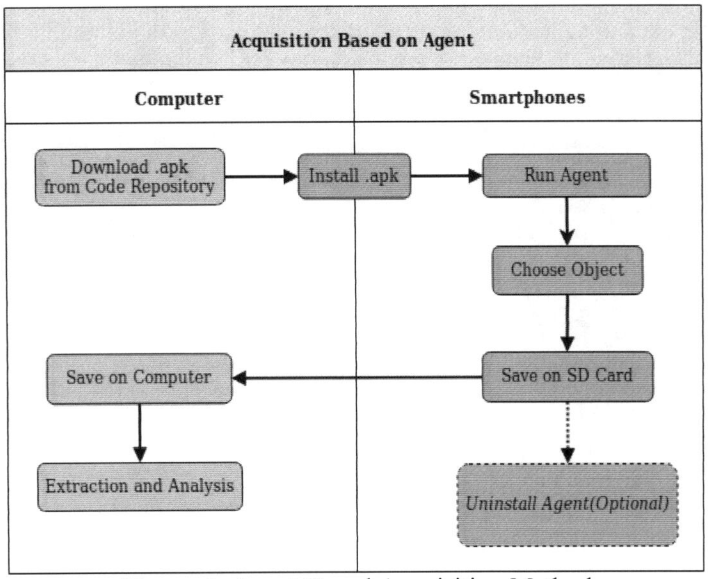

Figure 4. Agent-Based Acquisition Method

The acquisition process is divided into two parts, namely computer parts, and smartphones. In the computer agent section has been prepared in advance, namely AFLogical OSE and Laron in the form of files with extension .apk. The agent installation process can be installed via a smartphone or from a computer. If through a computer, the agent is uploaded to the smartphone first. The acquisition of agent-based digital evidence, the investigator will run the process on a smartphone. From the agent selected objects indicated as related to crime. Objects that are indicated as related to crime are copied on the external or internal memory. The selection of objects must be adjusted to the facts and the title of the case from a crime. After the acquisition process is complete, returning to the computer section uses ADB to copy the object. The final process, investigators carry out extraction and analysis of digital evidence obtained from the agent-based acquisition process. Software for extraction and analysis is adjusted to cases of crime that occur. The investigator is allowed to delete the agent installed on the smartphone, but this is optional.

4. Result and Discussion

The Agent used is AFLogical OSE and Laron has a license under the auspices Open Source Initiative [13]. AFLogical has a license GNU Public License, while Laron has a license MIT License. The agent is downloaded from the respective repository code available on the developer Github[34].

The main requirement for the implementation of the acquisition of agent-based digital evidence is that the smartphone is on, the screen is not locked, and ADB must be activated manually by the investigator. Optional requirements that can be done is to install the busy box and the cell phone is rooted, see Table 1. If these requirements have been met, the first step that must be prepared for acquisition is downloading AFLogical OSE and Laron.

Table 1. Comparison of the requirements of Agent AFLogical OSE and Laron

Agent	Smartphone ON	Unlock Screen	ADB Active	BusyBox	Rooted
AFLogical OSE	√	√	√	X	X
Laron	√	√	√	√	√

Install agent AFLogical and Laron uses command ADB, namely ADB install AFLogical-OSE_1.5.2.apk to install agent AFLogical OSE and ADB install Laron-app-release.apk to install agent Laron. The installation process is done on the computer by typing the command it. To get digital evidence on a smartphone agent AFLogical OSE and Laron must be run from a smartphone.

Digital evidence that can be obtained on smartphones using agent AFLogical OSE is Contact List, Call Log, SMS, and MMS, see Figure 5. Different from agent AFLogical OSE, evidence obtained or found using agent Laron is database file from the application installed on the smartphone, see Figure 6. For example, smartphone users who have installed Whatsapp then database files in the form of * .db can be acquired.

Figure 5. Acquisition Agent AFLogical OSE

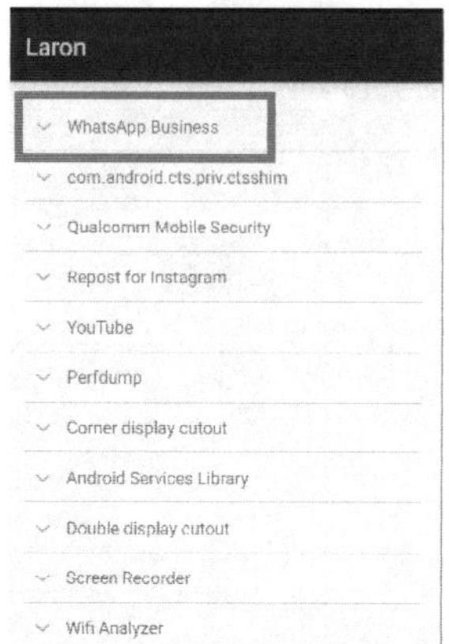

Figure 6. Acquisition Agent Laron

In the process of acquisition between agents AFLogical OSE and Laron, there are fundamental differences regarding checksums, agent AFLogical OSE does not provide checksum result information after the acquisition process, while the Laron agent hash values included as evidence of maintaining the integrity of the acquired evidence, see Table 2. Therefore, Laron agents have BusyBox installation requirements and rooted conditions to produce sha1 hash values.

Table 2. Comparison of Acquisition of Digital Evidence

Agent	Contact	Call Log	SMS	MMS	DB Apps	Hash
AFLogical OSE	√	√	√	√	X	X
Laron	X	X	X	X	√	√

The extraction results from the AFLogical OSE Agent are files *.csv dan *.xml, see Figure 7. Meanwhile, the extraction result from Laron Agent is a compressed database in the format tar.bz2 to avoid contamination and the text file contains sha1 hash values from the tar.bz2 file, see Figure 8. Since the extraction from the AFLogical OSE agent is a * .csv file, the analysis uses spreadsheet software while the Laron agent uses SQL viewer because digital evidence is an SQLite database.

Figure 7. Extraction Results of Agent AFLogical OSE

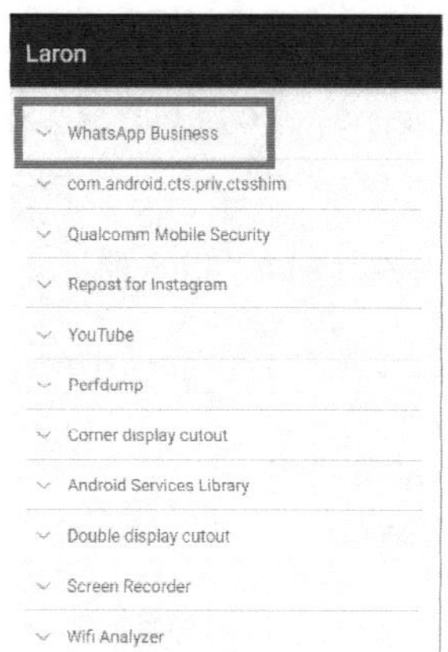

Figure 8. Extraction Results of Agent Laron

5. Conclusion

The acquisition of electronic evidence in the form of smartphones is different from computers, although it still pays attention to the prevailing digital forensic rules, namely minimizing interactions with electronic and/or digital evidence. The stages of logical acquisition are still allowed to speed up the process of investigating a crime. Installation of acquisition agents on smartphones can still be considered to obtain digital evidence because the acquisition process is following the applicable digital forensic rules and is right on target. According to the results of this study, the AFLogical agent OSE can acquire digital evidence in the form of Contact List, Call Log, SMS, and MMS, while the Laron agent can acquire proof of database applications installed on the smartphone. The two acquisition agents can be used by non-law enforcement personnel because they are freely available in the code repository and licensed under the auspices of the Open Source Initiative. So the two agents have their respective advantages and functions depending on the case faced and adjusted to the case title. Besides Aflogical OSE and Laron, there is still proprietary digital-based agent-based digital forensic software. It is hoped that further research can analyze the implementation of agent-based digital forensic software both proprietary and licensed under the Open Source Initiative.

References

[1] L. Vogel, "Getting started with Android development - Tutorial." [Online]. Available: https://www.vogella.com/tutorials/Android/article.html. [Accessed: 08-Jul-2019].
[2] D. Ashfield, "MOBILE DEVICE FORENSICS: DATA ACQUISITION TYPES," *CCL Solutions Group Ltd*, 2014. [Online]. Available: https://web.archive.org/web/20150319193302/http://www.cclgroupltd.com/mobile-device-forensics-data-acquisition-types/.
[3] R. Ayers, S. Brothers, and W. Jansen, "Guidelines on mobile device forensics (NIST Special Publication 800-101 Revision 1)," *NIST Spec. Publ.*, vol. 1, no. 1, p. 85, 2014.
[4] B. S. Nasional, *SNI 27037:2014 tentang Teknologi Informasi - Teknik Keamanan - Pedoman Identifikasi, Pengumpulan, Akuisisi dan Preservasi Bukti Digital*. Jakarta, 2014.
[5] R. Fasra and A. R. Nagoor Meeran, "Performing digital autopsy on an android device: An 'open source' approach," *Int. J. Control Theory Appl.*, vol. 9, no. 15, pp. 7111–7117, 2016.
[6] W. Buchanan, J. Graves, L. Saliou, H. Al Sebea, and N. Migas, "Agent-based forensic investigations with an integrated framework," *4th Eur. Conf. Inf. Warf. Secur. 2005, ECIW 2005*, no. May 2014, pp. 47–52, 2005.
[7] Z. R. Mair and A. Ashari, "Aplikasi untuk Identifikasi Short Message Service (SMS) Spam Berbasis Android," *J. Math. Nat. Sci.*, vol. 24, no. 3, pp. 257–262, 2014.
[8] N. Y. P. Lukito, F. A. Yulianto, and E. Jadied, "Comparison of data acquisition technique using logical extraction method on Unrooted Android Device," *2016 4th Int. Conf. Inf. Commun. Technol. ICoICT 2016*, vol. 4, no. c, 2016.
[9] D. Hariyadi and A. A. Huda, "Laron: Aplikasi Akuisisi Berbasis SNI 27037:2014 pada Ponsel Android," *Indones. Secure. Conf. 2015*, no. September, pp. 1–10, 2015.
[10] S. J. Yang, J. H. Choi, K. B. Kim, and T. Chang, "New acquisition method based on firmware update protocols for Android smartphones," *Digit. Investig.*, vol. 14, no. S1, pp. S68–S76, 2015.
[11] D. Quick and M. Alzaabi, "Forensic analysis of the Android file system Yaffs2," *Proc. 9th Aust. Digit. Forensics Conf.*, no. December, 2011.
[12] M. Taylor, J. Haggerty, D. Gresty, and D. Lamb, "Forensic investigation of cloud computing systems," *Netw. Secur.*, vol. 2011, no. 3, pp. 4–10, 2011.
[13] "The Open Source Definition (Annotated)." [Online]. Available: https://opensource.org/osd-annotated.

IOP Conf. Series: Materials Science and Engineering **771** (2020) 012025 doi:10.1088/1757-899X/771/1/012025

Rice Productivity Analysis by Province Using K-Means Cluster Algorithm

A Supriyatna[1,*], I Carolina[1], W Widiati[1], C Nuraeni[1]

[1]Universitas Bina Sarana Informatika, Indonesia

**Corresponding author e-mail :* adi.asp@bsi.ac.id

Abstract. Rice (*Oryza sativa L.*) is a very important food crop in the world after wheat and corn. It is also a staple food for most of the world's population, especially in Asia, like in Indonesia until now. In 2014 to 2018, rice productivity tended to change dynamically. In 2018, rice productivity in Indonesia was 51.92 (Ku/Ha). This research was conducted to classify rice productivity in 34 provinces in Indonesia in 2018. The data used were sourced from Statistics Indonesia. The method or approach used in this study is the K-Means cluster algorithm to classify rice productivity data by province in 2018. The results of the research are; (1) There are 19 provinces included in cluster 0 (Medium), (2) There are 4 provinces included in cluster 1 (High), and (3) There are 11 provinces that are included in cluster 2 (Low). Based on the results of the study, it was proven that there were 4 provinces in cluster 1 (High) they are West Java, Central Java, East Java and South Sulawesi, with the highest rice productivity. Three of them were on Java Island. It shows that Java still dominates the productivity of rice plants.

1. Introduction

Indonesia, as an agricultural country, enacts the agriculture sector as one of the main contributors in people's lives and becomes a major component in the government's strategy to reduce poverty. Rice *(Oryza sativa L.)* is a food crop that plays an important role because it produces rice which is a source of staple food. As in Indonesia, rice is the main commodity in supporting people's food in daily needs. Based on this background, various efforts were made to increase the productivity of rice plants in provinces in Indonesia. One way is to use an irrigation system. However, the fact that rice productivity has not met national needs. Figure 1 below shows a graph of the amount of rice productivity from 2014 to 2018, that is 51.92 (Ku / Ha).

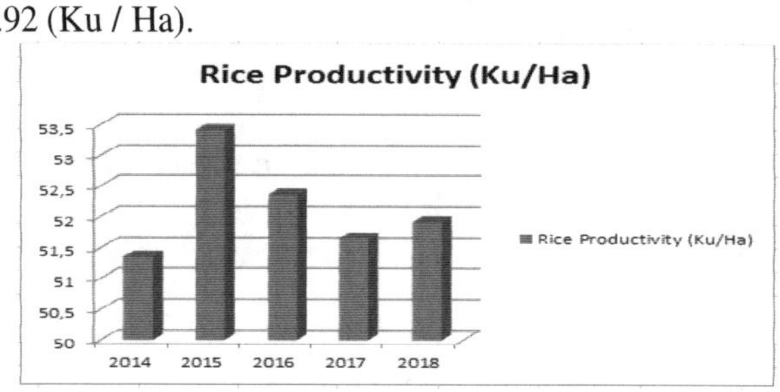

Source: Statistics Indonesia

Figure 1. Number of Rice Productivity

Agronomically, the increase in rice production is pursued through increased productivity rather than in rice planting area. Meanwhile, institutionally the program among others was realized through the implementation of Integrated Crop Field Schools / *Sekolah Lapangan Tanaman Terpadu (SL PTT)* with the aim of improving the quality of rice cultivation techniques carried out by farmers, increasing cropping intensity and rice productivity[1]. Based on data from Statistics Indonesia *(Badan Pusat Statistik/BPS)* rice productivity increased by 0.52% from 2017 to 2018[2]. In this study, the researchers conducted a study in 34 provinces in Indonesia with the purpose of classifying rice productivity in 2018 and data processed from the Statistics Indonesia. The technique that the researchers used was cluster analysis. It is a multivariant technique with the main objective of grouping objects based on the characteristics they possess[3][4]. Clustering analysis is a technology which can classify the similar sample points into the same group from a data set[5][6]. The K-Means algorithm is proposed based on the hybrid rice optimization algorithm with the aim to optimize the initial center of clustering as a good start for clustering. At the same time the optimization algorithm can improve the accuracy of clustering and the efficiency of the K-Means clustering algorithm[7]. The same study was conducted using data from 1993-2015, the results of clustering rice by province using Rapidminer, obtained high production clusters of 23 provinces, normal production clusters of 8 provinces and low production clusters of 3 provinces. The 3 provinces with the highest rice production are Central Java, East Java and West Java[8].

2. Research Method

2.1. Stages of the Research

In preparing the research, the researchers did the stages; they are:

1. Data collection.
   The data used in this study are data on harvested area, rice production and productivity according to 34 provinces in Indonesia. Data obtained from Statistics Indonesia.
2. Data Normalization.
   Normalised data used to scale the attribute values of the data so that they fit within a certain range. The formula used in data normalisation is the Z-Score which is a normalisation method based on the mean (average value) and standard deviation of the data.
3. Implement of the K-Means algorithm using Rapidminer tools.
   The normalised dataset is then processed using the K-Means Cluster algorithm with Rapidminer Studio.
4. Data analysis.
   Rice productivity data processing generated by Rapidminer tools in the form of cluster models and performance vectors, based on these results will group rice productivity by 34 provinces into three clusters.

2.2. K-Means Cluster

This study used the K-means algorithm to determine the best cluster. K-means clustering is a well-known partitioning method[7]. In these objects are classified as belonging to one of K- groups. The results of Partitioning method is a set of K clusters, each object of data set belonging to one cluster. In each cluster, there may be a centroid or a cluster representative. In a case where we consider real-valued data, the arithmetic mean of the attribute vectors for all objects within a cluster provides an appropriate representative; alternative types of centroid may be required in other cases[9]. Another expert stated that the K-Means clustering method, groups the data based on their closeness to each other according to the Euclidean distance[10].

The basic algorithm in K-means is:

a. Specify the number of clusters (k), set the cluster centre to arbitrary
b. Calculate the distance of each data centre cluster using the equation:

$$d_{ik} = \sqrt{\sum_{j=1}^{m} C_{ij} - C_{kj}{}^2} \tag{1}$$

c. Group data into cluster with the shortest distance using the equation:

$$Min \sum_{k=1}^{k} d_{ik} = \sqrt{\sum_{j=1}^{m} C_{ij} - C_{kj}{}^2} \tag{2}$$

d. Calculate the new cluster centre using the equation:

$$C_{kj} = \frac{\sum_{i=1}^{p} x_{ij}}{p} \quad (3)$$

e. Repeat steps two through four until no more data is moved to another cluster.

3. Result and Discussions

3.1. Dataset

The dataset used in this study is data on harvested area, rice production and productivity based on 34 provinces in Indonesia, as shown in table 1.

Table 1. Dataset of harvested area, production and productivity of rice

No.	Province	Harvested Area (ha)	Rice Productivity (ku/ha)	Production (ton)	No	Province	Harvested Area (ha)	Rice Productivity (ku/ha)	Production (ton)
1	Aceh	297294	57,11	1697756	18	West Nusa Tenggara	281013	49,8	1399495
2	North Sumatra	360716	52,89	1907725	19	East Nusa Tenggara	200877	39,87	800980
3	West Sumatra	318579	47,45	1511538	20	West Kalimantan	214877	28,95	622041
4	Riau	93755	38,96	365293	21	Central Kalimantan	202142	36,74	742758
5	Jambi	118408	42,23	500021	22	South Kalimantan	278853	40,76	1136511
6	South Sumatra	513209	51,57	2646566	23	West Kalimantan	58151	41,51	241398
7	Bengkulu	59455	42,76	254218	24	North Kalimantan	14133	32,07	45323
8	Lampung	397435	47,83	1901041	25	North Sulawesi	82051	44,69	366722
9	Bangka Belitung Island	8618	21,99	18951	26	Central Sulawesi	204158	46,77	954794
10	Riau Island	355	23,46	833	27	South Sulawesi	1145319	50,12	5740730
11	Dki Jakarta	690	57,83	3990	28	Southeast Sulawesi	131399	37,98	499007
12	West Java	1691725	56,39	9539330	29	Gorontalo	51765	46,74	241948
13	Central Java	1680406	56,61	9512434	30	West Sulawesi	67835	48,08	326169
14	DI Yogyakarta	92035	54,07	497599	31	Maluku	23317	38,98	90892
15	East Java	1828700	57,63	10537922	32	North Maluku	10343	34,19	35360
16	Banten	331444	48,38	1603550	33	West Papua	4851	28,69	13916
17	Bali	108794	59,77	650245	34	Papua	31133	41,99	130718

3.2. Normalisation Data

Normalisation of data is carried out to minimise the magnitude of the variable to avoid difficulties in the grouping or clustering process. The equation used for data normalisation is:

$$\textbf{Normalization} = \frac{(InitialValue - MinValue)}{(MaxValue - MinValue)} \quad (4)$$

The normalised dataset is shown in table 2 below:

Table 2. Data on Normalization of Harvested Area, Production and Rice Productivity

No.	Province	Harvested Area (ha)	Rice Productivity (ku/ha)	Production (ton)	No	Province	Harvested Area (ha)	Rice Productivity (ku/ha)	Production (ton)
1	Aceh	0,1624	0,9848	0,161	18	West Nusa Tenggara	0,1535	0,7709	0,1327
2	North Sumatra	0,1971	0,8613	0,181	19	East Nusa Tenggara	0,1097	0,4802	0,0759
3	West Sumatra	0,1741	0,7021	0,1434	20	West Kalimantan	0,1173	0,1607	0,059
4	Riau	0,0511	0,4536	0,0346	21	Central Kalimantan	0,1104	0,3886	0,0704
5	Jambi	0,0646	0,5493	0,0474	22	South	0,1523	0,5063	0,1078

						Kalimantan			
6	South Sumatra	0,2805	0,8227	0,2511	23	West Kalimantan	0,0316	0,5282	0,0228
7	Bengkulu	0,0323	0,5648	0,024	24	North Kalimantan	0,0075	0,252	0,0042
8	Lampung	0,2172	0,7132	0,1803	25	North Sulawesi	0,0447	0,6213	0,0347
9	Bangka Belitung Island	0,0045	-0,043	0,0017	26	Central Sulawesi	0,1115	0,6822	0,0905
10	Riau Island	0	0	0	27	South Sulawesi	0,6262	0,7802	0,5447
11	Dki Jakarta	0,0002	1,0059	0,0003	28	Southeast Sulawesi	0,0717	0,4249	0,0473
12	West Java	0,9251	0,9637	0,9052	29	Gorontalo	0,0281	0,6813	0,0229
13	Central Java	0,9189	0,9701	0,9027	30	West Sulawesi	0,0369	0,7205	0,0309
14	DI Yogyakarta	0,0501	0,8958	0,0471	31	Maluku	0,0126	0,4542	0,0085
15	East Java	1	1	1	32	North Maluku	0,0055	0,314	0,0033
16	Banten	0,1811	0,7293	0,1521	33	West Papua	0,0025	0,1531	0,0012
17	Bali	0,0593	1,0626	0,0616	34	Papua	0,0168	0,5423	0,0123

3.3. Centroid Data
Based on the data that has been normalised in table 2, then determine the number of clusters to be selected. The initial clusters selected are 3 (three) clusters based on Euclidean distances, as shown in Table 3 below:

Table 3. Centroid Initial Data

Cluster 1 (High)	1	1	1
Cluster 2 (Medium)	0,0281	0,6813	0,0229
Cluster 3 (Low)	0	0	0

Determination of initial centroid data points is done by taking the highest value as cluster 1 (C1) from the province of East Java, the middle value as cluster 2 (C2) from Gorontalo province and the lowest value as cluster 3 (C3) from the Riau Islands province.

3.4. K-Means Clustering Data Using Rapid Miner
Dataset of harvested area, production and rice productivity based on provinces with initial predetermined centroid data, then processed the data to get the results of grouping with K-Means cluster algorithm using Rapid miner Studio software.

Data processing with K-Means algorithm.

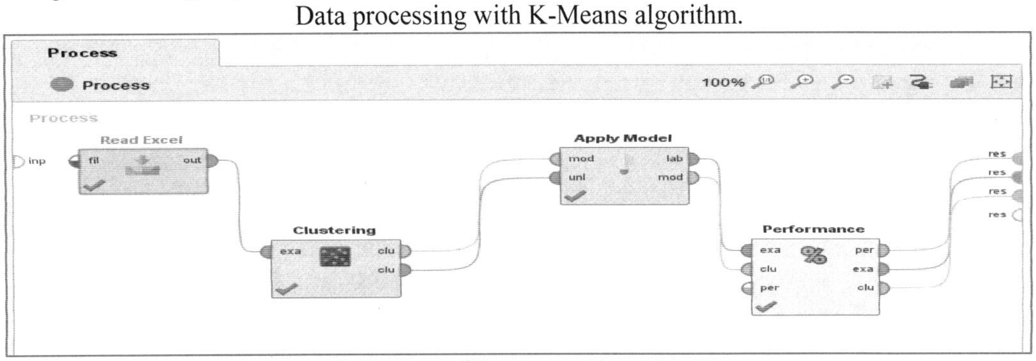

Figure 2. Design of K-Means Algorithm Operator

The design of data processing with the K-Means algorithm in Rapidminer studio uses 4 operators. Read Excel is used for reading the harvested area, production and rice productivity dataset based on the province that has been normalised. Next, the dataset is clustered or grouped using the K-Means clustering operator with a value of K = 3 and max runs = 10. The Apply Model operator is used to be able to read a clustered dataset based on data that has been studied. The operator of Cluster

Performance Distance is used to measure the performance of the K-Means clustering algorithm towards the dataset being read. The results of data processing with the K-Means algorithm in Rapidminer studio can be seen in the figure below:

Cluster Model

```
Cluster 0:  19 items
Cluster 1:  4 items
Cluster 2:  11 items
Total number of items:  34
```

Figure 3. Result of K-Means Cluster Model

The figure 3 above shows the results of processing rice productivity data using the K-Means cluster algorithm which is divided into 3 clusters, namely, 19 provinces included in the medium cluster, four provinces included in the high cluster and 11 provinces included in the low cluster.

Performance Accuracy of K-Means Cluster. The parameters used to measure the performance of the K-Means algorithm are done by counting avg._within_centroid_distance and davies_bouldin. avg._within_centroid_distance is the average in cluster distance calculated by the average distance between centroids and all cluster samples. Whereas davies_bouldin is an algorithm that produces clusters with low intra-cluster spacing (high level of intra-cluster similarity) and high inter-cluster spacing (low level of inter-cluster similarity) will have a low Davies-Bouldin index. The grouping algorithm that produces clusters with the smallest Davies-Bouldin index is considered as the best algorithm based on criteria. The results of the performance measurement of the K-Means cluster algorithm in grouping rice productivity data using the operator cluster performance distance in Rapidminer studio can be seen in Figure 5 below:

PerformanceVector

```
PerformanceVector:
Avg. within centroid distance: -0.039
Avg. within centroid distance_cluster_0: -0.038
Avg. within centroid distance_cluster_1: -0.058
Avg. within centroid distance_cluster_2: -0.035
Davies Bouldin: -0.615
```

Figure 4. Performance Vector K-Means Cluster

Figure 4. above shows PerformanceVector average within centroid distance of -0,039, average within centroid distance cluster 0 of -0,038, average within centroid distance cluster 1 of -0,058, average within centroid distance cluster 2 of -0,035 and PerformanceVector Davies Bouldin of -0,615.

3.5. Data Analysis

Grouping rice productivity data based on province in 2018 with the K-Means Cluster algorithm using Rapidminer studio has resulted in 3 clusters which can be seen in Figure 6 below. The following data is a grouping of 34 provinces into groups that are formed.

No	Provinsi	cluster	Luas Panen ...	Produktivita...	Produksi (ton)	No	Provinsi	cluster	Luas Panen ...	Produktivita...	Produksi (ton)
1	ACEH	cluster_0	0.162	0.985	0.161	18	WEST NUSA TENGGARA	cluster_0	0.153	0.771	0.133
2	NORTH SUMATRA	cluster_0	0.197	0.861	0.181	19	EAST NUSA TENGGARA	cluster_2	0.110	0.480	0.076
3	WEST SUMATRA	cluster_0	0.174	0.702	0.143	20	WEST KALIMANTAN	cluster_2	0.117	0.161	0.059
4	RIAU	cluster_2	0.051	0.454	0.035	21	CENTRAL KALIMANTAN	cluster_2	0.110	0.389	0.070
5	JAMBI	cluster_0	0.065	0.549	0.047	22	SOUTH KALIMANTAN	cluster_0	0.152	0.506	0.108
6	SOUTH SUMATRA	cluster_0	0.281	0.823	0.251	23	EAST KALIMANTAN	cluster_0	0.032	0.528	0.023
7	BENGKULU	cluster_0	0.032	0.565	0.024	24	NORTH KALIMANTAN	cluster_2	0.007	0.252	0.004
8	LAMPUNG	cluster_0	0.217	0.713	0.180	25	NORTH SULAWESI	cluster_0	0.045	0.621	0.035
9	BANGKA BELITUNG ISLAND	cluster_2	0.004	-0.043	0.002	26	CENTRAL SULAWESI	cluster_0	0.112	0.682	0.090
10	RIAU ISLAND	cluster_2	0	0	0	27	SOUTH SULAWESI	cluster_1	0.626	0.780	0.545
11	DKI JAKARTA	cluster_0	0.000	1.006	0.000	28	SOUTHEAST SULAWESI	cluster_2	0.072	0.425	0.047
12	WEST JAVA	cluster_1	0.925	0.964	0.905	29	GORONTALO	cluster_0	0.028	0.681	0.023
13	CENTRAL JAVA	cluster_1	0.919	0.970	0.903	30	WEST SULAWESI	cluster_0	0.037	0.721	0.031
14	DI YOGYAKARTA	cluster_0	0.050	0.896	0.047	31	MALUKU	cluster_2	0.013	0.454	0.009
15	EAST JAVA	cluster_1	1	1	1	32	NORTH MALUKU	cluster_2	0.005	0.314	0.003
16	BANTEN	cluster_0	0.181	0.729	0.152	33	WEST PAPUA	cluster_2	0.003	0.163	0.001
17	BALI	cluster_0	0.059	1.063	0.062	34	PAPUA	cluster_0	0.017	0.542	0.012

Figure 5. Result of ExampleSet (Clustering)

Based on the results of data processing with the K-Means Cluster algorithm, 19 provinces are included in cluster 0. They are Aceh, North Sumatera, West Sumatra, Jambi, South Sumatra, Bengkulu, Lampung, DKI Jakarta, DI Yogyakarta, Banten, Bali, West Nusa Tenggara, South Kalimantan, East Kalimantan, North Sulawesi, Central Sulawesi, Gorontalo, West Sulawesi and Papua. There are four provinces included in cluster 1, i.e. West Java, Central Java, East Java and South Sulawesi. While 11 provinces are included in cluster 2, they are Riau, Bangka Belitung Island, Riau Island, East Nusa Tenggara, West Kalimantan, Central Kalimantan, North Kalimantan, Southeast Sulawesi, Maluku, North Maluku and West Papua.

4. Conclusion

Based the research results, classification of rice productivity data based on provinces in Indonesia with the K-Means cluster algorithm using Rapidminer Studio produces 3 clusters, namely high clusters there are 4 Provinces, medium clusters there are 19 provinces and low clusters there are 11 provinces. 4 provinces in cluster 1 with the highest rice productivity, 3 of them are located on the island of Java, this shows that the island of Java still dominates the productivity of rice plants.

References

[1] V. D. Sari and B. M. Sukojo, "Analisa Estimasi Produksi Padi Berdasarkan Fase Tumbuh Dan Model Peramalan Autoregressive Integrated Moving Average (Arima) Menggunakan Citra Satelit Landsat 8," *Geoid*, vol. 10, no. 2, p. 194, 2015.

[2] Badan Pusat Statistik, "Luas Panen, Produksi dan Produktivitas Padi," Jakarta, 2018.

[3] Asroni and R. Adrian, "Penerapan Metode K-Means Untuk Clustering Mahasiswa Berdasarkan Nilai Akademik Dengan Weka Interface Studi Kasus Pada Jurusan Teknik Informatika UMM Magelang," *J. Ilm. Semesta Tek.*, vol. 18, no. 1, pp. 76–82, 2015.

[4] A. P. Windarto, "Implementation of Data Mining on Rice Imports by Major Country of Origin Using Algorithm Using K-Means Clustering Method," *Int. J. Artif. Intell. Res.*, vol. 1, no. 2, pp. 26–33, 2017.

[5] C. Y. Chen and F. Ye, "Particle swarm optimization algorithm and its application to clustering analysis," *2012 Proc. 17th Conf. Electr. Power Distrib. EPDC 2012*, no. 1, pp. 789–794, 2012.

[6] Ediyanto, M. N. Mara, and N. Satyahadewi, "Pengklasifikasian Karakteristik Dengan Metod K-Means Cluster Analysis," *Bul. Ilm. Mat. Stat. dan Ter.*, vol. 02, no. 2, pp. 133–136, 2013.

[7] C. Liu, C. Wang, J. Hu, and Z. Ye, "Improved K-means algorithm based on hybrid rice optimization algorithm," *Proc. 2017 IEEE 9th Int. Conf. Intell. Data Acquis. Adv. Comput. Syst. Technol. Appl. IDAACS 2017*, vol. 2, no. 2, pp. 788–791, 2017.

[8] Sudirman, A. P. Windarto, and A. Wanto, "Data mining tools | rapidminer: K-means method on clustering of rice crops by province as efforts to stabilize food crops in Indonesia," *IOP Conf. Ser. Mater. Sci. Eng.*, vol. 420, no. 1, p. 012089, Oct. 2018.
[9] N. Kaur, J. K. Sahiwal, and N. Kaur, "Efficient K-Means Clustering Algorithm Using Ranking Methode In Data Mining," *Int. J. Adv. Res. Comput. Eng. Technol.*, vol. 1, no. 3, pp. 85–91, 2012.
[10] P. Arora, Deepali, and S. Varshney, "Analysis of K-Means and K-Medoids Algorithm for Big Data," *Int. Conf. Inf. Secur. Priv.*, vol. 78, pp. 507–512, 2016.

IOP Conf. Series: Materials Science and Engineering **771** (2020) 012026 doi:10.1088/1757-899X/771/1/012026

The Stemming Application on Affixed Javanese Words by using Nazief and Adriani Algorithm

Dewi Soyusiawaty[1,*], **Anna Hendri Soleliza Jones**[1], **Nora Novita Lestari**[1]

Informatics Engineering, Ahmad Dahlan University, Yogyakarta, Indonesia.

*dewi.soyusiawaty@tif.uad.ac.id, annahendri@tif.uad.ac.id

Abstract. The Javanese language has various forms of affixed words. It makes people from other regions find it difficult to understand the Javanese language. One of the ways to understand the Javanese language is by studying and understanding Javanese vocabulary in the Javanese language dictionary or learning directly from ones who master the Javanese language. Both offline and online dictionaries have not been able to translate affixed words. This research used Nazief and Adriani Algorithm to find out the stems/root words in the Javanese language. The objective of this research is to apply the Nazief and Adriani Algorithm in stemming affixed words in the Javanese language. The subject discussed in this research is the application of stemming affixed Javanese words by using Nazief and Adriani algorithm. The stages started with identifying the affixes in the Javanese language. Nazief and Adriani algorithm has five steps those are detecting words that have affixes, cutting words that contain inflection suffixes, derivation suffixes and derivation prefixes, and last recording. The result of this research is a stemming application for affixed Javanese words that applies Nazief and Adriani algorithm method. Based on the accuracy test, it appears that 93.6% of words are accurate from the total of 219 words tested.

1. Introduction

Language is used as a primary means of daily communication, so understanding language is essential. One of the languages that is familiar to Indonesian society is the Javanese language used by societies living in Central Java, East Java, and Special Region of Yogyakarta. The Javanese language has several varieties which are Ngoko Javanese (low), Madyo Javanese (middle), and Kromo (high refined). Quoted from the book "Kawuruh Basa Jawa Pepak", it explains that affixed words in the Javanese language are divided into three those are *tembung lingga oleh ater-ater* (prefixed words), *tembung lingga oleh seselan* (infixed words), and *tembung lingga oleh panambang* (suffixed words) [1]. The way to find the stems or root words in the Javanese language is by using the process of stemming. In Natural Language Processing (NLP), there are several algorithms that can be used in the stage of stemming such as Porter Algorithm, Arifin Setiono Algorithm, Vega Algorithm, Tala Algorithm and Nazief and Adriani Algorithm [2]. This research used Nazief and Adriani algorithm because it uses the best algorithm in the process of stemming [3][4]. Nazief and Adriani algorithm has additional rules for reduplications, prefixes, and suffixes to improve the accuracy of each word. This research is to find out the application of Nazief and Adriani algorithm and its accuracy level in finding Javanese root words.

One of the techniques used in *Natural Language Processing* to restore the form of a word to be its root (root word) is the stemming technique. Stemming is removing prefixes and suffixes from the data of affixed words. *Stemming* is a process or technique used to find out the stem or root word of an affixed word [5]. According to Tala, stemming is used to change the form of a word to be its root based on the correct morphological structure of the language. There are three methods of stemming in Figure 1, which are [6][7]:

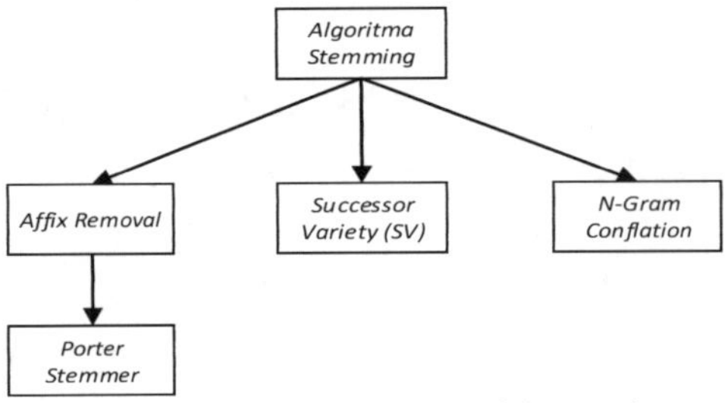

Figure 1. Diagram of Stemming

1. Affix Removal: removing suffixes and prefixes from a term to be its stem. The algorithm commonly used is the Porter algorithm because it has a simple model and is efficient.
   a. If a word ends with "ies" but not "eies" or "aies", "ies" is replaced with "y".
   b. If a word ends with "es" but not "aes" or "ees" or "oes", "es" is replaced with "e".
   c. If a word ends with "s" but not "us" or "ss", "s" is replaced with "NULL".
2. *Successor Variety (SV)*: emphasizes the structure of the letter in the word than the consideration on phonemes. The examples are: corpus, able, accident, ape, about, result in SV for the word "apple:
   a. Since the first letter of the word "apple" is "a", so the word group that has substring "a" followed by "b", "x", "c", "p" is called SV from "a" so "a" has 4 SV.
   b. Since the first two letters of the word "apple" are "ap" so the word group that has substring "ap" is only followed by "e" and called SV from "ap" so "ap" has 1 SV.
3. *N-Gram Conflation*: the basic idea of this technique is by grouping the words together based on the characters (substring) identified as N character.

Nazief and Adriani algorithm is a stemmer that was first developed by Bobby Nazief and Mirna Adriani. Nazief and Adriani algorithm is the best algorithm in the process of stemming because it has additional rules of reduplications, prefixes, and suffixes in improving the accuracy of each word. This algorithm is based on the comprehensive morphological rules of the Indonesian language, collected to be one group and encapsulated on allowed affixes and disallowed affixes. This algorithm uses the stem/ root word dictionary and supports recoding, which is restructuring the words that experienced the over-stemming process.[8]

The algorithm developed by Bobby Nazief and Mirna Adriani has the following stages [9]:

1. Find the word to be stemmed in the dictionary. If it is found, it can be assumed that the word is a stem or root word, so the algorithm stops.
2. The inflection of suffixes ("-lah", "-kah", "-ku", "-mu", or "-nya") are removed. If it is a particle "-lah", "-kah", "-tah" or "-pun"), this stage is repeated to remove the possessive pronoun ("-ku", "mu", or "nya"), if it exists.
3. Remove suffix derivation ("-i", "-an" or "-kan"). If the word is found in the dictionary, the algorithm stops. If it is not found, it goes to stage 3a.
   a. If "-an" has been removed and the last letter of the word is "-k", so "-k" is also removed. If the word is found in the dictionary, the algorithm stops. If it is not found, it goes to stage 3b.
   b. The suffixes to be removed ("-i", "-an" or "-kan") is restored and continue to stage 4.
4. Remove prefix derivation. If in stage 3, a suffix is removed, it then goes to the stage 4a. If no, it goes to stage 4b.
   a. Check the table of prefix and suffix combination which is not allowed. If it is found, the algorithm stops, if not, it goes to the stage 4b.
   b. In this stage, the iteration is carried out three times. Decide the prefix type, and then remove the prefix. If the root word has not been found, go to stage 5. If it is found, the algorithm stops.

Note: if the second prefix is similar to the first prefix, the algorithm stops.

5. Recoding

6. If all stages have ended, but it still fails, the word is assumed as the root word, then the process ends.

The type of prefix is decided through these following steps:
1. If the prefixes are: "di-", "ke-", or "se-", the type of prefix is successively "di-", "ke-", or "se-".
2. If the prefixes are "te-", "me-", "be-", or "pe-", it needs an additional process to decide the type of the prefixes.
3. If the two first characters are not "di-","ke-","se-", "te-", "be-", "me-", or "pe-" so it stops.
4. The type of prefixes is none, so it stops.

The Javanese language is the language used by the society living in Central Java, East Java, and the Special Region of Yogyakarta. The Javanese language has several courtesies those are Ngoko Javanese (low), Madyo Javanese (middle), and Kromo Javanese (high refined). The use of the three courtesy is different; for example, Ngoko is used to communicate with peers and the elder to the younger, Kromo is usually used to communicate with the elder or respectful people, while Madya is the mix between Ngoko and Kromo. Affixed words in the Javanese language are divided into four types those are [10]:

a. *Prefixes*. Verbs added with prefixes before the root words. The prefixes could be n-, ma-, ka-, ke-, ko-, dak-, di-, pi-. Example:
Makarya (ma + karya "work") work
Digawa (di + gawa "bring") being brought
b. *Infixes*. Verbs inserted with infixes in the middle of the root words. The infixes could be –um, -in, -el, -er. Example:
Gumuyu (guyu "laugh" + -mu-) laugh
Tumindak (tindak "act" + -um-) act
c. *Suffixes*. Verbs added with suffixes after the root words. The suffixes could be –ake, -na, -ne, -ana, -ing, -mu. Example:
Jupukake (jupuk "take" + ake) take it
Gawakna (gawa "bring" + na) bring it
c. *Confixes*. Verbs added with prefixes before the root words and suffixes after them. The affixes could be di-ake, n-i, n-ake, in-an. Example:
Digawakake (gawa "bring" + di-ake) being brought
Njupuki (jupuk "take" + n-i) take
Several examples of affixed words in the Javanese language are in the following Table 1.

Table 1. The Examples of Affixed Words in Javanese and Indonesian Language

Javanese Root Word	Indonesian Root Word	Affixed Javanese Word	Affixed Indonesian Word
Gawa (Bring)	Bawa	Gawak***na***	Bawakan
Guyu (Laugh)	Tawa	G***um***uyu	Tertawa
Jupuk (Take)	Ambil	Jupuk***ake***	Ambilkan
Adoh (Far)	Jauh	***Ng***adoh	Menjauh
Itung (Calculate)	Hitung	***Ng***itung	Menghitung

2. Research Method

System flowchart is the figure of the stemming stages carried out by the system. Figure 2 shows the design of the system flowchart.

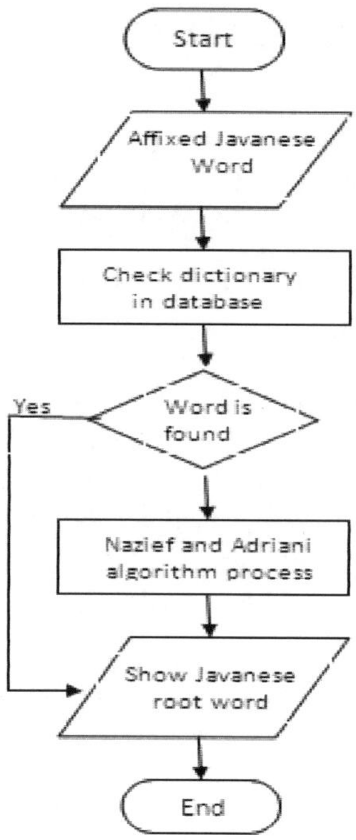

Figure 2. System Design

3. Result and Discussion

Data need analysis is the stage of deciding the limitation of data used. Data needed is data of affixes and root words in the Javanese language. Data grouping of Javanese affixed words of each category is shown in Table 2.

Table 2. Affixed Word Grouping

Category		Word Forming Affixes
inflectional suffixes	Particle	-a
		-ana
		-en
	inflectional suffixes possessive pronoun	-e
		-ku
		-mu
		-ne
Derivation Suffix		-ake
		-an
		-ke
		-na
		-ing
		-i
Derivation Prefix		Di-
		Dak-
		Kok-
		Nge-
		A-
		Ma-
		Ka-
		Ke-
		Sa-

	Pa-
	Pi-
	Pra-
	Pari-
	Kuma-
	Kami-
	Kapi-
	Tar-
	Nya-
	Ng-
	N-
	M-

A sample case of stemming Javanese based on the flowchart of Nazief and Adriani algorithm is shown in Figure 3.

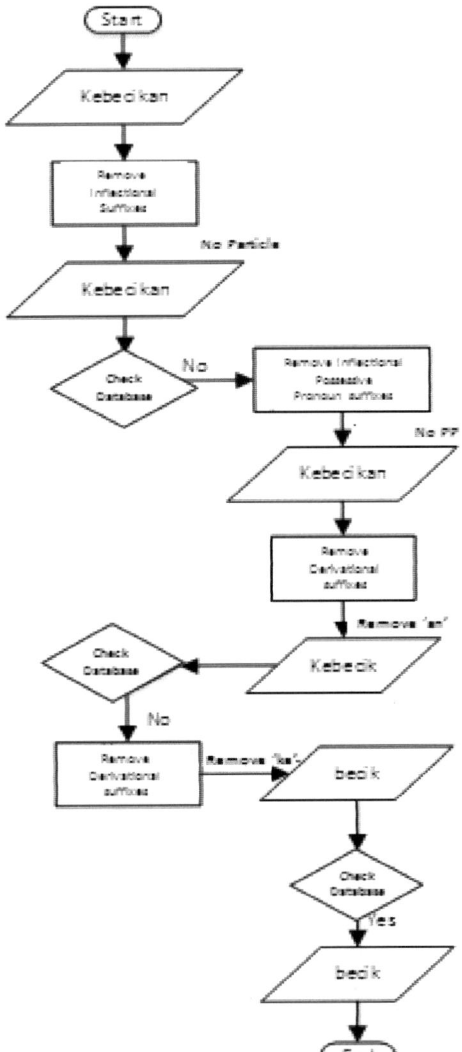

Figure 3. The Sample of Word Search Using Nazief and Adriani Algorithm

Nazief dan Adriani algorithm has 4 categories of affixes data grouping.

IOP Conf. Series: Materials Science and Engineering **771** (2020) 012026 doi:10.1088/1757-899X/771/1/012026

Algorithm and System *Pseudo-code* used in this research

a. Algorithm

Process 1: input the affixed Javanese word.

Process 2: check the root word dictionary on the database.

Process 3: the abstraction process of the root words and the affixes by using Nazief and Adriani algorithm.

Process 4: if the word is found in the database, the word inputted is assumed as the root word, and the stemming process will stop. However, if the word is not found, the system will go back to the input page.

b. *System Pseudo-code*

The following listing is Pseudo-code made based on the system flowchart as in Figure 4.

```
Algorithm (Stemming to decide the root word from the affixes in Javanese)
Declaration :
Root word: string
Result: string
Description :
Start
Read (affixed Javanese word)
Check the root word dictionary
Write "kamus kata dasar" ("root word dictionary")
If "root word is found" then
Print "root word"
Else If "root word is not found" then
Write "Jalankan Proses Algoritma Nazief dan Adriani" ("Run the process of Nazief and Adriani algorithm")
End.
```

Figure 4. Pseudo-code of system flowchart

The following Figure 5 is Pseudo-code made based on Nazief and Adriani algorithm flowchart.

```
Algorithm (Stemming root word by using Nazief and Adriani algorithm)
Declaration :
Root word : string
Result : string
Description:
Start
Read (Javanese affixed word)
Process of Nazief and Adriani Algorithm
Write "Hapus (Delete) Inflectional suffixes"
Write "Hapus (Delete) inflectional possessive pronoun suffixes"
Write "Hapus (Delete) Derivational Suffix"
If "root word is found" then
Print "root word"
Else If "root word is not found" then
Write "Hapus (Delete) Derivational Prefix"
If "root word is found" then
Print "root word"
Else If "root word is not found" then
Write "Recording"
Print "root word"
End
```

Figure 5. Pseudo-code of Nazief and Adriani Algorithm flowchart

Database which used is root words and affixes in Javanese language as in Table 3.

Table 3. Root Words

Name	Type of Data	Length	Explanation
Id_katadasar	Int	10	Primary key
Katadasar (Root word)	Text	-	NULL
Tipe_katadasar	Text	-	NULL

The result of system implementation is as shown in the following Figure 6. The stemming result page shows the stemming process by using Nazief and Adriani algorithm and the root word as the result of the affixed Javanese word inputted.

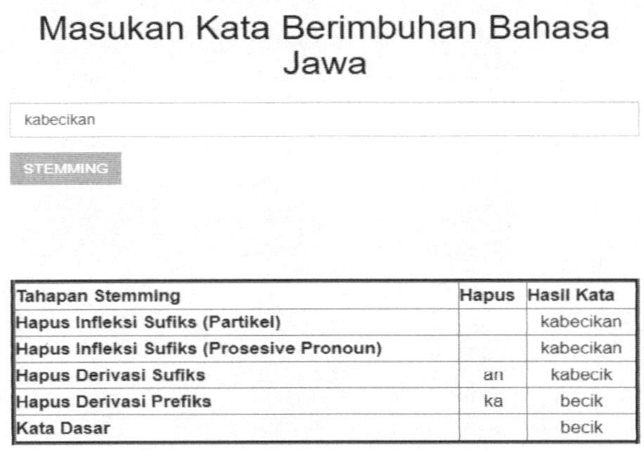

Tahapan Stemming	Hapus	Hasil Kata
Hapus Infleksi Sufiks (Partikel)		kabecikan
Hapus Infleksi Sufiks (Prosesive Pronoun)		kabecikan
Hapus Derivasi Sufiks	an	kabecik
Hapus Derivasi Prefiks	ka	becik
Kata Dasar		becik

Figure 6. Stemming Result Page

System testing is done to find out the capability of the system in finding the Javanese root words as in Table 4. This research used accuracy testing that is a test to measure the accuracy level of the system in finding the root words. Data used in this research was the affixed Javanese words.

Table 4. The Result of System Testing

	Affix	Affixed word	Root word	Root word on stem	Correct	Incorrect
Particles	-a	Maraa	Mara	Mara	✓	
		Sarapana	Sarapan	Sarap		✓
	-en	Panganen	Pangan	Pangan	✓	
		Gawanen	gawa	gawanen		✓
	-ana	Pakanana	Pakan	Pakan	✓	
Possesive Pronouns	-ku	Bukuku	Buku	Buku	✓	
		Panganku	Pangan	Pangan	✓	
	-mu	Ibumu	ibu	Ibu	✓	
		bapakmu	bapak	bapak	✓	
	-ne	Nyatane	Nyata	Nyata	✓	
		mejane	meja	meja	✓	
	-e	Pitike	Pitik	Pitik	✓	
		dhuwite	dhuwit	dhuwit	✓	
Derivasi	-ake	Jupukake	Jupuk	Jupuk	✓	

sufixes		Nulisake	Nulis	Nulis		
	-an	Gawean	Gawe	Gawe	✓	
		Takonan	Takon	Takon	✓	
	-i	Kemuli	Kemul	Kemul	✓	
		Mbalangi	Balang	Balang	✓	
	-ke	Jupukke	Jupuk	Jupuk	✓	
		Mbayarke	Bayar	Bayar	✓	
	-na	Gawakna	Gawa	Gawan		✓
		Golekna	Golek	Golek	✓	
	-ing	Pupusing	Pupus	Pupus	✓	
		Kawruhing	Kawruh	Kawruh	✓	
Prefixes	di-	Dipangan	Pangan	Pangan	✓	
		Dipikir	Pikir	Pikir	✓	
	Ke-	Kebecikan	Becik	Becik	✓	
		Kebutuhan	Butuh	Butuh	✓	
	Kok-	Kokgawe	Gawe	Gawe	✓	
		Kokpangan	Pangan	Pangan	✓	
	m-	Macul	Pacul	Pacul	✓	
		Mikul	Pikul	Pikul	✓	
	Dipun-	Dipunbukak	bukak	bukak	✓	
		dipunjagongi	jagong	jagong	✓	
	Ka-	Katabrak	Tabrak	Tabrak	✓	
		kapinteran	pinter	pinter	✓	
	Dak-	Dakpangan	Pangan	Pangan	✓	
		dakbukak	bukak	bukak	✓	
	Nge-	Ngecat	Cat	Cat	✓	
		ngepit	pit	Pit	✓	
		Prasekawan	Sekawan	Sekawan	✓	
		Pragen	Pragi	Prag		✓
	Tar-	Tarmulah	Mulah	Mulah	✓	
		tartantu	tantu	tantu	✓	
	Pari-	Parigawe	Gawe	Gawe	✓	
		Paripih	Pripih	Paripih		✓
	Kuma-	Kumakaruh	Karuh	Karuh	✓	
		kumayu	ayu	yu		✓
	Kami-	Kamigilan	Gila	Gilan		✓
		kamitenggengen	tenggeng	tenggeng	✓	
	Kapi-	Kapilare	Lare	Lare	✓	
		kapitunan	tuna	tunan		✓
	Ma-	magempuran	Gempur	Gempur	✓	
		manjel	anjel	Manjel	✓	
	Sa-	Sawengi	Wengi	Wengi	✓	
		salawang	lawang	lawang	✓	
	Pa-	Palilah	Lilah	Lilah	✓	
		Padhusunan	Dhusun	Dhusun	✓	
	a-	Apangan	Pangan	Pangan	✓	
		anganggo	anggo	anggo	✓	
	Pi-	Pitutur	Tutur	Tutur	✓	
		Piweling	Weling	Weling	✓	
	Ny-	Nyapu	Sapu	Sapu	✓	
		Nyunggi	Sunggi	Sunggi	✓	

Ng-	Ngombe nggoreng	Ombe goreng	Ombe goreng	✓
N-	Nambal	Tambal	Tambal	✓
	Nutup	Tutup	Tutup	✓

Accuracy Calculation:
Number of words: 219

Incorrect words: number of incorrect words = $\frac{number\ of\ incorrect\ words}{number\ of\ words\ tested}$ 100%

: number of incorrect words = $\frac{14}{219}x$ 100% = 6.4%

Correct words: number of correct words = $\frac{number\ of\ correct\ words}{number\ of\ words\ tested}x$ 100%

: number of correct words = $\frac{205}{219}x$ 100% = 93.6%

In this accuracy testing, several words could not be processed as in Table 5.

Table 5. Examples of Affixes that could not be processed by the system

Affix	Example	Explanation
Kami-	Kamigilan->gilan	In the word "kamigilan", the output from the system is "gilan"; the system does not remove affix "n" because in the Javanese affixes there is no affix "n", so the system does not remove it.
-na	Gawakna->Gawak	In the word "Gawakna" the output from the system is "gawak", the system removes suffix "na", but does not remove "k" because the system could not cut infix.

The percentage of the system accuracy is 93.6%, and 6.4% is the incorrect result.

4. Conclusion and Recommendation

This result of this research is a stemming application for affixed Javanese words that can be used to find root words of Ngoko Javanese affixed words by using Nazief and Adriani algorithm. The accuracy test that tested 219 words shows that the accuracy percentage of the system in finding root words is 93.6%.

Some recommendation for this research :

a. Affixes that still experienced incorrect processes, such as suffix "-na" and infixes could not be processed. Recommendation for further development is to use other methods to process the affixes that could not be processed in this research.
b. The assimilation of affixes at the end of the words could not be processed, such as "pa+turu+an" that becomes "paturon". The system could not find the root word.
c. It needs further research on prefixes and suffixes that are allowed and disallowed in the Javanese language.
d. This system can be developed with mobile technology to ease the users in accessing the system.

References

[1] Raharajo, S. H. (2010). *Kawruh Basa Jawa Pepak.* Semarang: Widya Karya.
[2] Simarangkir, M. S. (2017). STUDI PERBANDINGAN ALGORITMA - ALGORITMA STEMMING. *Jurnal Inkofar*, 40-46.

[3] Wibowo, J. (2016). Aplikasi Penentuan Kata Dasar dari Kata Berimbuhan pada Kalimat Bahasa Indonesia dengan Algoritma Stemming. *Jurnal Riset Komputer (JURIKOM)*, 346-350.
[4] Pramudita, H. R. (2011). PENERAPAN ALGORITMA STEMMING NAZIEF & ADRIANI DAN SIMILARITY . *Jurnal Ilmiah DASI* , 15-19.
[5] Asmara, D. A., Khairani, D., & Masruroh, S. U. (2013). Penerapan Algoritma Paice atau Husk untuk Stemming pada Kamus Bahasa Inggris ke Bahasa Indonesia. *Jurnal Teknik Informatika.*
[6] Amin, F. (2012). Sistem Temu kembali Informasi dengan metode Vector Space Model. *Fakultas Teknologi Informasi, Universitas Sitkubank.*
[7] Bunyamin, H. (2008). Aplikasi Information Retrieval(IR) CATA Dengan Metode Generalized Vector Space Model. *Jurnal Informatika* , 29-38.
[8] Wahyudi, D., Susyanto, T., & Nugroho, D. (n.d.). Implementasi dan Analisis Algoritma Stemming Nazief & Andriani dan Porter pada Dokumen Berbahasa Indonesia. *Jurnal Ilmiah SINUS*, 49-56.
[9] Nazief, B., & Adriani, M. (2007). Stemming Indonesian: A Confix-Stripping Approach. *ACM Transactions on Asian Language Information Processing* , 13-33.
[10] Wedhawati, & dkk. (2006). *Tata Bahasa Jawa Mutakhir.* Yogyakarta: Kanisius.

Diet Calorie Determination System using Case-Based Reasoning

Hindayati Mustafidah[1], Aprilia Olivia Yosi[1], Ardhista Shabrina Fitri[2], Vita Arfiana Nurul Fatimah[3], Yolla Arinda Nur Fitriana[4]

[1] Faculty of Engineering and Science - Universitas Muhammadiyah Purwokerto – Indonesia
[2] Faculty of Pharmacy - Universitas Muhammadiyah Purwokerto – Indonesia
[3] Faculty of Medicine, Public Health, and Nursing – Universitas Gadjah Mada – Indonesia
[4]Faculty of Industry Technology - Universitas Ahmad Dahlan - Indonesia

**Corresponding author e-mail:* fida.mustafidah@ump.ac.id

Abstract. The balance of food consumption plays a crucial role in the health and quality of metabolism in the body, especially for people with diabetes, because the balance of nutrients in the body is not balanced. But even for healthy people, it is necessary to maintain nutrition in the framework of a healthy lifestyle. To determine a healthy diet is not easy, especially food, which is a source of energy. By building a case-based reasoning (CBR) system, it can solve this difficulty. The flow of this system development includes 4R (retrieve, reuse, revise, retain). The work system of CBR is a new case having similarities with situations that have already existed (previous problems), so the solution has similarities with the earlier issues. The attributes used are age, gender, weight, height, and activity. This system will significantly help the community in determining the right calorie diet so that the body's health is maintained.

1. Introduction

Food is the fulfilment of the most basic needs of life because energy is obtained from food to be able to do daily activities. Foods that contain carbohydrates, fats, and proteins will produce a source of energy. Rice, sweet potatoes, wheat, and cassava will produce carbohydrates. Animal or vegetable, for example, eggs, chicken, tofu, and tempeh, will generate proteins, while avocado oil, cooking oil, olive oil, and so on can produce fat [1]. Fats and carbohydrates are needed by the human body as fuel to obtain energy and heat, which is used to maintain body condition, especially when the weather is cold. Carbohydrates provide most of the strength that the body needs. On average, there are 2100 calories per person of energy per day [2]. Calorie balance is the leading supplier of power required by the body, must be considered. If the amount of calories is excessive, it can cause a variety of diseases, such as coronary heart disease, high blood pressure, and diabetes [3]. One effort to balance calorie needs is to do a diet program.

Diet is a rule of food, specifically for health [4]. The cultural background adopted by specific communities has strongly influenced this type of diet. Although humans are omnivores, a community group usually has preferences or restrictions on certain types of food. Diets can increase the body's metabolism, which can reduce or increase body weight. Diet does not mean that you cannot eat at all, but only limits your calorie intake a little. Your body's needs must calculate a proper diet. As stated by [5] compiled from various sources, namely [6][7][8][9][10] that the average woman needs an intake of 2000 calories per day, while men need an average of 2500 calories per day. To lose weight, ideally only need to reduce 500 calories every day, so that women need 1500 calories, and men need 2000 calories per day. With a reduction of 500 calories per day, your body weight will decrease by 0.5-1 kg per week. This diet calorie is a unit of the human need for a diet or weight reduction. Meager diet calories are at risk for health, especially if run for more than 12-16 weeks. Running a diet program for more than 12 weeks is not recommended [11].

Balance of food consumption plays a vital role in the health and quality of metabolism in the body. Unhealthy eating patterns, such as eating prepared food or junk food, contain too many calories (unbalanced). It can trigger various diseases, such as cholesterol, diabetes, high blood pressure (hypertension), and so on. For example, diabetes mellitus is a metabolic disorder syndrome that results in abnormal conditions at high blood sugar levels (hyperglycemia). Diabetes management (also known as Sugar Disease or DM) requires sufferers to adopt a healthy lifestyle. This article aims to control diabetes and blood sugar levels and to prevent long-term complications that can be caused by diabetes. [12]. Based on this, it is considered challenging to determine the right and fast calorie diet to maintain health according to a healthy lifestyle. A computer-based system can overcome this problem.

Case-based reasoning/CBR is one of the developments in computer software in the field of intelligent systems in addition to Fuzzy Logic, Expert Systems, and Artificial Neural Network (ANN). CBR can identify cases based on existing facts. New problems are solved by reusing solutions that have been used to solve previous problems [13]. The assumptions used are if new cases have similarities with situations that have already existed (past issues), then the solution also has similarities to previous problems [14]. If necessary, solutions are adapted to current facts so that it becomes the right solution to solve the current problem [15]. CBR performs its duties using four cycles, namely retrieval, reuse, revise, and retain or what is known as 4R [16]. Similarity measurement in CBR case representation uses the feature or attribute approach of the case. Aggregation of differences in attributes between two cases will calculate a measure of similarity [17]. Research on CBR has matured as a case-based problem-solving method [18][19]. Therefore, with this CBR, it is expected can help those difficulties in determining the right calorie diet.

Intelligent system technology developed an expert system in 2012 to solve health problems, especially those related to calories for diet. This expert system uses certainty to help determine the daily calorie needs for diabetics [20]. Meanwhile, [21] developed an expert system application for the diagnosis of diabetes mellitus, which was produced later by [22]. A rule-based system for consulting dietary rules for rule-based diabetes mellitus patients was developed by [23]. Reference [24] developed a fuzzy logic-based application to diagnose diabetes mellitus and was further developed by [25].

2. Method

Research data obtained from documents related to the calculation of diet calories. The source of this document is print and electronic media. Some activities carried out in determining diet calories are representing cases, indexing cases, case retrieval, and making case adaptations. Reference [16] has illustrated the process of finding solutions in the CBR method, as in Figure 1. The figure explains, when there are new cases in the form of user conditions that are similar to cases stored in a database (case-based), the system will adapt and compare the suitability or similarity of the solution with the solution stored on case-based. If the answer is suitable, then that solution is the solution to the case.

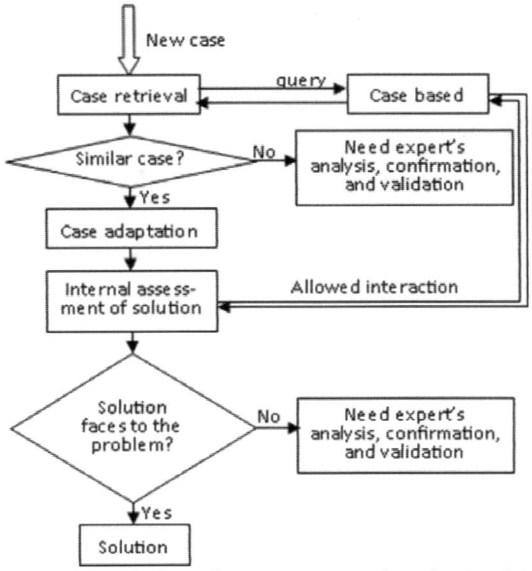

Figure 1. Solution discovery procedure in the CBR

3. Results and discussion

The variables in this study consisted of input variables, as in Table 1 and the output variable in the form of the number of calories needed per person.

Table 1. List of input variables

No.	Variable	Class Code	Variable Type
1	Age	1	10-14
		2	15-19
		3	20-24
		4	25-29
		...	...
2.	Gender	1	Female
		2	Male
...	...	...	...
7.	Activity	1	Mild
		2	Moderate
		3	Severe

The data in Table 1 are the basis for developing a CBR system. The CBR system designed contains several relevant pages, namely the main page (Figure 2), the login page (Figure 3), and the consultation page (Figure 4-7).

This CBR system has two groups based on users who can access the system. Both groups are end-users and admins. Besides being based on group access by users, this system consists of two groups of pages that can be accessed. These pages are consultation pages that can be accessed by end-users and admins, and administration pages that can only be accessed by the admin via the login page. The main page is the page that first appears when the application starts. On the main page, there are home, data input (Input Data –in Bahasa), and calorie calculation (Perhitungan Kalori –in Bahasa) buttons, as in Figure 2.

Figure 2. The main page of the CBR diet calorie system

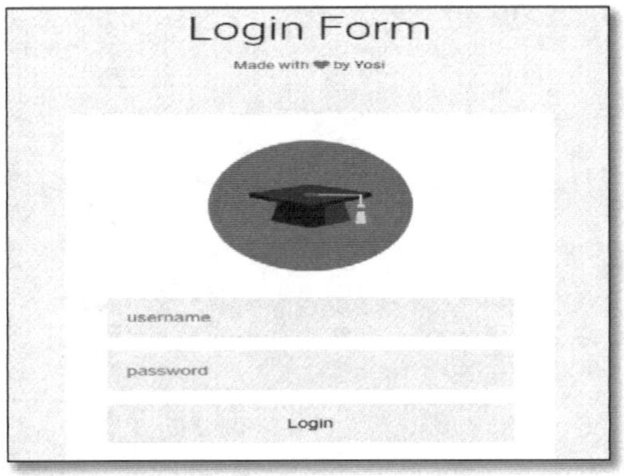

Figure 3. Login page

The login page in Figure 3 can only be used by admin or expert to process data masters and food data masters. If there is a new case, the admin can enter data with approval and validation from the expert. Some facilities on this page can be used to manage case-based. Experts can revise, save (retain) or delete the case-based data.

The consultation page, accessed through the "calorie calculation" menu on the main page, is used to enter new cases that will be processed by the system to find the number of calories needed per day according to user needs based on input variables (Figure 4). The solution provided by the system is the number of calories needed and examples of food according to the caloric content (Figure 5-7).

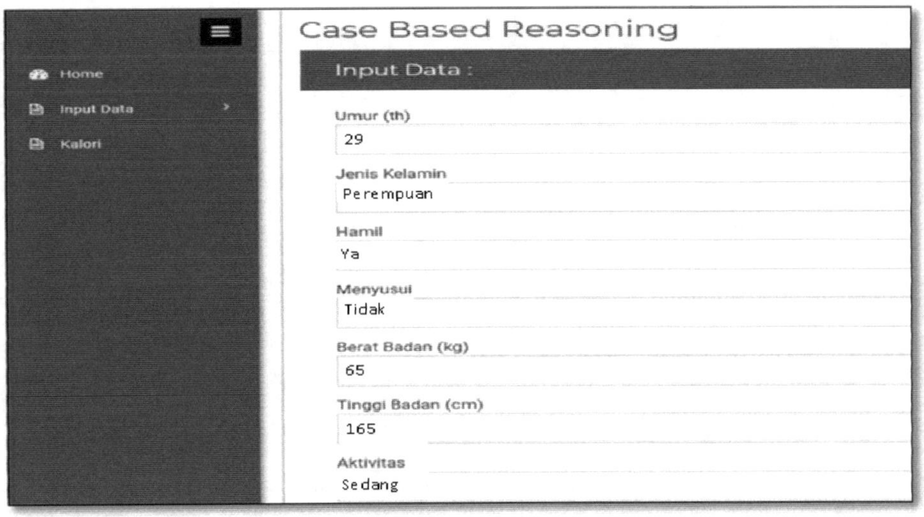

Figure 4. Page of a new case entering

New case entry pages with activity parameters have three choices, namely mild, moderate, and severe (Figure 5). From the three options, the user can choose with his daily activities, and on this page, there is an explanation of what the work is according to daily activities. Age and gender variables are the first factor in calorie calculation because a generation has a different function between children, adolescents, adults, and parents, while gender also determines that women need fewer calories than men.

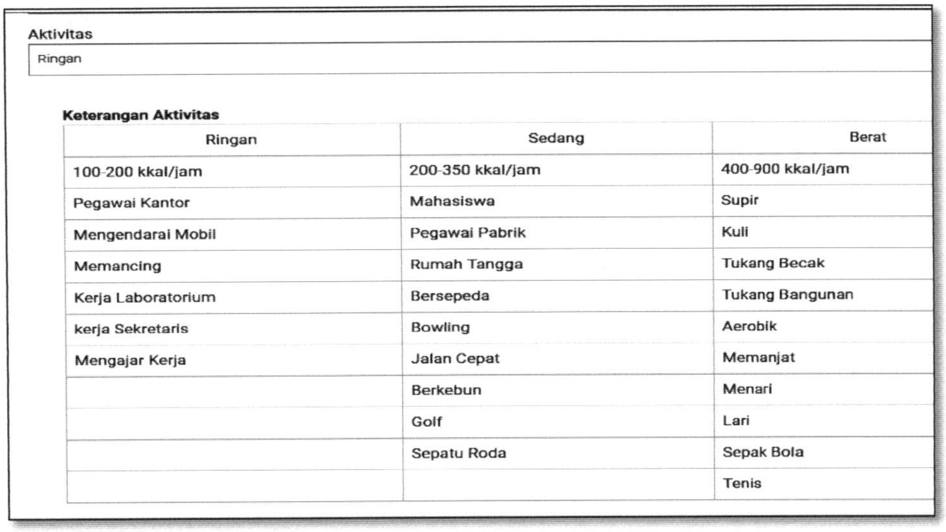

Figure 5. Display of new case entries for activity parameters

After the user enters the required parameters, the system will provide information on the number of calories needed and any food solutions that match the number of calorie calculations, as in Figure 6. Examples of these food solutions for one meal a day are divided into several groups. If there is an unwanted food in the solution, users can look at the bottom of this page. In this section, there is a list of exchange foods for each calorie needed by each person (Figure 7). Fig. 7 is a page listing the exchange of food items and is divided into each group to be a food solution.

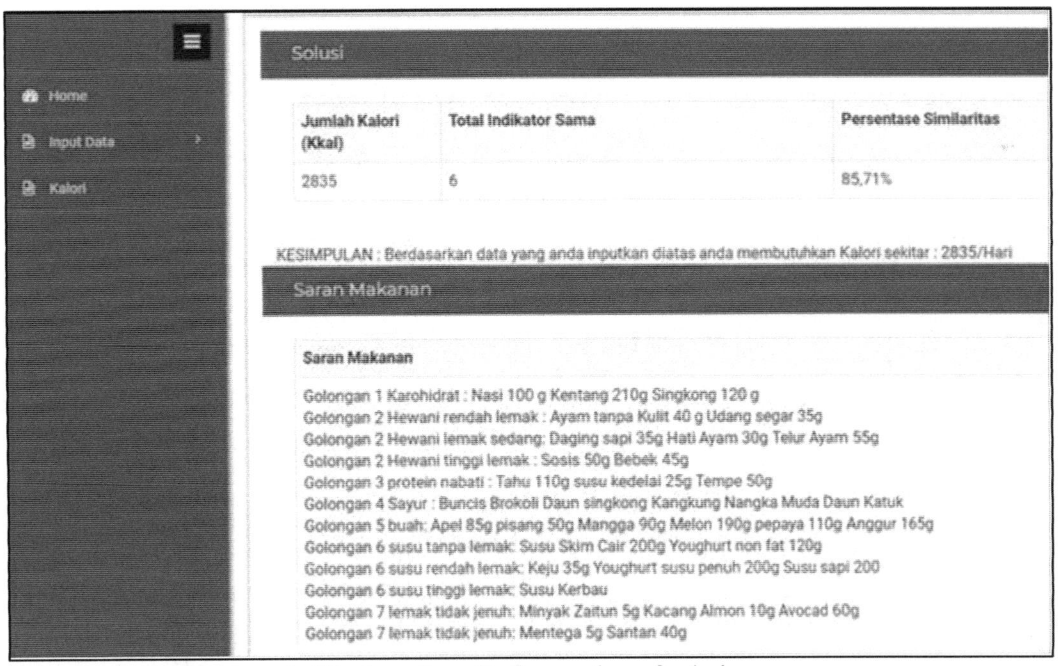

Figure 6. Display of results the number of calories per person

IOP Conf. Series: Materials Science and Engineering **771** (2020) 012027 doi:10.1088/1757-899X/771/1/012027

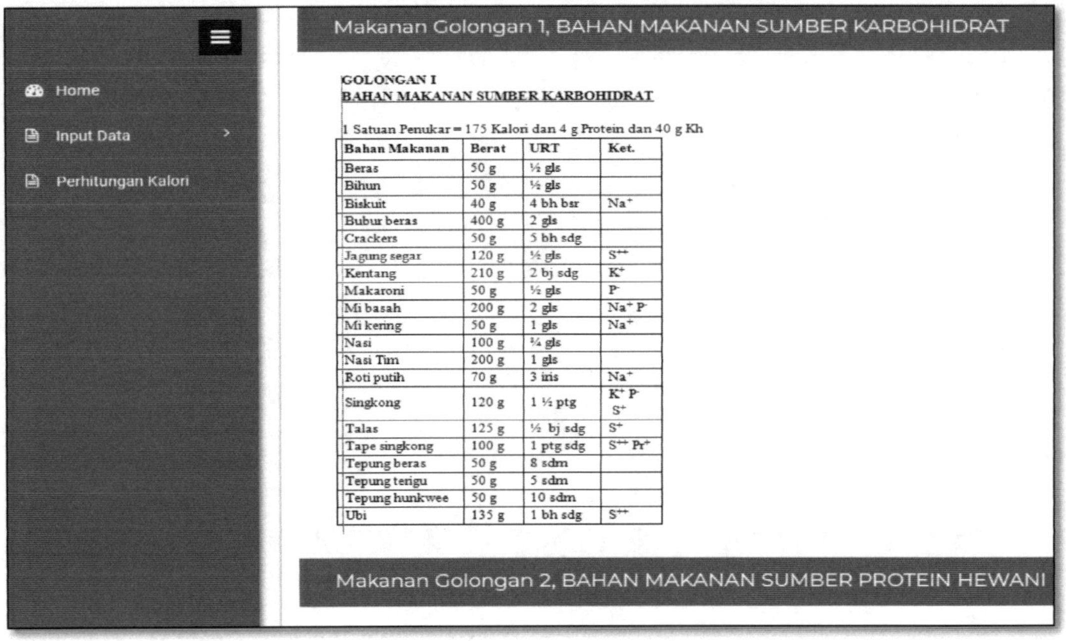

Figure 7. Display of list of exchanger feeding materials

Case-based stored some similar cases recommended to users of the results of the calculation of the number of calories needed. After getting a similar case search based on the selected root node, it does the similarity calculation. By tracing a similar case with comparing the input value with case-based will result in the same situation. The data in Table 2 is a new case as test case data.

Table 2. New case

Input Variable	Item of Input Variable	Item Class of Input Variables
Age	29	20-59
Gender	F	F
Pregnant	Yes	Yes
Breastfeeding	No	No
Weight	65	62-100
Height	165	165-200
Activity	Moderate	Mild

From the input entered by the user, the next process is to calculate a similar variable from the base of the case based on the root node selected by the system. Table 3 presented the results of calculating related variables.

Table 3. Results from similar case searches based on a new case

No.	Age (year)	Gender	Pregnant	Breast-feeding	Weight (kg)	Height (cm)	Activity	Solution
1	29	F	Yes	No	62-100	165-200	Mild	2535
2	16	M	No	No	50-53	153-155	Mild	2500

The next step is to calculate the number of input similarities by comparing inputs with the results of calculating similar variables. A similar case will give value 1. Otherwise, it will provide a value of 0, then the calculation of the total number of similarities will be calculated. Similarity calculation between new cases and cases stored in the case-based used equation 1. Table 4 presented the results of the similarity calculation. While the comparison between the original case with case-based to produce similarity values given in Table 4, is shown in Figures 8 and 9.

$$similarity(e, e_i) = \frac{common}{common + different} \quad (1)$$

Table 4. Calculation of similarity

No.	Age (year)	Gender	Pregnant	Breastfeeding	Weight (kg)	Height (cm)	Activity	Number of similarity	Similarity	Solution
1	1	1	1	1	1	1	0	6	0.857	2835
2	0	0	0	1	0	0	0	1	0.143	2500

Case Based / *New Case*

Input Variable	Class of Variable Item		Input Variable	Class of Variable Item
Age (year)	20-59	← 1 →	Age (year)	29
Gender	F	← 1 →	Gender	F
Pregnant	Yes	← 1 →	Pregnant	Yes
Breastfeeding	No	← 1 →	Breastfeeding	No
Weight (kg)	62-100	← 1 →	Weight (kg)	65
Height (cm)	165-200	← 1 →	Height (cm)	165
Activity	Mild	← 0 →	Activity	Moderate
Solution	2835		*New Case* based on the table	

Figure 8. Comparison of the new case with case-based for data 1

Case Based / *New Case*

Input Variable	Class of Variable Item		Input Variable	Class of Variable Item
Age (year)	16-19	← 0 →	Age (year)	29
Gender	M	← 0 →	Gender	P
Pregnant	No	← 0 →	Pregnant	Yes
Breastfeeding	No	← 1 →	Breastfeeding	No
Weight (kg)	50-53	← 0 →	Weight (kg)	65
Height (cm)	153-155	← 0 →	Height (cm)	165
Activity	Mild	← 0 →	Activity	Moderate
Solution	2500		*New Case* based on the Table	

Figure 9. Comparison of the new case with case-based for data 2

According to Figure 8 and 9, case similarity calculations can be calculated using the equation one as follows:

Data 1:
similarity(e,e_i)=((1+1+1+1+1+1))/((5+2))=6/7=85.7%

Data 2:
similarity(e,e_i)=((1))/((5+2))=1/7=14.3%

Based on these calculations, the system will display the highest value of all existing data on a case-based, which is 85.7%, then the value is used as a solution candidate for the given problem, namely 2835 Kcal. In Figure 7, the calories that are the solution to the needs of each person have a food solution as indicated by several examples of food exchanges which are divided into seven groups as follows:

- group 1: food ingredients as a source of carbohydrates
- group 2: food ingredients as a source of animal protein (low fat, medium fat, high fat)
- group 3: food ingredients for vegetable protein
- group 4: vegetables
- group 5: fruit and sugar

- group 6: milk
- group 7: oil and fat

As consideration, that some preliminary studies conducted by [26] stated that a low-calorie diet and metformin could reduce weight higher than a low-calorie or metformin diet in obese patients and [27] have developed a decision support system to calculate calorie diet values for diabetics. This research was conducted to determine the amount of calorie intake per day for people with diabetics to make it easier for diabetes to regulate their diet so that the threat of a drastic increase in blood sugar could receive them. Otherwise, reference [28] have also conducted research that refers to calorie labels influenced by various foods for diet, but not for free diets. The results show that the rush to provide calorie information does not prove to be the best approach to combat the obesity epidemic. Determination of the menu of healthy food in people with diabetes mellitus using an expert system has also been developed by [29].

4. Conclusions

A system of determining a calorie diet per day has been developed using Case-Based Reasoning. This Case-Based Reasoning system is a system that can decide diet calories for each person per day with similarity values based on case-based. The number of cases that are on a case-based will make the system better in conducting case searches. Researchers can develop this CBR system into an online system so that it can be utilized further wherever and whenever it is needed.

Acknowledgement

We want to thank the Research and Community Service Board – Universitas Muhammadiyah Purwokerto, for providing research funding support.

References

[1] DietSehat, "3 Cara Menghitung Kebutuhan Kalori per Hari untuk Wanita dan Pria," 2017. [Online]. Available: https://dietsehat.co.id/cara-menghitung-kebutuhan-kalori. [Accessed: 28-Oct-2017].

[2] BKKBN, "Gizi dan Kehidupan Manusia, Pusat Jaringan Informasi dan Dokumentasi Program KB Nasional," Jakarta, 1988.

[3] DietSehat, "13 Makanan Diet untuk Penderita Diabetes yang Wajib Dikonsumsi," 2017. [Online]. Available: https://dietsehat.co.id/makanan-diet-untuk-penderita-diabetes. [Accessed: 28-Oct-2017].

[4] M. E. Beck, *Ilmu Gizi dan Diet Hubungannya dengan Penyakit-penyakit untuk Perawat dan Dokter*. Yogyakarta: ANDI, 2011.

[5] A. Veratamala, "Berapa Anjuran Jumlah Kalori di Setiap Menu Makan Saat Sedang Diet?," 2017. [Online]. Available: https://hellosehat.com/?s=kalori+diet. [Accessed: 28-Oct-2017].

[6] Women's Health, "How Many Calories Do You NEED While Dieting?," 2013. [Online]. Available: http://www.womenshealthmag.com/weight-loss/counting-calories. [Accessed: 17-Oct-2016].

[7] H. L. Petersen, "Lose Weight: How Many Calories Should I Eat Per Day?," 2014. [Online]. Available: http://www.healthline.com/health/lose-weight-calories#LosingWeight3. [Accessed: 17-Oct-2016].

[8] G. B. Kris, "How Many Calories Should You Eat Per Day to Lose Weight?," 2014. [Online]. Available: https://authoritynutrition.com/how-many-calories- per-day/. [Accessed: 17-Oct-2016].

[9] M. T. Jacobsen, "How Many Calories Do You Really Need?," 2016. [Online]. Available: http://www.webmd.com/diet/guide/calories-chart. [Accessed: 17-Oct-2016].

[10] K. M. Zelman, "Weight Loss and Very Low-Calorie Diets," 2014. [Online]. Available: http://www.webmd.com/diet/low-calorie-diets#2. [Accessed: 17-Oct-2016].

[11] J. W. Gandy, A. Maden, and M. Holdsworth, *Gizi & Dietetika*. Jakarta: Kedokteran EGC, 2016.

[12] Web Kesehatan, "Diet Diabetes: Tips Pola Makan Sehat untuk Mengontrol Diabetes," 2017. [Online]. Available: https://www.webkesehatan.com/diet-diabetes-tips-pola-makan-sehat-mengontrol-diabetes/. [Accessed: 25-Oct-2017].

[13] S. K. Pal and S. C. K. Shiu, *Foundations of Soft Case-based Reasoning*. Hoboken, New Jersey: John Wiley & Sons, Inc., 2004.

[14] E. Hullermeier, *CASE-BASED APPROXIMATE REASONING*. Dordrecht, The Netherlands: Springer, 2007.

[15] R. L. D. M. Mántaras *et al.*, "Retrieval, Reuse, Revision, and Retention in Case-Based Reasoning," 2008. [Online]. Available: https://openair.rgu.ac.uk/bitstream/10059/54/1/ker05-mantarasetal.pdf. [Accessed: 06-Aug-2008].

[16] I. Watson, *Applying Case-Based Reasoning : Techniques for Enterprise Systems*. San Fransisco: Morgan Kauffman Publishers, Inc., 2001.

[17] H. Nu´n˜ez *et al.*, "A Comparative Study on the Use of Similarity Measures in Case-based Reasoning to Improve the Classification of Environmental System Situations," *Environ. Model. Softw. XX*, vol. Elsevier L, 2003.

[18] H. Mustafidah and S. Suwarsito, "Fish Diseases Control System Using Case-Based Reasoning," *IEEE Xplore Digit. Libr.*, pp. 1–5, 2011.

[19] S. Montani and L. C. Jain, *Successfull Case-Based Application-2*. Verlag Berlin Heidelberg: Springer, 2014.

[20] S. S. Kumala, "Sistem Pakar Pedoman Kebutuhan Kalori Harian untuk Penderita Diabetes," Semarang- Indonesia, 2012.

[21] A. Wuryandari and D. Trisnawati, "APLIKASI SISTEM PAKAR UNTUK DIAGNOSA PENYAKIT DIABETES MELITUS MENGGUNAKAN METODE DHEMSTER SHAFER," *MAGISTRA*, vol. 25, no. 85, pp. 73–79, 2013.

[22] A. K. Wardani, N. Santoso, and R. A. Asmara, "SISTEM PAKAR DIAGNOSA PENYAKIT DIABETES MELITUS," *J. Inform. Polinema*, vol. 1, no. 1, pp. 65–71, Nov. 2014.

[23] R. I. Perwira, "Purwarupa Sistem Pakar Untuk Menentukan Jumlah Kalori Diet Bagi Penderita Diabetes Mellitus," *Telematika*, vol. 10, no. 2, pp. 79–90, 2014.

[24] F. Masykur, R. Gernowo, and M. Suhartono, "IMPLEMENTASI SISTEM PAKAR DIAGNOSIS PENYAKIT DIABETES ELLITUS MENGGUNAKAN METODE FUZZY LOGIC BERBASIS WEB," Universitas Diponegoro, 2012.

[25] Z. Niswati, A. Paramita, and F. A. Mustika, "Aplikasi Fuzzy Logic dalam Diagnosa Penyakit Diabetes Mellitus pada PUSKESMAS di Jakarta Timur," *J. Teknol. dan Sist. Inf.*, vol. 2, no. 3, pp. 21–30, 2016.

[26] S. Suwandani, "Diet Rendah Kalori dan Metformin Menurunkan Berat Badan Lebih Tinggi Daripada Diet Rendah Kalori atau Metformin Saja pada Pasien Obesitas," Denpasar, 2010.

[27] F. Aldyningtyas, T. Pinandita, and Harjono, "Sistem Pendukung Keputusan Penghitung Kalori Diet bagi Diabetes," *JUITA (Jurnal Inform.*, vol. II, no. 2, pp. 145–157, 2012.

[28] L. Gilz, J. Polivy, C. P. Herman, and H. Lee, "The effects of Calorie Information on Food Selection and Intake," *Int. J. Obes.*, pp. 208–225, 2012.

[29] T. Lusiani and A. Qoiriah, "Sistem Pakar Untuk Menentukan Menu Makanan Sehat pada Penderita Diabetes Mellitus," *J. Ilmu Komput. dan Ilmu Sains Terap.*, vol. 5, no. 1, pp. 9–23, 2014.

IOP Conf. Series: Materials Science and Engineering **771** (2020) 012028 doi:10.1088/1757-899X/771/1/012028

Stress Detection from Multimodal Wearable Sensor Data

Fitri Indra Indikawati[1,*], Sri Winiarti[1]

[1]Informatics Engineering, Universitas Ahmad Dahlan, Ringroad Selatan, Tamanan, Bantul, DI Yogyakarta, Indonesia.

*fitri.indikawati@tif.uad.ac.id, sri.winiarti@tif.uad.ac.id

Abstract. Stress can be recognised by observing changes in physiological responses on the human body. Wearable sensors for stress detection are becoming more prominent in recent years due to their functionality and non-intrusive nature. By utilising data from wearable sensors, we have developed a personalized stress detection system. Our system performs classification on stress level using multimodal data from wrist-worn device Empatica E4 wearable sensor. We implemented three different classification algorithms: Logistic Regression, Decision Tree, and Random Forest and used four-class classification conditions: baseline, stress, amusement, and meditation. By evaluating the performance of the system, we demonstrate that our system can perform the best and consistent personalized stress detection using Random Forest classifier with the accuracy of 88%-99% on 15 subjects.

1. Introduction

Stress is an interesting affective state to explore because of its effects on the human body, both physically and mentally. Long-term stress can cause another health problem to appear, ranging from mild symptoms like headaches, weight gain or weight loss, sleeping trouble, until severe one such as cardiovascular disease [1].

In the earlier days, textual information from questionnaires and interview are employed to assess people's affective state. However, this method is intrusive because it interrupts the task that is being carried or there can be bias from the interviewer. Several kinds of research have been done in affective state and stress detection, i.e., using audio-visual data [2][3], writing [4], or by observing physiological responses [5]–[11].

Stress can be detected by analysing physiological changes in the human body. When people get stressed, the sympathetic nervous system (SNS) triggers physiological responses by releasing cortisol and adrenaline. The release of these hormones can cause increased in heart rate, breathing, sweat, or muscle tension. These physiological changes can then be used to detect stress in a person.

Most stress detection system uses generalized model [6], [8], [10]. However, physiological changes for each person is different from one another. Shi et al. [9] and Smets, et al. [12] create a general and personalized model but only classifies stress into two classes.

In this study, we implemented a personalized classification by using only personal physiological data and classify affective states into four-class classification conditions: baseline, stress, amusement, and meditation. We implemented three classical machine learning algorithm for classification task: Logistic Regression, Decision Tree, and Random Forest.

The remainder of this paper is as follows: Section 2 describes the dataset, exploratory analysis of the data from Empatica E4, and classification task. Then, we discuss the experimental result of our classification system in Section 3. Finally, we conclude our study in Section 4.

IOP Conf. Series: Materials Science and Engineering **771** (2020) 012028 doi:10.1088/1757-899X/771/1/012028

2. Research Method

In this section, we describe the data preprocessing and classification process in our stress detection system.

2.1. Dataset and Exploratory Data Analysis

WESAD dataset [6] is collected from 15 subjects, consists of 13 male and 2 female participants. Each subject participates in two hours of data collection protocol with four types of conditions: baseline, amusement, stress, and meditation condition. We used these four conditions data as the classes for the classification task.

WESAD provides data from two kinds of sensors: chest-worn (RespiBAN) and wrist-worn device (Empatica E4). In this study, we use only wrist-worn device data because it is not as intrusive as a chest-worn device, and they state that wrist-worn device data is quite promising to use for classification process [6].

Empatica E4 records six raw physiological and motion data: 3-axis accelerometer (ACC), blood volume pulse (BVP), electrodermal activity (EDA), temperature (TEMP), heart rate (HR), inter-beat interval (IBI). HR and IBI is a derivative data from BVP data, and we can ignore it in the dataset. We exclude ACC data because ACC data leads to low classification score compared to results using physiological data (BVP, EDA, TEMP) [6]. We used BVP, EDA, and TEMP data as features in the classification process. We can see the details for physiological data from Empatica E4 in Table 1.

Table 1 Empatica E4 data modalities

Data	Sampling	Description
ACC	32Hz	Accelerometer data consists of 3 axis channels
BVP	64Hz	Data from a photoplethysmograph (PPG).
EDA	4Hz	Consists of a tonic (referred to as skin conductance level (SCL)) and a phasic (skin conductance response (SCR)) component.
TEMP	4Hz	Temperature in °C

From Table 1, we can see that each data have different sampling rate. BVP data has a higher sampling rate than EDA and TEMP data. Therefore, the number of BVP data is significantly higher than the other data. We applied downsampling to BVP data into the window size of 0.25 seconds, similar to the EDA and TEMP data.

Figure 1 shows the histogram sample for BVP (a), EDA (b), and TEMP(c) data from one of the test subject. We can see the distribution for each raw data, and there is no outlier presents in the sample data.

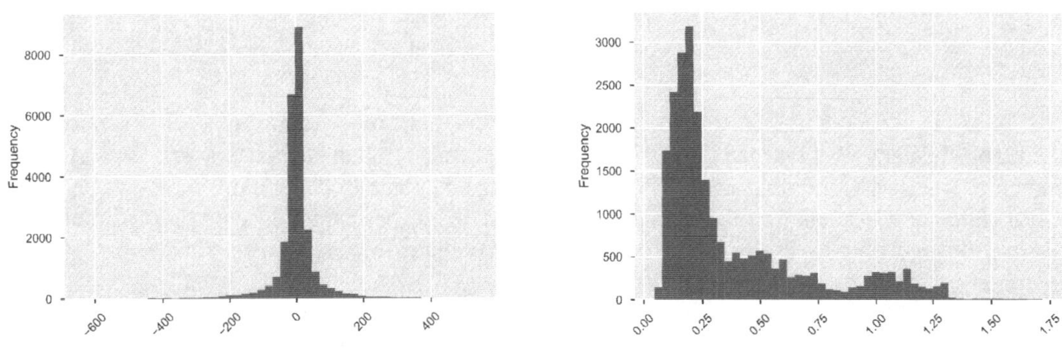

(a) (b)

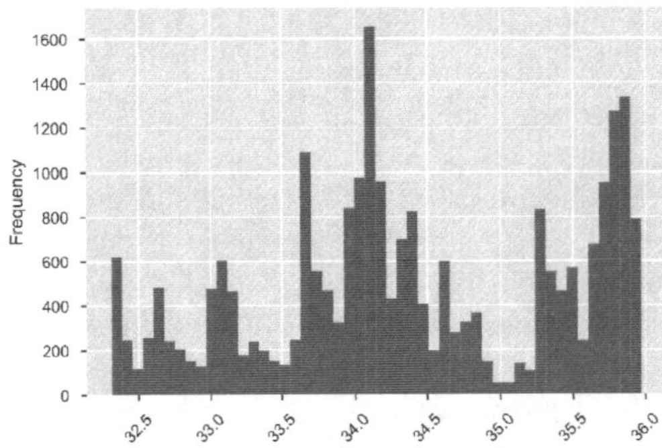

(c)

Figure 1 Histogram sample for Empatica data: (a) BVP (b) EDA (c) TEMP

2.2. Classification

The data considered in this study belong to four conditions (baseline, stress, amusement, and meditation) amount to approximately 180,000 records from a total of approximately 250,000 records or approximately 12,000 records per subject. From these records, ±40% belongs to the baseline condition, ±22% belongs to the stress condition, ±12% belongs to the amusement condition, and the remaining ±26% belongs to the meditation condition.

The experiments in WESAD [6] use binary class (stress, non-stress) and three- class (baseline, stress, and amusement) with accuracy result of 92% and 78%, respectively. Because the meditation condition records are significant, we included meditation condition as one of our four-class classification tasks. Markovic et al. performed an experiment using the dataset from the chest-based device but only use some part of the data in random [10].

Both experiments [6] [10] uses a generic model by combining all subjects data for constructing a training model. However, each person has a different degree of physiological changes in their body when faced with stress and amusement conditions. Because of this reason, we train the model for each subject separately in order to improve the classification accuracy.

We implement classic classification algorithms: Logistic Regression[13], Decision Tree[14], and Random Forest[15] using PySpark on top of Apache Spark. In our experiments, we divide data into 60% of training data and 40% of test data. We performed the experiments 10 times for each classification algorithm and each subject and randomised training and test data in each iteration.

3. Result and Discussion

In this section, we report the experimental evaluation of our classification system on a real dataset from a wearable sensor.

3.1. Experimental Setup

3.1.1. Environments. We implemented our classification system on Apache Spark [16] version 2.4.4 by utilizing PySpark (Python Spark Machine Learning Library). All experiments were conducted on a commodity machine with Intel Core i7-4500U 1.80 GHz CPU and 8 GB memory running a Windows 7 64-bit operating system. To ensure the reliability of the experiment results, we performed each classification algorithm for each subject ten times and averaged all the reported results.

3.1.2. Wearable Sensor Dataset. For evaluating our classification system, we used a real dataset named WESAD (Wearable Stress and Affect Detection) dataset [6]. The dataset is collected from two kinds of wearable sensor: RespiBAN and Empatica E4, a chest-worn and wrist-worn device, respectively. In this experiment, we only use Empatica E4 data because the authors concluded that data from the wrist-worn device is quite promising to use and because of the non-intrusive nature of the wrist-worn device.

In total, data were analysed and collected from 17 subjects. However, only data from 15 subjects are available because two subjects data has to be discarded due to sensor malfunction. WESAD dataset consists of millions of rows of raw sensor data which is collected from approximately 2 hours of the observation period. The total size of the synchronised data from the two sensors is more than 12 gigabytes.

3.2. Experiment Result

To show the performance of our classification system, we conducted experiments based on WESAD dataset using three multiclass classification algorithms: Logistic Regression, Decision Tree, and Random Forest. Table 2 shows the accuracy of the three classification algorithms on the four-class classification task: baseline, stress, amusement, and meditation.

Table 2 Accuracies result per subject on the four-class classification task

Subject	Logistic Regression (%)	Decision Tree (%)	Random Forest (%)
S2	75.89	86.69	93.08
S3	60.77	84.82	98.35
S4	73.85	79.40	91.60
S5	85.00	78.69	96.48
S6	89.81	80.41	99.57
S7	77.22	76.68	94.04
S8	62.21	94.44	99.21
S9	94.41	85.36	98.30
S10	85.41	92.65	97.02
S11	74.66	86.10	97.01
S13	66.36	74.13	88.88
S14	71.97	82.57	99.07
S15	87.73	85.05	99.76
S16	87.77	86.39	98.25
S17	88.71	87.75	99.61

We can see from Table 2 that Random Forest classifier produce the best and consistent classification accuracy compared for all of the subjects ranging from 88% to 99% classification accuracy. Compared to previous experiments which use generic model [6] [10], we can improve accuracy for our classification system by using a four-class classification problem and more personalized model by training the data per subject. Because we build our model by training the data for each subject, the accuracy result can be significantly different for a different subject. The result for S13 generally has lower accuracy for all three classification algorithms compared to the other subjects. Because each person has a different degree of physiological changes in their body therefore we cannot generalized the model for all subjects.

Figure 2 shows the confusion matrix for best prediction result (a) and worst prediction result (b) using Random Forest classifiers on the test data. In Section 2.2, we explained that we divided the dataset into train and test data, 60% and 40% respectively. Therefore we have approximately 4,000 records for test data per subject. In the best prediction scenario, the ratio for the true positives is ±40% for baseline condition, ±22% for stress condition, ±12% for amusement condition, and 26% for meditation condition. This ratio is consistent with the original dataset ratio explained in Section 2.2. False prediction happened when it tried to distinguish between baseline and meditation condition. It classifies meditation as a baseline condition. The main reason for this is because physiological data for baseline and meditation is stable compared to high physiological changes in high stimuli conditions (stress and amusement). In the worst-case scenario, it fails to classify the meditation condition. Also, it misclassified amusement condition as a baseline condition.

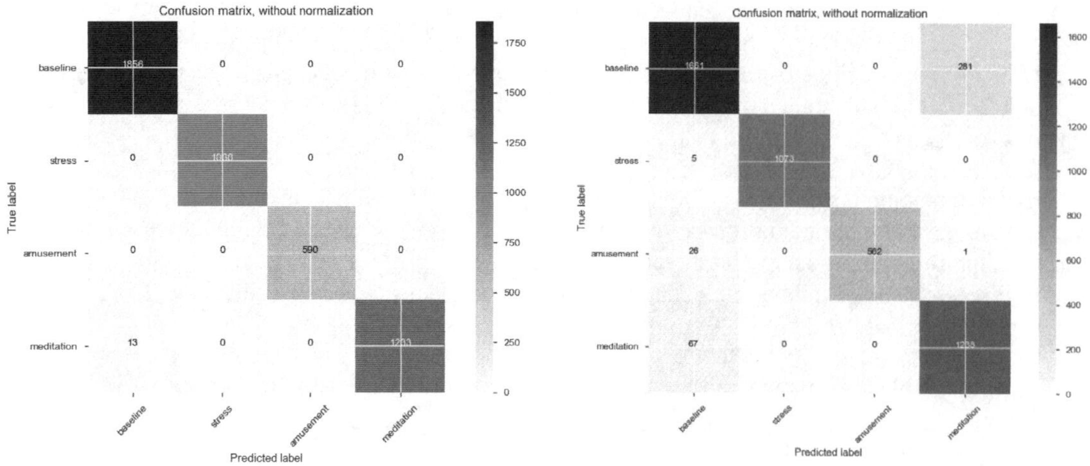

(a) (b)

Figure 2 confusion matrix for best prediction (a) and worst prediction (b) scenario

4. Conclusion

In this paper, we have implemented a personalized stress detection system from real wearable sensor dataset. We implemented three classification algorithms to the dataset and used four-class classification: baseline, stress, amusement, and meditation. Instead of using the generalized model, we train our data for each subject because physiological changes for each person is different and we need to make a personalized model for the classification. Our classification system can make a reasonable prediction by utilising physiological data from a wrist-worn device. The best accuracy is shown when we apply Random Forest classifier into the dataset with an 88-99% accuracy. Further work is required to create a more personalized model by utilising questionnaires available in the dataset. Moreover, we can also classify different stress level (low stress, moderate stress, and high stress) and amusement level (happy, sad, angry).

5. Acknowledgement

This work was funded by Universitas Ahmad Dahlan through Lembaga Penelitian dan Pengabdian Kepada Masyarakat (LPPM UAD) under research grant PDP-077/SP3/LPPM-UAD/2019.

References

[1] B. S. McEwen and E. Stellar, "Stress and the Individual: Mechanisms Leading to Disease," *Archives of Internal Medicine*, vol. 153, no. 18, pp. 2093–2101, 1993.

[2] S. Mirsamadi, E. Barsoum, and C. Zhang, "Automatic speech emotion recognition using recurrent neural networks with local attention," in *ICASSP, IEEE International Conference on Acoustics, Speech and Signal Processing - Proceedings*, 2017, pp. 2227–2231.

[3] P. Tzirakis, J. Zhang, and B. W. Schuller, "End-to-end speech emotion recognition using deep neural networks," *ICASSP, IEEE International Conference on Acoustics, Speech and Signal Processing - Proceedings*, vol. 2018-April, no. 8, pp. 5089–5093, 2018.

[4] A. Gangemi, V. Presutti, and D. Reforgiato Recupero, "Frame-based detection of opinion holders and topics: A model and a tool," *IEEE Computational Intelligence Magazine*, vol. 9, no. 1, pp. 20–30, 2014.

[5] Q. Xu, T. L. Nwe, and C. Guan, "Cluster-based analysis for personalized stress evaluation using physiological signals," *IEEE Journal of Biomedical and Health Informatics*, vol. 19, no. 1, pp. 275–281, 2015.

[6] P. Schmidt, A. Reiss, R. Duerichen, and K. Van Laerhoven, "Introducing WeSAD, a multimodal dataset for wearable stress and affect detection," in *ICMI 2018 - Proceedings of the 2018 International Conference on Multimodal Interaction*, 2018, pp. 400–408.

[7] N. Keshan, P. V. Parimi, and I. Bichindaritz, "Machine learning for stress detection from ECG signals in automobile drivers," in *Proceedings - 2015 IEEE International Conference on Big Data, IEEE Big Data 2015*, 2015, pp. 2661–2669.
[8] M. Gjoreski, M. Luštrek, M. Gams, and H. Gjoreski, "Monitoring stress with a wrist device using context," *Journal of Biomedical Informatics*, vol. 73, pp. 159–170, 2017.
[9] Y. Shi *et al.*, "Personalized Stress Detection from Physiological Measurements," in *Second International Symposium on Quality of Life Technology*, 2010, pp. 28–29.
[10] D. Marković, D. Vujičić, D. Stojić, Ž. Jovanović, U. Pesović, and S. Randić, "Monitoring System Based on IoT Sensor Data with Complex Event Processing and Artificial Neural Networks for Patients Stress Detection," in *2019 18th International Symposium INFOTEH-JAHORINA, INFOTEH 2019 - Proceedings*, 2019, pp. 1–6.
[11] B. Kikhia *et al.*, "Utilizing a wristband sensor to measure the stress level for people with dementia," *Sensors (Switzerland)*, vol. 16, no. 12, p. 1989, 2016.
[12] E. Smets *et al.*, "Comparison of machine learning techniques for psychophysiological stress detection," in *Communications in Computer and Information Science*, 2016, vol. 604, pp. 13–22.
[13] R. E. Wright, "Logistic regression.," 1995.
[14] J. R. Quinlan, "Induction of decision trees," *Machine learning*, vol. 1, no. 1, pp. 81–106, 1986.
[15] L. Breiman, "Random forests," *Machine learning*, vol. 45, no. 1, pp. 5–32, 2001.
[16] Apache Software Foundation, "Apache Spark." [Online]. Available: https://spark.apache.org/docs/latest/api/python/pyspark.ml.html.

IOP Conf. Series: Materials Science and Engineering **771** (2020) 012029 doi:10.1088/1757-899X/771/1/012029

Classification of Farmer's Eligibility as Recipients of Subsidized Fertilizer Assistance with C4.5 Algorithm

Diah Nur Yunita[1], Anna Hendri Soleliza Jones[2*], Dewi Soyusiawaty[3]
Department of Informatics, Universitas Ahmad Dahlan, Indonesia

[1]diahnuryunita@gmail.com, [2,*]annahendri@tif.uad.ac.id,
[3]dewi.soyusiawaty@tif.uad.ac.id,

Abstract. Farmers are the main actors in agricultural development, have a major contribution in realizing Indonesia's agricultural vision. To support the realization of agricultural production and productivity targets, the government's supports is by issuing policies to provide subsidized fertilizer assistance. The criteria for farmers who are eligible to receive subsidized fertilizer assistance are having land less than 1.9 hectares, have been members of farmer groups, and having work as a farmer or a private employee. Data collection was carried out using the method of observation and interviews with farmer groups in Gunung Kidul. Data classification is done by applying the C4.5 algorithm to 300 data of prospective farmers receiving subsidized fertilizer assistance. Application development includes interface design, making data flow diagrams, implementation and testing. From the research that has been done by applying the C4.5 Algorithm for farmer data, there are 300 data with 113 training data. Produce a data classification of farmers who are eligible to receive assistance at 95,58% accurate and targeted to the needs of farmers. This means that the C4.5 algorithm can be used as a classification for determining the eligibility of farmers to receive subsidized fertilizer.

1. Introduction

Farmers are one of the main actors in agricultural development that has a major contribution in realizing production and productivity goals to achieve the target of managing agricultural products in Indonesia. Agricultural development policies continue to be reviewed and improved, one of the government's policies to improve the quality of agricultural products is to provide subsidized fertilizer assistance to farmers. Based on the Regulation of the Minister of Agriculture No. 82 / Permentan / OT.140 / 8/2013 concerning the Development of Farmers Groups and Farmers Group Combinations, guidelines for the preparation of the Definitive Needs Group (RDKK) have been arranged. RDKK is made through the coordination of Gapoktan (Joint Farmers Group).

According to Regulation of the Minister of Agriculture No. 273 / Kpts / ot.160 / 4/2007 concerning guidelines for institutional development of farmers, Gapoktan is a collection of several farmer groups that join and work together to increase economies of scale and business efficiency. The existence of Gapoktan so that farmer groups can be more efficient and effective, and provide agricultural production facilities, improvement, capital, or expansion of farming for farmers and farmer groups from the upstream and downstream sectors, as well as increased cooperation and product marketing. And through the RDKK instructions for subsidized fertilizer are prepared as a reference for farmer groups, farmers, officers and agricultural service officials in the distribution of subsidized fertilizer. RDKK as the basis for redemption of fertilizer to retailers is not yet fully implemented as expected. Although subsidized fertilizer assistance has been implemented for a long time, the problems found in the implementation of subsidized fertilizer assistance are very complex in terms of data collection and distribution/distribution, so that often farmers who should get subsidized fertilizer products do not receive them. RDKK is not yet fully operational. Farmers have difficulty in obtaining the subsidized fertilizer product because the price of non-subsidized fertilizer is quite expensive for farmers.

Based on the background that has been described and previous studies that have been carried out using the C4.5 algorithm in determining the classification of data such as the Khairul Zaman study in 2016 on how to determine the eligibility of beneficiaries for social rehabilitation assistance for uninhabitable homes, Muhammad Arif Rahman's 2015 study was determining scholarship recipients using the C4.5 method, Wisti Dwi Septiani's research in 2014 the application of the C4.5 algorithm for prediction of hepatitis and 2017 Hikma Widayu research on data mining to predict the types of customer transactions in savings and loan cooperatives with the C4.5 algorithm. Based on the problems that have been discussed, the author tries to find the decision tree rules using the C4.5 algorithm to the data of prospective fertilizer recipients in Gunungkidul Regency in a study entitled "Classification of Determination of Fertilizer Aid Recipients with the C4.5 Algorithm Technique". The knowledge generated is expected to help the authorities in making policies, so that in the future the recipients of fertilizer assistance are right on target.

2. Method

2.1. Preprocessing Text

The process of data classification is the process of sorting data according to similarities or differences in the data held. Before the classification is done on the data, it is necessary to preprocess the data. Preprocessing data is to take steps to clean up the data so that the results of data mining will be more accurate. The data preprocessing process has the following steps: [1]

a. Data cleaning
Data cleaning is to get rid of inconsistent or noise data, one of the stages in data mining in the data cleaning process is very important because inconsistent data can cause invalid mining results. In general, the data obtained from the database and the results of experiments contained imperfect entries and there are data attributes that are not relevant to the data mining hypothesis that they have.

b. Data transformation (data is converted into a form suitable for mining)
Data is modified or combined into a format suitable for processing in data mining. Data in the form of continuous numeric numbers need to be divided into several intervals. (Han, 2006).

c. Application of data mining techniques
The main process in the method applied to find valuable and hidden knowledge from data.

d. Evaluate the patterns found
The final stage of the data mining process is to introduce patterns to the results of the algorithm. Pattern recognition is a discipline that studies ways to classify objects into classes or categories and recognizes data trends.

2.2. Algorithm C4.5

C4.5 algorithm is a tree structure where there are nodes that describe the attributes that are tested, and each leaf represents a class. C4.5 algorithm recursively visits each decision node, selects the optimal division, until it cannot be subdivided. C4.5 algorithm uses the concept of information gain or entropy reduction to choose the optimal division. Here is the gain and entropy formula:

2.2.1. *Gain (S, A)* is the acquisition of information from attribute A relative to the data output of S. The acquisition of information is obtained from the output data or the dependent variable S which is grouped by attribute A, denoted by the gain (S, A). [4]

$$\text{Gain}(S,A) = \text{Entropy}(s) - \sum_{i=1}^{n} - \frac{|si|}{|s|} * \text{entropy}(si) \qquad \ldots\ldots\ldots\ldots (1)$$

Information:
S: case set
A: attributes
n: number of attributes A
|Si|: number of cases on the i-th partition
|S|: number of cases in S

2.2.2. *Entropy*(S) is the estimated number of bits needed to be able to extract a class (+ or -) from a random amount of data in the sample space S. Entropy can be said to be the need for bits to express a class. the smaller the value of Entropy, the more Entropy is used in extracting a class. Entropy is used to measure the authenticity of S. [4]

$$\text{Entropy}(S) = \sum_{j=1}^{k} - pj \log 2\ pj \qquad \text{.............. (2)}$$

Information:
S: The set of cases
k: number of partitions S
Pj: Probabilities derived from the number (yes/no) divided by the total cases

After calculating the gain and entropy, the following are the steps of the decision tree construction using the C4.5 algorithm.

2.2.3. *Confusion Matrix*

Accuracy shows the closeness of the value of the measurement results with the actual value. To determine the level of accuracy, it is necessary to know the true value of the measured parameter, then the data is known to what level of accuracy. To analyze performance, a matrix is used by comparing the original class data with predictions from the input data or called a confusion matrix. [9]

To do confusion matrix calculations use the following formula:

1. *Recall* is the proportion of positive cases that are correctly identified.

$$Recall = \frac{d}{(c+d)} \qquad \text{...................................... (3)}$$

2. *Precision* is the proportion of cases that are correctly identified with the positive result

$$Precision = \frac{d}{(b+d)} \qquad \text{................................ (4)}$$

3. *Accuracy* is a comparison of correctly identified cases with a total number of cases.

$$Accuracy = \frac{(a+d)}{(a+b+c+d)}$$

Attribute Description:
a: If the prediction result is positive and the data is actually positive.
b: if the prediction result is positive while the actual value is negative.
c: if the result of the prediction is negative while the actual value is positive.
d: if the predicted result is negative and the actual value is negative

3. Results and discussion

3.1. *Result and Discussion*

The data in this research were obtained directly from the chairman of the Gunungkidul Farmers Group. There are 300 farmers data, with 3 criteria as support in determining the decision of whether farmers are eligible or not eligible for receiving subsidized fertilizer assistance. The 3 criteria used are; Land Area (Land), Occupation and the membership of farmer groups or not. Table 1 is a sample of data obtained.

Table 1. Sample Data of The Farmers.

No	Name	Land Area (Land)	Occupation	Member of Farmers Group
1	Kirnorejo	0,1	CS	No
2	Pariyem	0,2	Farmer	Yes
3	Nurhayadi	1,9	CS	Yes
4	Pujo. W	0,3	Farmer	No
5	Pujiyanto	0,5	Employee	No
6	Lasiyanto	0,3	Farmer	Yes
7	Turiyo	0,5	Farmer	Yes
8	Sukismo	0,1	Farmer	Yes
9	Tentrem	0,1	Farmer	Yes
10	Dedi Martanto	1,7	Farmer	Yes

Data processing is done by grouping the criteria used in a more efficient form. Criteria are grouped as follows:

1. Land

Land criteria are grouped based on the wide-area of land owned by farmers. The area will be divided into 3 groups, such as S for Small at a range of 0 – 0,9 Ha, M for Medium at a range of 1 – 1,9 Ha and L for large at more than 2 Ha.

2. Occupation

This criterion explains the variables that will be used to classify prospective recipients of subsidized fertilizer with the definition of job criteria. The variables are "Employee" for the farmer who has other jobs aside from a farmer, variable "Farmer" to describe a farmer and variable "CS" to describe a farmer who their main job as a civil servant.

3. Member of Farmers Group

This criterion explains the variables that will be used to classify prospective subsidized fertilizer recipients with the definition of the criteria whether the farmer is a member of the farmer group or does not belong to the farmer group, the variable will be "Yes" for the member of farmer group and "No" if not a member of farmer group.

After defining the next variable this research will classify the data by applying the C4.5 algorithm, by finding the highest gain value of the 300 existing data using Equation 2 and Equation 1 which have been described in Chapter 3. The calculation steps to find the gain are as follows:

Step 1: Calculate the total entropy. After we get the total entropy next, do calculate the entropy for each criterion, as examples:

a. Entropy (Total) = $\left(-\frac{30}{113} * log\left(\frac{30}{113}\right)\right) + \left(-\frac{83}{113} * log\left(\frac{83}{113}\right)\right)$

$= \left(-\frac{30}{113} * 0{,}2654\right) + \left(-\frac{83}{113} * -0{,}7345\right)$ = 0,5080 + 0,3269= 0,8349

b. Entropy (Land area, Large) = $\left(-\frac{5}{9} * log\left(\frac{5}{9}\right)\right) + \left(-\frac{4}{9} * log\left(\frac{4}{9}\right)\right)$

$= \left(-\frac{5}{9} * 0{,}8481\right) + \left(-\frac{4}{9} * 1{,}1700\right)$ = - 0,4711 + - 0,52= 0,9911

c. Entropy (Land area, Medium) = $\left(-\frac{11}{20} * log\left(\frac{11}{20}\right)\right) + \left(-\frac{9}{20} * log\left(\frac{9}{20}\right)\right)$

$= \left(-\frac{11}{20} * -0{,}8624\right) + \left(-\frac{9}{20} * -1{,}1520\right)$

= - 0,47432 + - 0,5184 = 0,99272

d. Entropy (Land area, small) = $\left(-\frac{14}{84} * log\left(\frac{14}{84}\right)\right) + \left(-\frac{70}{84} * log\left(\frac{70}{84}\right)\right)$

$= \left(-\frac{14}{84} * -0{,}7369\right) + \left(-\frac{70}{84} * -0{,}2630\right)$

= - 0,1228 + - 0,2191 = 0,6500

Repeat the calculation for each of criterions.

Step 2: Calculate the total gain. Use formula 2 to get the total gain, the example can be seen below:

a. Gain (total, Land area)=Entropy (total) $\sum_{i=1}^{n} \frac{|land\ area|}{|total|}$ * Entropy (Land area)

$= 0{,}8349 - \left(\frac{9}{113} * 0{,}9910\right) + \left(\frac{20}{113} * 0{,}9927\right) + \left(\frac{84}{113} * 0{,}6500\right)$

= 0,8349 – (-0,0789) + (-0.1756) + (-0,4831)= 0,0970

b. Gain (total, member of farmer group)=Entropy (total) $\sum_{i=1}^{n} \frac{|Member\ of\ farmer\ group|}{|total|}$*Entropy (member of farmer group)

$= 0{,}8349 - \left(-\frac{84}{113} * 0{,}2222\right) + \left(-\frac{29}{113} * 0{,}3620\right)$

= 0,8349 – (-0,1651) + (-0,0929) = 0,5767

c. Gain (total, Occupation)=Entropy (total) $\sum_{i=1}^{n} \frac{|Occupation|}{|total|}$ * Entropy (Occupation)

$= 0{,}8349 - \left(-\frac{14}{113} * 0\right) + \left(-\frac{79}{113} * 0{,}0979\right) + \left(-\frac{20}{113} * 0{,}8112\right)$

= 0,8349 – 0 + 0,0684 + 0,1435= 0,6228

The results of the calculation of entropy and gain values can be seen in Table 2.

Table 2. Calculation of algorithm C4.5

Node			Total Cases	No	Yes	Entropy	Gain
1	**Total**		300	85	215	0,8599	
	Land Area	Large	21	14	7	0,9183	0,0456
		Medium	89	30	59	0,9219	
		Small	190	41	149	0,7524	
	Member of farmer group	Yes	227	18	209	0,3997	0,5575
		No	73	68	5	0,3603	
	Occupation	CS	38	38	0	0	0,5420
		Farmer	207	8	199	0,2360	
		Employee	55	40	15	0,8453	

From the calculation of Entropy and Gain values, as shown in table 3.2, it can be seen that the attribute that has the highest Gain is the work attribute that is equal to 0.6228. The work attribute will be the root node which will be the basis of the classification of data in the form of a decision tree. Decision tree resolution is obtained by finding the largest gain from the rest of the data that has been used in the first node decision tree search. Gain search will continue to be repeated until there is no more data that can still be broken down to look for gain. Thus, the results of the decision tree in this study can be seen in Figure 1.

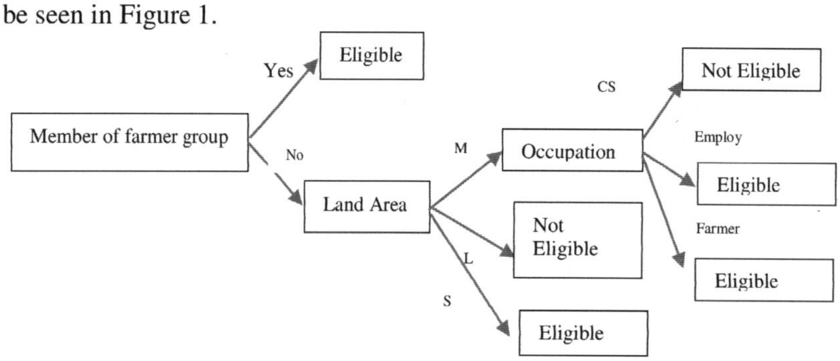

Figure 1. Decision Tree Classification of Farmers.

With the production of the classification tree of farmers, the search for the determination of farmers who are not suitable or not eligible to receive subsidized fertilizer can be done more efficiently.

3.2. System Testing

Accuracy testing is performed using 300 training data and 113 test data. Test results obtained:

Table 3. Accuracy of test result

Accuracy	True: Eligible	True: Not Eligible	Precision
Eligible Prediction	82	4	95,35%
Not Eligible Prediction	1	26	96,30%
Recall	98,80%	86,67%	

From Table 3, it was found that with 300 training data and 113 precision test data the decision of farmers who were eligible to receive subsidized fertilizer assistance was 95.35% with an accuracy value of 95.58%.

IOP Conf. Series: Materials Science and Engineering **771** (2020) 012029 doi:10.1088/1757-899X/771/1/012029

4. Conclusion

From the research that has been done, the result of C4.5 Algorithm has an effect when applied in the data of farmers to determine whether farmers are eligible or not eligible to receive subsidized fertilizer assistance. By using 300 farmer data as training data and 113 test data. The study resulted in the classification of data of farmers who met the requirements to receive assistance with the accuracy of the target accuracy of 95.58% subsidized fertilizer recipients. This means that the C4.5 algorithm can be used as a classification to determine the eligibility of farmers to receive subsidized fertilizer.

Reference

[1] Anwar, N., Pranolo, A., & Kurnaiwan, R. (2018). *Grouping the community health center patients based on the disease characteristics using C4.5 decision tree. IOP Conference Series: Materials Science and Engineering, 403, 012084.* doi:10.1088/1757-899x/403/1/012084

[2] ARIBOWO, Agus Sasmito; CAHYANA, Nur Heri. Feasibility study for banking loan using association rule mining classifier. **International Journal of Advances in Intelligent Informatics**, [S.l.], v. 1, n. 1, p. 41-47, mar. 2015. ISSN 2548-3161. Available at: <http://ijain.org/index.php/IJAIN/article/view/8%7Cto_array%3A0>. Date accessed: 09 oct. 2019. doi:https://doi.org/10.26555/ijain.v1i1.8.

[3] J. A. Suyatno, F. Nhita and A. A. Rohmawati, "Rainfall Forecasting in Bandung Regency Using C4.5 Algorithm," *2018 6th International Conference on Information and Communication Technology (ICoICT)*, Bandung, 2018, pp. 324-328. doi: 10.1109/ICoICT.2018.8528725

[4] L. Dongming, L. Yan, Y. Chao, L. Chaoran, L. Huan and Z. Lijuan, "The application of decision tree C4.5 algorithm to soil quality grade forecasting model," *2016 First IEEE International Conference on Computer Communication and the Internet (ICCCI)*, Wuhan, 2016, pp. 552-555.doi: 10.1109/CCI.2016.7778985

[5] R. K. Amin, Indwiarti and Y. Sibaroni, "Implementation of decision tree using C4.5 algorithm in decision making of loan application by debtor (Case study: Bank pasar of Yogyakarta Special Region)," *2015 3rd International Conference on Information and Communication Technology (ICoICT)*, Nusa Dua, 2015, pp. 75-80.doi: 10.1109/ICoICT.2015.7231400

[6] S. Das, S. Dahiya and A. Bharadwaj, "An online software for decision tree classification and visualization using c4.5 algorithm (ODTC)," *2014 International Conference on Computing for Sustainable Global Development (INDIACom)*, New Delhi, 2014, pp. 962-965. doi: 10.1109/IndiaCom.2014.6828107

[7] Siahaan H., Mawengkang H., Efendi S., Wanto A. and Windarto A. P. 2018 Application of Classification Method C4.5 on Selection of Exemplary Teachers *IOP Conference Series* 1-6

[8] SIA, Florence; ALFRED, Rayner. Tree-based mining contrast subspace. International Journal of Advances in Intelligent Informatics, [S.l.], v. 5, n. 2, p. 169-178, july 2019. ISSN 2548-3161. Available at: <http://ijain.org/index.php/IJAIN/article/view/359%7Cto_array%3A0>.Date accessed: 09 oct. 2019. doi:https://doi.org/10.26555/ijain.v5i2.359.

[9] S. Visa, Brian. Ramsay, A. Ralescu, and E.V.D. Knaap, "Confusion matrix-based feature selection," Midwest Artificial Intelligence and Cognitive Science Conference, vol. 710, April 2011.

[10] Wahyuni, Sri & Saputra S, Kana & Iswan, Mochammad. (2017). THE IMPLEMENTATION OF DECISION TREE ALGORITHM C4.5 USING RAPIDMINER IN ANALYZING DROPOUT STUDENTS.

[11] Xuefei Wang and Yan Shi, "Design and implementation of targeting advertising system based on C4.5 algorithm," *2015 4th International Conference on Computer Science and Network Technology (ICCSNT)*, Harbin, 2015, pp. 669-672.doi: 10.1109/ICCSNT.2015.7490833

IOP Conf. Series: Materials Science and Engineering **771** (2020) 012030 doi:10.1088/1757-899X/771/1/012030

Detection of Mad Lazim Harfi Musyba Images Uses Convolutional Neural Network

A Noeman[1,*] and D Handayani[1]

[1]Departement of Information Technology, Faculty of Engineering, Universitas Bhayangkara Jakarta Raya, Jakarta, Indonesia

[*]*Corresponding author e-mail*: achmad.noeman@dsn.ubharajaya.ac.id

Abstract. Reading of Al Quran is an obligation for all Muslims. Lack of knowledge about the knowledge of recitation in reading the Al Quran is certainly a problem. The purpose of this study makes it easier for everyone to learn the law of recitation, especially Madd Lazim Harfi Musyba in the Al Quran verses. The method used is one of the deep neural network methods, namely Convolutional Neural Network (CNN), as real-time detection of the law of Madd Lazim Harfi Musyba, implementation of the method using the help of Tensorflow GPU (Graphic Processor Unit) library. The results of trials with the Deep Convolutional Neural Network model show the detection performance of 9 verses with an average accuracy of 93.25%. The conclusion is that the training data on the CNN model is very reliable in detecting the Mad lazim harfi musyba law. Therefore this system can be used to assist in applying the Mad Lazim Hafi Musyba legal recitation while reading the Al Quran.

1. Introduction

Developing computing technology is Artificial Intelligence[1][2]. Convolutional Neural Network model is a variation of the Multilayer Perceptron (MLP) model [3] which has good classification capabilities on image data and has a high degree of accuracy for detecting images [4] from the recitation of mad lazim harfi musyba from Al-Quran verses.

The results of this system research have an average detection rate of 84.85%. This research can be continued with input in the form of image reading patterns of reading-law on the Qur'an. This application makes it easy for people who want to learn the law reading mad lazim harfi musyba on the Al-Quran [5].

The purpose of this study is to detect the percentage of legal reading of Mad lazim harfi musyba on the Qur'an with the Convolutional Neural Network and to detect the reading of the recitation of the verses of the Qur'an with the Convolutional Neural Network.

2. Theoretical

2.1 Convolutional Neural Network (CNN)

Convolutional Neural Network is a development of the Multilayer Perceptron (MLP) which aims to process two-dimensional data. CNN is included in the Deep Neural Network [6]type and is widely applied to image data. MLP has a disadvantage in storing spatial information on image data so that it produces bad images. An MLP in Figure 1 has a layer (red and blue circle) with each circle layer containing ji neurons. MLP accepts one-dimensional input data and propagates the data on the network to produce the output[7]. Each relationship between neurons on two adjacent layers[8] has a one-dimensional weight parameter that determines the quality of the model. Each input data on the layer is carried out linearly with the existing weight value, and then the computational results will be transformed using a non-linear operation called the activation function.

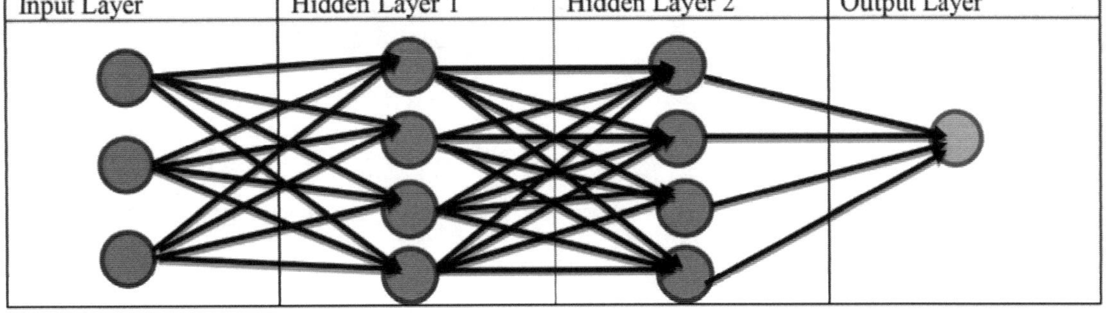

Figure 1. Simple Multilayer Perceptron (MLP) architecture

Data propagated on the network are two-dimensional data, so the linear operation and weight parameters on CNN are different. In CNN, linear operations use convolution operations, whereas weights are no longer just one dimension, but in the form of four dimensions which are a collection of convolution kernels like Figure 2, the weight dimensions on CNN are input neurons x output neurons x height x width. Because of the nature of the convolution process, CNN can only be used on data that has a two-dimensional structure such as image and sound.

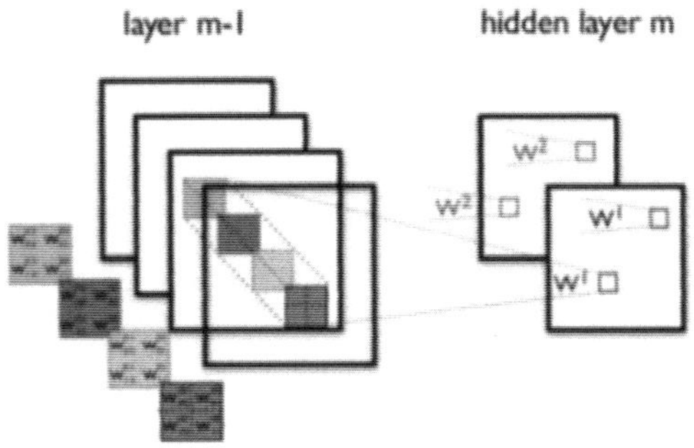

Figure 2. Convolution Process in Convolutional Neural Network

2.2 Convolution Layer
Convolution Layer performs convolution operations at the output of the previous layer. This layer is the main process that underlies a CNN. Convolution is a mathematical term that means to apply a function to the output of another function repeatedly. The purpose of convolution in image data is to extract features from the input image. Convolution will produce a linear transformation of the input data according to spatial information in the data[9].

2.3 Subsampling Layer
Subsampling aims to improve the position invariance of features. Convolution layer into several small grids and then take the maximum value from each grid to arrange the reduced image matrix.

2.4 Fully Connected Layer
The Fully Connected Layer aims to transform the dimensions of the data so that the data can be classified linearly. Each neuron in the convolution layer needs to be transformed into one-dimensional data before it can be put into a fully connected layer. Because it causes data to lose spatial information and is not reversible, the fully connected layer can only be implemented at the end of the network.

IOP Conf. Series: Materials Science and Engineering **771** (2020) 012030 doi:10.1088/1757-899X/771/1/012030

2.5 Activation Function

The activation function is to transform input data into a higher dimension so that simple hyper lane cuts can be made which enable classification. There CNN activation function is used, namely the sigmoid function. The sigmoid function transforms the range of values from input x to between 0 and 1 with the shape of the distribution function, as shown in Figure 3.

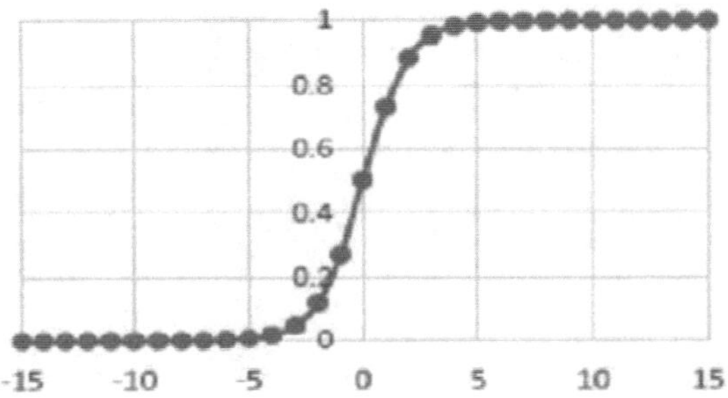

Figure 3. Distribution of Sigmoid Functions

2.6 Mad lazim harfi musyba

in the reading of the Al-Quran, there is tajweed science which is the knowledge to know how to pronounce the Alquran verses[10]. Mad lazim harfi musyba occurs only at the beginning of the surah in the Al-Quran. The following are hijaiyah letters on the AlQuran that have punctuation marks on mad lazim harfi musyba, ie. : كم ل، س، ع، ص، ق، ن، ،
Examples of surah contained in the recitation of Mad lazim harfi musyba on Al-Quran such as the name of Al-Quran Al-baqarah:1, name of AlQuran Yassin:1, name of Al-Quran Qof:1, Maryam:1, name of Al-Quran Al-qalam:1.

In reading the Mad lazim harfi musyba is 6 harakat (3 alifs). As an example in Al-Baqarah verses one الٓمٓ, Al-A'raf verses one الٓمٓصٓ, Al-Qasas verses 1 طسٓمٓ.

3. Method

In developing this system, there are system components, namely obtaining sample data, data labelling, training set, training data and data testing.

3.1 Samples Data

Data are taken from the Al-Quran; this data that will be used in this research is a verse from the Al-Quran, which includes punctuation marks from the mad lazim harfi musyba.

3.2 Data Labelling

There are two stages of the labelling process in this system, namely, making a rectangle box and labelling each training data will be a training data set. A rectangle box is to make boundaries on objects that will become training data. Making a rectangle box as shown in Figure 4.

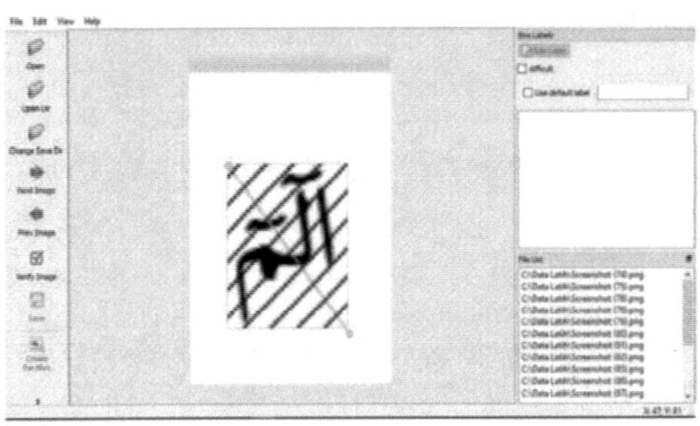

Figure 4. Making Rect Box

The rectangle box has a label for each of their training data to be trained. Labelling on each the rectangle box aims to have a label for training data as shown in Figure 5.

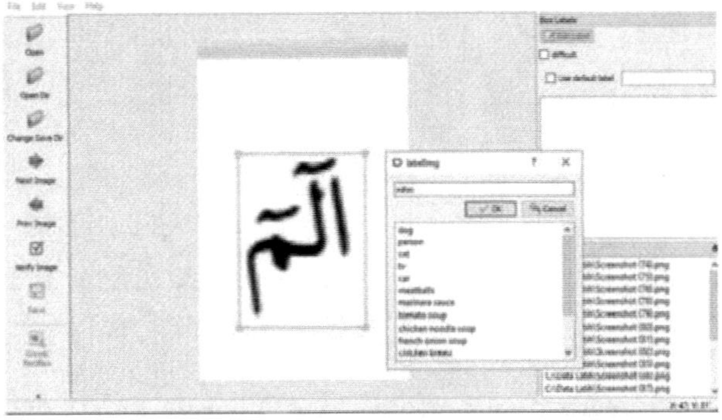

Figure 5. labelling for training data

3.3 Training Set

The training set is the data needed to do the training data process, then first to collect training data[11]. in the study find 27 surahs that there are 59 mad law lazim harfi musyba, in table 1 is the name of Al-Quran, there is a mad lazim harfi musyba punctuation mark :

Table 1. Punctuation Mark of Mad Lazim Harfi Musyba

No	Name of Surah	Verses	Mad lazim harfi musyba Images
1	Al-Baqarah	1	الم
2	Ali-Imran	1	الم
3	Al-A'raf	1	المص
4	Yunus	1	الر
5	Hud	1	الر
6	Ar-Rad	1	المر
7	Ibrahim	1	الر
8	Al-Hijr	1	الر
9	Maryam	1	كهيعص
10	Asy-Syu'ara	1	طسم
11	An-Naml	1	طس
12	Al-Qasas	1	طسم
13	Al-Ankabut	1	الم
14	Ar-Rum	1	الم
15	Luqman	1	الم
16	As-Sajdah	1	الم
17	Yaasiin	1	يس
18	Shad	1	ص
19	Ghafir	1	حم
20	Fussilat	1	حم

21	Asy-Syura	1 and 2	حم عسق
22	Az-Zukhruf	1	حم
23	Ad-Dukhan	1	حم
24	Al-Jasiyah	1	حم
25	Al-Ahqaf	1	حم
26	Qof	1	ق
27	Al-Qalam	1	ن

3.4 Training Data

This system uses the Tensorflow library as a deep learning library for conduct training data with the convolutional neural method network, this system conducts training data with as many epochs 56140, with the value of loss or error stable at 0.01 as shown Figure 6.

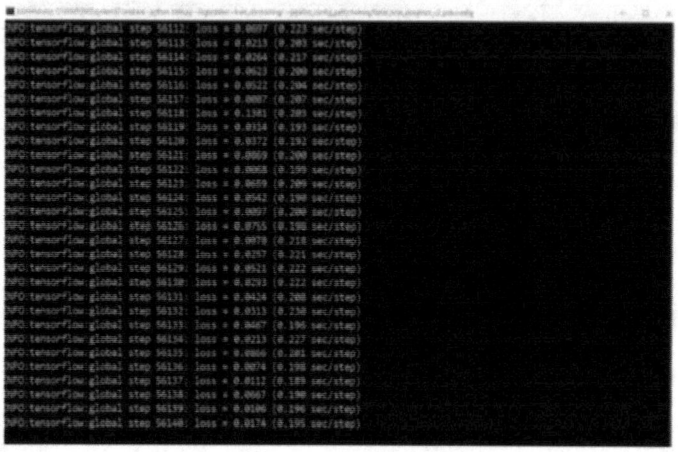

Figure 6. labelling for training data

Results of training data conducted on this system, by including the results of total loss that is visualised in the graphic, as shown in Figure 7.

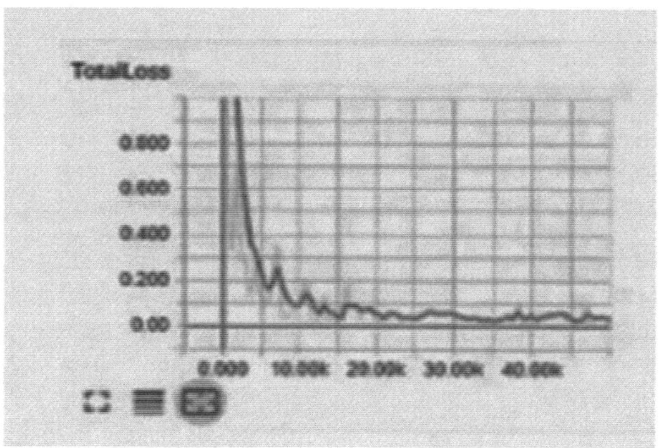

Figure 7. The result from the total-loss of training data

3.5 Data Testing

Data processing is the stage where data is collected, labelled, and trained, to produce data that is ready to be tested data[12] then result from detection system images of mad lazim harfi musyba.

4. Result and Discussion

4.1. Training Set

This Training-Set will later be used to make machine learning models, and Training set to collect author summarises only a few surah which can represent the pattern in each of the common mad lazim harfi musyba letters. shown in Figure 8.

Letter of Mad lazim harfi musyba	ل	م	ص	ع	ق	س	ك	ن
Mad lazim harfi musyba on Al-Quran	آلٓ	مٓ / مٓ	صٓ / صٓ	عٓ / عٓ	قٓ / قٓ	سٓ / سٓ	كٓ	نٓ
Training Set	الٓمٓ الٓمٓصٓ صٓ عٓصٓ عٓسٓقٓ سٓ قٓ كٓ نٓ							

Figure 8. Training Set to collect letters of mad lazim harfi musyba images

4.2 Data Testing on the System

In the process of detect mad lazim harfi musyba then there is a colour mark on the reading that contains mad lazim harfi musyba. Shown in Figure 9.

No	Name of AlQuran	Training Set	Testing data	Response time	Accuracy
1	Al-Baqoroh:1	الٓمٓ		2 detik	97%
2	Al-'Arof:1	الٓمٓصٓ		2 detik	97%
3	Shood:1	صٓ		2 detik	98%
4	Maryam:1	كٓ		4 detik	62%
5	Maryam:1	عٓصٓ		2 detik	97%

Figure 9. Detecting images on Al-Quran containing mad lazim harfi musyba

Testing data above using a sample of 9 verses from 8 surah in the Al-Quran, the total surah contained law tajweed madd lazim harfi musyba there are 27 surah out of 114 surah of the Al-Quran, but of the 27 surah many have the same pattern so that it can be collected and represented on 8 surah, on testing this system produces an average detection accuracy of 93.25%. However, there are still some law tajweed madd lazim harfi musyba detected less than 80%, and many other letters that are not letters from the recitation of law tajweed madd lazim harfi musyba are detected.

5. Conclusion

Detection System of law tajweed Mad lazim harfi musyba by using Deep Convolutional Neural Network has an average detection accuracy of 93.25% in the detection of the verses of the Al-Quran that contain the law madd lazim harfi musyba.

References

[1] Plotnikov D, Sopov E and Panfilov I 2018 An approach for automating the design of convolutional neural networks *IOP Conf. Ser. Mater. Sci. Eng.* **450**

[2] Arrofiqoh E N and Harintaka H 2018 Implementasi Metode Convolutional Neural Network Untuk Klasifikasi Tanaman Pada Citra Resolusi Tinggi *Geomatika* **24** 61

[3] Zufar M and Setiyono B 2016 Convolutional Neural Networks Untuk Pengenalan Wajah Secara Real-Time *J. Sains dan Seni ITS* **5** 72–7

[4] Hu H and Cheng Y 2019 Video Logging Casing Damage Image Recognition Based on Improved Convolutional Neural Network *J. Phys. Conf. Ser.* **1237** 032025

[5] Ahsiah I, Noor N M and Idris M Y I 2013 Tajweed checking system to support recitation *2013 Int. Conf. Adv. Comput. Sci. Inf. Syst. ICACSIS 2013* 189–93

[6] Cheng G and Guo W 2017 Rock images classification by using deep convolution neural network *J. Phys. Conf. Ser.* **887**

[7] Zainudin Z, Shamsuddin S M and Hasan S 2019 Deep Learning Layer Convolutional Neural Network (CNN) Scheme for Cancer Image *IOP Conf. Ser. Mater. Sci. Eng.* **551**

[8] Plotnikov D, Sopov E and Panfilov I 2018 An approach for automating the design of convolutional neural networks *IOP Conf. Ser. Mater. Sci. Eng.* **450**

[9] Zheng H, Gu N and Zhang X 2018 An Efficient and Slight Convolutional Neural Network for Vehicle Type Classification *J. Phys. Conf. Ser.* **1069**

[10] Rizal, Fadlisyah, Muhathir and Akfal A M 2015 Detection System Tajwid Al Quran on Image Using Bray Curtis Distance *Int. J. Comput. Technol.* **2** 293–300

[11] Sugata T L I and Yang C K 2017 Leaf App: Leaf recognition with deep convolutional neural networks *IOP Conf. Ser. Mater. Sci. Eng.* **273** 012004

Convolutional Neural Network for Brain Tumor Detection

D C Febrianto[1,*], I Soesanti[1] , H A Nugroho[1]

[1]Department of Electrical Engineering and Information Technology, Faculty of Engineering, Universitas Gajah Mada, Yogyakarta, Indonesia

[*]e-mail: dany.candra@mail.ugm.ac.id

Abstract. Magnetic resonance imaging (MRI) is the imaging technique used to diagnosing brain tumor disease. Early diagnosis of brain tumors is an essential task in medical work to find out whether the tumor can potentially become cancerous. Deep learning is a handy and efficient method for image classification. Deep learning has been widely applied in various fields including medical imaging, because its application does not require the reliability of an expert in the related field, but requires the amount of data and diverse data to produce good classification results. Convolutional Neural Network (CNN) is the deep learning technique to perform image classification. In this paper, we compared two model CNN find the best model CNN to classify tumours in Brain MRI Image and at the end, we have trained CNN and obtained a prediction accuracy of up to 93%.

1. Introduction

According to the Indonesian Ministry of Health, in 2018 Indonesia became the number of 8th cancer patient in Southeast Asia and ranked 23rd in Asia[1]. Brain tumors are the second cause of death in cancer cases after breast cancer. From all cases, it is known that women are more affected by brain tumors than men. Brain tumors have continued to increase in incidence for a decade last in several countries [2]. Medical imaging is key in diagnosing brain tumors and can help prevent more virulent diseases. MRI is imaging techniques that researchers rely on to detect brain tumors[3]. MRI is one of the most widely used medical imaging techniques for brain tumors because it does not use ionizing radiation [4].

There are many studies about the detection of brain tumors, Parveen et al. [5], using a combination of Fuzzy C-Means and SVM methods. In this study, using Fuzzy C-Means to segment between parts of the brain and the parts indicated to have tumors, the use of Fuzzy C-Means in this paper shows good results. After segmentation, the Gray Level Run-Length Matrix (GLRLM) is used to feature extraction. The purpose of feature extraction is to find relevant features of an image to facilitate the classification process. The classification technique used in the method is SVM. In the method used, the accuracy achieved was 83.33%.

Avizenna et al. [6] using the fluid-attenuated inversion recovery (FLAIR) method. This study proposed to classify MRI images into two classes, normal brains and abnormal brains. BRATS database data 2017 used in this study. Process classification brain MRI images using multinomial logistic regression models with ridge estimators, and scoring using sensitivity, specificity, accuracy with cross-fold validation.

However, in unsupervised method-based segmentation, as before the extraction feature is not obtained implicitly, there is a need for human touch [7]. Then, to segment the brain image there are still many obstacles, there are problems such as the shape or location that is different from each patient.

According to Işin A et al [8], the devices used for scanning vary greatly so as to produce images that have different characteristics in the dataset.

In this paper, the CNN method is proposed to solve the problem of data complexity. CNN is able to extract features without removing the spatial information from the input data. CNN is a machine learning method to process two-dimensional data. CNN has two methods: classification using feedforward and learning using backpropagation.[7]. But apart from the advantages of CNN architecture, which is simple for users, but on the development process, CNN requires amount data for the learning process and consume a lot of learning time compared to unsupervised learning approaches. In this research, 2 CNN models were made as a comparison to find the best model for classification.

2. Literature Review

2.1. Convolutional Neural Network

CNN is a neural network that aims to process data that has a grid structure. Convolution is an operation in convolution layer that is based on a linear algebra operation that multiplies matrix of the filter in the image to be processed [9]. The convolution layer is the primary layer that is most important to use. Another type of layer that is commonly used is the pooling layer, which is a layer that is used to take the maximum value or the average value of the pixel portions of the image.

CNN has the ability to learn complicated features by forming a feature map. The convolution layer kernel is wrapped around the input sample to calculate several feature maps. Features are detected from input samples than represented by small boxes on the feature map. These maps are forwarded to the maximum collection layer, which preserves relevant features and discards the rest. The features of the max-pooling layer are converted to a one-dimensional feature vector in the fully connected layer, which is then used to calculate the output probability. The configuration of CNN is shown in Figure 1[10].

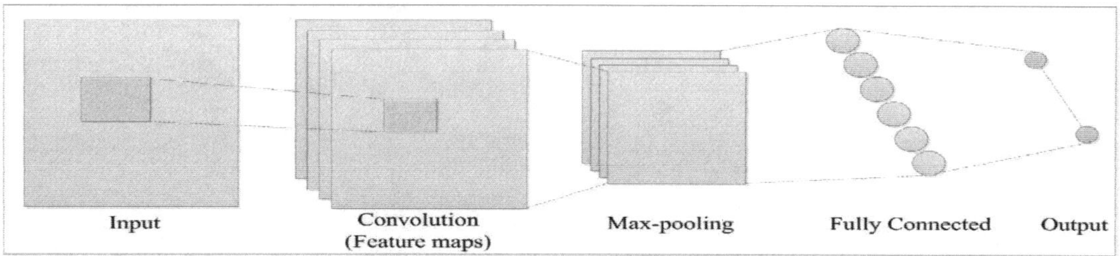

Figure 1. CNN architecture [10]

2.1.1. Convolution Layer

Convolution Layer is the core layer in the CNN method which aims to extract features from the input. Convolution performs linear transformations of input data without changing spatial information in the data. Convolution kernels are determined from the weight of the layer so that the convolution kernels can process the input data training on CNN.

2.1.2. Subsampling Layer

Subsampling aims to reduce the size of image data and to increase the invariance of feature positions. CNN uses Max Pooling as a subsampling method. The way Max Pooling works is to divide the output of the convolution layer into several smaller grids and then take the maximum value from each grid to produce a smaller image matrix. With a small image size will make it easier to process the next convolution layer.

2.1.3. Fully Connected Layer

The Fully Connected Layer changes the dimensions of the data so that it can be classified linearly. In the convolution layer, each neuron must be transformed into one-dimensional data before being inserted into another layer that is connected as a whole[11]. This process is caused by data losing its spatial information and at the end of the Fully Connected Layer network is applied.

3. Method

3.1. Dataset

The dataset used in this study is Brain MRI Images for Brain Tumor Detection obtained from kaggle.com. The dataset consists of 253 images grouped into 2 groups, 155 brain images that have tumors, and 98 brain images that do not have tumors. Figure 2 shows images in the dataset.

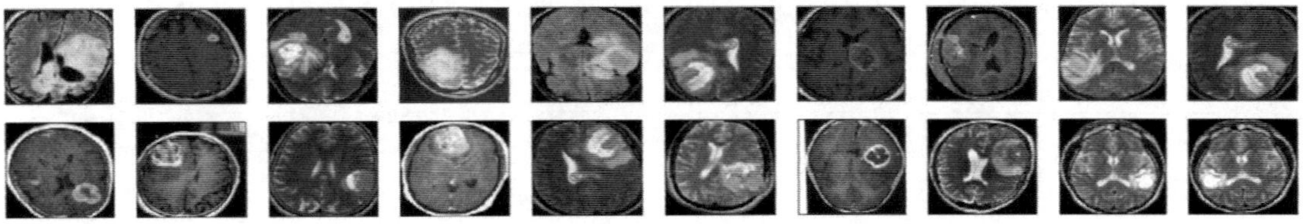

Figure 2. samples of the dataset

3.2. Proposed Method

This paper uses CNN for the automatic detection of brain tumors. This study uses input images labeled (yes/no) from the raw data and then uses these patterns to distinguish between tissues that do not contain tumors and those that contain tumors. CNN was trained to use 2065 sample images consisting of 1085 images containing tumors and 980 not containing tumors. Therefore, the proposed system is illustrated in Figure 3 method proposed in this study.

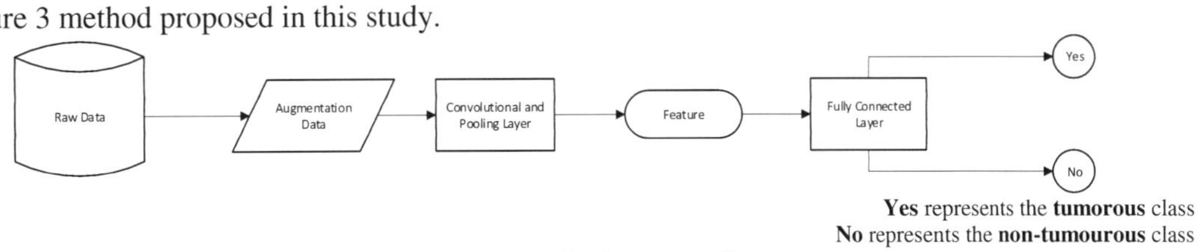

Figure 3. Method proposed

3.3. Data Augmentation

The amount of data in the dataset is not enough to be used as training data for CNN. Therefore the augmentation method is used to overcome the imbalance of issues. Augmentation is an algorithm that can utilize statistical data information and form an integrated model. This algorithm can produce a number of two-dimensional images of various poses and sizes. The application of augmentation to obtain image variants can improve the accuracy of CNN segmentation[12]. In this paper, each image with a tumor is segmented into 6 images, and an image with no tumor is segmented into 9 images. After data augmentation, the dataset consists of 1085 samples containing tumors (53%) and 980 samples not containing tumors (47%), bringing a total of 2065 images.

3.4. Image Pre-processing

Pre-processing performs to create smooth training because there are different variants of intensity, contrast, and size in images [13]. Input image will be processed into the first pre-process that is the process of wrapping and cropping. In wrapping, the input image is checked against the edge of the main object in the image. From the edge of the image, the maximum edge is determined so that when the results of cropping, the object in the image remains intact. After cropping, resize the image to shape (240, 240, 3) = (image_width, image_height, number of channels) because the images in the dataset have different sizes. Apply normalization: to scale pixel values to the range 0-1 to facilitate the learning process.

3.5 Model CNN

In this research, the CNN model contains several layers, namely the convolution layer, the pooling layer, the flatten layer, the dropout layer, and the dense layer. In addition to the layers used in the CNN process, there is also an activation function in this study using rule activation. In this research, 2 CNN models were made as comparison material. The CNN model design can be seen in table 1 and table 2. An image in the form of a number interweaving the first convolution, image size of 240x240 pixels. Kernels that have a size of 3x3 with a thickness of 3 in accordance with the channel of the image data and filters are used as many as 32. After getting the results of the operation, the model will perform the activation and pooling data functions. The Pooling layer process serves to reduce the dimensions of the feature map. The result of the convolution process is a feature map that is used for the subsequent convolution process repeatedly. The next step is a flallten feature map in vector form to carry out a fully-connected layer process to produce a classification of images.

Table 1. First CNN model.

Layer (type)	Output Shape	Param #
conv2d (Conv2D)	(None, 240, 240, 32)	896
max_pooling2d (MaxPooling2D)	(None, 60, 60, 32)	0
max_pooling2d_1 (MaxPooling2D)	(None, 30, 30, 32)	0
max_pooling2d_2 (MaxPooling2D)	(None, 15, 15, 32)	0
flatten (Flatten)	(None, 7200)	0
dense (Dense)	(None, 256)	1843456
dense_1 (Dense)	(None, 1)	257
Total params: 1,844,609 Trainable params: 1,844,609 Non-trainable params: 0		

Table 2. Second CNN model.

Layer (type)	Output Shape	Param #
conv0 (Conv2D)	(None, 240, 240, 32)	896
max_pooling2d (MaxPooling2D)	(None, 60, 60, 32)	0
dropout (Dropout)	(None, 60, 60, 32)	0
conv2d_1 (Conv2D)	(None, 60, 60, 32)	9248
max_pooling2d_1 (MaxPooling2D)	(None, 15, 15, 32)	0
dropout_1 (Dropout)	(None, 15, 15, 32)	0
flatten (Flatten)	(None, 7200)	0
dense (Dense)	(None, 256)	1843456
dropout_3 (Dropout)	(None, 256)	0
dense_1 (Dense)	(None, 1)	257
Total params: 1,863,105 Trainable params: 1,863,105 Non-trainable params: 0		

4. Result and Discussion

Experiments in this article were carried out on 2065 images consisting of 1085 samples containing tumors and 980 samples containing no tumors. The data is further divided into 70% of the data as data training, 15% of the data as data validation, 15% of the data as data testing. The data is run 10 times, each using the CNN model that has been made before, each experiment using 25 epochs and 32 batches. The results are then compared using standard deviations, the mean and mean of the loss, accuracy and f1 score. In table 3 and table 4 can be seen from the results of the experiments conducted in this study. In the first CNN model, 1 convolution is used, and the average accuracy value is 94% and has an average loss value of 0.14181 on the training data, but there is a significant difference with the test data results, the average test data accuracy value is 85% and an average loss value of 0.44037. In the second CNN model using 2 convolution gets better results, the training data obtained an accuracy value of 96% and a loss value of 0.10046, then the test data obtained an accuracy value of 93% and a loss value of 0.23264. The f1 score on the second model is 92% but has a longer training time compared to the first model.

Table 3. Result of First CNN Model

No	Train			Val		Test		
	Time Training	Loss	Acc	Loss	Acc	Loss	Acc	f1
1	00:01:30	0,1895	0,9343	0,3797	0,8452	0,3938	0,8645	0,8742
2	00:01:31	0,1159	0,9592	0,3751	0,8774	0,4381	0,8483	0,8458
3	00:01:29	0,1793	0,9349	0,3908	0,8323	0,4013	0,8581	0,8682
.	.	.	.	.	.	.	.	.
.	.	.	.	.	.	.	.	.
8	00:01:29	0,0259	0,9722	0,3674	0,8903	0,5348	0,8709	0,8802
9	00:01:26	0,1221	0,9619	0,3762	0,8581	0,4632	0,8387	0,8538
10	00:01:26	0,2707	0,9183	0,3961	0,8355	0,4405	0,8322	0,8479
Averae	00:01:29	0,14181	0,9496	0,38254	0,85452	0,44037	0,85098	0,86048
Mean	00:01:29	0,14215	0,946	0,37795	0,8532	0,4293	0,8502	0,8575
STD	1,95511e-05	0,07132	0,0225	0,01134	0,02188	0,04233	0,01175	0,01119

Table 4. Result of Second CNN Model

No	Train			Val		Test		
	Time Training	Loss	Acc	Loss	Acc	Loss	Acc	f1
1	00:01:40	0,1625	0,9329	0,2765	0,9065	0,2655	0,9193	0,9158
2	00:01:39	0,0455	0,9862	0,2245	0,9258	0,2272	0,9419	0,9415
3	00:01:40	0,1105	0,9585	0,2104	0,9129	0,2258	0,9322	0,9302
.	.	.	.	.	.	.	.	.
.	.	.	.	.	.	.	.	.
8	00:01:41	0,1123	0,9578	0,1959	0,9032	0,2532	0,9258	0,9225
9	00:01:45	0,0789	0,9744	0,1874	0,9355	0,2172	0,9354	0,9328
10	00:01:46	0,1215	0,9543	0,2156	0,9129	0,2412	0,9291	0,9256
Average	00:01:42	0,10046	0,96402	0,22538	0,91485	0,23264	0,93221	0,92996
Mean	00:01:42	0,09965	0,96435	0,2181	0,9129	0,2269	0,9338	0,9315
STD	2,7411e-05	0,03143	0,01506	0,0317	0,00988	0,01629	0,00625	0,00725

5. Conclusion

Convolutional Neural Networks are good enough to diagnose brain tumors on MRI images. This study resulted in accuracy of 93% and a loss value of 0.23264. The number of convolution layers affects the quality of classification, more convolution layers increase the accuracy results, but more number of convolution layers will require more time for training. The process of image augmentation can improve the variants of existing datasets, thereby increasing the classification results. Finally, for future suggestions, more images can be used to improve classification results. Future studies can also classify certain types of tumors.

References

[1] "Penyakit Kanker di Indonesia Berada Pada Urutan 8 di Asia Tenggara dan Urutan 23 di Asia." [Online]. Available: http://p2p.kemkes.go.id/penyakit-kanker-di-indonesia-berada-pada-urutan-8-di-asia-tenggara-dan-urutan-23-di-asia/.

[2] S. EDY, W. I, and W. A, "Clinical Characteristics and Histopathology of Brain Tumor at Two Hospitals in Bandar Lampung," *Fac. Med. Lampung Univ.*, vol. 69, pp. 48–56, 2014.

[3] H. Mohsen, E.-S. A. El-Dahshan, E.-S. M. El-Horbaty, and A.-B. M. Salem, "Classification using deep learning neural networks for brain tumors," *Futur. Comput. Informatics J.*, vol. 3, no. 1,

pp. 68–71, 2018.

[4] A. Wadhwa, A. Bhardwaj, and V. Singh Verma, "A review on brain tumor segmentation of MRI images," *Magn. Reson. Imaging*, vol. 61, no. January, pp. 247–259, 2019.

[5] Parveen and A. Singh, "Detection of brain tumor in MRI images, using combination of fuzzy c-means and SVM," *2nd Int. Conf. Signal Process. Integr. Networks, SPIN 2015*, pp. 98–102, 2015.

[6] M. H. Avizenna, I. Soesanti, and I. Ardiyanto, "Classification of Brain Magnetic Resonance Images Based on Statistical Texture," *Proc. - 2018 1st Int. Conf. Bioinformatics, Biotechnol. Biomed. Eng. BioMIC 2018*, vol. 1, pp. 1–5, 2019.

[7] S. Dabeer, M. M. Khan, and S. Islam, "Cancer diagnosis in histopathological image: CNN based approach," *Informatics Med. Unlocked*, vol. 16, no. May, p. 100231, 2019.

[8] A. Işin, C. Direkoğlu, and M. Şah, "Review of MRI-based Brain Tumor Image Segmentation Using Deep Learning Methods," *Procedia Comput. Sci.*, vol. 102, no. August, pp. 317–324, 2016.

[9] K. P. Danukusumo, Pranowo, and M. Maslim, "Indonesia ancient temple classification using convolutional neural network," *ICCREC 2017 - 2017 Int. Conf. Control. Electron. Renew. Energy, Commun. Proc.*, vol. 2017-January, pp. 50–54, 2017.

[10] S. Hussain, S. M. Anwar, and M. Majid, "Segmentation of glioma tumors in brain using deep convolutional neural network," *Neurocomputing*, vol. 282, pp. 248–261, 2018.

[11] R. Refianti, A. B. Mutiara, and R. P. Priyandini, "Classification of melanoma skin cancer using convolutional neural network," *Int. J. Adv. Comput. Sci. Appl.*, vol. 10, no. 3, pp. 409–417, 2019.

[12] Z. Tang, K. Chen, M. Pan, M. Wang, and Z. Song, "An augmentation strategy for medical image processing based on Statistical Shape Model and 3D Thin Plate Spline for deep learning," *IEEE Access*, vol. 7, pp. 1–1, 2019.

[13] T. Zhou, S. Ruan, and S. Canu, "A review: Deep learning for medical image segmentation using multi-modality fusion," *Array*, vol. 4, no. July, p. 100004, 2019.

IOP Conf. Series: Materials Science and Engineering **771** (2020) 012032 doi:10.1088/1757-899X/771/1/012032

QOS Analysis of Routing Protocol Babel on Mobile Ad-Hoc Network (MANET)

A Wijayanto[1,*], R Adhitama[2]

[1]Department of Software Engineering, Faculty of Industrial Technology and Informatics, Telkom Institute of Technology, Purwokerto, Indonesia
[2]Department of Software Engineering, Faculty of Industrial Technology and Informatics, Telkom Institute of Technology, Purwokerto, Indonesia

[*]aditya.wijayanto@ittelkom-pwt.ac.id, rifki@ittelkom-pwt.ac.id

Abstract. The unequal access to internet communication infrastructure in most parts of Indonesia can cause difficulties for people in the area to communicate. As an alternative communication, the mobile ad-hoc network requires a reliable routing algorithm. Quality of service is vital to address the problem of communications in the mobile ad-hoc network. This study analyzes the quality of service routing protocol babel on a mobile ad-hoc network. Test carried out using 3 scenarios, namely testing feasibility condition, self-healing, and delay, jitter, packet loss, and throughput on multihop condition. Values delay and jitter generated by the scenario testing conducted according to the standards TIPHON into the category of a very good and good. In the packet loss data capture value generated by the scenario testing conducted according to the standards TIPHON into the category of good and bad, but most of the values obtained into the bad category. At throughput data capture value generated by the scenario testing conducted according to the standards TIPHON into the category of good, medium and bad but mostly the value obtained in the category of good.

1. Introduction

Based on a survey from the Indonesian Internet Service Providers Association (APJII), internet user penetration in Indonesia reached 64.8%, where the percentage includes 143.26 million users from 262 million Indonesians. However, the majority of internet user contributors are still found in Java (55%) and Sumatra (21%). Whereas for regions outside Java and Sumatra, they still have low penetration (<10%). The unequal access to internet communication infrastructure in most parts of Indonesia can cause difficulties for people in the area to communicate. On the other hand, disasters can also cripple or destroy communication infrastructure, be they earthquakes, floods, or tsunamis, where people exposed to such hazards will find it difficult to communicate outside the affected area. One solution that can overcome these communication problems is to create a wireless communication infrastructure that can be used in rural areas or extreme conditions such as natural disasters and war as alternative communication media. One alternative form of communication is the Mobile Ad-Hoc Network (MANET).

Development of wireless communication networks Began since the discovery of Maxwell's equation which shows that data transmission can occur without the need for cables. Signals emitted in the air can be a medium for delivering information, even being a cheaper medium than cable media[1]. In reality, the use of MANET can be assumed to be a "relationship anywhere and anytime", which is why the implementation of MANET is beneficial in various fields of application such as disaster recovery, victim search and rescue, communication on the battlefield, educational needs, entertainment, robots, and sensor networks [2]. The rapid development of wireless technology, the smaller volume of batteries in

IOP Conf. Series: Materials Science and Engineering **771** (2020) 012032 doi:10.1088/1757-899X/771/1/012032

accommodating the electric current, and the development of wireless chips that are smaller but with good signal strength, Mobile Ad-Hoc Network (MANET) emerged as a result of the development of the superior technology. MANET is one of the infrastructures of the Internet OF Things[3]. By utilizing a single board computer as a node in MANET, MANET can be built as a mobile network [4]. MANET can be a collection of mobile communication devices running on inadequate infrastructure and the absence of a predetermined liaison or router [5].

Even though manet is flexible as secondary communication, the MANET network has some lack of Routing, Security and Reliability, Quality of Service, Internetworking, and Power Consumption. Nodes in MANET generally always move, existing wireless communication will often be disconnected and then reconnected. Apart from that, most of the nodes in MANET have limitations in computational capability and power source resources such as batteries. Therefore it is necessary to adjust the routing protocol for nodes in MANET. For optimal utilization, MANET requires a reliable routing algorithm so that the data packets sent by the source node can be received in full by the destination node. Adaptation to network conditions that are not too fixed can be handled by reactive, proactive, or hybrid routing methods. BABEL is included in the proactive routing protocol. Babel is a distance vector routing protocol that has fast convergence and fast route repair time performance [6]. BABEL is designed to make flexible route choices. Babel uses a feasibility condition that guarantees no routing loop. The router will ignore the renewal of the route that does not meet the feasibility condition [7]. However, despite its convergence speed, Babel has the most significant frequency of route changes than BATMAN and OLSR routing[8].

Babel is designed to make flexible route choices. Babel uses a feasibility condition that guarantees no routing loop. The router will ignore the renewal of the route that does not meet the feasibility condition [9]. Babylon can also be called a hybrid routing protocol, in the sense that it can carry routes for several network-layer protocols (IPv4 and IPv6) [6]. Quality of Service (QoS) is essential to overcome communication problems in MANET. QoS refers to the ability of networks to provide better services on specific network traffic through different technologies. The primary purpose of the QoS parameters is to provide priorities including appropriate bandwidth, control jitter, latency and correct loss characteristics [10]. Implementing QoS on an ever-changing network is a challenge. Resources owned by MANET are limited. Therefore QoS is needed to measure the quality of communication with these limited resources [11].

2. Research Method

In this study, we measure the Quality of Service in the MANET design using the Babel routing protocol that is using raspberry as nodes in the MANET network. The general process of the proposed model described in Figure 1.

Figure 1. Proposed model

3. Analysis software and hardware requirements for building MANET nodes

In this section, we will describe the analysis of software and hardware used in our research to build the MANET network. This section divided into three subsections, i.e., system planning, hardware design, and software design.

3.1. System Planning

In general, a system must have an input, process, and output. As shown in Figure 2, as input functions are data traffic, as the processing device is Single Board Computer (SBC) and Notebook with Babel routing protocol installed, and the output is Quality of Service analysis.

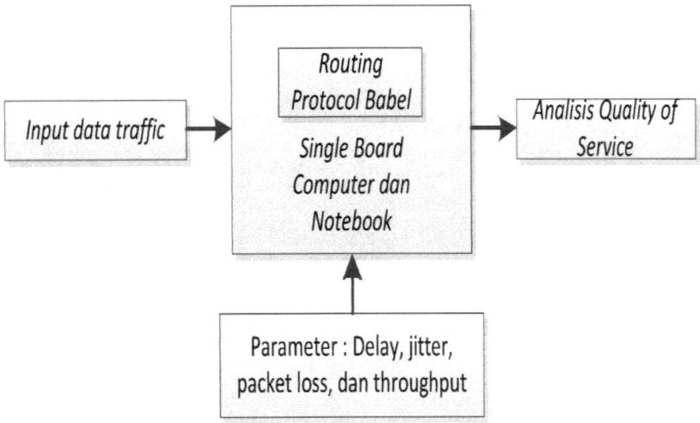

Figure 2. System planning for building MANET nodes

3.2. Hardware Design

As mentioned before, this section will describe the hardware design. The hardware used in this research are Notebooks, Single Board Computer, Micro SDHC Card, Wifi Dongle Ralink RT5370, and Power Bank. The notebook used to store data and operating systems, and the notebook also has a wireless interface that used as a transceiver, receiver, and router. Single Board Computer requires a Micro SDHC Card to store operating systems and data storage. Single Board Computer nodes are installed as a wireless device to build a MANET network that functions to connect with other nodes. Single Board Computer as a dynamic node in MANET requires resources that can take anywhere, Power Bank as resource storage has a flexible nature that can be taken anywhere so that it is suitable for use in this test.

3.3. Software Design

Software design section will describe the software used in this research, such as babel routing protocol.

3.3.1 Babel routing protocol

The process of installing the babel routing protocol requires packages that must be installed first. Babel routing protocol is suitable for use on Single Board Computers because it only requires a small CPU and memory requirements. The following algorithm from the Babel routing protocol:

- Detect all links that are connected inside the node
- Record all link information in the routing table
- The routing table owned by each router will contain the network address that is directly connected to the router.
- Periodically, each router exchanges information so that the contents of the routing table are entirely filled (converged)
- If there is a change in network topology, the router will immediately update the routing information.
- Periodically update every router.
- If the location of the router is far or remote, it will take a long time to receive information about network changes at a location.
- Tell each link connected in the node to the neighbour node.

IOP Conf. Series: Materials Science and Engineering **771** (2020) 012032 doi:10.1088/1757-899X/771/1/012032

4. Experiments and Results

4.1. Experimental Setup

After designing the MANET node, which consists of hardware and software design, then the formation of MANET, if the device was located in the wireless signal coverage area of the device, will be synchronized. This experiment conducted in three scenarios i.e., three nodes, four nodes and five nodes, which each scenario will send 2000, 2500, and 3000 bytes of data. The formation of each scenario can be seen in Figure 3.

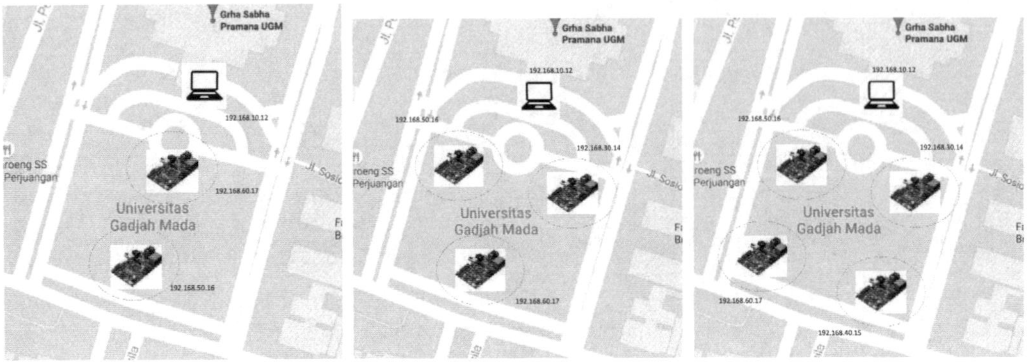

Scenario I Scenario II Scenario III

Figure 3. Block diagram of the formation of MANET

4.2. Experimental scenario

After the system is prepared, the next process is to test. Each node has a wifi dongle installed, and each node is connected to other nodes within the coverage signal range of 10 to 15 meters. Quality of Service parameters that will be observed is a delay, jitter, packet loss, and throughput. Delay is a time delay in the process of transferring packets from the source packet to its destination that passes through a series of nodes. Jitter is the variation of delay between packets that occur on an IP network. Packet loss can be defined as packet loss in the network[12]. Throughput is the rate (rate) of effective data transfer, measured in bps. This throughput can be measured after data transmission occurs[13]. Testing delay, jitter, packet loss, and throughput in a multihop condition where not all nodes are connected directly to all other nodes but must go through individual nodes that are considered as neighbours with the best link quality. In the testing of delay, jitter, packet loss, and multihop throughput conditions which is divided into three scenario, namely multihop testing with three nodes, four nodes, and five nodes. This test is essential because to see the ability of the Babel routing protocol in multihop conditions. The nodes for first scenario are node IP 192.168.10.12, node IP 192.168.50.16, and node IP 192.168.60.17, second scenario are IP node 192.168.10.12, IP node 192.168.30.14, IP node 192.168.50.16, and IP node 192.168.60.17, and the third scenario are IP node 192.168.10.12, IP node 192.168.30.14, IP node 192.168.30.14, IP node 192.168.50.16, and IP node 192.168.60.17.

4.3. Experiments results and Analysis

The results of this research based on the TIPHON standard is obtained from the analysis process on the results of the average UDP (User Datagram Protocol) data values that have been given varying network loads using predetermined network topology scenarios. It is known that the TIPHON (Telecommunications and Internet Protocol Harmonization Over Networks) standard has a standard with QoS parameters as a whole[14]. The results of the analysis are in Table 1, which has been obtained from the tests that have been carried out compared with the TIPHON standard provisions so that conclusions can be drawn from the tests that have been done. In Table 1, it can be seen that several parameters are included in the category of very good and good according to TIPHON standards.

Table 1. Quality of Service results with four-parameter (Delay, Jitter, Packet Loss, and Throughput).

Parameter QoS	Scenario	*Workload (byte)*	Average results	*TIPHON Standart*
Delay (ms)	1 (3 nodes)	2000	3.53	Very Good
		2500	4.76	Very Good
		3000	4.49	Very Good
	2 (4 nodes)	2000	3.77	Very Good
		2500	4.63	Very Good
		3000	6.53	Very Good
	3 (5 nodes)	2000	3.83	Very Good
		2500	4.93	Very Good
		3000	10.63	Very Good
Jitter (ms)	1 (3 node)	2000	4.67	Good
		2500	5.06	Good
		3000	6.43	Good
	2 (4 nodes)	2000	4.70	Good
		2500	5.06	Good
		3000	6.50	Good
	3 (5 nodes)	2000	6.67	Good
		2500	13.10	Good
		3000	20.28	Good
Packet loss (%)	1 (3 node)	2000	3.85	Good
		2500	13.86	Good
		3000	25.39	Bad
	2 (4 nodes)	2000	7.95	Good
		2500	35.14	Bad
		3000	45.40	Bad
	3 (5 nodes)	2000	35.25	Bad
		2500	66.60	Bad
		3000	74.05	Bad
Throughput (%)	1 (3 nodes)	2000	80.58	Good
		2500	60.88	Medium
		3000	50.28	Medium
	2 (4 nodes)	2000	47.66	Bad
		2500	59.16	Good
		3000	79.98	Good
	3 (5 nodes)	2000	56.38	Good
		2500	39.28	Bad
		3000	35.32	Bad

The comparison results shown in Table 1 is the Babel QoS routing protocol on the Mobile Ad-hoc Network with the TIPHON standard; the results are as follows:

- In scenarios 3, 4, and 5 nodes with a load of 2000 bytes, 2500 bytes, and 3000 bytes, the delay is included in the very good category.
- In scenarios 3, 4, and 5 nodes with a load of 2000 bytes, 2500 bytes, and 3000 bytes, jitter is included in the good category.
- In the scenario of 3 nodes with a load of 2000 bytes and 2500 bytes, packet loss is included in the good category. In the scenario of 3 nodes with a load of 3000 bytes, packet loss is included in the ugly category. In the scenario of 4 nodes with a load of 2000 bytes, packet loss is included in the good category. In the scenario of 4 nodes with a load of 2500 bytes and 3000 bytes, packet

loss is included in the ugly category. In the scenario of 5 nodes with a load of 2000 bytes, 2500 bytes, and 3000 bytes, packet loss is included in the ugly category.

- In the scenario of 3 nodes with a load of 2000 bytes, throughput is included in the good category. In the scenario of 3 nodes with a load of 2500 bytes and 3000 bytes, throughput is included in the medium category. In a scenario of 4 nodes with a load of 2000 bytes, throughput is included in the ugly category. In the scenario of 4 nodes with a load of 2500 bytes and 3000 bytes, throughput is included in the good category. In the scenario of 5 nodes with a load of 2000 bytes, throughput is included in the good category. In the scenario of 5 nodes with a load of 2500 bytes and 3000 bytes, throughput is included in the ugly category.

5. Conclusion

The results of the throughput, delay, jitter, and packet loss above are good because the loop-free characteristics of the babel also influence to avoid unnecessary waste of bandwidth. The characteristics of fast route repair also cause the results of good throughput, delay, jitter, and packet loss.
Following the characteristics of MANET, i.e., bandwidth limitations, links on wireless networks tend to have low capacity when compared to wired networks. The effect that occurs on low capacity networks is congestion; variations will significantly influence the bad value of throughput, delay, jitter, and packet loss in traffic load and the amount of congestion in IP networks. The higher the traffic load on the network will cause, the higher the chance of congestion so that the value of throughput, delay, jitter, and packet loss will be increased. The characteristic of MANET, which has a dynamic topology, also causes the distance of connection points that are far away and requires other nodes to communicate or multihop, allowing the weak value of throughput, delay, jitter, and packet loss. Furthermore, in the MANET characteristic is autonomy, so that it burdens the node's performance. Even so, testing the actual conditions, of course, various other factors can affect the reduced throughput, delay, jitter, and packet loss obtained, especially concerning the tools and conditions of the test site.

6. References

[1] W. Stallings, *Wireless Communications and Networking*. 2005.
[2] J. R, “Mobile Ad Hoc Network,” *J. Terap. Teknol. Inf.*, pp. 601–607, 2011.
[3] F. Bonomi, R. Milito, J. Zhu, and S. Addepalli, “Fog Computing and Its Role in the Internet of Things,” vol. 231, no. 1. pp. 266–277, 2017.
[4] M. Affandes, “[2013] Muhammad Affandes - SNTIKI 5.” 2013.
[5] P. Aggarwal and H. Aggarwal, “Comparative analysis of routing protocols in MANETS,” 2014.
[6] J. Chroboczek, “1/29/13 RFC 6126 - The Babel Routing Protocol,” no. April 2011. pp. 1–90, 2016.
[7] K. Vareldzhian and V. Fitsov, “Babel Routing Protocol Research using Quagga Software Router.” pp. 118–123.
[8] M. Abolhasan, B. Hagelstein, and J. C. P. Wang, “Real-world performance of current proactive multi-hop mesh protocols,” *2009 15th Asia-Pacific Conf. Commun. APCC 2009*, no. October, pp. 44–47, 2009.
[9] J. Nguyen and W. Yu, “An SDN-based approach to support dynamic operations of multi-domain heterogeneous MANETs,” in *Proceedings - 2018 IEEE/ACIS 19th International Conference on Software Engineering, Artificial Intelligence, Networking and Parallel/Distributed Computing, SNPD 2018*, 2018.
[10] Cisco, *Cisco 3200 Series Mobile Access Router*, no. 6387. 2004.
[11] P. Latha and R. Ramachandran, “QoS Constrained Multicast Routing For Mobile Ad Hoc Networks,” *J. Comput. Sci.*, vol. 9, no. 7, pp. 66–70, 2009.
[12] J. F. Kurose, *Computer Networking - A Top-*. .
[13] A. A. Abror, M. Z. S. Hadi, I. Winarno, M. Politeknik, E. Negeri, and J. T. Telekomunikasi, “Rancang Bangun Dan Analisa Qos Audio Dan Video,” pp. 1–9, 2015.
[14] ETSI, “Tr 101 329,” 1999.

IOP Conf. Series: Materials Science and Engineering **771** (2020) 012033 doi:10.1088/1757-899X/771/1/012033

Hanoman adventure game based on Android

Sigit Sugiyanto[1,*], Achmad Fauzan[1], Lahan Adi Purwanto[1], Dimara Kusuma Hakim[1]

[1]Department of Informatics Engineering, Universitas Muhammadiyah Purwokerto, Jawa Tengah, Indonesia.

[*]sigitsugiyanto@ump.ac.id,

Abstract. The use of smartphones to play mobile games is increasingly being used by everyone from children to adults. Along with the rapid development of the game industry, game engine software is now popping up, one of which is Construct 2. Construct 2 is an HTML5-based game maker software that is devoted to the 2D platform. Games can be a medium for cultural preservation in a country. One of them is Indonesian culture that needs to be preserved is wayang culture. This adventure game will raise a cultural puppet character named Hanoman, Hanoman is a figure in the form of a white monkey who acts as a character in helping Rama to save Shinta. This Hanoman adventure game in carrying out missions on each level will face the enemy characters who must be defeated to be able to open the next level. In this game, there are 3 levels where each level has different difficulties so that it can provide challenges for its users.

1. Introduction

A smartphone is a communication tool that is not only used for telephones or sending text messages but also has other supporting features such as games, music, internet, and others. The use of smartphones to play mobile games is increasingly being used by everyone from children to adults. Games are all forms of activities that require thinking, intellectual agility and achievement of certain targets. [1]. Games are also a participatory, or interactive, entertainment [2].

The development of the game industry in Indonesia has developed very rapidly and has various types of game variations, one of which is an Android-based game. The number of game applications on smartphones with the Android operating system because it provides an open platform for developers to create their own applications. Compared to other operating systems on smartphones, the Android operating system most supports developers to create their own applications [3].

Creating a game application requires game engine software to make it easier to build a game. The game engine is the nuts and bolts that are responsible behind the making of each video game. Starting from graphic arts to mathematical calculations that determine each frame on the screen is determined by the engine. Starting from the rendering process - the process of displaying graphics on the screen, and integrating control methods and rules that will be applied in the game - engine is something that is made by developers to make a game [4]. The game engine software that will be used in building Construct 2, which functions as a powerful and very user-friendly application. This application has been around for more than 5 years, has been downloaded more than 3.5 million times. And has an active community and responsive development team. The output from Construct 2 can also be exported to desktop PCs, Macs and Linux using Node Webkit. [5].

Games can be a medium of cultural preservation in a country, one of which is an adventure game. The adventure game is an adventure game that tells the journey of a character [6]. This adventure game will raise a puppet culture figure named Hanoman, who is a tangible white ape character who acts as a character in helping Rama to save Shinta. Currently, there is a game that raises the character of the Hanoman puppet, but the game is still very simple in appearance, action, obstacles and there is no element of cultural delivery in it. With the several Hanoman themed adventure games above, a better Hanoman adventure game will be developed in which there are attractive designs, more challenging games and cultural elements. This Hanoman Adventure Game consists of 3 levels, namely wilderness, Sempati beach, and Alengka royal park in which there are enemy characters that must be defeated to complete missions at each level.

2. Method

2.1 Multimedia Development Life Cycle (MDLC)

This research method in building Android-based Hanuman adventure games uses the Multimedia Development Life Cycle (MDLC) method, which consists of 6 stages: concept, design, material collecting, assembly, testing, and distribution [7]. Figure 1 is the MDLC stages as follows:

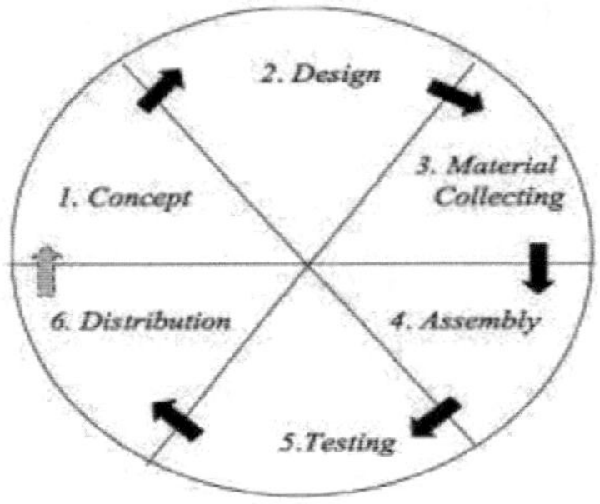

Figure 1. Multimedia Development Life Cycle (MDLC)

2.2 Flowchart Game

Builds an Android-based Hanuman adventure game requires a flowchart to guide the game workflow before the game will be built and implemented. Flowchart system workflow when the user in playing the game Hanoman Adventure, starting from the user taking action by pressing the Start button then the initial level will appear to the final level. The following picture is a flowchart:

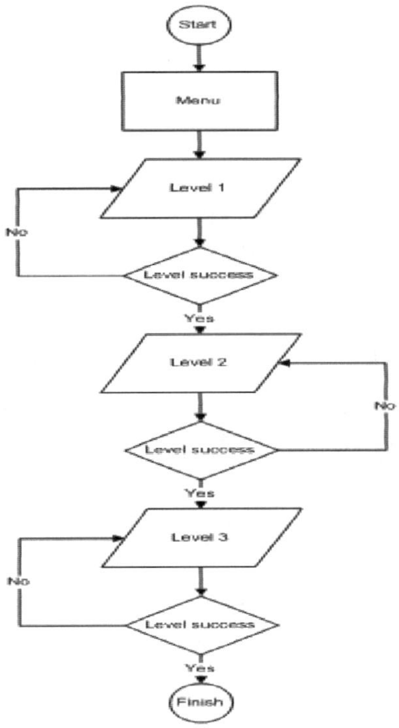

Figure 2. Flowchart of Hanoman adventure game

2.3 Design Character

In the process of designing the characters that become the reference is the story script which then records the needs of any character that will appear in the story. In this case, the writer is in charge of making eight central characters. After knowing what characters will be made, the next process is to study the character description either through literature, existing studies from similar films or historians. After knowing the description of each character, the next process is to determine the appropriate archetype to strengthen the character's personality [8]. The types of character design in building a game are as follows [9] :

1. Player Character
    a. A player or several players control the character.
    b. In some games one player controls one character, several one players control the character (strategy, sport)
2. Non-Player Character
    a. Divided into allies, facilitators, and decorations
    b. Showing limited reaction, but generally not hostile except in provocation
    c. Examples of sports: cheerleaders, judges, announcers, etc.
3. Enemies
    a. Divided into rivals, aggressive enemies, and passive elements
    b. An NPC that prevents players from winning the game
    c. The enemy can be smart and cunning, or aggressive
4.Finally Boz
    a. Included in the character of enemies
    b. Has a strong character, smarter and more difficult to beat.

3. Result and Discussion

In building an Android-based Hanuman adventure game with the Multimedia Development Life Cycle (MDLC) method which will be explained in detail at each stage of concept, design, material collecting, assembly, testing and distribution.

3.1 concept

This Hanoman adventure game is an adventure game that emphasizes the player's reaction. Adventure while fighting is an action in this genre. The blow-kick action, the use of swords and several other tools become entertainment for players. Opponent players can be anything and usually quite a lot. Judging from the actions that are quite a lot this requires quite a lot of control too [10]. This game has a 2D action element that tells the story of a puppet character, Hanoman, who was assigned to rescue Dewi Shinta who was kidnapped by Rahwana. In building Hanoman adventure games using Construct2 which is a game engine to create game applications especially for 2D5 based HTML5 games, which in calling functions in Construct 2 only uses the Events settings that have been provided. Events are choices of actions and conditions that will become lives in the game and the game can run as we want [11].

The storyline of this adventure game does not use the high score system because this adventure game uses a story or story model in which there are missions at each level. This adventure game has 3 levels namely wilderness, Sempati beach, and Alengka royal park. At the beginning of opening the Hanoman adventure game will display loading, when loading is complete it will be directly connected to the main menu display (main menu) main menu display there are several buttons namely the start button, how to play, settings, and exit. The button starts directing the player directly into the game level that is started by the prologue scene.

3.2 Design

3.2.1 Use Case Diagram

Use Case diagrams are used to make it easier to design applications because the usecase diagram is clearly visible when the user performs an application, the following is a Use Case diagram in the game Hanoman Adventure:

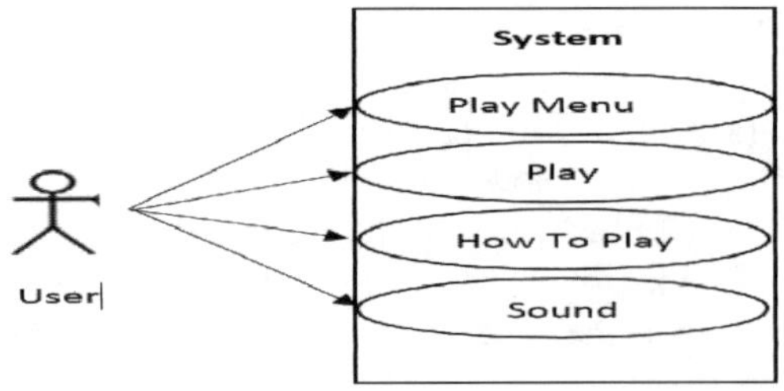

Figure 3. Use Case Diagram

3.2.2 Storyboard

Storyboard illustrates a scenario that is made in stages, that is every scene in the game. This storyboard will explain the composition of the game material created. The following are the scene tables in the Hanoman Adventure game:

Table 1. Storyboard

Scene	Menu Name	Information
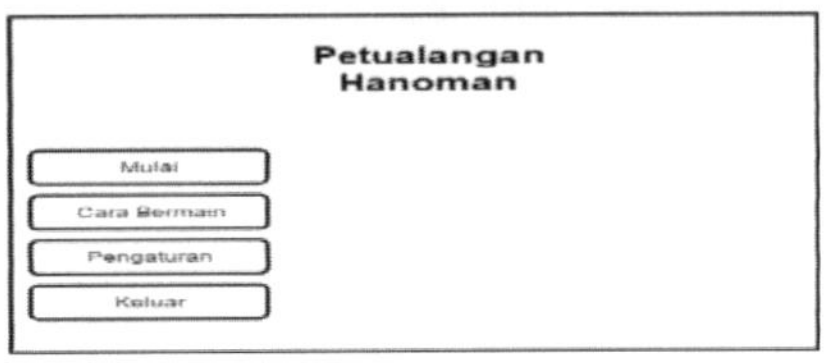	Playing Menu	Display the main menu of the Hanoman Adventure game
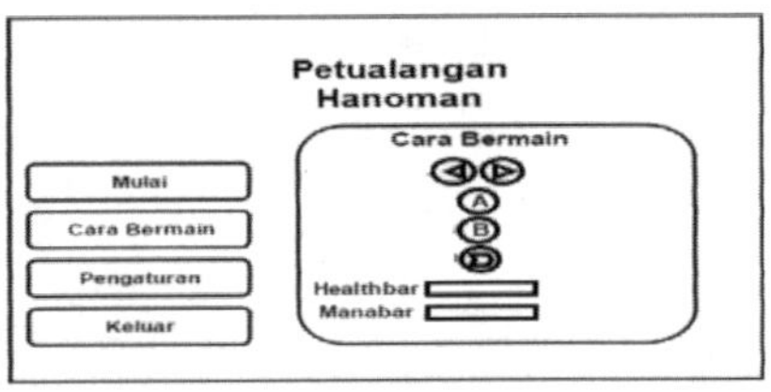	How to play	Showing how to play the Hanoman Adventure game
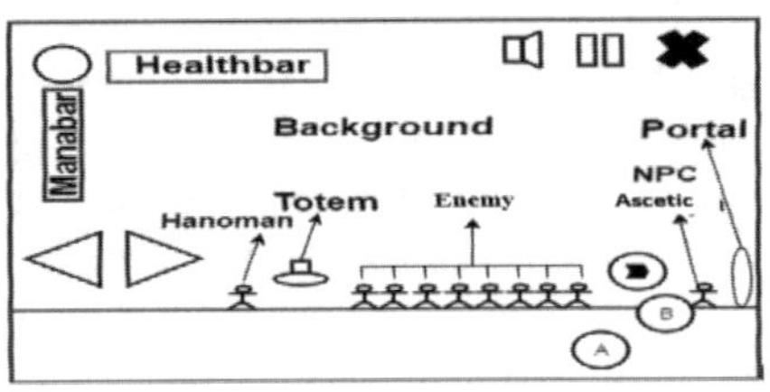	The mission of *Level* 1, Level 2 and Level 3	Featuring the main actor Hanoman to be able to complete each mission traversed from scenes level 1, level 2 and level 3 to save Dewi Shinta.

3.2.3 Design Character

The process of character design in Hanoman adventure game is made with the characters in the direction of the anime design character if the original character of the puppet will be less interesting because the character will look old-fashioned. Anime is a key part of popular visual culture. In the midst of the huge role of the mass media (society) in Indonesian society, anime and manga become an inseparable part of people's lives which are currently very visual. Anime plays an important role in the formation of global mediascape, both print and electronic [12]. Next is the character design of the Hanoman adventure game:

3.2.3.1 Main Player Karakter

The Hanoman Adventure Game uses one main character, Hanoman, who is a clever, strong, and clever white ape. Hanoman characters are made as attractive as possible so that they can be accepted by all ages, both children and adults. (Figure 4) is a Hanoman character design in the Hanoman Adventure game:

Figure 4. Hanoman character design

3.2.3.2 Non-Playable Character (NPC)

NPCs are characters that cannot be played by users. In the Hanoman Adventure game, there are two NPCs whose role is to provide information to the user. The NPCs are ascetics, Sempati and Dewi Shinta. Here is an NPC in the Hanoman Adventure game. Next is the NPC character design:

Figure 5. Design of Non-Playable Character

3.2.3.3 Enemy

The enemy is a character whose role is to prevent the user from achieving goals in a game. In the Hanoman Adventure game, there are five types of enemies namely Archer, Orc, Troll, Dragon and Knight. Here are pictures of enemies in the Hanoman Adventure game:

Figure 6. Design of Enemy Character

3.2.3.4 Final Boss

The final boss is the boss who is near the end of the game, with the completion of the storyline the game usually follows victory in battle. The final boss in this Hanoman Adventure game is Ravana. Rahwana is the king of the Alengka kingdom who has a red body figure and long hair. Rahwana has a weapon in the form of a deep red sword called the Chandrahasa sword. The character Ravana is made as attractive as possible so that it can be accepted by all ages both children and adults. (Figure 7) is a Hanoman character design in the Hanoman Adventure game:

Figure 7. Design of Rahwana Character

3.3 Material Collecting

Material collecting is the stage used to collect materials according to the needs carried out. The materials needed are as follows (Table 2):

Table 2 Material collecting

Game Ingredients	Type
Text	Fellowmaiden, Pristina, Impact
Image	Button, UI, Sprite, Parralax, Tileset
Sound	Ogg, Wav

3.4 Assembly

The assembly stage is the stage of making all objects or multimedia material. Making an application based on the design stage, such as storyboards, flowcharts, navigation structures as follows :

3.4.1 Scene Menu

The Scene Menu will display the main menu in the Hanoman Adventure game which helps the user to start the game consisting of start, how to play, settings, and game exit buttons. Here (Figure 8) is a picture from the scene menu:

Figure 8. Scene Menu

3.4.2 Scene How to Play

This scene serves to show the user how to play the Hanoman game. It can be seen in Figure 9 that in this scene will display a dialogue box instructions on how to play the Hanoman Adventure game and there is a cross button to close the dialogue box.

Figure 9. Scene how to play

3.4.3 Scene Game Level

The game level scene is a scene in which there is a mission that Hanomun must complete to find the whereabouts of the Goddess. At each level that will be taken to have different difficulties in completing its mission. Each can complete the mission at each game level, it will be able to know the whereabouts of the Goddess Shinta. In the left corner, there is a picture of the health bar and manabar that indicate the life and indication of the style of the Hanoman. If Healthbar reaches zero then Hanoman will repeat from the beginning of the

Scene level being pursued. Manabar serves as an indication of the style, if Manabar is full then Hanoman can release a style by pressing the shoot button. In this scene there are 8 buttons namely pause, mute, exit, right, left, A button, B button and Shoot button.

Figure 10. Scene Game Level

3.5 Testing

At this stage the black box method is tested for functions in the game by running each menu and button to see whether there are errors or not. Table 3 shows the results of the black box testing:

Table 3. Black Box Testing

Testing Name	Input	Output	Status
starting the *game*	press the start button	starting game	ok
how to play the game	press the how to play button	display the how to play dialog box	ok
settings	press the setting button	display the setting dialog box	ok
mute sound	push the mute button	sound does not appear	ok
exit game	push the exit button	exit the game	ok
player movement	push the right button	player moves to the right	ok
player movement	push the left button	player moves to the left	ok
player movement	push the 'a' button	player jumps	ok
player movement	push the 'b' button	player hits	ok
player movement	push the kick button	player make a move/power	ok

3.6 Distribution

At this stage, the Hanoman Adventure game project has been completed then the build process is made into a .apk file (android file), the Hanoman Adventure game is ready to be published for users.

4. Conclusion

The results of research carried out in building Hanoman adventure games with the game engine Construct 2 will make it easier to make the game look and very user-friendly and in making Hanoman adventure game characters adopt characters into anime designs, with anime character designs able to attract the attention of many people from among children and even adults so that this game application will be more interesting and fun to play. This game is included as an adventure genre game that elevates puppet characters in Indonesian culture. With this game in addition to playing also in it can provide elements of cultural delivery, in the form of Hanoman figures who help

Raja Rama to find Goddess Shinta who has been kidnapped by King Rahwana. The storyline at each level of the game is made according to the story of Hanoman's character searching for Dewi Shinta.

References

[1] A. Ismail, *Education Games* Yogyakarta: Pro U Media, 2009.
[2] A. Rollings, and E. Adams, *Game Design*. USA: New Riders Publishing, 2003
[3] H. Syariati, *Makalah Sistem Operasi Android* Kuningan: Fakultas Ilmu Komputer Universitas Kuningan, 2012.
[4] W. Goldstone, *Unity Game Development Essentials*. Birmingham: Packt Publishing Ltd, 2009
[5] L. Stemkoski, and E. Leider, *Game Development with Construct* Springer Science New York, 2017
[6] T. Dillon, *Adventure Game for Learning and Story Tellin,* Future Lab the UK, 2005
]7] I. Binanto, *Multimedia Digital Dasar teori dan Pengembangannya* Yogyakarta: Andi, 2010
[8] C. Boeree, G. Carl Jung, *Archived from the original* on 6 February 2006. [Accessed:09-Sep-2019]
[9] L. Pardew, *Beginning Illustration and Storyboarding For Game,* Thomson Course Technolog PTR, 2005
[10] I. C. Sibero, *Langkah Mudah Membuat Game 3D,* Mediakom Yogyakarta, 2009
[11] Scirra, *Construct 2* http://www.scirra.com. [Accessed: 09-Sep-2019]
[12] M. William, *Japanese Visual Culture,* New York: M.E. Sharpe, 2008

Design of Pond Water Quality Monitoring System Based on Internet of Things and Pond Fish Market in Real-Time to Support the Industrial Revolution 4.0

A Junaidi[1,*] and C Kartiko[2]

[1] Department of Informatics, Institut Teknologi Telkom Purwokerto, Indonesia
[2] Department of Software Engineering, Institut Teknologi Telkom Purwokerto, Indonesia

*apri@ittelkom-pwt.ac.id, condro.kartiko@ittelkom-pwt.ac.id

Abstract. Pond cultivation is one of the contributors to foreign exchange and employment in Indonesia. The results of pond aquaculture production in Indonesia, according to statistical data in 2016, amounted to 3,012,467 tons, and the area of fish ponds in Indonesia in 2016 was 674,135 ha. The factors that lead to crop failure are illness, theft, poor water quality, and natural disasters such as floods. This research provides a solution called the Internet of Things Ponds (I-Tamb). I-Tamb applies the Internet of Things (IoT) technology to monitor water quality and the pond environment by installing several sensors. This study also aims to reduce crop failure and make it easier for farmers to market their crops. The cloud server receives sensor data in real-time. I-Tamb Apps produces information that can be used by farmer farmers as material for decision making. I-Tamb also facilitates the sale of crops in the form of a marketplace called the I-Tamb Marketplace. The output of this study is the design of an IoT-based pond water quality monitoring system on mobile and web applications.

1. Introduction

Ponds are an average artificial pond in the coastal area for aquaculture. Ponds are used to saving, grow and harvest results in a controlled environment [1]. Pond cultivation is one of the contributors to foreign exchange and employment in Indonesia. The results of pond aquaculture production in Indonesia, according to statistical data in 2016, amounted to 3,012,467 tons, and the area of fish ponds in Indonesia in 2016 was 674,135 ha. Activities to produce aquatic plants and animal species are called aquaculture. Field activities consist of Training, Knowledge, and Techniques for Breeding aquatic plants and several animal species. This activity has a significant interest in economic development and food production. Continuous monitoring of the physical, chemical and biological parameters of pond water helps not only to predict and control adverse aquaculture conditions but also to prevent environmental damage and the collapse of the production process. Monitoring physical and chemical variables such as oxygen, temperature, and pH in water is very important for adequate safety requirements and avoiding unwanted problems that can cause the collapse of the aquaculture system [2].

Water quality is the sum of all the physical, chemical, biological, and aesthetic characteristics of water that affect its use. Technology can be used to measure all the physical, chemical, biological, and aesthetic aspects of water. Any water characteristics in a production system that affect the survival, reproduction, growth, and production of aquaculture species, influence management decisions, cause environmental impacts or reduce product quality and safety can be considered as water quality variables [3].

Failure to harvest, which is often experienced by pond farmers, is one indication of the occurrence of water quality degradation [4]. Based on [5] the factors that lead to crop failure are illness, theft, poor water quality, and natural disasters such as floods [6]. The influence of the state of water is indeed significant because there are several influential water parameters such as water temperature, oxygen content, pH, and salinity [7]. This requires special attention to be able to increase production in each pond in several ways, one of them by paying attention to water quality. Poor water quality affects the slow growth of fish. In the aquaculture business, the availability of water and water quality is one of the factors that determine success in the fish farming business [8].

Internet of things (IoT) can optimize several tools such as media sensors, radio frequency identification (RFID), wireless sensor networks, and other smart objects that allow humans to interact with all equipment connected through the internet network easily [9]. IoT can help in many ways, such as monitoring water quality and environmental health [10].

The use of technology in aquaculture ponds provides breakthroughs to monitor and control pond water quality, such as by installing several sensors around the lake. The sensor gets the data then is processed into information about the pond water quality. With IoT technology that can be used by pond farmers, infrastructure support such as the internet, solar technology to support electricity energy-saving programs. Some sensors that can be used to measure pond water quality, such as Dissolved Oxygen (DO) [11] or also called dissolved oxygen, Potential Hydrogen (pH) [12], water temperature, soil temperature, salinity, CO2, ammonia, nitrite, nitrate, orthophosphate, plankton abundance, and soil texture [13].

2. The Design

This research has prospects to increase market demand, improve pond environmental control, reduce production costs, improve crop quality. I-Tamb was developed based on IoT, which uses a realtime database so that it can provide information about the state of water quality through several sensor nodes installed around the pond. Besides, I-Tamb has prospects for fishpond owners as well as the ease of getting information through mobile phones and websites.

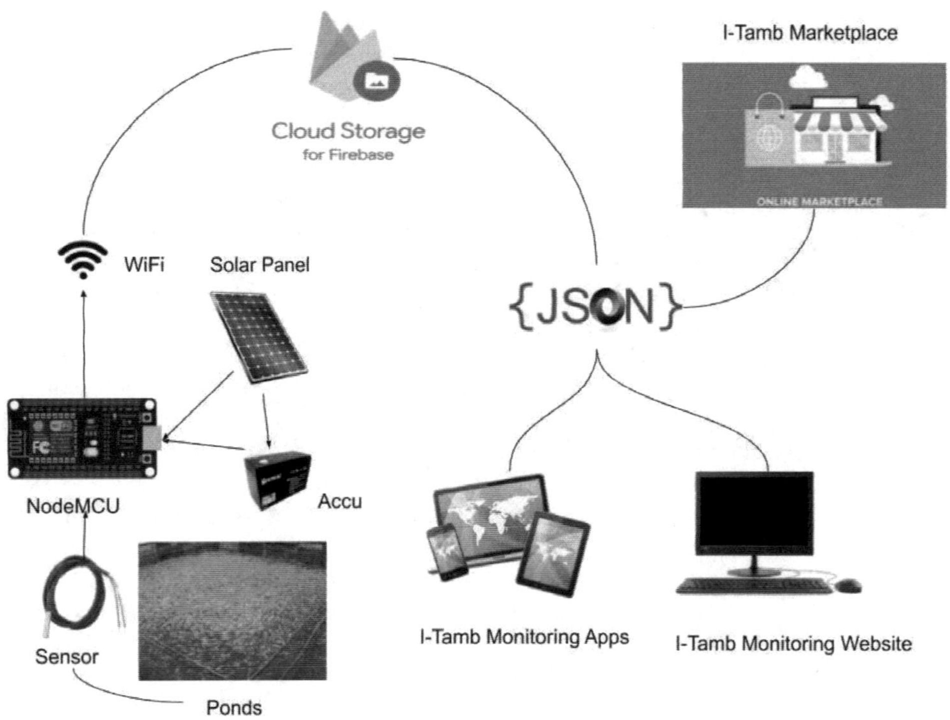

Figure 1. I-Tamb Framework

Based on Figure 1. can be explained the workings of the I-Tamb Framework: the data obtained from sensors around the pond is sent to the cloud database, in real-time using WiFi. WiFi is chosen as the preferred communication technology. WiFi is an IEEE 802.11 standard for wireless local area networks (WLAN) [14]. The cloud database used is Firebase, a Google-owned multi-service cloud-computing solution for mobile and web developers. The service includes a real-time NoSQL database management system that enables fast queries execution. Each database is consolidated in a single JavaScript Object Notation (JSON) file, where tabular relations are used rather than traditional relational databases. This allowed highly-responsive data synchronization across system modules with latency in the range of milliseconds. Note that the use of Firebase does not ensure connection reliability, rather, real-time data transmission/retrieval is procured when a stable WiFi connection is available. Furthermore, these data produce information about water quality. Farm owners can access data through the smartphone as material for decision making. The website can also manage data. Marketplace applications can accommodate the marketing of ponds so that people can access the harvest directly. This marketplace application can make buying and selling transactions wherever they are. The series of industrial revolution 4.0 processes in this study is censorship collecting pond data and then generating data management through the I-Tamb application and accumulation of pond crop sales in the marketplace.

Synergy is carried out with the Software Development Life Cycle (SDLC) software development method and explains the problem through data retrieval from various predetermined sources. The research flow diagram, in terms of developing mobile and web applications, is illustrated in Figure 2 :

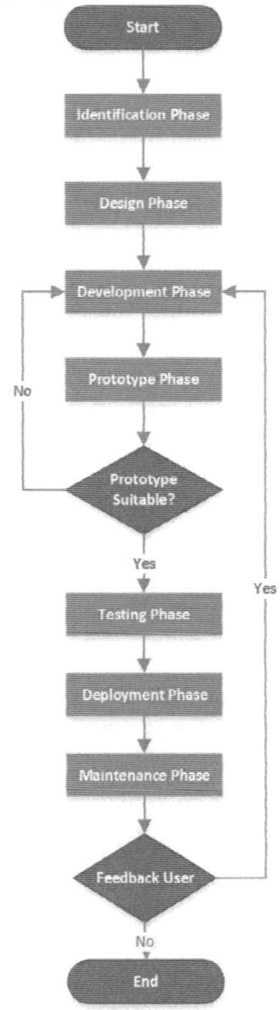

Figure 2. Research Flowchart

2.1. Identification Phase

In this phase, the developer collects ideas that come from potential users or developers themselves. The idea is analyzed and described. Look for applications that are considered the same, then study, compare, and document the results of the comparison. The developer also determines the time needed to develop the application.

2.2. Design Phase
In this phase, the developer determines what platform to use. Furthermore, the developer specifies the release of the application as a free or paid version. Then, the developer defines the functions used as modules and prototypes. The developer plans the functionality and architecture requirements. The most important part of this design phase is that the developer creates a storyboard for user interface interactions. The developer carries out the coding process after the documentation results in the design phase are complete.

2.3. Development Phase
In this phase, the developer does the coding. The first coding is done in parallel and refers to the design documentation.

2.4. Prototype Phase
This stage analyses each functional prototype requirement; the prototype is tested and sent to prospective users for feedback. After feedback, the changes needed are back in the development phase. When the second prototype is ready, integrate with the prototype, then test it and send it back to the prospective user. This stage is the phase that iterates through prototyping and testing until the final prototype is ready.

2.5. Testing Phase
In this phase, it is one of the most critical aspects of each development life cycle model. This test is carried out on an emulator or simulator, followed by testing on a real device. The testing phase uses an actual method, usually done using several versions of the operating system and some android specifications. All test results need documentation.

2.6. Deployment Phase
This phase discusses the stages of deploying applications. Before researchers distribute applications to users, researchers need to prepare application release documents. The most important thing to do in this preparation process is [15]:

1. Gathering Materials and Resources
   Researchers prepare cryptographic keys, this key is a certificate owned by the developer. Next is to make sure the icons used to meet the suggested guidelines. The next process is the end-user license agreement (EULA), this process can help protect individual, organizational and intellectual property developers.
2. Configuring Applications for Release
   The researcher chooses the package name that is suitable for the application's lifetime, then turns off logging and debugging. Researchers clean the project directory; this process aims to facilitate the compilation process.
3. Building Application
   Researchers integrate with Android Studio to build APK files that are ready for release.
4. Test the Application
   Make sure the user interface elements have the right size, performance, and battery efficiency. The stages in testing are listed in [16].
5. Releasing Application
   After the application passes the application release preparation process, the researcher distributes the application by uploading the application that is ready for Google Playstore.

2.7. Maintenance Phase
This phase, the application has been uploaded to Google Playstore. After the user downloads the app, the user conducts a review of the app by rating and commenting on the application on Google Playstore. The results of the feedback are to make improvements and updates.

3. Conclusion And Future Work
The impact and benefits of this research are as follows:

1. Knowledge / Education
   Educating the public about the use of IoT in pond management.
2. Economy
   Increase the income of pond farmers and local government with the I-Tamb marketplace that makes it easy to market the farm's harvest that can be accessed directly by the public through a smartphone.
3. Technology
   I-Tamb produces data obtained through sensors mounted around the pond; this data can be processed with data mining techniques, so it supports for various business and management purposes.
4. Management
   It makes it easy to control and monitor pond aquaculture activities.

References

[1] *UNDANG-UNDANG REPUBLIK INDONESIA NOMOR 31 TAHUN 2004 TENTANG PERIKANAN,* 2004.

[2] C. Encinas, E. Ruiz, J. Cortez and A. Espinoza, "Design and implementation of a distributed IoT system for the monitoring of water quality in aquaculture," in *2017 Wireless Telecommunications Symposium (WTS)*, Chicago, 2017.

[3] C. E. Boyd and C. S. Tucker, Pond aquaculture water quality management, Springer Science & Business Media, 2012.

[4] M. N. Suparjo, "Daya Dukung Lingkungan Perairan Tambak Desa Mororejo Kabupaten Kendal," *Saintek Perikanan: Indonesian Journal of Fisheries Science and Technology,* vol. 4, no. 1, pp. 50-55, 2008.

[5] R. H. Bosma and M. C. Verdegem, "Sustainable aquaculture in ponds: principles, practices and limits," *Livestock science,* vol. 139, no. 1, pp. 58-68, 2011.

[6] K. K. Saguin, "States of hazard: Aquaculture and narratives of typhoons and floods in Laguna de Bay," *Philippine Studies: Historical and Ethnographic Viewpoints,* vol. 64, no. 3, pp. 527-554, 2016.

[7] A. Kristianto, I. Setiawan and S. Sumardi, "Pengendalian pH Air Dengan Metode PID Pada Model Tambak Udang," *Transmisi Teknik Elektro Undip,* vol. 14, no. 4, pp. 119-126, 2008.

[8] F. Mas'ud, "Pengaruh Kualitas Air Terhadap Pertumbuhan Ikan Nila (Oreochromis sp.) Di Kolam Beton Dan Terpal," *Grouper Faperik,* pp. 1-6, 2014.

[9] A. Junaidi, "Internet of Things , Sejarah , Teknologi Dan Penerapannya : Review Internet of Things , Sejarah , Teknologi Dan Penerapannya : Review," *Jitter,* pp. 62-66, 2016.

[10] M. V. Ramesh, K. V. Nibi, A. Kurup, R. Mohan, A. Aiswarya, A. Arsha and P. R. Sarang, "Water quality monitoring and waste management using IoT," in *2017 IEEE Global Humanitarian Technology Conference (GHTC)*, San Jose, 2017.

[11] W. Wang, C. Deng and X. Li, "Soft sensing of dissolved oxygen in fishpond via extreme learning machine," in *Proceeding of the 11th World Congress on Intelligent Control and Automation*, Shenyang, 2014.

[12] Y. Y. Maulana, G. Wiranto and D. Kurniawan, "Online Monitoring Kualitas Air pada Budidaya Udang berbasis WSN dan IoT," *Jurnal Informatika, Sistem Kendali, dan Komputer (INKOM),* vol. 10, no. 2, pp. 81-86, 2016.

[13] M. N. Suparjo, "Daya Dukung Lingkungan Perairan Tambak Desa Mororejo Kabupaten Kendal," *Jurnal Saintek Perikanan,* vol. 4, no. 1, pp. 50-55, 2008.

[14] C. Cordeiro, D. Akhmetov and M. Park, "IEEE 802.11Ad: Introduction and Performance Evaluation of the First Multi-gbps Wifi Technology," in *Proceedings of the 2010 ACM international workshop on mmWave communications: from circuits to networks*, New York, 2010.

IOP Conf. Series: Materials Science and Engineering **771** (2020) 012034 doi:10.1088/1757-899X/771/1/012034

[15] Google Developers, "Menyiapkan aplikasi untuk dirilis | Android Developers," Google, 2019. [Online]. Available: https://developer.android.com/studio/publish/preparing?hl=id#publishing-resources. [Accessed 10 November 2019].

[16] Google Developers, "Kualitas Aplikasi Inti | Android Developers," Google, 2019. [Online]. Available: https://developer.android.com/docs/quality-guidelines/core-app-quality?hl=id#tests. [Accessed 10 November 2019].

IOP Conf. Series: Materials Science and Engineering **771** (2020) 012035 doi:10.1088/1757-899X/771/1/012035

Build Design of Information Media Application in Purwokerto Based on Multimedia

Tito Pinandita[1,*], Silvia Deviani Saputri[1]

[1]Informatic Engineering Department of Universitas Muhammadiyah Purwokerto
Jl. Raya Dukuh Waluh Kembaran, Purwokerto, Jawa Tengah, Indonesia, 53182

[*]e-mail: titop04@gmail.com

Abstract. Transportation is a major problem for newcomers in certain places. Types of transportation and trails make newcomers confused when they want to visit a certain place. There are several media that can solve the problem, but the limitations of the media make it difficult for newcomers to find the right track in the city, for example in Purwokerto city. One of the media that can be used is multimedia-based media. In this case, it can be created using Adobe Flash CS5. This software can be used to create animations about urban transport vehicle lines in Purwokerto. Multimedia applications make users know information that is accurate, precise, fast and interesting.

•

1. Introduction

Information technology can help us to provide fast, precise, accurate and attractively. In this case, the use of media to support multimedia applications very user as a means of delivering information. This multimedia application allows users to get information quickly and accurately.

The terminal is a bus after cessation and will operate[1]. Typically, the first time in a city, for example, Purwokerto and do not understand where the actual dismissal, must always choose the terminal as a place to stop. Not infrequently, the new people who come to Purwokerto confusion regarding the place to be visited. So that the person will ask other people, who sometimes do not know or many other reasons. If someone wants to get to the next place after the terminal, many people do not know what to use transportation like, as well as the way to go about anything.

In this case, the information needs are felt, which is faced with a situation that is filled with the urgent need to decide on a policy, in order to travel as planned. In this case, the multimedia application is needed, because by using the application display which highlighted the route that will be passed can be seen easily. Therefore, it should be made Media City Transport Information[4]-Based Multimedia Purwokerto.

2. Methods

The method of research in making applications Purwokerto city transportation information media multimedia based are as follows[2]:

1. Data Collection
    a. Interview
    b. Literature review

2. Design System Structure [3].
   a. Conceptual Design
      Here is a flowchart of this application (Figure 1).

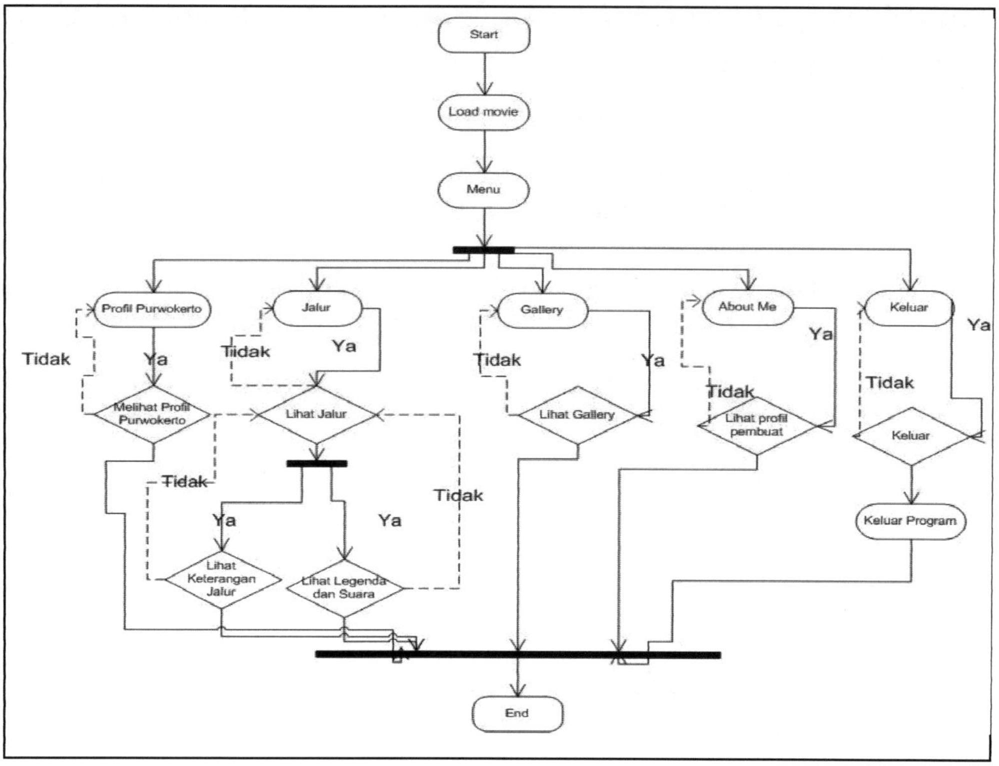

Figure 1. Flowchart Transport Applications These Purwokerto

   b. Physical design
      Design Image Shape Design About any menu that will be in this application that public transportation [5] will be built.
3. Making Applications
4. Testing
5. Implementation

3. Results and Discussion

3.1. Hardware and Software specifications.

One unit of the computer used as a tool to use application with minimum specifications as follows:

a. Hardware
- Intel Core Duo Processor T6600 2.2GHz
- RAM: 1GB
- HDD 160 GB
- Monitor LED / LCD

b. Software
- Adobe Flash Professional CS5 that has been published with an .exe file format.

3.2. User Interface

Application of this system is implemented in the form of interface, such as the following pictures:

a. Loading Form. This form is shown in Figure 2. In this form, it only shows a simple animation before heading into the user menu.

Figure 2. Form Image Loading

b. Form Home, this form is shown in Figure 3. In this form, the user will be shown an animation in which there is a public transportation driver who is talking and there is a menu button to get to the next menu

Figure 3. Form Home

c. Form Menu. This form is shown in Figure 4. In this form, the user can view and select the menu on this program. The program has three menus that each menu item will present information on the menu that have been selected.

Figure 4. Form Main Menu

d. Form Menu Profile. This form is shown on Figure 5. On the menu this profile, the user can see at a glance about profile Purwokerto

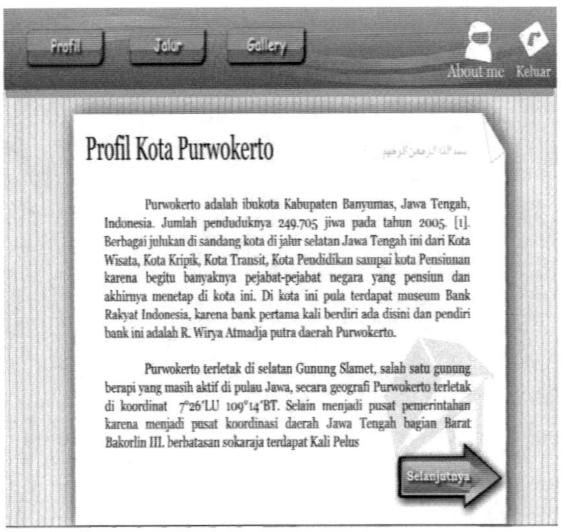

Figure 5. Form Menu Profile

e. Form Menu Line. This form is shown in Figure 6. In this form, the user can choose the path that has been provided. Each cursor is pointing to each track will be shown the route to be traversed by the freight. If the user has determined will be the means of transport which, users just click on that path

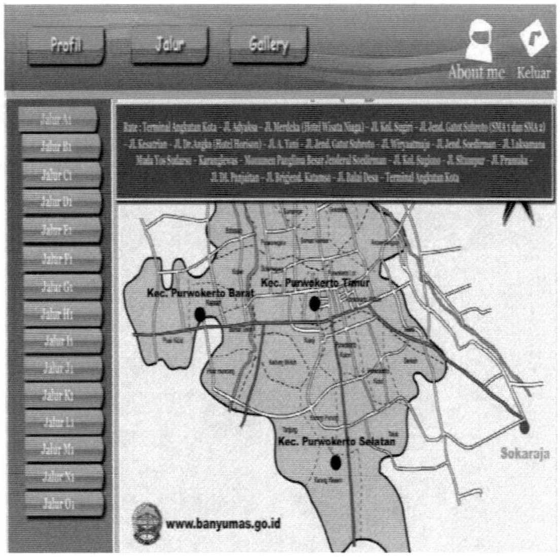

Figure 6. Menu Line.

f. Form menu each selected path is shown in Figure 7. The menu of each of these paths the intention is to have 16 Purwokerto city transit lines, where each track and city transportation have different routes, the program has provided a choice of paths A1 to P1. Where the user can select one of the lines to see the route of the city transport chooses. In the picture below, will be shown in one of the lines, ie Line A1. Route of city transport A1 indicated by a yellow stripe and there can also be seen there is animation car that is moving through the route.

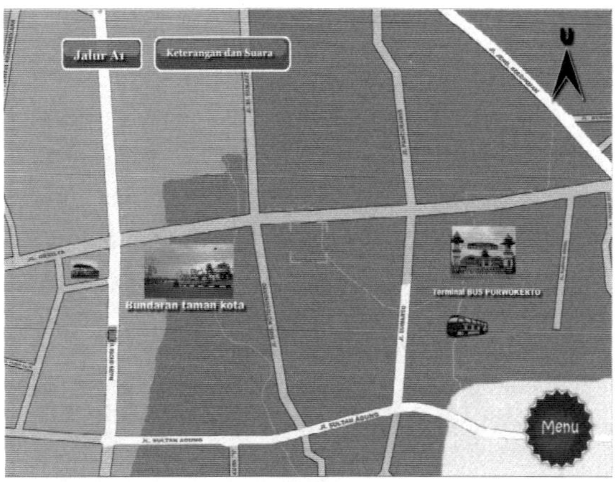

Figure 7. Form menu of every stripe

g. Form of each of the selected path. In the picture above can be seen that there are 16 menu button Line A1 (the selected path) and the menu button Description and Voice. Here is a screenshot of the Line Description and Specification and Sound. Shown in Figure 8 and Figure 9.

Figure 8. Description of the Form The selected route of Line

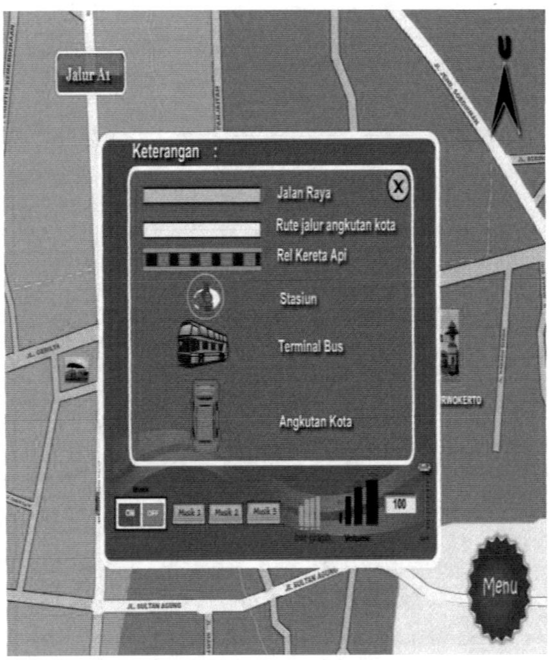

Figure 9. Form regarding the information on maps and sound settings

h. Form Gallery. In the form shown in Figure 10 contains about drawings or photographs for a map of Purwokerto city[6]. So that visitors to the city of Purwokerto can imagine how the state of it. The image also comes with a description of the place name.

3.3. Due Diligence Questionnaire Results

After performing the division and filling the questionnaire, then to calculate the value of the average scores used the formula described previously, namely:

IOP Conf. Series: Materials Science and Engineering **771** (2020) 012035 doi:10.1088/1757-899X/771/1/012035

$$\text{Average value} = \frac{\text{Total score of answers used}}{\textit{Total validator}}$$

The total score of answers used is the average total score. It can be calculated as follows:

The average value $= \frac{42{,}9}{10} = 4.29$ rounded up to 4.3

From the calculation can be concluded that the average value obtained reaches the average value is expected, which is 4.3. It can be concluded that the application of information media Purwokerto city transportation can already be said to be good enough for use in the terminal town of Purwokerto.

4. Conclusion

The conclusion that can be drawn from this is the final report of this information the media serves to use the terminal Purwokerto visitors to be aware of the route and the transit lines that existed during this Purwokerto. The program is expected to help visitors with car use public transportation so that visitors do not get lost later on, because visitors will not be confused again about the route and the path to take to get to a place. According to the questionnaire that has been analyzed, it can be concluded that the application is fit for use in Purwokerto city terminal so that the visitors can get the information quickly and accurately.

References

[1] Juanita & Iskahar. 2012. Optimization of Total Fleet Public Transport in Purwokerto. Scientific articles. Faculty of Engineering, Universitas Muhammadiyah Purwokerto. Purwokerto.

[2] Chandra. 2012. Actionscript Flash CS5 for the Layman. Palembang: Maxikom.El Revelation's Blog. 2010. Adobe CS5 was released 2010. http://elwahyu.wordpress.com/2010/03/25/adobe-cs5-dirilis-2010/, accessed on October 23, 2013.

[3] Kadir, A. 2003. Introduction to Information Systems. Yogyakarta: Andi Offset.

[4] Nasution, M. N. 2004. Transport Management. Jakarta: Ghalia Indonesia.

[5] Juanita, Pinandita, 2015, Analisis Pelayanan Angkutan Umum dalam Kota Purwokerto Berdasarkan Metode Importance Performance Analysis dan Customer Satisfaction Index, Techno, Volume 12 No.2, Oktober 2015

[6] Perbup [2] [3] No.35 th 2007_Trayek lengkap.pdf

IOP Conf. Series: Materials Science and Engineering **771** (2020) 012036 doi:10.1088/1757-899X/771/1/012036

Case-Based Reasoning for Mortality Prediction of Catfish Farming Based on Water Quality

Suwarsito[1,*], H Mustafidah[2], S Mujiono[2], M A Fitriani[2]
Department of Geography[1], Department of Informatics[2], Universitas Muhammadiyah Purwokerto, Central Java, Indonesia

*e-mail: suwarsito.ito@ump.ac.id

Abstract. Catfish (Clarias gariepinus) is a species of freshwater fish that has good prospects for cultivation. This type of fish has a fast growth rate, is able to adapt to a bad environment and is easily cultivated. Water quality is the main element that influences the level of fish life. However, catfish deaths often occur when water quality is in certain conditions. Fish mortality can be predicted using an appropriate system so that it does not greatly affect the cultivation process. This research has succeeded in developing a prediction system for catfish mortality as a result of the water quality used for catfish farming. This system is based on case-based reasoning. The matching process starts by entering water quality condition variables such as Temperature, pH, Ammonia, Nitrite, and Dissolved Oxygen. Features entered by the user will be referred to the knowledge base. The system will look for a base case that has been indexed before. The similarity calculation with the highest value is used as a prediction of the effect of water quality on the mortality rate of catfish farming. The similarity of the test cases shows that the new case has a similarity level of up to 80%.

1. Introduction

Aquaculture is a sector whose growth can continue encouraged, given the potential use of the existing land is still low compared to the area of land which is available. The potential of aquaculture land in Indonesia is estimated to reach 17,744,303 Ha while the total usage has only reached 1,125,597 Ha. Existing conditions of potential pool use recorded at 22.6% [1]. Water quality greatly affects the type of fish and the level of fish life [2]. The water of good quality and in accordance with the nutritional needs of fish will enhance the results of cultivation and vice versa.

Catfish (Clarias gariepinus) is a species of freshwater fish that has good prospects for cultivation. This fish has a fast growth rate, is able to adapt to the environment that is not good and easily cultivated, besides it is popular with the wider community because it has good taste, tasty, soft texture, and has fairly high nutrition. Recently, Clarias Gariepinus has extensively cultivated throughout Indonesia, mainly in Java, Sumatra, Bali, and Kalimantan. Utilization of the catfish in Indonesian aquaculture was initiated in 1985 through the introduction of the species from Taiwan. Its fast-growing performance is well appreciated by fish farmers, led to the rapid development of aquaculture. This species was very popular and locally named as "Lele Dumbo" [3].

The high demand for catfish and the lack of supply of catfish, fish farming is one of the right ways to increase fish supply. Catfish farming is an activity to produce aquatic organisms (organisms) for profit [4]. Hatchery activities begin with the preparation of the hatchery media unit, management, or management of a good parent, spawning, to hatching eggs into seeds or larvae. According to Standard Nasional Indonesia (2000), in rearing nursery 1, namely rearing larvae up to 1-3 cm in size, the survival rate is 60%. This is related to various factors that affect growth and survival, both internal and external factors. Internal factors that influence are genetic, age and parasitic, while external factors are food availability and water quality which includes temperature, dissolved oxygen, carbon dioxide, water pH, nitrite, and ammonia. The availability of foods that have high nutrition is very needed larvae for the development of organs that are still simple to perfection [6].

Water as the main media for fish farming must have a quantity and quality standard that is in accordance with the fish's living requirements. Water must meet indicators in order to maintain the viability of aquaculture [7]. To predict catfish mortality rates, a case-based reasoning (CBR) system was developed. CBR is one branch of artificial intelligence that uses the knowledge-oriented problem approach of the past to solve current problems [8]. CBR is a methodology for solving a problem that uses previous data or memorized problem situations called cases. The processes of the CBR system proceed in four main steps such as retrieve, reuse, revise, and retain [9]. The new case starts at the top of the stage, where input is entered into the system. The previous case is compared to the new case and start to retrieve step. In a practical CBR system is a comparison between all the cases in the system and a new case, then the result will list the ranking of similar cases [10].

CBR development has been carried out previously, namely to control fish diseases [11]. This research will apply case-based reasoning to help farmers prepare the best water criteria in catfish farming so that the percentage of deaths can be minimized. This system was built using C# programming language as a reliable programming language [12].

2. Research Method

2.1. Data Collection

Water quality data were obtained from the Biofloc Laboratory of Universitas Muhammadiyah Purwokerto.

2.2. Case-Based Reasoning Process

Figure 1 shows the scheme of CBR. New cases can be added according to variables that have been determined by the system. The variables in this system consist of water temperature, water pH, dissolved oxygen nitrite, and ammonia. Retrieve in the system will search into an indexed case base using the C.45 Algorithm. The next retrieval process is to look for closeness between cases within the case base. The similarity calculation has been calculated using Equation 1.

$$similarity(e, ei) = \frac{common}{common + different} \quad (1)$$

Similarity (e, e_i) : a degree of similarity between e to e_i.
Common : represents a number of features that have the same feature value between e to e_i..
Different : represent a number of features that have different feature values between e to e_i..

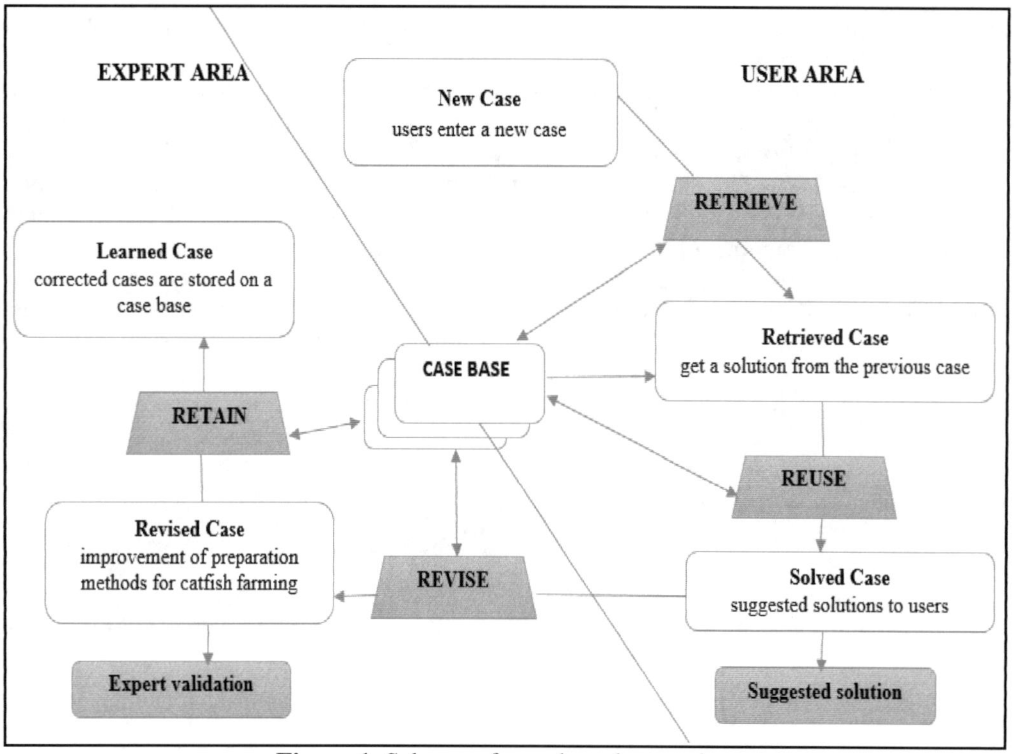

Figure 1. Scheme of case-based reasoning

3. Result and Discussion

Water quality is a picture of the fertility of water. Although sometimes catfish can live and grow in minimum quality conditions, Some water parameters that can be used as indicators in assessing the quality of water are presented in Table 1. Based on the results of the analysis that has been done, the data requirements include input data and output data can be seen in Table 2 and Table 3.

Table 1. Water Quality

Parameter	Value Range for fish farming	Measurement equipment
Physical Aspects		
Temperature	20 – 30 °C	Thermometer
Brightness	> 10 cm	Secchi Disc
Turbidity	25 - 400 JTU	Turbidity meter
Salinity	Freshwater 0 – 5 °/oo	Salinometer
	Brackish water 6 – 29 °/oo	Refractometer
	Freshwater 30-35 °/oo	
Water discharge	Heavy water 50 l/sec	Current meter
	Calm water 0,5 – 5 l/sec	
Chemical Aspects		
Dissolved oxygen	5 – 6 ppm	DO meter/ Winkler Method
Carbon dioxide	Max 25 ppm	CO meter/ Titration Method
pH	6,5 – 8	pH meter/ Litmus Paper
Alkalinity	50 – 500 ppm $CaCO_3$	
Hardness	3 – 15 dH	dH meter
Ammonia	< 1,5 ppm	Spectrophotometer
H_2S	< 0,1 ppm	Spectrophotometer
Nitrite	< 0,2 ppm	Spectrophotometer
Nitrate	0 – 1,5 ppm	Spectrophotometer

Phosphate	< 0,02 ppm	Spectrophotometer
Biological Aspects		
Plankton abundance	As needed	Plankton net
Bhentos abundance		Haemocytometer
Periphytons abundance		Ekman Dredge

Table 2. Sample input

Variable	**Class**	**Sample input value**	**Variable**	**Class**	**Sample input value**
Ammonia	1	0,01040	pH	1	7,5
	2	0,00980		2	7,4
	3	0,01490		3	7,3
	4	0,01420		4	7,2
			...	...	...
			Dissolved oxygen	1	7,7
				2	7
				3	6,4
				4	6,2

Table 3. Sample output

Class	**Sample output value**
1	0%
2	0,1%
3	0,333667%
4	1,339136257%
...	...
49	0,299850075%

The similarity is calculated based on the new case added by the user. An example of the new case can be seen in Table 4. Before doing the calculation, the system will search for cases with similar criteria in the case-based that can be seen in Table 5. After getting cases with similar criteria, the next process is to calculate the degree of similarity of the new case.

Table 4. New case

Variable	**Value**
Ammonia (NH_3)	0,01490
Water pH	7,4
Temperature (°C)	27
Nitrite (NH_2)	0,0400
Dissolved Oxygen (DO)	5,4

IOP Conf. Series: Materials Science and Engineering **771** (2020) 012036 doi:10.1088/1757-899X/771/1/012036

Table 5. Cases that have similar criteria value to the new case

Case_Id	Ammonia (NH_3)	Water pH	Temperature (˚C)	Nitrite (NH_2)	Dissolved Oxygen (DO)	Distance
6	0,01490	7,4	29	0,0400	5,4	0,3337
13	0,01490	7,2	28	0,0400	5,4	0,3478

Table 6. Comparison of new cases with Case_Id = 6

Case_Id = 6			New case	
Variable	**Value**	**Similarity**	**Variable**	**Value**
Ammonia (NH_3)	0,01490	1	Ammonia (NH_3)	0,01490
Water pH	7,4	1	Water pH	7,4
Temperature (˚C)	29	0	Temperature (˚C)	27
Nitrite (NH_2)	0,0400	1	Nitrite (NH_2)	0,0400
Dissolved Oxygen	5,4	1	Dissolved Oxygen	5,4

Table 7. Comparison of new cases with Case_Id = 13

Case_Id = 13			New case	
Variable	**Value**	**Similarity**	**Variable**	**Value**
Ammonia (NH_3)	0,01490	1	Ammonia (NH_3)	0,01490
Water pH	7,2	0	Water pH	7,4
Temperature (˚C)	28	0	Temperature (˚C)	27
Nitrite (NH_2)	0,0400	1	Nitrite (NH_2)	0,0400
Dissolved Oxygen	5,4	1	Dissolved Oxygen	5,4

Using eq.1, the similarity value of new case based on Case_Id = 6 can be seen as follow:

$$similarity(e, ei) = \frac{(1+1+1+1)}{4+1} = \frac{4}{5} = 0.8 * 100\% = 80\ \%$$

The similarity value of new case based on Case_Id = 13 can be seen as follow:

$$similarity(e, ei) = \frac{(1+1+1)}{3+2} = \frac{3}{5} = 0.6 * 100\% = 60\ \%$$

The comparison of new case values by Case_Id = 6 can be seen in Table 6, while the comparison of new case values by Case_Id = 13 can be seen in Table 7. By using Equation (1) the similarity value of the new case with Case_Id = 6 and Case_Id = 13 respectively obtained similarity values of 80% and 60%. The case that has the highest similarity value will be used as a search solution. This result will help the user in estimating the percentage of catfish mortality.

4. Conclusion

Based on the basis of the case, there are two cases that had similarity criteria with the new case, those cases will be the suggested solution for new cases. The similarity criteria value from each case can be seen that: 1) the similarity criteria of Case_Id = 6 is 0.3337 and 2) the similarity criteria of Case_Id = 13 is 0.3478. The calculation shows that both are respectively obtained similarity values of 80% and

60%. Therefore, the best result that will be the suggested solution for the new case is using Case_Id = 6 with a similarity value 80%.

References

[1] Bappenas, "Kajian Strategi Pengelolaan Perikanan Berkelanjutan," *Kementeri. PPN/Bapenas Direktorat Kelaut. dan Perikan.*, 2014.

[2] S. Suwarsito and H. Mustafidah, "Determination of Appropriate Fish Culture Method Based on Water Quality Using Expert System," *Adv. Sci. Lett.*, vol. 24, no. 12, pp. 9178–9181, 2018.

[3] B. Iswanto, I. Imron, R. Suprapto, and H. Marnis, "MORPHOLOGICAL CHARACTERIZATION OF THE AFRICAN CATFISH (Clarias gariepinus BURCHELL, 1822) STRAINS INTRODUCED TO INDONESIA," *Indones. Aquac. J.*, 2015.

[4] I. Effendy, *Budidaya Perikanan*. Jakarta: Universitas Terbuka, 2009.

[5] Standar Nasional Indonesia, "Produksi Benih Ikan Lele Dumbo (Clarias gariepinus x C. fuscus) Kelas Benih Sebar." SNI, 2000.

[6] N. B. W. Mufidah, B. S. Rahardja, and W. H. Satyantini, "Pengkayaan Daphnia spp. Dengan Viterna Terhadap Kelangsungan Hidup Dan Pertumbuhan Larva Ikan Lele Dumbo (Clarias gariepinus)[Enrichment Of Daphnia spp. With Viterna To Survival And Growth Of African Catfish (Clarias gariepinus) Larvae]," *J. Ilm. Perikan. dan Kelaut.*, vol. 1, no. 1, pp. 59–66, 2019.

[7] A. S. Oladimeji, S. O. Olufeagba, V. O. Ayuba, S. G. Sololmon, and V. T. Okomoda, "Effects of different growth media on water quality and plant yield in a catfish-pumpkin aquaponics system," *Journal of King Saud University - Science*, 2018.

[8] R. L. D. M. Mántaras *et al.*, "Retrieval, Reuse, Revision, and Retention in Case Based Reasoning," 2008. [Online]. Available: https://openair.rgu.ac.uk/bitstream/10059/54/1/ker05-mantarasetal.pdf. [Accessed: 06-Aug-2008].

[9] A. Agnar and E. Plaza, "Case-Based reasoning: Foundational issues, methodological variations, and system approaches," *AI Commun.*, 1994.

[10] P. Chantamit-O-Pas and M. Goyal, "A case-based reasoning framework for prediction of stroke," *Adv. Intell. Syst. Comput.*, vol. 625, no. October, pp. 219–227, 2018.

[11] H. Mustafidah and S. Suwarsito, "Fish Diseases Control System Using Case-Based Reasoning," *IEEE Xplore Digit. Libr.*, pp. 1–5, 2010.

[12] P. Deitel and H. Deitel, *C# 2010 for Programmers*, 4th ed. Boston: Pearson Education, 2011.

IOP Conf. Series: Materials Science and Engineering **771** (2020) 012037 doi:10.1088/1757-899X/771/1/012037

Preservation potential of lemon basil essential oil on tofu: Development of a natural food preservative

A Hamad[1,*], M B Ramadhan[1], D Y S Dewi[2], A D Djalil[2], D Hartanti[2]

[1] Department of Chemical Engineering, Faculty of Engineering and Science, Universitas Muhammadiyah Purwokerto, Jl. Raya Dukuhwaluh PO Box 202, Banyumas, Indonesia
[2] Faculty of Pharmacy, Universitas Muhammadiyah Purwokerto, Jl. Raya Dukuhwaluh PO Box 202, Banyumas, Indonesia

[*]e-mail: alwanihamad@ump.ac.id; hamadalwani@gmail.com

Abstract. The antimicrobial activity of lemon basil (*Ocimum* × *africanum* Lour.) has been described. This activity can be elaborated further and possibly applied for natural foods preservative. In this study, the constituents of lemon basil essential oil were analysed, and its preservation potential on the tofu was evaluated. The essential oil was distilled from dried plant materials using a steam and water distillation process. The constituents of essential oil of lemon basil were analysed by Gas Chromatography-Mass Spectroscopy (GC-MS) technique. Its application for preservation of tofu was evaluated based on its capacity in inhibiting the growth of bacteria on the tofu as well as the physical observation of the tofu during ten days of preservation at room temperature. The lemon basil essential oil constituted of 21 constituents, with a high fraction of oxygenated compounds, including neral and geranial, were detected. Lemon basil essential oil at an optimum concentration of 3.125 mg/ml inhibited the bacterial growth on the tofu during storage and improved its shelf life up to 4 days.

1. Introduction

Tofu is made from precipitation of water extract of whole soya beans and mainly consumed in East- and Southeast Asia. It typically contains high amounts of water (84.55%) and proteins (8.08%). This nature makes it is easily spoiled by microorganisms and hence has a relatively short shelf life [1]. A tofu preservation process is needed to prolong its shelf life as well as minimise the risk of food poisoning.

Spices and herbs, particularly those with profound antimicrobial activity, are the interesting sources of natural food preservative. The presence of antimicrobial compounds in those spices enabled the bacterial growth inhibition capacity toward a given food matrix [2]. Empirically, spices have been applied to food, and hence we have been familiar with their taste and aroma. This aspect might benefit us with better acceptability when they are used as a food preservative.

Lemon basil (*Ocimum* × *africanum* Lour., *kemangi* in *Bahasa Indonesia*; Lamiaceae) is a popular spice commonly used in savoury meals, including tofu-based ones. Lemon basil essential oil has demonstrated antimicrobial activities against both Gram-positive and Gram-negative bacteria [3–5]. In this study, the chemical constituents of lemon basil essential oil were analysed, and its application to be used as a natural tofu preservative was evaluated. The preservative potential of the oil was evaluated based on its capability in inhibiting the bacterial growth on the tofu as well as improving the shelf life of the tofu.

2. Materials and Method

2.1. Plant materials

Aerial parts of lemon basil were collected from Banyumas, Central Java. The plant materials were dried under direct sunlight. The identity of the plant was determined in the Laboratory of Taxonomy, Faculty of Biology, University of Jenderal Soedirman, Purwokerto, Indonesia.

2.2. Distillation of lemon basil essential oil

Lemon basil essential oil was distilled from the dried plant materials using a steam and water distillation process as previously described [6].

2.3. Identification of chemical constituents of lemon basil essential oil

Lemon basil essential oil was analysed using Shimadzu QP2010S SE Gas Chromatograph-Mass Spectrometer. The separation was performed on the AGILENT HP 5MS column (30mx0.25mmx0.25μm). The oven temperature was programmed from 50°C and raised to 260°C for 8 minutes, and then held for 9.5 minutes. The carrier gas was helium with a flow rate of 1ml/min. The injection ratio was 1:50. The injector and detector temperatures were 280 and 300°C, respectively. The volume of sample injection was 1μl at a concentration of 100ppm with n-hexane as the solvent. The solvent readout time was 2 minutes, and the sample analysis was run for 60 minutes. The ionisation voltage used was 70eV. The identification of the compounds in lemon basil essential oil was performed by comparing the mass spectrum of respective compounds with those available in Wiley ver. 9.0. Library.

2.4. Evaluation of the preservative potential of lemon basil essential oil

The preservative potential of lemon basil essential oil on the tofu was evaluated by following a previously described method with a minor modification [7]. In brief, after surface sterilisation and under sterile condition, the cubes of tofu were placed in the manually homogenised essential oil in water with to obtain the concentration of 0.125, 0.625, and 3.125 mg/ml, respectively. Sterilised water was used as a negative control. The preserved tofu was stored at room temperature. On day 2, 4, 6, 8, and 10, the bacterial growth on the tofu and their physical characteristic were evaluated. Each cube of tofu was put in 25 ml of sterile nutrient broth (NB) medium and then homogenised for a minute. One ml of suspension was transferred into 9 ml of sterile NB and then incubated in a temperature of 37°C for 24 hours. The enumeration of bacterial growth was conducted by recording the optical densities of cultured bacterial suspensions with UV-Vis spectrophotometer at a wavelength of 600 nm. The colour, odour, texture, and formation of slime on the surface of tofu were used as the evaluated physical characters to determine the prolonged shelf life of the tofu. All the works were replicated 3 times.

2.5. Data analysis

The total area under curve day 2-10 (AUC_{2-10}) of the optical density of cultures of tofu preserved with lemon basil versus the preservation time was calculated using the trapezoid method. Means separation of the optical densities of cultured bacterial suspensions in NB and AUC of the curve, as mentioned above, were evaluated using by Duncan's tests. Significance was evaluated at p-value <0.05. Statistical analysis was conducted by the general procedures of SPSS Statistics v.20 (SPSS Inc.)

3. Results and Discussion

There were 21 compounds identified in the lemon basil essential oil with major constituents of geranial, neral, α-humulene, β-caryophyllene, β-caryophyllene oxide, and α-bergamotene (Table 1). The constituents and their respective ratio in the essential oil of a given plant species are widely varied as the result of the phytochemical infraspecific variation. However, geranial and neral were found to be the typical main character constituents of lemon basil essential oil [8]. For example, the oil obtained from lemon basil collected in Chiang Mai, Thailand mainly constituted of estragole, geranial, and neral [9]. Another report mentioned linalool, α-bisabolene, β-caryophyllene, and geranial as the characteristic constituents of lemon basil essential oil originated from northern India [10].

An equal amount of oxidised and hydrocarbon compounds were found in the lemon basil essential oil, respectively (Table 1). Oxygenated compounds, including those with phenolic, aldehyde, ester, hydroxyl, and ether functional groups, are considered possessing a higher antimicrobial activity than those of hydrocarbons [11]. Hence, geranial, neral, β-caryophyllene oxide, neryl acetate, methyl heptenone, nerol oxide, bisabolol oxide A, and geranyl acetate might be the compounds responsible for lemon basil essential oil antibacterial activity and further tofu preservative potential activity.

Table 1. Chemical constituents of lemon basil

Retention time (min)	Compounds	Area (%)
15.685	Methyl heptenone	2.25
23.433	Nerol oxide	1.17
24.619	2-Methyl-4,5-nonadiene	0.95
27.181	Neral	15.60
28.398	Geranial	22.79
31.506	Geranyl acetate	0.41
32.065	α-Copaene	1.01
32.234	Neryl acetate	3.24
32.425	β-Bourbonene	0.45
33.747	β-Caryophyllene	11.74
34.139	α-Bergamotene	6.11
34.717	β-Farnesene	2.19
34.850	β-Selinene	3.42
35.749	Germacrene D	4.40
36.037	Zingiberene	0.95
36.375	α-Farnesene	0.60
36.475	β-Bisabolene	1.14
36.993	γ-Caryophyllene	1.54
37.660	α-Humulene	12.56
39.111	β-Caryophyllene oxide	6.28
39.858	Bisabolol oxide A	1.15
Total		99.95
Total oxidised compounds		52.89
Total hydrocarbon compounds		47.12

The optical density of the culture of the tofu in NB was used as the parameter of the capability of lemon basil essential oil in inhibiting the growth of bacteria in tofu. The lower optical density indicated the presence of fewer bacteria on the tofu, and vice versa. In general, the optical densities of the culture of tofu were gradually increased from day 2 and reached the maximum value on the final day of observation. The optical densities of tofu cultures in day 2 and 4 were not significantly different and might indicate the lag phase of microbial growth, while the considerably highest optical density change observed from day 4 to 6 might refer to the log phase. The optical density of the culture of tofu treated with negative control was consistently higher than those treated with lemon basil essential oils. However, its pattern was only similar to that of lemon basil essential oil at a concentration of 0.125 mg/ml and different from those of essential oil in higher concentrations. Hence, particularly in day 2, 6, 8, and 10, lemon basil essential oil at a concentration of 0.625 and 3.125 mg/ml was capable of inhibiting the growth of bacteria on the preserved tofu (Figure 1).

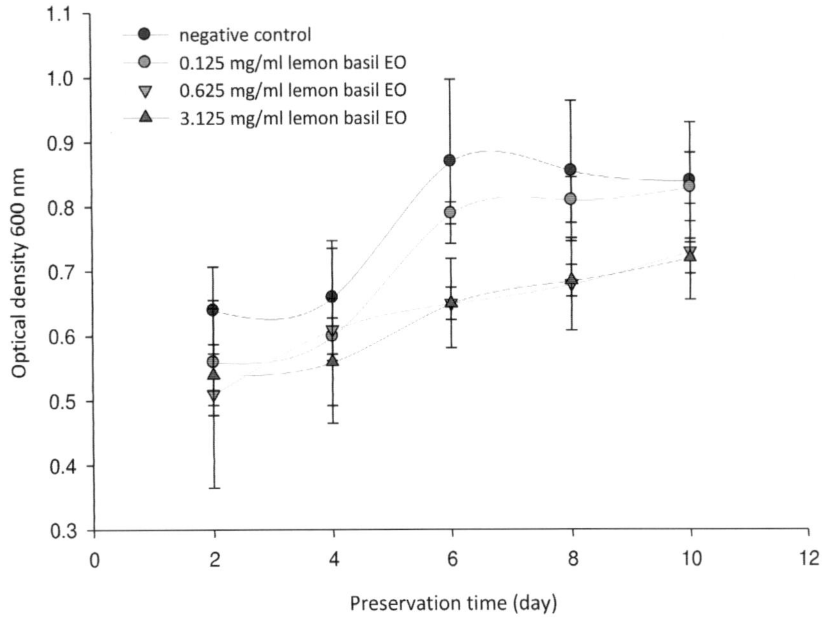

Figure 1. The profile of the optical density of cultures of tofu preserved with lemon basil essential oil

AUC_{2-10} of the optical density of cultures of tofu preserved with lemon basil versus the preservation time represented the overall bacterial growth in day 2-10. The lowest total AUC_{2-10} value indicated the best inhibitory activity against the bacteria grown on the tofu during ten day-preservation. The total AUC_{2-10} of tofu preserved with lemon basil essential oil was in a dose-dependent manner, which the higher concentration used resulted in, the lower total $AUC_{2-10.}$ However, only tofu treated with lemon basil essential oil at a concentration of 3.125 mg/ml demonstrated a significant difference in the value of total AUC_{2-10} compared to the negative control (Figure 2). Hence, only the treatment with the highest concentration of the oil that was capable of inhibiting the growth of bacteria on the tofu during preservation.

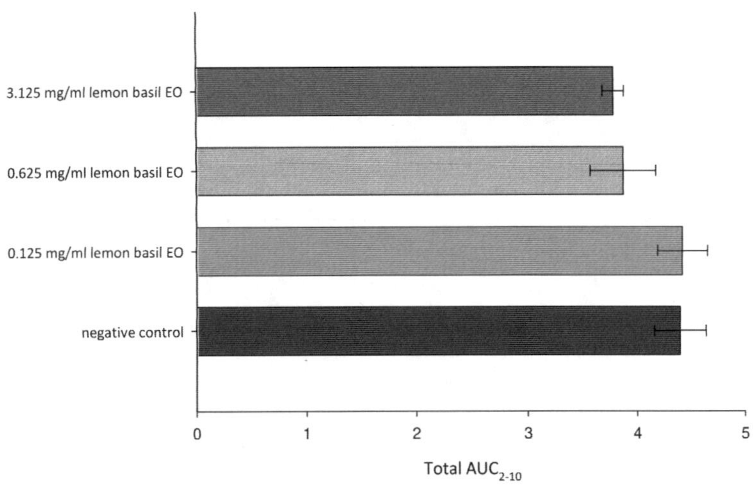

Figure 2. The AUC_{2-10} of the profile of the optical density of cultures of tofu preserved with lemon basil essential oil

Lemon basil essential oil has shown antimicrobial properties that support the result of our study. It inhibited the growth of *Bacillus subtilis* and *Staphylococcus aureus* with MIC value of 31.25 µg/mL and 125 µg/mL, respectively [3]. Another report mentioned that its minimum bactericidal (MBC) value against those two bacteria was 0.5 and 0.25% v/v, respectively [4]. The lemon basil essential oil was identified as the most active oils against *Enterococcus faecalis, Enterococcus faecium, P. vulgaris, S. aureus* and *S. epidermis* compared to 5 others essential oils from seven Ocimum taxa [5].

Physical characters of preserved tofu were evaluated to determine the capability of lemon basil essential oil in extending the shelf life of tofu. The bacterial growth on tofu caused the changes in the colour, odour, texture, and also slime formation on it. Improvement of tofu shelf life by treatment with the lemon basil essential oil was obtained from the comparison of time of each parameter changing between those treated with the oil and the negative control. The colour of the tofu was initially white. The colour of tofu treated with negative control was changed into yellowish in day 6, while those in lemon basil essential oil in all given concentrations started to be yellowish in day 10. The odour of unpreserved tofu started to be deteriorated in day 4, and the use of lemon basil essential oil effectively delayed it in a dose-dependent manner. The higher the concentration used, the stronger the hint of lemon basil aroma and the longer the first deteriorated odour time were observed. The same trend was also demonstrated in texture and slime formation parameters. At the concentration of 3.125 mg/ml, lemon basil essential oil was capable of maintaining the initial texture and inhibiting the slime formation until the final day of observation. The capability of lemon basil in improving the tofu shelf life was presented in Figure 3. The improvement of tofu shelf life was in a dose dependent manner in regards to odour, texture, and slime formation. When all parameters were taken into account, the improvement of tofu shelf life by treatment of lemon basil essential oil at lower, intermediate, and higher concentrations was 0, 2, and 4 days, respectively.

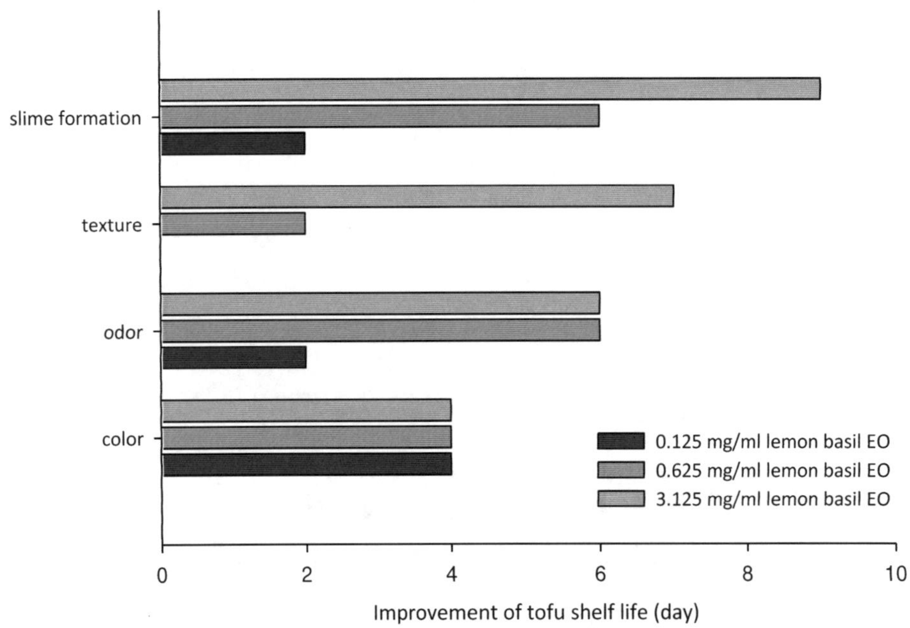

Figure 3. Improvement of tofu shelf life based on the changing of physical characters

However, the preservation potential of lemon basil essential oil in this study is considered low, that only the highest concentration showed the bacterial growth inhibitory activity. It might due to the unsuitability model used in this study. The essential oils were immiscible with the water as the medium, which minimize the contact between the essential oils with tofu. The essential oil would immediately separate from the water, form film in the upper surface of the water and was possibly evaporated from the medium afterwards [14]. For a better experimental model, the use of co-solvents or emulsifiers to increase the solubility of essential oil in medium might be applied. The use of

IOP Conf. Series: Materials Science and Engineering **771** (2020) 012037 doi:10.1088/1757-899X/771/1/012037

cetylpyridinium chloride to enhance the stability of cinnamon essential oil in the water had benefited the preservation of minimally processed basil leaves [15]. This strategy was also successfully applied to the preservation of Nile tilapia fillets, in which propylene glycol and tween were utilized to enhance the solubility of essential oils of rosemary and thyme in water medium [16].

4. Conclusion

Lemon basil essential oil at an optimum concentration of 3.125 mg/ml inhibited the bacterial growth on the tofu during storage and improved its shelf life up to 4 days. This preservation activity might be due to the presence of the oxygenated volatile compounds, including geranial and neral.

Acknowledgements

The authors acknowledge the Directorate General of Higher Education, Ministry of Research, Technology, and Higher Education, the Republic of Indonesia for financial support of this study.

References

[1] USDA. *National Nutrient Database for Standard Reference Legacy Release: Tofu.* Washington DC, https://ndb.nal.usda.gov/ndb/foods/show?ndbno=16427 (2019).
[2] Liu Q, Meng X, Li Y, Zhao CN, Tang GY and Li HB 2017 *Int. J. Mol. Sci.* **18** 1283.
[3] Hamad A, Nuritasari A and Hartanti D 2017 *der Pharm. Lett.* **9** 109-16.
[4] Maryati, Fauzia RS and Rahayu T 2007 *J. Penelit. Sains. Teknol.* **8** 30-8.
[5] Carovic´-Stanko K, Orlic´ S, Politeo O, Strikić F, Kolak I, Milos M and Satovic Z 2010 *Food Chem.* **119** 196–201.
[6] Hartanti D, Haqqi MZU and Hamad A 2018 *Adv. Sci. Lett.* **24** 91-4.
[7] Hamad A and Hartanti D 2015 *Farmasains* **2** 289-94.
[8] Pisutthanan N and Pisutthanan S 2009 *Naresuan Univ. J.* **17** 269-74.
[9] Tangpao T, Chung HH and Sommano SR 2018 *Foods* **7** E175.
[10] Padalia RC, Singh VR, Bhatt G, Chauhan A, Upadhyay RK, Verma RS and Chanotiya CS 2018 *J. Essent. Oil. Res.* **30** 437-43.
[11] Chouhan S, Sharma K and Guleria S 2017 *Med.* **4** 58.
[12] Widati RR, Suranto and Pangastuti A 2006 *Biofarmasi* **4** 22-6.
[13] Aristawati AT, Hasanuddin A and Nilawati J .2016 *J. Sains. Teknol. Tadulako* **5** 7–15.
[14] Van de Vel E, Sampers I and Raes K 2017 *Crit. Rev. Food. Sci. Nutr.* **30** 1–22.
[15] Park JB, Kang JH and Song KB 2018 *Food Sci Biotechnol* **27** 47–55.
[16] Albarracín WH, Alfonso CA and Sánchez ICB 2012 *Vitae Rev. Fac. Química. Farm.* **19** 34-40.

IOP Conf. Series: Materials Science and Engineering **771** (2020) 012038 doi:10.1088/1757-899X/771/1/012038

The Effect of Olive Oil Composition on The Staining Quality of Klowong Batik Wax

Dulmalik[1*], A Chafid Z[1], H Setiadi[1], and F N Khaliq[1]

[1]Department of Chemical Engineering, Faculty of Industrial Technology, Islamic University of Indonesia, Yogyakarta, Indonesia.

[*]815210102@uii.a.id

Abstract. Batik is visual art that integral to the cultural identity as well as illustrates the life values upon which the life of the community is laid. It uses a technique of wax-resist dyeing applied to whole cloth. This research aims to know the composition of olive oil that required on batik Klowong and to know its effect on the results. The research used both qualitative and quantitative methods such as ingredients measurement technique, questionnaire distribution and voter calculation. The wax for batik Klowong consists of olive oil, resin, and white paraffin. The beeswax is an independent variable that is varied in the composition for five different samples (10 grams, 20 grams, 30 grams, 40 grams, 50 grams). The sharpness of the motifs is visually analysed by six respondents. The results showed that sample D and E have better quality than the other samples. While sample A has the highest level of motifs sharpness.

1. Introduction

Batik is a cultural heritage of Indonesia is undoubtedly authenticity, as evidenced by the award of batik as one of the world cultural heritage produced by the Indonesian nation by UNESCO on September 28, 2009 [1]. The term Batik comes from the Javanese vocabulary that is Amba and Titik. Amba means cloth and point is a way of giving motifs on fabric by using liquid wax by means in dots [2]. Batik work is basically cover the surface of the cloth with liquid (wax) so that when the cloth is dyed into the dye liquid, the cloth closed the wax does not come into contact with the color [3]. If the process of making batik motif is done by "written" by using a tool called canting, then batik is called batik [5].

The batik wax is a material used to cover the surface of the fabric according to the image of Batik Motif so that the closed surface has the properties of resist or reject the color given on the fabric. Initially, the material used to cover the cloth is a slurry of glutinous rice, and the fabric is made of this called Simbut cloth [6]. After the discovery of batik candles, glutinous rice is not used anymore. The key ingredients of wax are Gondorukem, cat's resin, paraffin (white and yellow), Micro wax, animal fat, coconut oil, candle (evening) wasp, and candle lancing. The number of basic materials used and comparison are various, in their experience. So the batik candle is already a combination of waxes. In the beginning batik candles made only from wasp wax only (people call it Javanese batik wax or wasp wax), then because of increased experience then mixed with Gondorukem and resin cat eyes [7]. Then to relax or lower the melting point then mixed with animal fat or coconut oil [4].

So far, people only know the process of batik and batik candles alone without knowing the influence of the composition of batik wax itself to the results of batik. For that, this research is done so that people know more about the influence of batik wax composition and more focused on the influence of the wasp wax which is one of the batik waxes [7], [8]. The purpose of this research is to know the composition of the wax needed on batik Klowong and what influence the composition of wax on the quality of batik result [9], [10].

2. Method

2.1. Tools

The tools used for the experiment are:

Making batik patterns consisting of pencil, Mori cloth commonly used for batik. Then, making a sample wax experiment consisting of the wok (to heat the wax), stove, mixer, place/mold to accommodate wax samples. After that, making batik with tools Canting Klowong, Pan, Stove. And then, make batik coloring with dye Basin (bucket) for the dye container, Large frying pan, Glass beaker.

2.2. Materials

Materials used for this experiment are:

Sample recipe:

Sample A :
- 20 gr cat's eye resin.
- 10 gr white *paraffine*.
- 40 gr Gondorukem.
- 30 gr wasp wax.

Sample B :
- 20 gr cat's eye resin·
- 10 gr white *paraffine*.
- 40 gr Gondorukem.
- 10 gr wasp wax.

Sample C :
- 20 gr cat's eye resin·
- 10 gr white *paraffine*.
- 40 gr Gondorukem.
- 20 gr wasp wax.

Sample D :
- 20 gr *resin* cat's eye·
- 10 gr white *paraffine*.
- 40 gr Gondorukem.
- 40 gr wasp wax

Sample E :
- 20 grams of cat's eye resin.
- 10 gr of white paraffin.
- 40 gr Gondorukem.
- 50 gr wasp wax.

To create 5 (five) prescription samples wax is then required materials as much as:
- 100 g of cat's eye resin (@sample = 20 gr).
- 200 g Gondorukem (@ samples = 40 grams).
- 50 gr of white paraffine (@sample = 10 gr).
- 150 gr wasps wax (different each sample). *Naphtol AS. OL* and *Scarlet Salt GG* for coloring
- Coldwater

2.3. Experiment Steps

The step of the experiment is the steps taken from the preparation of raw materials to be something that can be observed in accordance with the desired results. In each process, there are individual stages. For more details about the experimental steps of this study can be seen in Figure 1 below.

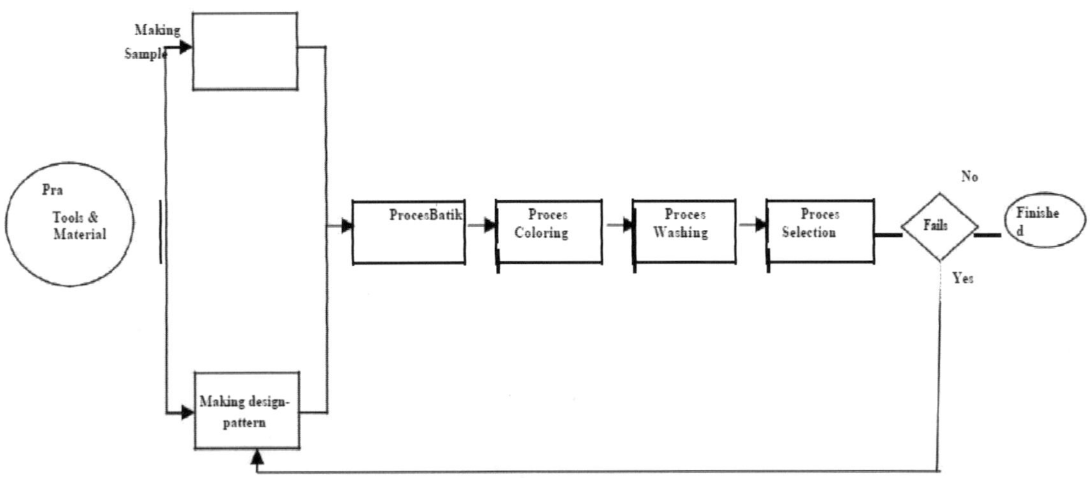

Figure 1. Flow Diagram of the Research Process

3. Results and Discussion

3.1. Test of Candle Samples

This study aims to obtain a sample with the right composition for the type of batik klowong with the pattern of lines and dots. The parameter used in this research is the composition of wasp night. To obtain the objective it must create several sample compositions with different wasp tonnes - each sample. The samples made in this research are 5 (five) samples. The materials used to make an evening consist of cat's resin, gondorukem, white parade and wasp night. Each material has its own characteristics and functions. Resin of a cat's eye when viewed from its shape like a slightly brownish and textured glass is a bit hard as a rocksugar. If heated, the cat's resin will quickly melt and form a thick brownish liquid texture.

Sample A Wasp Wax 30 gr

- if the fire is too large, the wax becomes too liquid and too hot to make the wax come out of the motive and make a mess of batik. besides, the candle can also penetrate the back of the fabric
- the size of the fire must always be maintained so that the wax is not too thick or liquid so that it can be easily used in the batik process

Sample B Wasp Wax 10 gr

- cannot be used if the fire is too large or small. This sampell must be lifted from the top of the stove after it is liquid so that the wax does not overheat. because this wax is very easy to melt

Sample C Wasp Wax 20 gr

- treatment for this sample is almost the same as sample 1 because the composition of the night wasp is not much different. However this sample can still be applied to a small stove fire

Sample D Wasp Wax 40 gr

- this sample must be treated on medium fire which is not too large or small. This is because this sample has properties that are easily thickened if the flame is too small

Sample E Wasp Wax 50 gr

- the treatment for this sample is almost the same as sample 3 but this sample can still be applied if on a large flame because the nature of this sample is very thick so it must always be heated

Figure 2. Special Treatment of Candle Samples.

While Gondorukem when viewed from its shape like clear glass yellow and very hard textured like glass. If heated gondorukem same as cat's eyes resin that will form a slightly thick yellow liquid. Furthermore, white paraffine which, when viewed from its shape like a normal wax white and slightly textured soft. If heated it will be a little liquid and clear like oil. The latter is a wasp wax which, when viewed from its shape like a textured tree sap, is very soft and sticky brown almost to the yellow. If heated the evening wasps will form a slightly thick liquid brown. The wasp's night is a bit hard to melt because of its soft, sticky texture.

Each wax sample has special treatment during batik process so that the result is as expected. Based on the observation, special treatment to be done on the wax sample can be seen in Figure 2 above. The treatment on each sample should be adjusted to the properties of each material. To obtain a good batik result, special treatment should be done so that the quality of candle samples remain good and not easily damaged. If the wax samples are treated incorrectly, the wax sample may not be good anymore, so it will affect the result of the batik.

One characteristic of a waxy sample that is not good can be seen from the color that is different from the first usage is blackish brown. Wax samples are continuously heated using a large fire can cause the color change. Wax samples that are not good anymore can still be used but will look the difference in the process of batik. From the wax sample test it can be concluded that the wax sample E is the best wax sample among the others. This is because the wax sample has a composition.

Table 1. Non-microwax Color Differences Test Result (R %)

No.	Sample code Non Microwax	Test Value Color Difference L*	a*	b*	dE*ab
1		100,23	0,11	0,05	0,00
2	A	82,25	4,09	9,99	13,12
3	B	98,59	1,02	6,41	7,03
4	C	89,83	7,52	10,53	14,29
5	D	98,51	4,20	6,51	9,34

Test the color difference with the composition without using Microwax (non microwax) can be seen in Table 1, In L
* of the A-D sample has a temporary increase in a * of the A-D sample the test results are up and down. While the b
* of the A-D sample also experienced the same thing that is the result up and down and for dE* ab from the A-D
sample of color difference test results have increased. Therefore, batik wax is considered as the main material to make batik, especially for handmade batik in which the pattern and carve applied directly by the artisans . Thus the results showed that non microwax can be used to substitute the funktion of beeswax with approopriate composition. The generaly test value color difference result relative minim but have been excellent because full dynamics and fluctuatif.

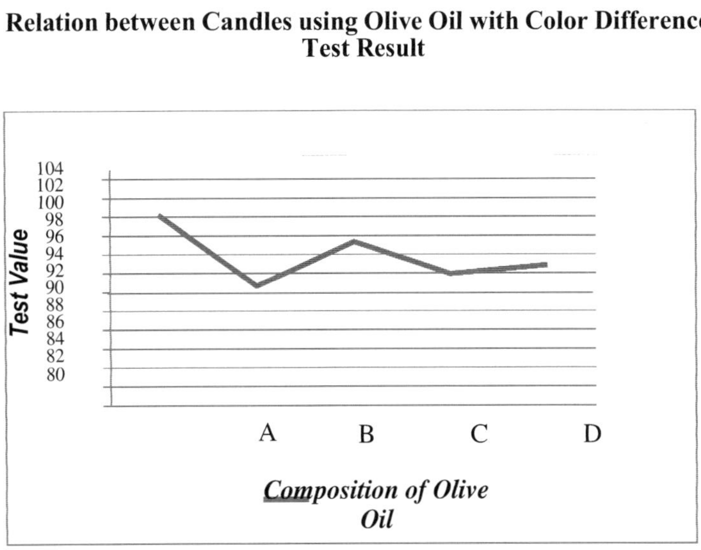

Figure 3. Color Differences Test Result

Figure 3 above shows the results of testing different colors by using micro wax. From the graph shows the value of different color test decreased, the decline in the numbers here show the color tested darker as the decline in test numbers. Testing the color difference there is a scale that is from the scale of 0-100, the number approaching the number 100 means brighter so vice versa close to the number 0 means the dark. This shows the more micro wax used by the different color values are getting lower/darker.

4. Conclusion

From research and study done then it can be concluded that:
From the results of our research can be drawn conclusion as follows:

1. The more the olive oil composition is used, minimizing the temperature of the melting point during the process of melting the raw materials into the wax.
2. The more olive oil composition used in the sample the clearer and softer the evening results when viewed visually.
3. From the test results, the more the composition of olive oil does not make the better the quality of the wax. The best wax quality is obtained on samples with the least amount of olive oil.
4. The composition of olive oil should be balanced with other ingredients in order to produce the best quality

References

[1] A. Kusrianto, *Filosofi of Batik, Motif and Utility,* CV. ANDI OFFSET, Yogyakarta, 2013.

[2] H. S. Gumbolo, *Batik and Natural Dyes,* Ardana Media, Yogyakarta, 2015.

[3] A. Wulandari, *Nationality of Batik, Experience of Filosofis, Manufacturing Methods and Industrial Batik*, CV. ANDI OFFSET, Yogyakarta, 2011.

[4] S. Susanto, *The Art of Batik Craft in Indonesian,* BPKB, Yogyakarta, 1975.

[5] Hamzuri, *Classical of Batik,* Balai Pustaka, Jakarta, 2000.

[6] Research Team, *The Batik wax mix composition,* BPKB, Yogyakarta, 1975.

[7] Bilal et.al, The Effect Micro wax Composition on Coloring Quality of Batik, *Skripsi* , Chemical Engineering Departement, Faculty Of Industrial Technology, Islamic University of Indonesia, 2016.

[8] Retno et.al, The Effect Wasp Wax Composition on Wax Quality for Batik, *Skripsi*, Chem.Eng. Dept, Faculty of Industrial Technology, Islamic University of Indonesia, 2015

[9] Malik et al, The effect of Micro wax Composition on the Staining Quality of Klowong Batik Wax, *Proceeding ICET4SD 2017-603*, Yogyakarta, Indonesia, 2017.

[10] Khaliq, et.al, The use of hydrogen peroxide and sky stabilizer agent in the bleaching process of textile fabrics, *AIP Conference Proceeding 2138, 050011(2019),* https://doi.org/10.1063/1.5121116, 2019

IOP Conf. Series: Materials Science and Engineering **771** (2020) 012039 doi:10.1088/1757-899X/771/1/012039

Mass Transfer Coefficient Study of Shogaol Extraction in Ginger Using Subcritical Water

M E Yulianto[1], R Amalia[1], V Paramita[1], I Hartati[2], Q A K Nisa'[1,*]

[1] Department of Technology Industry, Vocational School of Diponegoro University Semarang 50239, Indonesia
[2] Chemical Engineering Department, Universitas Wahid Hasyim, Semarang 50236, Indonesia

*qurrotunayuni@student.undip.ac.id

Abstract. Ginger (*Zingiber officinale*) has bioactive components such as shogaol and zingerone which are able to mediate cardiac contractions, are antioxidant, antiproliferative and apoptosis. To increase the economic value, it is necessary to develop a process of extracting ginger active compounds through subcritical water. The superiority of hydrothermal extraction by using water is easy to get, cheap, abundant availability, high purity, non-toxic and easy to handle. This study aims to obtain data on the mass transfer coefficient of the extraction process of ginger bioactive compounds such as shogaol through subcritical water. These data are very much needed in the design and scale-up of the extractor tool. The results of the study show that hydrothermal extraction using subcritical water can increase the content of Shogaol compounds, compared to conventional extraction processes such as soxhletation, percolation, supercritical fluid extraction, microwave extraction, and ultrasonic extraction. Mass transfer coefficient (K_{la}) in the model mass transfer extraction of shogaol, ginger using subcritical water respectively 3,7503 s^{-1}, 3,6912 $^{s-1}$, 0.31435 $^{s-1}$.

1. Introduction

The need for raw materials for national drug production in Indonesia, 90% is still met through imports with a value reaching 11.4 trillion in 2018 [1]. Efforts to increase independence in the health sector include developing the production of medicines and raw materials based on Indonesia's biodiversity plant diversity [2] [3]. Ginger *(Zingiber officinale*) which is one of the medicinal plants is known to have various pharmacological effects such as anticancer, antioxidant, anti-carcinogenic and antiosteoporosis [4] [5]. The main bioactive components of ginger that show pharmacological properties include shogaol, and zingerone [6] [7] [8] [9] [10].

Although 6-gingerol and 6-shogaol have similar chemical structures, and both are known to have beneficial bioactive properties, 6-shogaol has a stronger active power. Numerous studies have shown that 6-shogaol has greater effectiveness than 6-gingerol in many cases, such as in antioxidant activity [8] [11] [12] [13], anti-inflammatory activity [8] [14] [15], anti-platelet aggregation effect [16], inhibits muscle contraction [17] [18], inhibiting colorectal cancer [14], ovarian cancer [19] [20], breast cancer [21], and lung cancer [22] [23].

Therefore, efforts are needed to develop 6-shogoal production through the 6-gingerol extraction and conversion process. The development of the 6 shogaol extraction process resulting from 6-gingerol conversion can be done through several separation techniques, including: (i) soxhletation [24], (ii) subcritical extraction [25], (iii) supercritical fluid extraction [26], (iv) hydrotropic extraction [27], (v) microwave extraction [27] [28], (vi) enzymatic extraction [29], (vii) ultrasonic extraction {30] [31], and ionic liquid extraction [32]. The extraction process is carried out using various solvents such as ethanol, methanol, acetone, dichloromethane, hexane, water, CO_2 and *1-decyl-3-methylimidazolium bromide.*

This research was conducted to determine the effect of time to shogaol mass transfer into the subcritical water phase.

2. Methods

2.1. Study of Mass Transfer Events and Computational Processes

The modeling activity begins by compiling a mathematical equation about the mass transfer events of the bioactive compound from the feed phase to the solvent phase which may be based on theoretical studies and based on many previous studies, which were carried out at the Laboratory of Chemical Engineering Process Computation at Diponegoro University Vocational School Semarang. The model is made based on the theory of mass transfer at 1 stage extraction. The postulated model is then derived to obtain an equation which will later be tested using data obtained from the experimental.

Experiments were conducted to obtain data that are useful in determining the extraction rate coefficient parameters that are modeled in empirical equations. The data that has been measured is used as a tool to validate the postulations that have been set. Data measurements were carried out at the Laboratory of Chemical Engineering Operations at the Diponegoro University Vocational School. Thus, validated models and empirical equations will be obtained.

Research materials in the form of wet ginger, nitrogen, and chemicals needed to support the hydrothermal extraction process such as N_2. Materials for product analysis include: acetonitrile, nitrogen, methanol, phosphate buffer, ethanol, distilled water, sodium hydroxide, potassium hydroxide, hydrochloric acid and phenolphthalein indicators. The main ingredients for the research are ginger rhizomes from the banyumanik market and ginger farmers in Tembalang District, Semarang. Chemicals bought at CV. Jurus Maju Semarang. Standard Shogaol and are buy from Sigma.

The main tool that will be used in this research is the hydrothermal extractor unit. While the supporting tools that will be used for analysis include centrifugation, magnetic stirrers, burette analytical balance, with a volume of 10 ml and a scale of 0.02, pycnometer, with a volume of 5 ml, viscosimeter, Erlenmeyer, with a volume of 250 ml, volume pipette, with volume 10 ml, beaker glass, with a volume of 500 ml and 100 ml, pipette drops, and papa capillaries.

The research variables varied were temperature, solvent-feed ratio, extractor pressure and extraction time. extractor temperature is set at 130, 140, 150, 160, 170, and 180°C because it is a temperature range with perfect solubility of shogaol to the subcritical solvent phase. Feed solvent ratios are set at 1: 1, 2: 1, 3: 1, 4: 1 and 5: 1. The extractor pressure is set at 2, 2.5, 3, 3.5 and 4 MPa. The extraction time is set at 10, 20, 30 and 40 minutes.

The ginger rhizome is cleaned of dirt and grinded to get the size of the powder. 75 g of ginger powder, put into a stainless tube with a lid and put into extraction cells. 700 ml of distilled water and varied to a certain ratio were added to the cell. The cell is closed with a stainless lid, N_2 gas is liquefied into the cell for 2 minutes to remove air and dissolved oxygen. The excess pressure is discharged through the valve. The temperature is set under certain conditions and heating takes 3-5 minutes to reach the desired temperature. The extraction process starts (t = 0) when the temperature is reached under certain conditions. All experiments are carried out under certain pressure. After the process is complete, the extractant is transferred to a cooling cell at 25°C, 1 Mpa for 1 minute for a brief cooling. During extraction, a number of samples are taken every 10 minutes. Extracts in the form of active compounds in the continuous phase (water) were analyzed for physical and chemical properties and their chemical composition was analyzed using HPLC-MS.

3. Result & Discussion

3.1. Mass transfer solid-liquid

The mechanism of mass transfer in the extraction process of a solution contained in a solid by a liquid until an equilibrium condition is reached takes place in three continuous stages, namely:

1. Solute phase changes occur when the solute changes from the solid phase to the liquid phase. Solutation of solutes occurs through the solid-liquid interface. This phenomenon is considered to occur quickly and does not affect the overall extraction rate
2. The solute diffuses into the solvents found in the pores of the solid. The process of moving solutes from inside solid particles to the surface takes place because of the concentration gradient between the solid-liquid interface and the external surface of the solid. Solvents in the pores remain stationary, therefore the process of moving the solute from the zone with greater

concentration to the outside will occur through a molecular diffusion process. The mass transfer rate of this stage is expressed in equation 1 :

$$N_{AS} = -D_e.A.\frac{dC}{dr} \quad (1)$$

with

N_{AS} = mass flux (kg. s^{-1})
De = solute diffusivity ($m^2.s^{-1}$)
A = diffusion contact surface area (m^2)
C = solute concentration in solution ($kg.m^{-3}$)
r = distance of mass transfer (m)

3. When the solute reaches the particle surface, the solute moves into the solution because there is a concentration gradient. The rate of mass transfer in the solution phase is expressed in equation 2:

$$N_{AS} = K_{La}V(C_S - C) = \frac{dM}{dt} \quad (2)$$

With:

M = mass of the solute being moved (kg)
V = solvent volume (m^3)
t = time (s)
Cs = solute concentration on a solid surface, which is equal to solute concentration in a saturated solution ($kg.m^{-3}$)
K_{La} = volumetric mass transfer coefficient (s^{-1})

3.2. Ginger Bioactive Hydrothermal Extraction Mass Transfer Model

The hydrothermal extraction events of gingerol, shogaol, ginger paradol are considered as a series of mass transfer events which include:

1. Diffusion of gingerol, shogaol, paradol from the solid to the surface of the solid.
2. Gingerol, shogaol, paradol mass transfer from the surface of the solid to the liquid solvent in the pores of the solid.
3. Diffusion of gingerol, shogaol, paradol in liquid solvents.

The volumetric mass transfer rate of gingerol, shogaol, paradol from the surface of the solid to liquid, where CS is the concentration of gingerol, shogaol, paradol on the surface of the solid which is in balance with the concentration of the solute in the saturated solution. Equation (2) can also be expressed in the form of:

$$N_{AS} = K_{La}V(C^* - C) \quad (3)$$

The correlation between solute concentration in a saturated solution and initial solute concentration in solid solute content in the remaining sample at an infinite time and the mass of the initial sample (m) can be stated in the equation:

$$m.(C^0 - C^\infty) = VC^* \quad (4.a)$$

$$\frac{m.(C^0 - C^\infty)}{V} = C^* \quad (4.b)$$

Gingerol mass balance, shogaol, total paradol in solution are expressed in the following equation

$$N_{AS} = \frac{d(VC)}{dt} \quad (5)$$

The total differentials of equation are:

$$N_{AS} = C\frac{d(V)}{dt} + V\frac{d(C)}{dt} \quad (6)$$

Since there is no change in the volume of the solvent, equation (6) can be simplified to:

$$N_{AS} = V\frac{d(C)}{dt} \tag{7}$$

Equation (7) and equation (4.b) are substituted into equation (3) to obtain:

$$V\frac{d(C)}{dt} = K_{La}V(\frac{m.(C^0 - C^\infty)}{V} - C) \tag{8}$$

Equation (8) can be rearranged into:

$$\frac{d(C)}{dt} + K_{La}.C = K_{La}\left(\frac{m.(C^0 - C^\infty)}{V}\right) \tag{9}$$

If, A = K_{La} and B = $K_{La}\left(\frac{m.(C^0 - C^\infty)}{V}\right)$

Then equation (9) can be changed to:

$$\frac{dC}{dt} + AC = B \tag{10}$$

Equation (22) is an ordinal differential equation of order 1. Analytical settlement with the boundary conditions when t = 0, C = 0 and when t = t, C = C is obtained:

$$C = \frac{B}{A} - \frac{B}{A}e^{(-At)} \tag{11}$$

The K_{La} value can be obtained through one variable optimization with the help of the Golden Section optimization method or MATLAB calculation software. Optimization is done to obtain the sum square of errors (SSE) or the sum of the squares of the difference between the C count (C_{hit}) with the smallest C experiment ($C_{perc.}$).

$$SSE = \sum (C_{hit} - C_{perc})^2 \tag{12}$$

3.3. Shogaol mass transfer to the subcritical water phase

The mass transfer model in the hydrothermal extraction of ginger shogaol was also verified with experimental data. Figure 2 shows that the hydrothermal extraction rate at the beginning of extraction is very fast. However, after the extraction time of more than 10 minutes, no increase in the extract was obtained. The proposed model matches the experimental data with constants A and B found are 3.6912 and 54,078. As shown in Figure 24, the straight line represents the proposed model and most of the data is attached to the line. The mass transfer coefficient (Kl_a) in the hydrothermal extraction mass transfer model of the Goga shogaol is 3.6912 s^{-1}. The use of green solvents in subcritical areas for the shogaol extraction process will be effective if it is at high temperatures. That is because, at high temperatures, the viscosity and surface tension of the water will decrease so that it will encourage the rate of mass transfer, absorption into the matrix particles and selectivity to [33] [34].

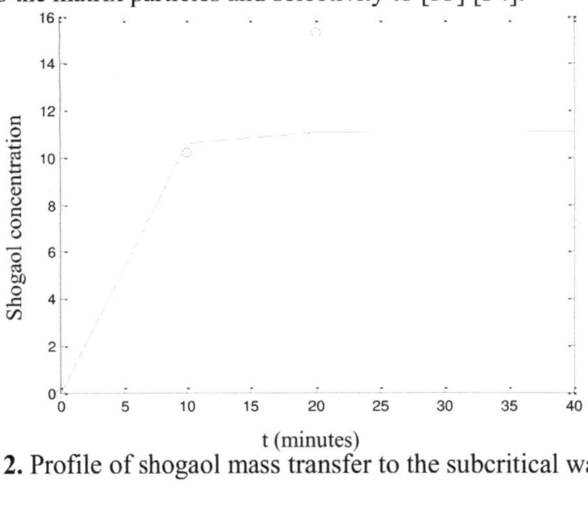

Figure 2. Profile of shogaol mass transfer to the subcritical water phase

4. Conclusion

Hydrothermal extraction using subcritical water can increase the content of shogaol compounds, compared to conventional extraction processes such as soxhletation, percolation, supercritical fluid extraction, microwave extraction, and ultrasonic extraction. The mass transfer coefficient (K_{la}) in the shogaol mass transfer model using subcritical water were 3,7503 s^{-1}, 3,6912 s^{-1}, 0.31435 s^{-1}, respectively.

References

[1] Ministry of Industry. 2018. Government Prepares Medicinal Plant. Industry News.
[2] Ministry of Agriculture. 2007. Prospects and Directions for the Development of Medicinal Plants Agribusiness.
[3] Ministry of Health. 2017. Regulation of the Minister of Health of the Republic of Indonesia No 17 of 2017 concerning the Action Plan for the Development of the Pharmaceutical Industry and Medical Devices.
[4] Dhanik J., Arya N., Nand V. 2017. A Review on Zingiber officinale. Journal of Pharmacognosy and Phytochemistry 2017; 6(3): 174-184.
[5] Mishra R.K., Kumar A., Kumar A. 2012. Pharmacological Activity of Zingiber officinale. International Journal Of Pharmaceutical And Chemical Sciences. Vol 1(3).
[6] Ali, B. H., Blunden, G., Tanira, M. O., & Nemmar, A. 2008. Some phytochemical, pharmacological and toxicological properties of ginger (Zingiber officinale Roscoe): A review of recent research. Food and Chemical Toxicology, 46, 409–420.
[7] Dhanik J., Arya N., Nand V. 2017. A Review on Zingiber officinale. Journal of Pharmacognosy and Phytochemistry 2017; 6(3): 174-184.
[8] Dugasani, S., Pichika, M. R., Nadarajah, V. D., Balijepalli, M. K., Tandra, S., & Korlakunta, J. N. 2010. Comparative antioxidant and anti-inflammatory effects of [6]-gingerol, [8]-gingerol, [10]-gingerol and [6]-shogaol. Journal of Ethnopharmacology, 127(2), 515–520.
[9] Gupta S.K., Kumar S. 2014. Medicinal properties of Zingiber officinale Roscoe - A Review. IOSR Journal of Pharmacy and Biological Sciences.
[10] Malhotra S., Singh A.P., 2003. Medicinal Properties of Ginger. Natural Product Radiance 2 (6),
[11] Bhattarai S., Tran V.H., Duke C.C. 2007. Stability of [6]-Gingerol and [6]-Shogaol in Simulated Gastric and Intestinal Fluids. Journal of Pharmaceutical and Biomedical Analysis. 45(4): 648-653.
[12] Cheng, X. L., Liu, Q., Peng, Y. B., Qi, L. W., & Li, P. 2011. Steamed ginger (Zingiber officinale): Changed chemical profile and increased anticancer potential. Food Chemistry, 129(4), 1785–1792.
[13] Pawar, N., Pai, S., Nimbalkar, M., & Dixit, G. 2011. RP-HPLC analysis of phenolic antioxidant compound 6-gingerol from different ginger cultivars. Food Chemistry, 126(3), 1330–1336.
[14] Pan, M. H., Hsieh, M. C., Kuo, J. M., Lai, C. S., Wu, H., Sang, S., & Ho, C. T. 2008b. 6-Shogaol induces apoptosis in human colorectal carcinoma cells via ROS production, caspase activation, and GADD 153 expression. Molecular Nutrition & Food Research, 52(5), 527–537.
[15] Sang, S., Hong, J., Wu, H., Liu, J., Yang, C. S., Pan, M. H., Ho, C. T. 2009. Increased growth inhibitory effects on human cancer cells and anti-inflammatory potency of shogaols from Zingiber officinale relative to gingerols. Journal of Agricultural and Food Chemistry, 57(22), 10645–10650.
[16] Shih, H. C., Chern, C. Y., Kuo, P. C., Wu, Y. C., Chan, Y. Y., Liao, Y. R., Wu, T. S. 2014. Synthesis of analogues of gingerol and shogaol, the active pungent principles from the rhizomes of zingiber officinale and evaluation of their anti-platelet aggregation effects. International Journal of Molecular Sciences, 15(3), 3926–3951.
[17] Kimura, I., Pancho, L. R., Koizumi, T., & Kimura, M. 1989. Chemical structural requirement in gingerol derivatives for potentiation of prostaglandin F2 alpha induced contraction in isolated mesenteric veins of mice. Journal of Pharmacobio-dynamics, 220–227.
[18] Pancho, L. R., Kimura, I., Unno, R., Kurono, M., & Kimura, M. 1989. Reversed effects between crude and processed ginger extracts on PGF2 alpha-induced contraction in mouse mesenteric veins. Japanese Journal of Pharmacology, 243–246.

[19] Kim, J. S., Lee, S. I., Park, H. W., Yang, J. H., Shin, T. Y., Kim, Y. C., Kim, D. K. 2008. Cytotoxic components from the dried rhizomes of Zingiber officinale Roscoe. Archives of Pharmacal Research, 31(4), 415–418.

[20] Rhode, J., Fogoros, S., Zick, S., Wahl, H., Griffith, K. A., Huang, J., & Liu, J. R. 2007. Ginger inhibits cell growth and modulates angiogenic factors in ovarian cancer cells. BMC Complementary and Alternative Medicine, 7(5), 1–9.

[21] Wu, C. H., Hong, B. H., Ho, C. T., & Yen, G. C. 2015. Targeting cancer stem cells in breast cancer: Potential anticancer properties of 6-shogaol and pterostilbene. Journal of Agricultural and Food Chemistry, 63(9). 2432-41.

[22] Hsu, Y. L., Hung, J. Y., Tsai, Y. M., Tsai, E. M., Huang, M. S., Hou, M. F., & Kuo, P. L. 2015. 6-Shogaol, an active constituent of dietary ginger, impairs cancer development and lung metastasis by inhibiting the secretion of CC-chemokine ligand 2 (CCL2) in tumor-associated dendritic cells. Journal of Agricultural and Food Chemistry, 63(6), 1730–1738.

[23] Warin, R. F., Chen, H., Soroka, D. N., Zhu, Y., & Sang, S. 2014. Induction of lung cancer cell apoptosis through a p53 pathway by [6]-shogaol and its cysteineconjugated metabolite M2. Journal of Agricultural and Food Chemistry, 62(6). 1352-62.

[24] Usman Y O., Abechi S.E., Benedict O.O., Victor O.,Udiba U.U., 2013. Anyahara S.E., Effect of solvents on [6]-Gingerol content of ginger rhizome and alligator pepper seed. Annals of Biological Research, 2013, 4 (11):7-13.

[25] Yulianto M.E., Kusumo P., Hartati I. 2017. Subcritical Water Extraction of Gingerol from Zingiber Officinale. Rasayan Journal of Chemistry. 10 (1): 738-743.

[26] Mesomo, M.C., Corazza, M.L., Ndiaye, P.M., Santa, O.R.D., Cardozo, L., and Scheer, A.D.P. 2013. Supercritical CO2extracts and essential oil of ginger (Zingiber officinale R.) : Chemical composition and antibacterial activity. The Journal of Supercritical Fluids 80 (2013): 44 – 49.

[27] Handayani D., Yulianto M.E., Ikhsan D. 2015. Mass transfer coefficient in ginger oil extraction using microwave hydrotropic solution. 2nd International Conference Chemical and Material Engineering.

[28] Purwanto, H Hartati I., Kurniasari L., 2010. Development of Microwave Assisted Extraction (MAE) in Ginger Oil Production High in Zingiberene. Momentum Vol 6 No. 2.

[29] Handayani D., Amalia R., Yulianto M.E., Murni. 2018. Production of Ginger Oil into Nanoemulsion from Enzymatic Extraction of Isolated Cow Rumen. AIP Conference Proceedings

[30] Kou, X., Li, X., Rahman, M.R.T., Yan, M., Huang, H., Wang, H., Su, Y. 2017. Efficient dehydration of 6-gingerol to 6-shogaol catalyzed by an acidic ionic liquid under ultrasound irradiation. Food Chemistry, 215:193–199.

[31] Supardan M.D., Fuadi A., Alam P.N., Arpi N., 2011. Solvent Extraction of Ginger Oleoresin Using Ultrasound. MAKARA, SAINS, VOL. 15, NO. 2, 2011: 163-167.

[32] Guo, J. B., Zhang, W. J., Wu, H., & Du, L. M. 2015. Microwave-assisted decomposition coupled with acidic food condiment as an efficient technology for ginger (Zingiber officinale Roscoe) processing. Separation and Purification Technology, 146, 219–226.

[33] Sarip M.S., Morad N.A., Ali N.A.M., Yusof Y.A.M., Yunus M.A.C., 2014. The kinetics of extraction of the medicinal ginger bioactive compounds using hot compressed water. Separation and Purification Technology.

[34] Teo C.C., Tan S.N., Yong J.W.H., Hew C.S., Ong E.S., 2010. Pressurized hot water extraction (PHWE). Journal of Chromatography A, 1217 (2010) 2484–2494.

IOP Conf. Series: Materials Science and Engineering **771** (2020) 012040 doi:10.1088/1757-899X/771/1/012040

Preliminary Study of Gaultherin Extraction from Gandapura Through Uv-Photo Extraction

Sutrisno[1,*], D Ariwibowo[1], M E Yulianto[2], N A Maulinda[2]

[1]Department of Mechanical Engineering, Vocational School of Diponegoro University
[2]Department of Chemical Engineering, Vocational School of Diponegoro University
Jl. Prof Soedarto SH, Tembalang, Semarang, Indonesia, 50239

*e-mail: didik.ariwibowo@live.undip.ac.id

Abstract. Gaultherin in gandapura can be extracted using hot water and calcium carbonate, followed by solvent extraction using acetic ester hydrate at a temperature of 100 °C. However, the process yields only 4 g/kg of fresh leaves. A technique of simultaneously gaultherase enzyme inactivation and gaultherin extraction with ethanol solvent should be studied. Preliminary studies of Gaultherin extraction with ethanol has been done and it is able to obtain gaultherin up to 14.46% at pH 8. This study will examine the development of the process photo-extraction-UV and osmosis dehydration simultaneously. The advantages of this process are (i) the gaultherase enzyme experiences unfolding, resulting in reduced hydrolysis reaction, (ii) increase in gaultherin yield, (iii) needs less energy, and (iv) purification of the product would be lighter. Study of the effect of solvent concentration and extraction time on gaultherin recovery was carried out for 2 months in the Laboratory of Chemical Engineering Operations at UNDIP Vocational School. Variables include dependent and independent variables. Dependent variables include the solvent-feed ratio of 5:1, pH 4.8, 75 rpm of mixer speed, chopper blades speed of 125 rpm, a drying temperature of 80 °C. Independent Variables include ethanol concentration at 80, 85, 90, and 95%. Extraction time for 10, 20, 30, 40 and 50 minutes. The result shows that gandapura extraction using ethanol had dual function, inactivate gaultherase enzymes and extract the active compounds gaultherin. Gaultherin yields was increased by increasing the duration of extraction. Whereas increasing the concentration of the solvent, gaultherin extracted was decreased.

1. Introduction

Gandapura is an essential oil-producing plant that in the list of commodities fostered of Directorate General of Plantation. Product diversification from gandapura can be a form of gaultherin production, which has a higher economic value (US $ 690 per 10 mg) compared with gandapura oil (US $ 22 per 15 ml). Theoretical analysis shows that gaultherin productivity 12 times higher than gandapura oil production (Yulianto et al., 2013). Gaultherin has properties that make it as a natural aspirin and anti-cancer (Ribnicky et al., 2003; Hoffman, 2007). As estimated, the aspirin consumption reaches 20-50 million pounds per year (West, 1998). The global pharmaceutical industry's demand for gaultherin is expected to increase in the coming years.

At this time, taking gaultherin from gandapura plant has not been effective, because during the extraction process gaultherin hydrolyzed into methyl salicylate components and disaccharides. The hydrolysis process is to be catalyzed by the enzyme gaultherase contained in the plant itself (Waters et al., 1931). Extracting method of Gaultherin plant when conditions gaultherase minimal activity or inactive need to be sought so that the hydrolysis reaction gaultherin be methyl salicylate and the disaccharide is not going to happen.

In 1928, gaultherin on Gaultheria procumbens extracted using hot water and the addition of calcium carbonate, and followed by solvent extraction using acetic ester hydrate at a temperature of 100°C. The process only produces the final yield of 4 g / kg of fresh leaves (Bridel and Grillon, 1928), Small yield caused by gaultherin contained has been hydrolyzed by gaultherase. It is believed that alcohol, methylene chloride, acetonitrile can inhibit gaultherase activity (US Patent No. 2002/0031562 A1; Yulianto et al., 2008; Kusumo et al., 2013; US Patent No. 7033618).

For gaultherin production of gandapura, enzyme inactivation techniques and extraction gaultherin gaultherase simultaneously with ethanol should be examined. Ethanol solvent will have dual function, i.e simultaneously extract gaultherin enzyme inactivation. Preliminary studies of recovery gaultherin through the extraction process have been carried out (Yuniastuti et al., 2008, Faizah et al., 2011, Kusumo et al., 2013). The results show that inactivation of the enzyme using ethanol gaultherase potential and prospective, because it can increase the active compound gaultherin acquisition of up to 14.46% at pH 8. However, gaultherin recovered can not be produced optimally. It is caused by: (i) the diffusion of ethanol into the cellular cytoplasm of cells retained, so that not all the enzymes gaultherase through a process of unfolding, and (ii) the hydrolysis reaction gaultherin be methyl salicylate and disaccharide still occur. Thus, it is need photo-extraction-UV process development and drying agent simultaneously, which is able to summarize gaultherase inactivation treatment, the extraction process, and the process of osmotic dehydration in a single stage. Ethanol solvent functions as inactivation of the gaultherase enzyme while extracting gaultherin. The UV light-based extractor functions as the inactivation of the enzyme gaultherase and cellular cell destruction, so that the gaultherin can optimally recovered. While the *drying agent* serves to minimize the reaction of hydrolysis gaultherin to methyl salicylate. UV Oxidizing properties makes UV as a disinfectant in drinking water treatment (US Patent No. 5.78086 million) and the degradation of the organic components of wastewater (US Patent No. 2006/0163168 A1 and US Patent No. 7279092), as well as purification and sterilization of food production (US Patent No.7.391.041 and US Patent No. 2007/0114465 A1). The use of ultraviolet light was also developed to inactivate the enzyme and simultaneously extract the bioactive compounds in the extractor (Rizki et al., 2018).

Scheme process photo of extraction-UV and dehydration osmosis simultaneously is expected to provide benefits i.e. (i) the enzyme gaultherase experienced unfolding as penetrated ethanol, resulting hydrolysis reaction gaultherin into methyl salicylate catalyzed by the enzyme gaultherase reduced, (ii) yield gaultherin increased, due to the degradation mobile cell by UV light, (iii) less energy requirements for enzyme inactivation and gaultherin extracting, (iv) lighter product purification load, and (v) less amount of waste.

For this reason, the research activities aimed at: (i) design and fabrication of a 5 liter/day capacity UV photo-extractor prototype for inactivating the enzyme gaultherase, reducing water (inhibiting the hydrolysis reaction) and extracting the active compound gaultherin, (ii) studies of the effect of adding *drying* magnesium *agent* sulfate, sodium sulfate, and calcium chloride to increase gaultherin productivity, (iii) the optimization of process parameters on productivity gaultherin, and (iv) techno-economic evaluation study for feasibility of investment. Therefore, production gaultherin can be one of the diversification options for the gandapura plants. Hopefully, by producing gaultherin may increase the economic value gandapura industry.

2. Method

Research on the inactivation of the enzyme gaultherase and the extraction of gaultherin simultaneously through UV-photobio extraction technology has been investigated experimentally. This UV-photo extractor is a multi-function, which is able to summarize three stages in one device, namely the inactivation treatment, the extraction process gaultherin and dehydration process of osmosis. Fotoektstraktor-UVberupa stirred tank equipped with blade/knife enumerators as in a blender and is also equipped with UV light. The series of UV-stage photo-extractor and osmotic dehydration devices used for recovery of active compounds of gaultherin are presented in Figure 1. This series of devices consists of UV-based enzymatic inactivation extractor equipped with a stirring motor, a chopper *blade,* temperature control and stirrer rotation.

3. Materials and Devices Research

Raw materials to be used in this study are the leaves and flowers of plants gandapura obtained from KT Sikunang, Wonosobo. The raw material is frozen in order to inhibit the enzyme activity gaultherase. Another necessary ingredient is a chemical of PT. Bratachem Semarang is used as a solvent such as ethanol, pH buffer, drying agent (gelatin, calcium chloride and sodium sulfate) and the standard material for the purposes of analysis gaultherin levels, levels of salicylic acid and methyl salicylate levels using HPLC-MS analysis.

The main equipment used in this research is UV-photo-extractor and one-stage osmotic dehydration which is used to inactivate enzymes, reducing water and extract gaultherin. Supporting tools needed include: filtration equipment, oven, centrifuge and heater to assist in determining the levels of salicylic acid and methyl salicylate levels. Identification and analysis gaultherin performed using a spectrophotometer and HPLC-MS.

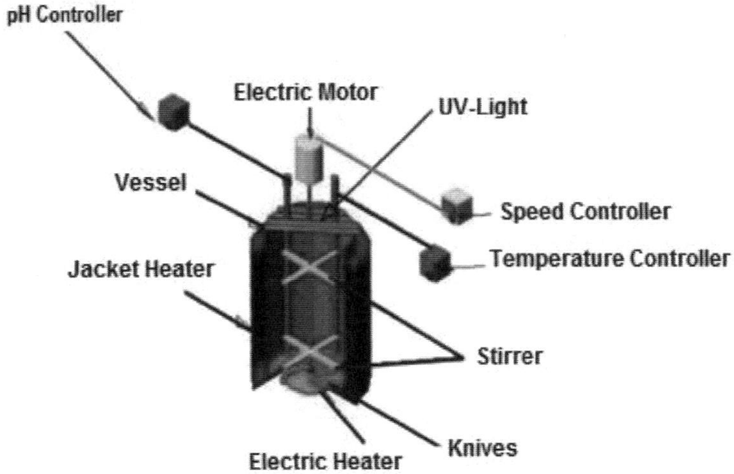

Figure 1. UV-photobio Extractor

4. Study Effect of Concentration and Time Solvent Extraction

Study of the effect of solvent concentration and extraction time on gaultherin recovery was carried out for 2 months in the Laboratory of Chemical Engineering Vocational School Operations UNDIP. Efforts to improve productivity including the effect of the addition of drying agent (sodium sulfate, calcium chloride, and calcium sulfate) to UV-photobio extraction. In general, the level of productivity gaultherin rather than with the addition of the drying agent. The use of *drying agent is* useful to bind, reduce or eliminate the water content in solution or in a mixture of water so that no hydrolysis reaction occurs by the enzyme gaultherase. It is expected that with the addition of the drying agent, the yield obtained in the production of gaultherin will be higher when compared to the production process without the addition of a *drying agent*.

5. Variable Experiment

Variables include the experiment remains variable and independent variables. Fixed variables include the solvent-feed ratio of 5: 1, pH 4.8, 75 rpm rotational speed mixer, chopper blades rotating speed of 125 rpm, a drying temperature of 80°C product. Free Variable includes concentration ethanol at 80, 85, 90, and 95%. Extraction time for 10, 20, 30, 40 and 50 minutes. Type drying agent sodium sulfate, calcium chloride, and calcium sulfate concentration of drying agent 5; 7.5; 10; 12.5% (w / w).

6. Experimental procedure

In this experiment, the leaves and flowers of the plant gandapura ice or frozen using liquid nitrogen. Extraction was carried out in a UV- enzyme photobio extraction enzyme with ethanol as a solvent. UV-photobio extraction is equipped with a *blade* or chopper knife at the bottom and UV light at the top. The ratio of solvent: feed, 5: 1. Ethanol is added at a certain concentration. The pH was maintained at pH 4.8 using a pH buffer. Speed of stirrer rotary and rotational chopper blades, respectively 75 and 125 rpm. Drying agent added according to experimental variables. Extraction was held for a certain time. The solids are then separated from the extract using a filter or centrifugation.

The extract was separated from the solids then added with chemicals or heated to remove the solvent. To get gaultherin in the form of a solution, extract that has been heated can be resuspended using a buffer or water. The extraction and then analyzed the content of gaultherin, methyl salicylate, salicylic acid using HPLC-MS analysis. HPLC-MS used were HPLC Alliance 2695 (Waters) with Photodiode Array Detector 2996 (Waters), column Symmetry C18 5 um, 4.6 mm x 150 mm (Waters).

7. Result and Discussion

Gandapura plant extraction using polar compound in the form of ethanol has a dual function, which inactivates enzymes gaultherase and extract the active compounds gaultherin. Diffusion of ethanol into gandapura leaves (Figure 2) intended to make the enzyme gaultherase are located in the cytoplasm penetrated with solvent, causing the enzyme activity to be inhibited. The statement also expressed by Poulev et al. (2003) that gaultherase activity inhibited by the addition of polar compounds. The next mechanism that the ethanol solvent will infiltrate the walls of the tonoplast membrane and phase contact occurs with the active compound gaultherin. The polar solvent will diffuse out of the leaf cell by carrying gaultherin. This is caused by differences in solubility.

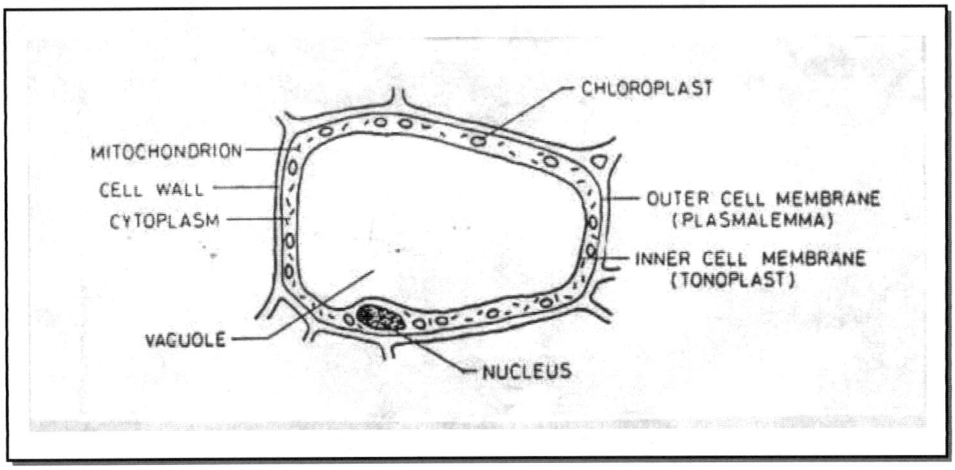

Figure 2. Gandapura leaf cell

Enzymes are giant molecules with a molecular weight that varies between 5000 Da-5 million Da. The enzyme belongs to a group that is larger macromolecular protein and consists of a series of linear chains of specific amino acids. In optimum conditions, the enzyme will through a folding process (Figure 3). The process of forming the composition of the enzyme folding is a spontaneous process that occurs in seconds (Bugg, 2004; Yang et al., 2004). Therefore, if the enzyme gaultherias is in a *folding* state, and damage to the tonoplast membrane also occurs, the enzyme catalyzes the reaction of the hydrolysis of the gaultherin compound to methyl salicylate. This cause acquisition of the active compound gaultherin is relatively low.

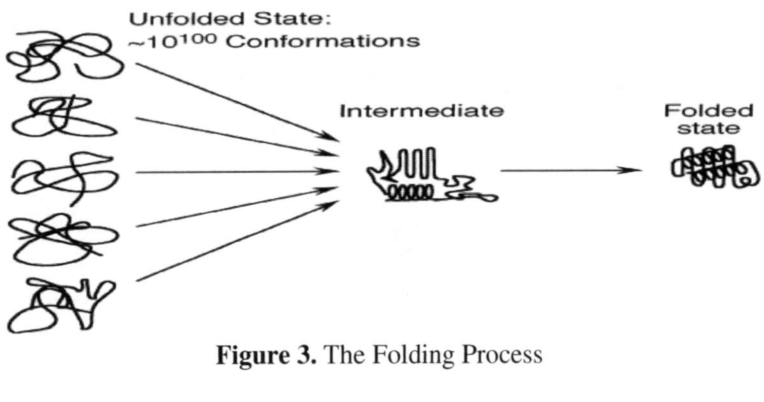

Figure 3. The Folding Process

The amino acid sequence of the enzyme will form a three-dimensional arrangement of certain, specific to each type of enzyme (tertiary structure). Part of the tertiary structure of the enzyme that is responsible for the catalytic activity of the enzyme called the active side. The number of active sides of an enzyme reaches 10-20% of the total volume of the enzyme (Bugg, 2004; Yang et al., 2004). The active side of an enzyme usually a hydrophilic gap consisting of a series of amino acid chains that binds the substrate (Figure 4.a) or bind a cofactor (Figure 4.b) and catalyzes the reaction.

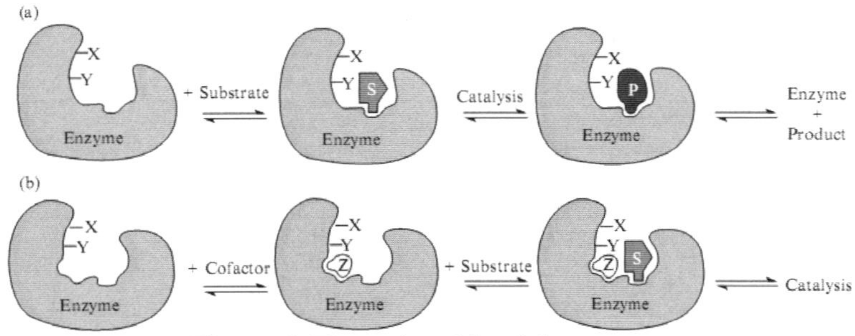

Figure 4. The active side of the enzymes

The *folding* process in an enzyme is a process that involves the inclusion of amino acid chains that are hydrophobic to the inner side of the enzymes and process out or shifting the amino acid chain that is hydrophilic to the exterior of the three-dimensional arrangement of the enzyme.

7.1. Effect of Extraction Time

Figure 5 presents the relationship gaultherin recovery to extraction time. The longer the extraction time, the gaultherin obtained increases, but after 10 minutes, the extraction approaches equilibrium. The longer the contact time between the phase dispersion phase (gaultherin) with the continuous phase (solvent phase), causing many dispersion phases to drag and cross the continuous phase based on differences in solubility. This is accordance with the statement Yuniastuti et al., (2017), the longer the time, the yield gaultherin has increased.

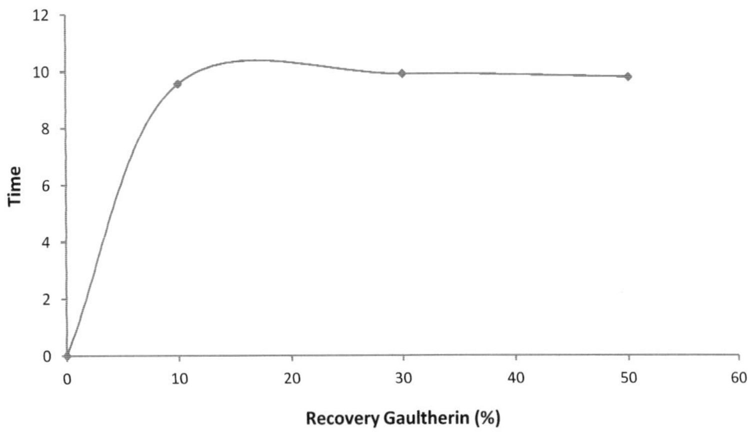

Figure 5 Gaultherin acquisition against extraction time graph

7.2. Effect of Solvent Concentration

Figure 6 presents a graph of the relationship between the ethanol concentration gaultherin levels. The greater the concentration of the solvent, gaultherin extracted decreases. It is possible, increasing concentrations of ethanol above 85%, causing some diluents such as methyl salicylate shipped to the continuous phase due to the increased solubility (figure 7). This is in contrast with the statement Yuniastuti et al., (2017), by increasing the concentration of solvent means enlarging the continuous phase, resulting in the volume fraction of the dispersed liquid phase and the smaller the particle diameter is also smaller. With decreasing particle diameter, it will expand inter-phase contact brought about by increased solute was dragged into the solvent phase.

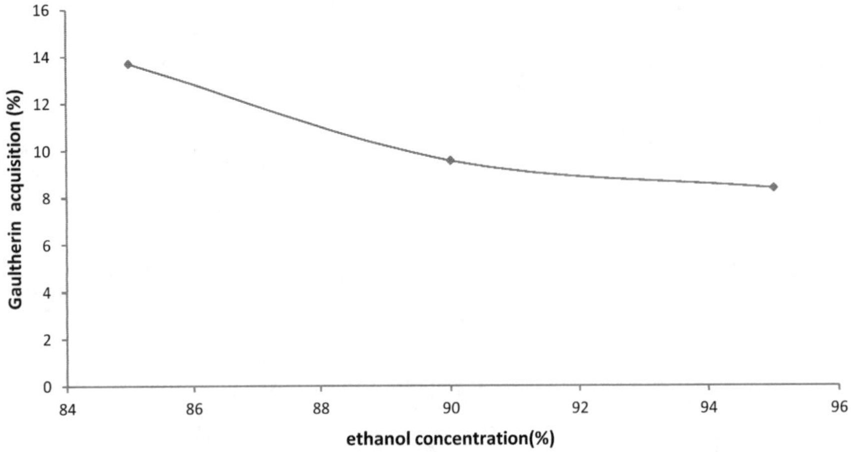

Figure 6. Gaultherin acquisition of ethanol concentration

It is inexplicable that gaultherase a type hydrolase enzyme, which has optimum activity around pH weak acid solution. Therefore, on the condition of a weak base bio-extraction condition cause enzyme gaultherase is unfolding, consequently it will reduce the hydrolysis reaction of gaultherin to methyl salicylate which is catalyzed by the enzyme gaultherase.

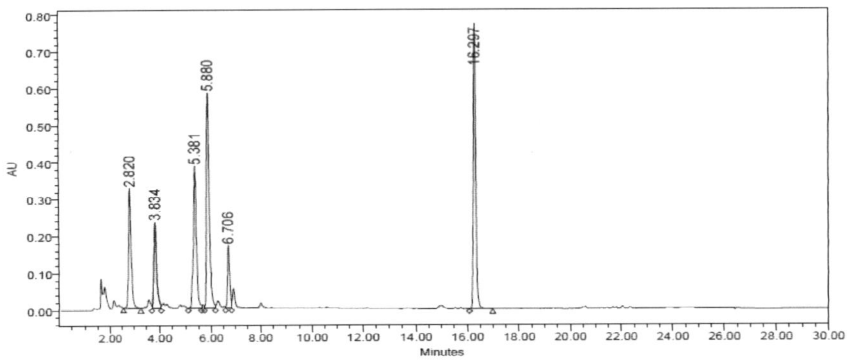

	RT	Area	% Area	Height
1	2.820	2441581	13.69	323410
2	3.834	1734364	9.72	230749
3	5.381	3073729	17.23	382806
4	5.880	4444864	24.92	582563
5	6.706	1095720	6.14	170395
6	16.297	5044594	28.29	769152

Figure 7. HPLC-MS chromatogram for gaultherin components and methyl salicylate in a 10-minute extraction using ethanol 85%.

8. Conclusion

Gandapura extraction using ethanol to dual function, which activates enzymes gaultherase and extracts the active compounds gaultherin. The longer the extraction time, the gaultherin obtained is increasing. The greater the concentration of the solvent, gaultherin extracted decreases.

References

[1] Yulianto ME, Ariyanto HD 2013. Per Gram US $ 98.1-million. Poster April issue. Thing. 124-125.

[2] Ribnicky, Poulev. 2003. *The determination of salicylates in gaultheria p for use as a natural*

[3] West. 1998. *Use of medications and polypharmacy are increasing among the elderly*. Journal of Clinical Epidemiology **55** (8): 809-817

[4] Walters A, Robertson. 1931. *Synthesis of gluycosides, the synthesis of monotroposides*. Journal Of American Engineering.

[5] Bridel MM, Grillon S. 1928. *Methyl salicylate on P. Gaultheria de Comptes* Rendus: 609-611.

[6] Yulianto ME, Hartati I, Anggun PS. 2008. *Inactivation of gaultherase on the gaultherin production from wintergreen by alcoholic solvent extraction*. National Seminar UKSW Salatiga.

[7] Kusumo, P., Yulianto, ME and Kasmiyatun, M. 2013. *Effect of drying agent on Extraction and Simultaneous Gaultherase Enzyme Inactivation of Gandapura (Gaultheria fragantissima.* National national conference on chemical engineering Kejuangan UPN Yogyakarta.

[8] Yuniastuti A, Yulianto ME, Hartati Gaultherin I. 2017. *Production From Gandapura (Gaultheria Fragantissima) By Enzymatic inactivation Of Gaultherase.* International Journal of Scientific and Technology Research Volume **6**, Issue 03.

[9] Faizah L, Yulianto ME. 2011. *Development of Enzymatic inactivation extractor For Gaultherin Production of Gandapura (Gaultheria fragantissima) As anticarcinogenic*. UNDIP Technology Development Research Report.

[10] Rizki, ILN, Yulianto, ME, Hartati, I., Paramita, V., Abidin, Z., Nisa, QAK 2018. *Optimization of simultaneous enzymatic inactivation and extraction of linamarin from cassava leaf by UV-assisted photobioextraction.* AIP Conference Proceedings 2026, 020 016;https://doi.org/10.1063/1.5064976

[11] PoulevJM O'Neal, S. Logendra, RB Pouleva, V. Timeva, US Garvey, D. Gleba, IS Jenkins, BT Halpern, R. Kneer, GM Cragg, and I. Raskin. Elicitation, *a new window into plant chemodiversity and phytochemicals drug discovery*. J. Med. Chem., Vol. **46**, 2542 - 25472003.

[12] Ribnicky., David, Poulev, Alexander, and Ilya Raskin,2002 "*Recovery of gaultherin from plants* ", US Patents No. 2002 /0031562 A1,

[13] Ribnicky., David, Poulev, Alexander, and Ilya Raskin, 2006, "*Methods of Administering gaultherin-containing Compositions*", US Patent No. 7033618,

[14] Faizah L, Yulianto ME. 2011. *Development of Enzymatic inactivation extractor For Gaultherin Production of Gandapura (Gaultheria fragantissima) As anticarcinogenic*. UNDIP Technology Development Research Report

[15] Adyana, MO, 1989. *Economic Analysis in Farming Systems Research*. Farming Systems Research Methodology exercises. IAARD. Jakarta. 15h.

[16] Backer CA, Bakhisizen RC, and Brink JR. 1965. *Flora of Java (Spermatophytes), NV P.Noordhoff, Groningen*. TheNetherland: 180-181.

[17] Bhirud, PR, & Sosulski, FW 1993. *Thermal inactivation kinetics of wheat germ Lipoxygenase*. Journal of Food Science, **58**. 1095-1098.

[18] Chua, MSM and S. Sunarti, 1999. *Essential Oil Plants. Plant Resources Of South East Asia*. 19: 110-114.

[19] Gittinger, JP, 1986. *Economic Analysis of Agricultural Projects*. Issue **2**. UI. Press-John Hopkins. Jakarta. 169 h.

[20] Hawley, GG, 1981. *The condensed chemical dictionary*. 10th Ed. Van Nostrand Reinhold Co., New York. 1135 p.

[21] Hener, U .; Kreis P. and A. Mosandl, 1990. *Enantiomeric distribution of α-pinene, β-pinene and limonene in essential oils and extracts. Part 2. Oils, perfumes and cosmetics*. Flavor and fragrance Journal. **5**: 201-204.

[22] Heyne, K., 1987.Tumbuhan berguna Indonesia. Volume II. Forestry Research, Jakarta: 1553-1554.

[23] Hernani. 2004. *Gandapura: processing, phytochemicals, essential oils and power herbicide*. TRO Bulletin XV (**2**). Center for Postharvest Research and Development Research.

[24] EJ Hoffman. 2007. *Cancer and the Search for Biochemical Selective Inhibitors*. Second Ed CRC Press.

[25] Lydon. J and SODuke, 1987. *Progress toward natural herbicides from plants*. The herb, spice and medicinal plant digest. **5** (4): 1-5.

[26] Manurung R. 2002. *Essential oils, the business opportunities that have not worked optimally*. General daily Sinar Harapan evening.

[27] Mauludi L. 2003. *Feasibility gandapura oil processing business in Wonosobo district of Central Java*. TRO Bulletin XIV (**2**). Research Institute for Spices and Medicinal Plants.

[28] Miliszkiez, D; P. Wieczorek; B. Lejezak; E. Kowalik and P. Kafarski, 1992. *Herbicidal activity of phosphonic and analogues and aspartic acid*. Pestic.Sci. **34**: 349-354.

[29] Mohammed W. 1985. *Biochemistry, Metabolism Energy, Carbohydrate and Lipid*. Publisher ITB. Bandung.

[30] Nezu, Y; M. Miyazaki; K. Sugiyama; N. Wada; I. Kajiwara and Miyazama, 1996. Dimethoxypirimidines *as novel herbicides. Part 2. Synthesia and herbicidal activity of O-pirimidines salycilates and analogues. Pestic*. Sci. **47**: 115-124.

[31] LPA Oyen, Dung NX. 1999. *Plants resources of South-East Asia: Essential Oil 19*. Prosea, Bogor. Indonesia: 110-114.

[32] TCV Penna., & Ishii, M. 2002. *Selective permeation and organic extraction of recombinant green fluorescent protein (gfpuv) from Escherichia coli*. BMC Biotechnology **2** (7), 11 pages.

[33] Raskin I, 1990, »*Salycilic acid in thermogenic Plant*» Annual Botany, 1990.

[34] Ravid, U; E.Putievsky and N. Sniv, 1983. *The volatile components of oleoresins and essential oils of Foeniculum vulgare in Israel*. J.Nat.Prod. **46** (6): 848-851.

[35] Shiva, MP; A. Lehri and A. Shiva, 1996. *Aromatic and medicinal plants: yielding essential oils for pharmaceutical, cosmetic industries and trade.* Vedams eBooks (P) Ltd. New Delhi.

[36] Sriwatanapongse, A., Balaban, M., & Teixera, A. 2000. *Thermal inactivation kinetics of bromelain in pineapple juice*. Transactions of the ASAE, **43**. 1703-1708.

[37] Wijlhuizen, AE, Kerkhof, PJAM & Bruin, S. 1979. *Theoretical study of the inactivation of phosphatase during spray drying of skim milk*. Chemical Engineering Science. **34**. 65160.

IOP Conf. Series: Materials Science and Engineering **771** (2020) 012041 doi:10.1088/1757-899X/771/1/012041

The Development of Hydrogel Polymer from Diapers Waste with The addition of Straw Nano Fibers as The Growing Media of Green Beans Plant

C L Imalia ,[1], G A Selviana[1], A Chafidz[2], and Haryanto[1,*]

[1]Chemical Engineering Department, Universitas Muhammadiyah Purwokerto, Purwokerto 53182, Indonesia

[2]Chemical Engineering Department, Universitas Islam Indonesia, Yogyakarta 55584, Indonesia

[*]Corresponding Author : harymsl@gmail.com

Abstract. The increase in diapers usage gives a serious contribution to environmental pollution. The hydrogel in diaper waste can be an alternative growing media. But, the growing media must have enough nutrition to fulfil the plant's need. Therefore, some amounts of filler are necessary to be added to the growing media of diapers waste hydrogel. This research aimed to find out the characteristics of diapers waste hydrogel, the effect of adding straw nanofiber towards the release constant rate, and the growth of green beans. The hydrogel of diapers waste was loaded with various concentration (0%, 1%, 5%, 10%, 15%, and 20%) of straw nanofibres and dried at 40°C for 3 days. The results show the morphology of the diapers waste hydrogel surface more dense with the addition of straw nanofibres. The release rate constant value increased with the increasing of straw nanofiber content. In this research, it can be proved that diaper waste hydrogel can be developed as the growing media, especially for green bean plant.

1. Introduction

Along with the development of technology, the use of disposable diapers for toddlers is increasing because it is considered more practical for both toddlers and housewives. The Increase use of disposable causes new problems, i.e. environmental pollution. Disposable diapers are discarded and not utilised. But in terms of technology and the economy diapers can be utilised as useful materials [1].

Based on that condition, the existence of diaper waste disturbing society. As happened in Surabaya, the governor of East Java received a notice from the environmental consultant and the lecturer of environmental engineering on the presence of alleged violations of the Law No. 32, 2009 about environmental management and constitution No. 18/2008 on Waste management. East Java Government is considered not to manage garbage properly, especially baby diapers [2].

Inside the diapers, there are the hydrogels. The hydrogel functions are to absorb and store water and nutrients for plants in large quantities. The hydrogel can decompose through decay by microbes so that the product is safe to use. The hydrogel is insoluble in water, but it only absorbs and releases the water and nutrients proportionally when needed by plants. Thus the plant will always have water supply and nutrients at all times. The hydrogel is able to absorb water as much as 100-200 times the weight of the hydrogel itself. This hydrogel can be used for soil replacement media and mixed planting media on potted plants, farmland, plantations, forests etc. Hydrogel helps to reduce the volume and frequency of plant watering.

As the development of technology, Hydrogel has been widely utilised in various ways, such as water-absorbent gel on diapers, media planting, soil improvement, fertiliser, drug release control, and so on [3,4]. Besides, the hydrogel from diapers waste can be an efficient and practical planting medium without having to cultivate routinely. Concerning this, the hydrogel is used as an alternative to planting media.

The production of green beans in Indonesia is still low and does not meet domestic needs. One of the causes of low green beans is the low level of soil fertility. One of the efforts that can be done to improve soil fertility is the supply of nutrients through fertilisation. Fertilisation is intended to meet the need for adequate amounts of nutrients to support vegetative and generative growth of crops. One effort to increase the production of green beans and soybeans is to add organic ingredients that can meet the nutritional needs of green beans. Straw fibre is an organic material that can be an additional material of planting hydrogel from waste diapers [5-7].

The utilisation of straw waste has not been optimal yet. Rice straw that is usually only burned in paddy fields often causes air pollution (CO_2, NOx, SOx) that can harm the environment [8]. The benefits of using straw waste include straw waste to be wasted, improving the physical properties and chemical soils, increasing water resistance so that soil moisture can be maintained, providing the microelements needed plant, and improve fertiliser efficiency (reduce the use of chemical fertilisers), suppress the cost of use of fertilisers and can ultimately increase production. The contents of some rice straw waste nutrients are: macro element Nitrogen (N) 2.11%; Phosphorus (P_2O_5) 0.64%; Potassium (K_2O) 7.7%; Calcium (Ca) 4.2%; and micro Elements Magnesium (Mg) 0.5%; Cu 20 ppm; Mn 684 ppm and Zn 144 rpm [8].

The hydrogel in diaper waste can be used as an alternative growing media. But, the growing media must have enough nutrition to fulfil the plant's need. Therefore, some amounts of filler are necessary to be added to the growing media of diapers waste hydrogel. This research aimed to find out the characteristics of diapers waste hydrogel, the effect of adding straw nanofibre towards the constant release rate, and the growth of green beans.

2. Materials and methods

This research was conducted in the laboratory of Polymer and Hydrogel Research Center, Universitas Muhammadiyah Purwokerto. The materials used in this research works are diapers, straw nanofiber, aquades, KOH 4%, and $NaClO_2$ 5%.

2.1. Preparation of straw nanofibres

Preparation begins by drying the rice straw under the sunlight. The dry weighed straw then be processed by cutting with a size of 1-2 cm. Then straw was grounded using a blender to get a powder. Then screened with a screening of 100 mesh. 10 gr of straw fibre that has been sieved then soaked and stirred in 200 ml solution of KOH 4% at 80 °C for one hour. The solution then was bleaching twice using $NaClO_2$ 5% at 70 °C for 1 hour while stirring. After washing, the treatment using KOH 4% at 80 °C was repeated, then washed again to remove the alkali residue. The final stage was mechanical treatment. The material was homogenised using a mixer at a speed of 22,000 rpm for 10 minutes.

2.2. Preparation of hydrogel of diapers waste

Various concentration of straw nanofibres was added into 2 grams of diapers that already dissolved in water. Afterwards, it was put into the oven and heated at 40°C for 3 days. The hydrogel was then analysed with various procedures.

2.3. Analysis of hydrogel

The various analysis was to know the characteristic of hydrogel from diapers waste. FTIR analysis, SEM morphology analysis, the effect of the addition of straw Nanofibres on the growth of green bean seeds and the kinetics of Straw nanofibres release in water were studied in this research.

3. Results and Discussion

Analysis of the chemical functional group of hydrogel using FTIR

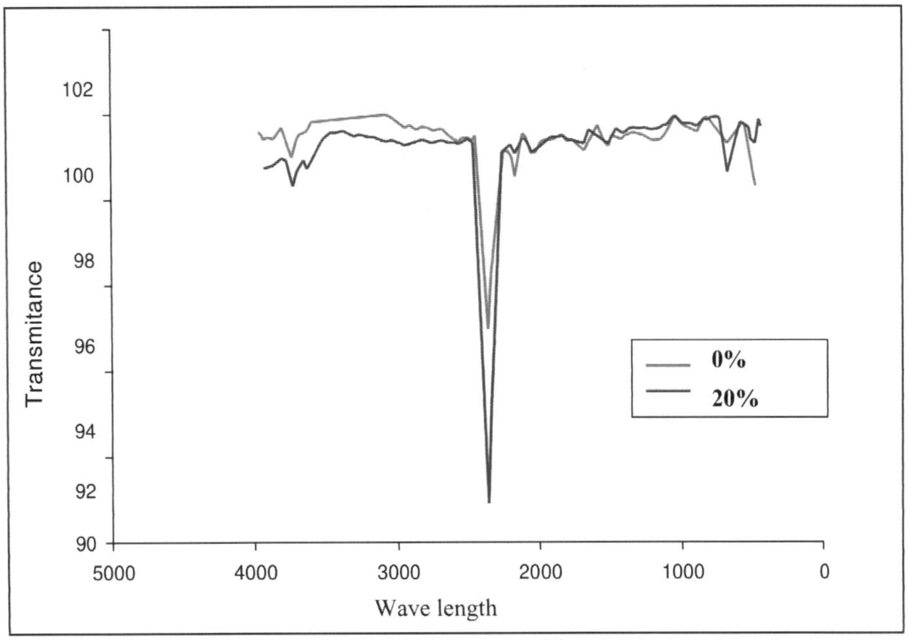

Figure 1. FTIR spectra of Straw nano fibre-loaded Hydrogel

The figure shows the FTIR Spectra of the waste-diapers hydrogel with the addition of straw nanofibre and the pure one. The diapers with the addition of straw nanofibres have a higher intensity compare with the pure one. It may be caused by the increasing of a hydroxyl group from straw nanofibres into the hydrogel.

3.1. Analysis of the morphology of waste diapers hydrogel

The morphology of waste diapers hydrogel can be seen from Scanning Electron Microscope (SEM) analysis as shown in Figure 2 and Figure 3. The morphology of straw nano fibers-loaded hydrogel looked more dense. This is likely due to the addition of the straw nano fibers can cover the structure formed on the hydrogel so that it looks more dense.

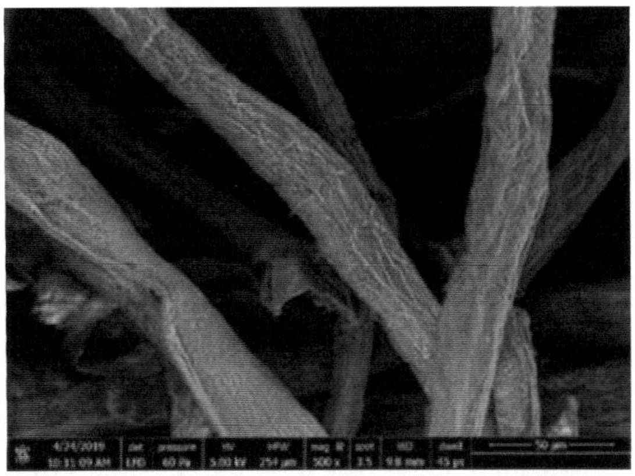

Figure 2. SEM Morphology of pure diapers waste hydrogel

Figure 3. SEM Morphology of diapers waste hydrogel with the addition of straw nanofibres

3.2. The kinetics of Straw nanofibres release in water

In a Nanofiber release system required kinetics equations to estimate the rate of release of Nanofibers on certain conditions/concentrations and estimating the nanofibers release mechanism. The rate of release of nanofibers into Aquadest is tested with the theoretical rate of Higuchi model kinetics which depicts the release of nanofibers from an insoluble matrix.

$$\mathbf{Q = kH\ t^{1/2}}$$

Where Q is the concentration and kH is the rate constant of release. The obtained data is shown in the table as the correlation between the concentration of straw nanofibres and constant rate.

Table 1. Rate constant of release according to Higuchi law

Concentration	**kH**
1%	0,0201
5%	0,1187
10%	0,1483
15%	0,2077
20%	0,2175

Based on Table 1, it can be seen that the higher the concentration of straw nanofibres, the higher constant rate of nanofiber release (kH). It may be caused by the bigger difference of concentration the bigger driving force for diffusion and moreover, the durability of the hydrogel structure is lower at higher concentration of straw nanofibers which diapers so nanofibers which indicated more easily leave the tissues.

3.3. Effect of the addition of straw Nanofibers on the growth of green bean seeds

Table 2. The growth of green beans seeds on hydrogel

Day	**The concentration of straw nanofibers/The growth of green beans seed**					
	0%	**1%**	**5%**	**10%**	**15%**	**20%**
1	-	-	-	0,5 cm	0,5 cm	0,5 cm
2	0,5 cm	0,5 cm	0,5 cm	0,5 cm	0,5 cm	0,5 cm
3	0,5 cm	0,5 cm	0,5 cm	0,5 cm	0,5 cm	0,5 cm
4	0,5 cm	0,5 cm	0,5 cm	0,5 cm	0,5 cm	0,5 cm

IOP Conf. Series: Materials Science and Engineering **771** (2020) 012041 doi:10.1088/1757-899X/771/1/012041

Based on the observation, on the planting media of diapers without mixing with soil, the seeds of green beans still live and grow for four days. It shows that the growing media fulfil the need for plants on adequate nutrient and mineral elements. The availability of nutrient and mineral elements in hydrogel affects the green bean growth process.

4. Conclusion

From the results and discussion, we can conclude that the addition of straw nanofibers on the diapers waste change the morphology of hydrogel and increase the rate constant according to the Higuchi equation. Moreover, it can be proved that the growing media can be developed by adding the straw nanofibers on the diaper waste.

Acknowledgement

This work was financially supported by the Ministry of Research, Technology and Higher Education. (T/140/E3/RA.00/2019).

References

[1]. Rohmawati, Dini. 2016. Making composit with MOL organic waste. Yogyakarta: Jurdik Kimia FMIPA UNY.
[2]. Riski, Petrus. 2017. Governour of East Java was sued by citizens on diaper waste problems. Voa Indonesia.
[3]. Haryanto, N. Aries Nurulita, E. Sundhani, P. Ho Huh, S. Cheol Kim, 2017 Polymer-Plastics Technology and Engineering, 57, 14, 1-7.
[4]. Yoshii, F. et al, 1999 Radiation Physics and Chemistry, 55, 133-138.
[5]. Adi, S H. 2011. Nanotechnology for Agriculture: Application of Hydrogel for Irrigation Efficiency. Jurnal Sumberdaya Lahan Vol 6, No 1.
[6]. Laila, Farida. 2010. Porous polyacrylamidehydrogel synthesis and characterization for urea absorption and release. Program Magister Kimia Fakultas MIPA Universitas Indonesia.
[7]. Sannino, A., C. Demitri, and M. Madaghiele. 2009. Biodegradable cellulose- based hydrogels: design and applications. Materials 2, 353.
[8]. Jannah, Rauzatul and Jannah, Rauzatul. 2018. The influence of rice straw composit on growth and yield of peanut crop (Arachis Hipogaea L). Agrotropika Hayati Vol 5, No 1.

IOP Conf. Series: Materials Science and Engineering **771** (2020) 012042 doi:10.1088/1757-899X/771/1/012042

Ultrafiltration of Laundry Wastewater Using Natural Zeolite-PVA Hybrid Membrane

A Ma'ruf [1,*], M A S Al Fathoni[2], A M Purnawanto[3] and I Astiyani[1]

[1]Chemical Engineering Department of Universitas Muhammadiyah Purwokerto
[2]Civil Engineering Department of Universitas Muhammadiyah Purwokerto
[3]Agrotechnology Department of Universitas Muhammadiyah Purwokerto
Jl. Raya Dukuh Waluh Kembaran, Purwokerto, Jawa Tengah, Indonesia 53182

[*]e-mail: anwarump@yahoo.com

Abstract. The laundry wastewater produced by laundry service becomes a pollutant of the water. The main characteristics of laundry wastewater are a high value of BOD dan COD. This research is aimed to study the utilization of ceramic hybrid membranes for laundry wastewater treatment. The ceramic hybrid membrane was produced from natural zeolite that coated with polyvinyl alcohol (PVA). The results show that the average percentage of the decreasing value of BOD reaches 83.37% and the average percentage of the decreasing value of COD is 89.86%. Based on turbidity, the higher rejection is obtained at the pressure of 5 bars with the value of rejection of 96.85%.

1. Introduction

Dukuh Waluh is an area where the University of Muhammadiyah Purwokerto is founded, which of course almost all houses become a boarding house for students. One of the business activities carried out by the Dukuh Waluh community is laundry services. Most of the material that requires a laundry process such as blankets, towels, sheets, clothes, and so forth. This shows that the laundry process plays an important role in the campus environment. The laundry process generally consumes a lot of water. The amount of water consumed and discharged is very dominant. A large amount of laundry wastewater discharged is closely related to the stages of the process in the laundry process that requires a lot of water.

The source of pollutants comes from the use of detergents as washing agents. Detergent is used because it has good washing power and is not affected by water hardness. However, it has a high phosphate content because phosphate is the main building material in detergents. The characteristics of laundry wastewater obtained from Slovenia have chemical oxygen demand (COD) values from 600 to 2500 mg O_2/l [1] and biochemical oxygen demand (BOD) values from 48 – 1200 mg O_2/l [2].

There are several methods to treat the laundry wastewater before it is discharged to the river or municipal irrigation. The conventional methods are using flocculation methods and adsorption methods[1]. The alternative methods for laundry wastewater are using membrane processes. Seo et al. [3] were using the ultrafiltration process that combined with ozone for domestic laundry wastewater reclamation and reuse. Turk et al. [1] were used two membranes process separation namely ultrafiltration and reverse osmosis. Polak et al. [4] were treated the laundry wastewater using the ceramic membrane that modified by carbon.

This research is aimed to study the utilization of ceramic hybrid membranes for laundry wastewater treatment. The ceramic hybrid membrane was produced from natural zeolite that coated with polyvinyl alcohol (PVA).

2. Materials and Methods

2.1. Materials

The Ceramic hybrid membrane was obtained from Polymer and Membrane Research Center (PMRC) of Chemical Engineering Department of Universitas Muhammadiyah Purwokerto, Indonesia. The Ceramic hybrid membrane was produced from natural zeolite with polyvinyl alcohol (PVA). Laundry wastewater was obtained from laundry service near Universitas Muhammadiyah Purwokerto, Indonesia. The characteristics of laundry wastewater are presented in Table 1.

2.2. Ultrafiltration Set-up

The ultrafiltration process was done by cross-flow proses (Figure 1). Laundry wastewater was filled in the feed tank. Then pumped to the flat sheet membrane at the constant pressure. The permeate was collected and weighed periodically. The retentate was recycled to the feed tank.

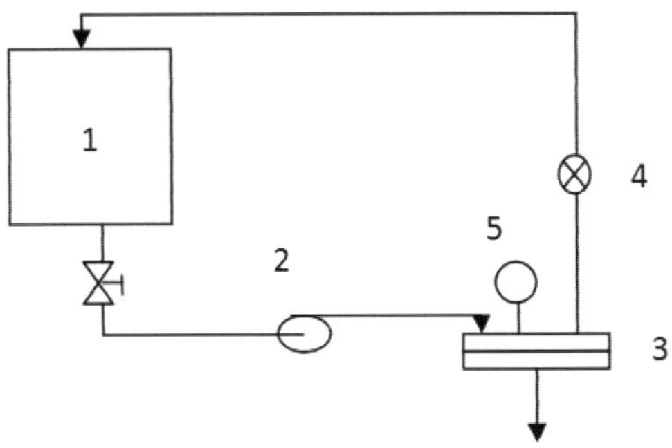

1. Feed tank
2. Pump
3. Membrane module
4. Valve
5. Pressure gauge

Figure 1. Experimental set-up

Table 1. The characteristics of laundry wastewater

Parameters	values
TDS, ppt	2.09
PH	10
Conductivity, mS	4.17
Turbidity, NTU	255
COD, mg/l	7480
BOD, mg/l	889.18

3. Results and Discussion

3.1. The Characteristics of Ceramics Hybrid Membrane (CHM)

The CHM that used was produced from natural zeolite that coated with PVA 7%. The characteristics of CHM are shown in Table 2. The Permeability test of CHM shows that the CHM is categorized as ultrafiltration.

Table 2. The characteristics of CHM

Parameters	Value
Density, g/cm^3	1.8
Porosity, %	32.2
Water absorption, %	17.91
Permeability, $l/m^2.h$	40.42 – 58.36

3.2. Effect of concentration of feed

Figure 2 shows the BOD of laundry wastewater feed and permeate. Based on BOD values, the decrease of BOD values of feed, the BOD values of permeate will be decreased. The average percentage of the decreasing value of BOD is 83.37%. Figure 3 shows the COD of laundry wastewater feed and permeate. Based on COD values, the decrease of COD values of feed, the COD values of permeate will be decreased. The average percentage of the decreasing value of COD is 89.86%. Ashfaq and Qiblawey [5] have reported at the process of ultrafiltration of laundry wastewater using polyethersulfone (PES) UF Membrane with MWCO of 5 kDa was achieved the percentage rejection of COD was 88.79% to 90.86 %. The percentage of rejection was decreased when pressure was increased. Islam et al. [6] used an aerobic membrane bioreactor (MBR) system for laundry wastewater. The result showed that the COD decreased varied from 87% at the first cycle to 95% at the fifth cycle.

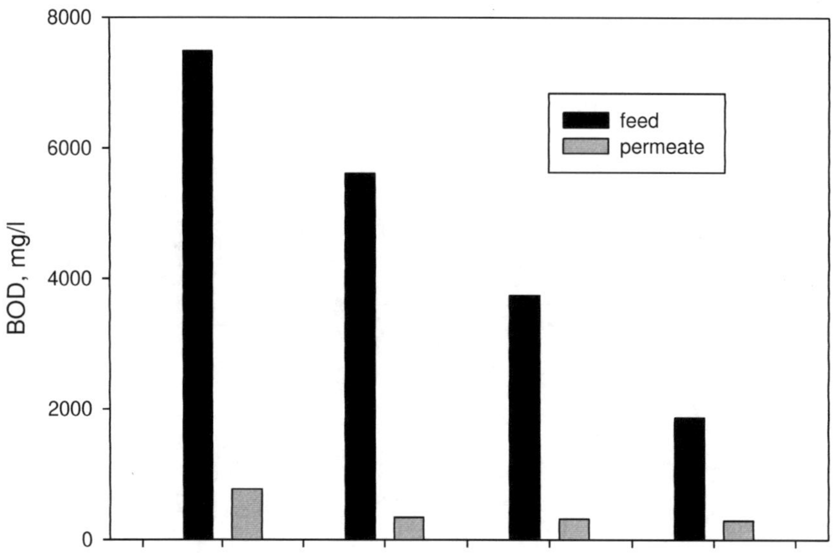

Figure 2. Characteristics of BOD of feed and permeate of laundry wastewater

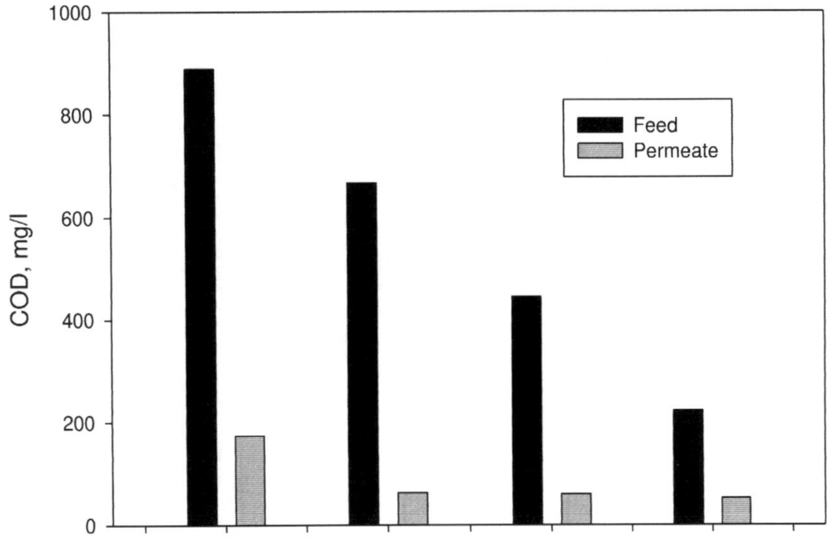

Figure 3. Characteristics of COD of feed and permeate of laundry wastewater

3.3. Effect of pressure

Figure 4 shows the flux of laundry wastewater at various pressure. The flux of the CHM increases with the pressure increasing. Figure 5 shows the rejection of CHM based on the turbidity of the laundry wastewater of feed and permeate. The higher rejection is obtained at the pressure of 5 bars with the value of rejection of 96.85%. Figure 6 shows the laundry wastewater before and after ultrafiltration process.

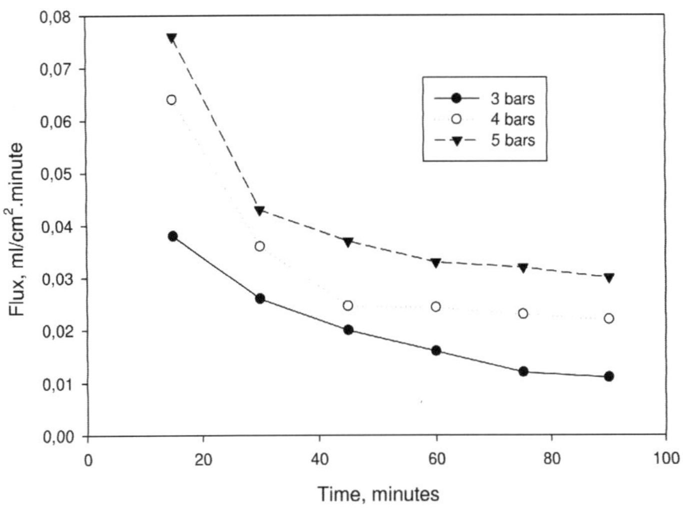

Figure 4. The flux of laundry wastewater at various pressure

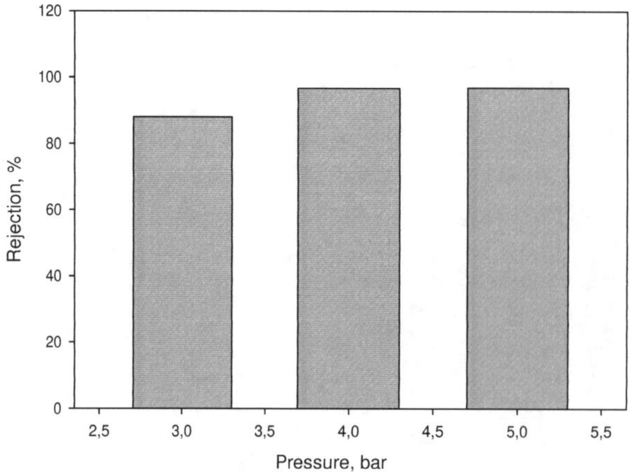

Figure 5. Rejection of laundry wastewater at various pressure

Table 1 shows the characteristics of permeate during the ultrafiltration of laundry wastewater. Overall, the parameters of laundry wastewater such as total dissolve solute (TDS), pH, conductivity and turbidity will decrease significantly.

Table 3. Characteristics of permeate during ultrafiltration of laundry wastewater

Parameters	Feed	Pressure, bar		
		3	4	5
TDS, ppt	2.09	1,49	1,34	0.685
pH	10	9	8.5	7.5
Conductivity, mS	4.17	3.10	2.75	1.370
Turbidity, NTU	255	30.9	8.57	8.02

Figure 6. Laundry wastewater: (a) before ultrafiltration; (b) after ultrafiltration

4. Conclusion

The CHM can be used for the ultrafiltration of laundry wastewater. The average percentage of the decreasing value of BOD reaches 83.37% and the average percentage of the decreasing value of COD is 89.86%. Based on turbidity, the higher rejection is obtained at the pressure of 5 bars with the value of rejection of 96.85%.

Acknowledgments
The authors acknowledge to Ministry of Research Technology and Higher Education of Indonesia for funding this research.

References

[1] S. Šostar-Turk, I. Petrinić, and M. Simonič, "Laundry wastewater treatment using coagulation and membrane filtration," *Resour. Conserv. Recycl.*, vol. 44, no. 2, pp. 185–196, 2005.
[2] O. Lade and Z. Gbagba, "Sustainable water supply : Potential of recycling laundry wastewater for domestic use," *J. Civ. Eng. Environmental Sci.*, vol. 4, no. 2, pp. 56–60, 2018.
[3] G. T. Seo, T. S. Lee, B. H. Moon, and J. H. Lim, "Ultrafiltration combined with ozone for domestic laundry wastewater reclamation and reuse," *Water Sci. Technol. Water Supply*, vol. 1, no. 5–6, pp. 387–392, 2001.
[4] D. Polak, M. Szwast, W. Fabianowski, and M. Rosiński, "Ceramic Membranes Modified by Carbon used for Laundry Wastewater Treatment," *Chem. Eng. Trans.*, vol. 74, no. March 2018, pp. 931–936, 2019.
[5] M. Y. Ashfaq and H. Qiblawey, "Laundry wastewater treatment using ultrafiltration under different operating conditions," *AIP Conf. Proc.*, vol. 2022, 2018.
[6] M. A. Islam, S. M. Muhtad, A. Anwar, M. M. Islam, M. R. Amin, M. Ar Islam and A. K. M. A. Haque., "Water filter-based MBR unit in laundry effluent treatment," *Indian J. Chem. Technol.*, vol. 21, no. 2, pp. 134–138, 2014.

Characterization and ultrasound-assisted activation of Indonesian natural zeolite from Cikalong West Java Province

Regawa Bayu Pamungkas[1,3,*], Mohamad Djaeni [2], Bakti Jos [2], Wancuh[1]

[1] Department of Chemical Engineering, Faculty of Engineering and Science, University of Muhammadiyah Purwokerto, Jl. Dukuhwaluh Raya, Purwokerto, 53182, Indonesia

[2] Department of Chemical Engineering, Faculty of Engineering, Diponegoro University Jl Prof H. Soedharto, SH, Tembalang, Semarang, Indonesia

[3] Postgraduate Program of Department of Chemical Engineering, Faculty of Engineering, Diponegoro University, Jl Prof H. Soedharto, SH, Tembalang, Semarang, Indonesia

*e-mail: rbayup11@gmail.com

Abstract. Natural zeolite activation from Cikalong, Tasikmalaya Regency, West Java Province, Indonesia, has been investigated experimentally. The activation process is carried out with a solution of sodium chloride (NaCl), which is utilized by ultrasound cavitation. The purposes of this research are to determine the effect of variation of process variables on the physical-chemical characteristics of zeolites. There were two variables studied, namely; sonication time and NaCl concentration. The study was conducted with the one factor at a time (OFAT) method, which is observing responses to changes in one variable at other constant. The range of sonication time of 20 - 60 minutes, while the NaCl concentration is 0.25N - 4N. Natural zeolites and activated zeolites are then characterized using SEM, XRF, FTIR, iod number, and cation exchange capacity to test their physical-chemical characteristics. Characterization results showed that activated natural zeolite had better quality than natural zeolite. The results of the SEM analysis showed the crystal type of natural zeolite minerals in the form of elongated crystals and others, which indicate the shape of quartz and mordenite. XRF analysis results prove that the elements of Ferrum, calcium, magnesium, and potassium are cations that can be exchanged with other metal elements. The results of the volumetric analysis showed an increase in iodine number and cation exchange capacity, along with an increase in sonication time and decreased along with the increasing of NaCl concentration. Based on these properties, zeolites have excellent performance as an adsorbent.

1. Introduction

Zeolite is a natural mineral that can exchange cations, adsorb molecules, or filter molecules so that they are widely used as adsorbents [1],[2], catalysts, and soil remediation [3]. They occur naturally in the environment but can also be synthesized in a laboratory [4]. According to Ramesh *et al.* (2011), natural zeolites are some of the most commonly occurring sedimentary rocks [5]. Therefore, they are abundant quantity, available easily, and cheaply [6]. There are now over 50 species of natural zeolite recorded [5,7,8], but only nine zeolite minerals are often found, such as analcime, chabazite, clinoptilolite, heulandite, erionite, ferrierite, laumontite, mordenite, and phillipsite. Indonesia has many types of natural zeolites that are spread across many islands. The Indonesian zeolite stockpile has estimated more than 400 million tons [8].

Zeolite is a hydrated porous mineral of alumina silicate crystal which has a three-dimensional skeletal structure, formed from tetrahedral $[SiO4]^{4-}$ and $[AlO4]^{5-}$ [9]. The structure of natural zeolite formed by alumina-silica atoms tetrahedral form three-dimensional structures that are connected through the sharing of oxygen atoms to form cavities and channels that contain water molecules and cations alkali and or alkaline [10]. In the zeolite, there is the partial substitution of Si^{4+} by Al^{3+}, which results in excess of negative charge [4],[6]. The negative charge is neutralized by alkali and alkaline earth cations (Na^+, K^+, Ca^{2+}, or Mg^{2+}) [1],[11]. Zeolite structure, divided into three components, namely the framework of alumina silicates, empty space interconnected containing metal cations and water molecules. The water contained within the zeolite pores can be removed by heating to push the water in the zeolite pores [12].

Natural zeolites have advantages, including cation exchange capacity, high surface area due to the structure of the porous, has structural stability, capable of neutralizing the acid solution through H^+ ion exchange of cations in the solution with the cations zeolite. Behind its leading, natural zeolite also has a disadvantage since its content of impurities (alkali and alkaline earth metals), relatively low ion-exchange capacities, and the water content that can reduce the absorption capacity. Therefore, they are commonly activated by chemical or physical methods before used [13]. Ion-exchange with sodium chloride solution is widely employed because sodium ions can readily remove other cations and therefore enhancing the adsorption capacity [14]. Hydrochloride acid (HCl) is another reagent used to activate the natural zeolite. The ability of HCl-activated natural zeolite was higher than the natural zeolite without activation. Hence, the HCl substance might be able to exchange cation in the framework and to dissolve undesired components such as metal oxides contained within the zeolite [15].

In the view of environmental sustainability, the use of NaCl is more attractive than HCl, because it does not dispose of waste that damages the environment. Unfortunately, the activation process with NaCl still requires quality improvement. This process requires a long time consuming on the activation process, and the cation exchange capacity obtained is still relatively low. The aim of this research is to study the activation of natural zeolite using a solution of NaCl, which is treated by ultrasonic cavitation. Variations of two factors were examined, namely sonication time and NaCl concentration in response to cation exchange capacity and surface area indicated by iodine numbers.

2. Method of Experiment

This section consists of two subsections, namely, materials and equipment, and experimental and analytical.

2.1. Materials and equipment

Natural zeolite materials used in this study were obtained from Cikalong, Tasikmalaya Regency, West Java Province, Indonesia. Sodium hydroxide, oxalic acid, hydrochloric acid, iodine, potassium bromide, silver nitrate, sodium thiosulfate were from Sigma Aldrich. Meanwhile, equipment used in this study are crusher, sieve shaker, oven dryer, ultrasonic homogenizer (TU-250Y probe 6 mm, power 250 W), and instrument analysis X-Ray Florence (EDXRF, Nex CG Rigaku), FTIR (IRTracer-100 Shimadzu), SEM (TM 3000 Hitachi), spectrophotometer uv-vis single-beam 1280 (Shimadzu).

2.2. Experimental and analytical

2.2.1. Activation of zeolite. Natural zeolite was cleaned, crushed, and sieved to obtain -80+100 mesh size and dried. The dried natural zeolites 20 g were added to 100 mL of sodium chloride on a certain concentration. The mixture was sonicated for certain minutes at room temperature on ultrasound homogenizer TU-250Y probe 6 mm, power 250 W by operation of 5 seconds on and 1 second off, and percent power 50 and filtered. The activated zeolites were then air-dried prior to further use. There were two factors of ultrasound-assisted activation have been investigated. Factor 1 is sonication time, and factor 2 is the concentration of NaCl. The study of factor 1 effects was carried out by activating of zeolite with 1N NaCl solution on the series of sonication time from 20,30, 40, 50, 60 minutes. The study of factor 2 was carried out by treating of zeolite on the variation of NaCl concentration 0,25N, 0,5N, 1N, 2N, 3N, 4N on the best sonication time.

2.2.2. Cation Exchange Capacity (CEC)
The method adopted in this study is the acid-base titration method. The dried sample 1 gram was added to a 10 mL solution of 5.5 N HCl and stirred for 2 hours with the speed of the stirring motor on a scale of 3. After the time stirring completed, the mixture separated by filtering and then washing the solid using distilled water to remove the remaining chlorine (white deposits) followed by washing with NaOH. The first washing was done by adding distilled water until the filtrate did not contain chloride. The test was done by putting $AgNO_3$ to the filtrate. The second washing was carried out by pouring carefully 250 ml of 1N NaOH, which has been standardized with oxalic acid. The cation exchange capacity was determined by acid-base titration of the last filtrate using 1 ml of 1N HCl with phenolphthalein (PP) indicator. The formula for testing the CEC value is to use the following equation:

$$\text{CEC meq/100 g} = ((N_1-N_2)\text{x}23\text{x}100)/(4\text{x}40),$$

where: N_1 = normality of NaOH for washing, N_2 = normality of NaOH after washing

2.2.3. Iod number measurement. Iod number measurement was carried out to determine surface area based on the result of the Mianowski's research [16]. Zeolite sample 0.5 g was added to 100 mL beaker glass, which contained 50 mL of 0.1 N iodine solution. The mixture was stirred at three scale magnetic stirrer for 10 minutes and filtered. Twenty mL filtrate was then titrated with 0.1 N sodium thiosulfate until pale yellow and added starch indicator, and titration continued until the blue color of solution disappeared. The volume of thiosulfate then recorded (Vs), and the titration is repeated for 20 ml of standard iodine solution (without the addition of zeolite) as blank. The volume of thiosulfate then recorded as Vb. The iodine number was calculated by the followed formula ;

$$\text{Iodine number} = \frac{V_{\text{fil}} - \left(\frac{V_{\text{tio}} \text{ x } N_{\text{tio}}}{N_{\text{iod}}}\right) \text{ x } W_{\text{iod}}}{\text{Mz}}$$

Where V_{fil} = volume of filtrate, V_{tio} = volume of thiosulfate, N_{tio} = concentration of thiosulfate in normality, N_{iod} = iodine normality, Mz = zeolite mass, W_{iod} = molecular weight of iodine

2.2.4. Chemical analysis. Dry natural and activated zeolite samples were crushed by milling machine for 1-5 min to a fine powder. 8.0 g of the powder samples and 2 g of binder wax were homogenized by mixing, pressed in a small aluminum cup under pressure of about 200 KN for 30 seconds to get a disc of the sample with 40 mm diameter and 4 mm thickness. The XRF instrument EDXRF, Nex CG Rigaku operated at 50 KV and 10 mA was used to analyze the major oxides in these samples.

3. Result and discussion

3.1. Zeolite Morphology
Zeolite morphology was obtained by instrument scanning electron microscopy (SEM) TM 3000 Hitachi. Figure 1 shows the SEM images of natural (a) and activated zeolites (b). SEM image shows the morphology of natural zeolite consists of amorphous and crystal materials with various shapes and sizes. The picture of natural zeolites shows various shapes of crystal, such as fiber-like, rod-like, prism-like, needle-like. That picture indicated that the zeolite mineral consists of mordenite and quartz, as stated by Widianti [17]. Sodium chloride (NaCl) activation induced by ultrasound

cavitation did not make any change on crystal forms. There were crystal in the prism formation still existed on both zeolite but activated zeolite illustrated cleaner crystal with bigger pore size. The bigger pore is about 6 μm for raw zeolite and 10 μm for activated zeolite, while the smaller pore is about 0,5 μm for raw zeolite and 1 μm for activated zeolite. Those images indicated that the amorphous structure materials had been diluted.

(a) (b)

Figure 1. SEM image natural zeolite (a) and activated zeolite with NaCl (b)

3.2. Chemical compositions

The EDXRF Nex CG Rigaku instrument was used to analyze the chemical composition of natural zeolite and activated natural zeolite. Table 1 indicates the content of the two types of zeolite. The XRF characterization shows that the silica and alumina are the main compositions of zeolites. Other elements that are in zeolites include magnesium, calcium, potassium. Theoretically, these elements can be exchanged with other metals. While other elements, which are impurities present in the zeolite pore, can be reduced by heating and chemical activation in the zeolite by acid, base, or salt [18]. Based on Table 1, it can be seen that the activation with sodium chloride solution (NaCl), which was induced by ultrasound cavitation on natural zeolite, leads to an insignificant change of almost all cations except Ca^{2+} and Fe^{3+}. The percentage of Ca element inactivated natural zeolite is smaller than the percentage in the raw natural zeolite, while the Fe element is the opposite. The percentage of Fe_2O_3 in activated zeolite increased by 7%, while the percentage of Ca_2O_3 decreased by 14% compared to raw natural zeolite. Natural zeolite pore cleansing has been carried out during activation.

Table 1 shows that the ratio of SiO_2/Al_2O_3 natural zeolites is 6.2 (79.5/12.8), and increased slightly to 6.5(79.2/12.2) in the activated natural zeolite. The composition of silica and alumina in activated natural zeolites decreased compared with the results of XRF natural zeolite. The decrease in the percentage of SiO_2 is much smaller than Al_2O_3, which is 0, 377 percent compared to 4.687 percent. This has happened might because there had been an exchange of Na^+ cations from the NaCl activator solution to Al^{3+} on alumina structure. Dealumination of zeolites on sodium chloride salt treatment is more effective than their desilication.

Table 1. Chemical analysis (EDXRF) of natural zeolite and natural zeolite activated

	SiO_2 (%)	Al_2O_3 (%)	K_2O (%)	CaO (%)	Fe_2O_3 (%)	TiO_2 (%)	SrO (%)	MnO (%)	others (%)
Natural zeolite	79.500	12.800	3.720	1.940	1.600	0.190	0.095	0.030	0.125
Activated natural zeolite	79.200	12.200	3.610	1.650	1.720	0.189	0.097	0.029	1.305

3.3. Effect of sonication time

Figure 2 indicates that the effect of sonication time on the response of cation exchange capacity (CEC) and an iodine number value (IN). The value of iodine number (IN) related to surface area (S_{BET}) [16], which is analyzed by BET methods [19]. The relation could be applied on the range 0 – 800 IN with formula S_{BET} (m^2/gr) ≅ 0.986 IN (mg/ gr) ≅ 1 IN (mg/ gr). Based on Fig. 3, it can be understood that the longer sonication time will increase the value of the cation exchange capacity and the value of the iodine number to a certain value, then a further increase of sonication time causes a decrease in the value of both responses. The maximum response occurs at 50 minutes of sonication time.

The longer the sonication time, the more bubbles are formed, and the more bubble bursts occur. This bubble burst energy is capable of producing free radicals, which can accelerate the process of adsorption or reactions related to cation exchange. The longer the activation time, the bubble growth (bubble) due to the cavitation effect gained to optimal. Mass transfer occurs when the bubble that carries Na^+ ions broken into the natural zeolite tetrahedral framework and exchange Al^{3+} ions. The optimal CEC value is obtained at the activation time of 50 minutes. While in the 60 minutes sonication, the CEC value has decreased, this is because the bubble growth phase (bubble) has passed the optimal growth time so that ion exchange does not occur and only generates heat.

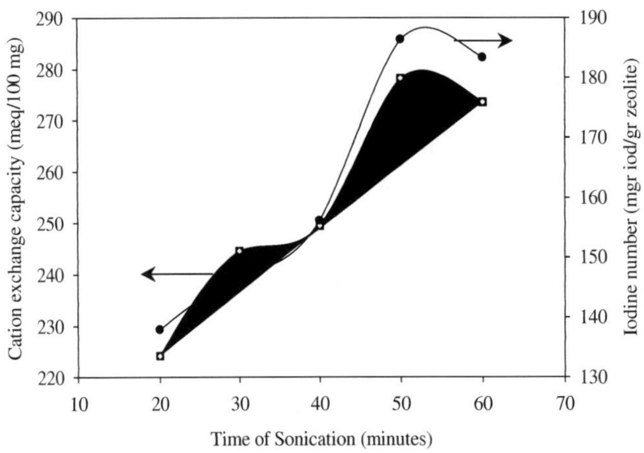

Figure 2. Effect of sonication's time on cation exchange capacity and iodine number

3.4. Effect of concentration NaCl.

Figure 3 shows the cation exchange capacity and iodine number values at various activator concentrations (NaCl). Based on Fig. 3, it can be seen that an increase in NaCl concentration will increase the value of cation exchange capacity and the value of iodine number to a certain value, but increasing NaCl concentration after 0.5 N causes a sharp decrease in the value of both responses. The decrease of CEC is because the active site in the zeolite framework has been saturated, where the amount of Al^{3+} ions was exchanged for Na^+ ions from NaCl activating solutions. The Na^+ ion, which cannot exchange Al^{3+} ions in the tetrahedral structure, will attach to the outer surface of the zeolite. The decrease in the value of the iodine number with increasing NaCl concentration after the 0.5 N because of the increasing number of Na^+ ions, which are not exchanged with other ions in the tetrahedral zeolite structure. These ions accumulated together with other ions that might be formed in the sonication process and stick to the surface of zeolite so that it can cover the zeolite pores. Therefore, the amount of iodine entering the zeolite pore is only small.

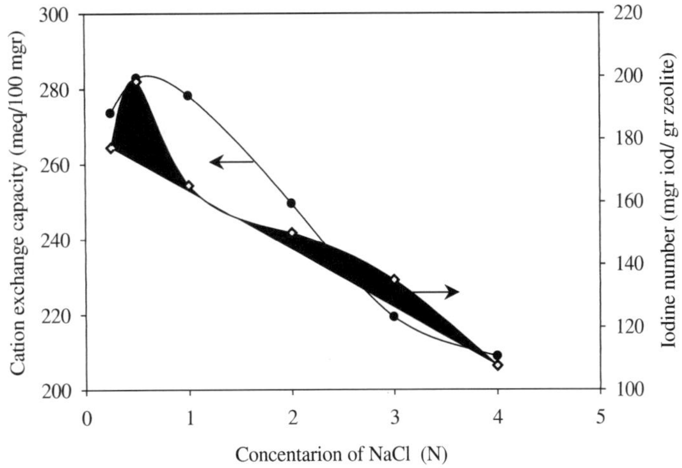

Figure 3. Effect of concentration NaCl on cation exchange capacity and iodine number

4. Conclusion

An active zeolite could be prepared by activation using sodium chloride (NaCl) solution induced by ultrasonic cavitation. The effect of sonication time on the quality of active zeolite could be explained that the iodine number and cation exchange capacity increase along with the increase of sonication time before 50 minutes. On the other hand, the iodine number and cation exchange capacity decrease along with the increase of NaCl concentration for greater than 0.5 N. This condition stated that the highest quality was found on 50 minutes irradiation for the fixed NaCl concentration and 0,5 N concentration for the fixed sonication time.

References

[1] Erdem E, Karapinar N, and Donat R 2004 *J. Coll. Interface Sci.* **280** 309
[2] Motsi T, Rowson N A, and Simmons M J H 2009 *Int. J. Mineral Process.* **92** 42
[3] Wang S and Peng Y 2010 *Chem. Eng. J.* **156** 11
[4] Colella, C., Mumpton, F.A. 2000 Natural Zeolites for the Third Millennium. ICNZ, *International Committee on Natural Zeolites*. A. De Frede.
[5] Ramesh, K., Damodar Reddy, D., Kumar Biswas, A., Subba Rao, A. 2011 Zeolites and their potential uses in agriculture *Adv. Agron.* **113**, 219-241
[6] Bowman, R.S. 2003 Applications of surfactant-modified zeolites to environmental remediation. *Microporous Mesoporous Mater.* **61**, 43-56
[7] Kowalczyk, P., Sprynskyy, M., Terzyk, A.P., Lebedynets, M., Namie_snik, J., Buszewski, B., 2006 Porous structure of natural and modified clinoptilolites *J. Colloid Interface Sci.* **297**, 77-85.
[8] Kusdarto, 2008, Indonesian zeolite potential *Indonesian Zeolite Journal* **Vol. 7** No. 2. November 2008 (in the Indonesian language)
[9] Setiawan Y., F. Widhi Mahatmanti, and Harjono 2018 Preparation and characterization of Nanozeolites from Gunungkidul Natural Zeolites with Top-Down Method *Indo. J. Chem Sci* **7** (1); 43-49 (in Indonesian language)
[10] Mirzaei N, Hadi M, Gholami M, Fard R.F., and Aminabad M.S. 2016 *J. the Taiwan Inst. Chem.Eng.* **59** 186
[11] Doula, M.K. 2006 Removal of Mn^{2+} ions from drinking water by using Clinoptilolite and a Clinoptilolite-Fe oxide system *Water Research* **40** (17) 3167–3176

[12] Lestari D Y 2010 Proceedings of the National Seminar on Chemistry and Chemistry Education Yogyakarta 1 (in Indonesian language)

[13] Inglezakis, V.J., Hadjiandreou, K.J., Loizidou, M.D., Grigoropoulou, H.P. 2001 Pretreatment of natural clinoptilolite in a laboratory-scale ion-exchange packed bed *Water Research* **35** (9), 161–2166.

[14] Günay, A., Arslankaya, E., Tosun, I. 2007 Lead removal from aqueous solution by natural and pretreated clinoptilolite: Adsorption equilibrium and *kinetics Journal of Hazardous Materials* **146** (1–2), 362–371.

[15] Ola, P.D., and Nesimnasi, A.M. 2012 Utilization of Natural Zeolites to Improve Quality of Used Cooking Oil *Journal of Mathematics and Natural Sciences,* **Vol. 12**, No. 1 59-68 (in Indonesian language)

[16] Mianowski, M. Owczarek & A. Mareck 2007 Surface Area of Activated Carbon Determined by the Iodine Adsorption Number *Energy Sources, Part A*, 29:839–850

[17] Widianti, T. 2006 Testing the zeolite cation exchange capacity as a natural cation exchange for industrial wastewater treatment, *Annual meetings on testing and quality*, Indonesian Institute of Sciences, Research Center for Quality Systems and Testing Technology (in Indonesian language)

[18] Margeta K, Logar N , iljeg and arkas A 2013 *Natural zeolites in water treatment –howeffective is their use (Water Treatment)* ed W Elshorbagy and R K Chowdhury (Rijeka: InTech) chapter 5 pp 81-112

[19] Brunauer, S., Emmett, P. H., and Teller, E. J. 1938 Surface Area Measurements of Activated Carbons, Silica Gel, and Other Adsorbents. *J. Am. Chem. Soc*. **30**:309–315.

IOP Conf. Series: Materials Science and Engineering **771** (2020) 012044 doi:10.1088/1757-899X/771/1/012044

Analysis of Flood Discharges of Serayu River at Serayu Barrage using HSS Gama I and Nakayasu Models

Teguh Marhendi[1,*] and Fathurohman[1]

[1] Department of Civil Engineering, Universitas Muhammadiyah Purwokerto, Indonesia.

Corresponding author: *tmarhendi@gmail.com, afthur.rokhman@gmail.com

Abstract. Serayu River, located in the central of the Java island, is one of the largest rivers in the island. The length of the river to the control point of the Serayu barrage is 118.276 km and the basin areas is 3001.73 km^2. The aim of the research was to analysis flood discharge in the Serayu river at Serayu Barrage. The analysis of flood discharges on the Serayu River would use the HSS Gama I and Nakayasu models. The result of research shown that f flood discharges using the Nakayasu model Qp = 18.833 m3/sec and HSS Gama I model Qp = 59.953 m3/sec. by using rainfall intensity of 7.431 mm / hour, the peak discharge with the Nakayasu model Qp = 139.156 m3/sec and HSS Gama I model Qp = 445.509 m3/sec.

1. Introduction

The river is an important part of human life. The river is a large and elongated water flow that flows continuously from upstream to downstream. The river besides being used for human needs is also for meeting irrigation and environmental needs. Rivers can be formed naturally, through processes carried out by nature, and can also be formed due to the process of the human intervention[14].

One part that needs to be analyzed from river flow is the amount of flood [8] [17]. Flood is a condition where water overflows outside the normal limits of river, lake and seawater flow. The occurrence of floods is not only influenced by natural phenomena, but more dominant is the impact of human activities. Flood disaster is a routine phenomenon in the rainy season in various watersheds in various parts of Indonesia.

Serayu River is one of the largest rivers on Java that needs to be considered the magnitude of the flood. The average debit of the Serayu River in the control point of the Serayu Barrage in 2007-2016 ranged from 87.28 m3 / sec in August and 377.70 m3 / sec in December as in Table 1 and Figure 1. The river crosses several districts in the Province Central Java includes the Regencies of Wonosobo, Banjarnegara, Purbalingga, Banyumas and Cilacap. Serayu River has a length of 118,276 km2 with an area of Watershed reaching 3001.73 km2. The Serayu River Basin has experienced environmental degradation and pollution which has resulted in a decline in the quality of Serayu River Water [1] [2] [3] [11] [15]. To reduce the risk of damage due to overflowing river water flood control efforts are needed. Flood control planning can be done well if it is known that the flood discharge [4] [5] [7] [8]. In this study, the analyzes of flood discharge in the Serayu River will be using the HSS Gama I and Nakayasu methods [5] [11] [12] [13]. The calculation is intended to find a large difference in flood discharge using these two methods [11].

Table 1. Serayu River average Discharge at the Serayu Barrage control point in 2007-2016 [2]

Month	Jan	Feb	Mar	Apr	Mey	Jun	Jul	Aug	Sep	Oct	Nov	Dec
Discharge	339,13	312,92	337,58	308,80	198,85	147,49	135,01	87,28	108,89	206,72	322,29	377,70

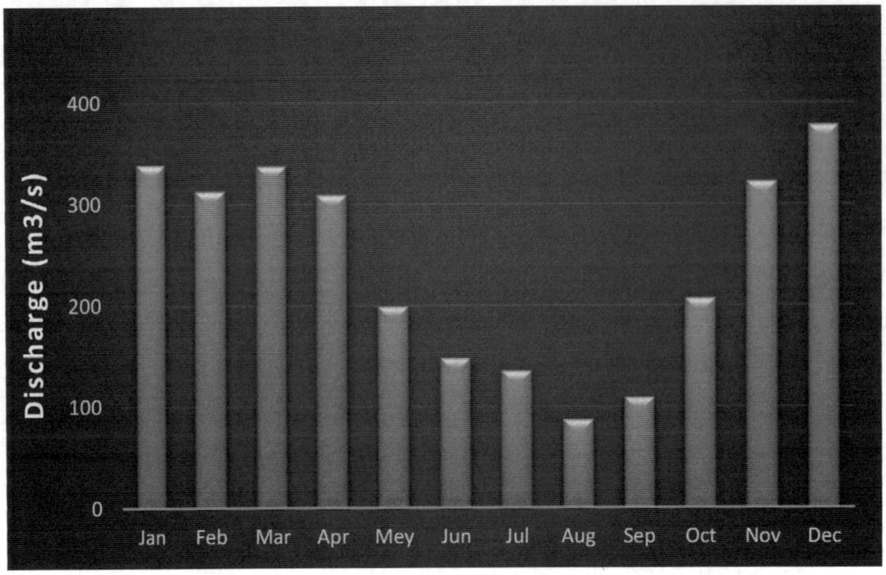

Figure 1. Serayu River average Discharge at the Serayu Barrage control point in 2007-2016 [2]

2. Methodology

2.1. Research Sites

The location of this research is located on the Serayu River, with the control point in the Serayu Barrage, exactly in Kebasen Village, Rawalo District, Banyumas Regency, Central Java Province.

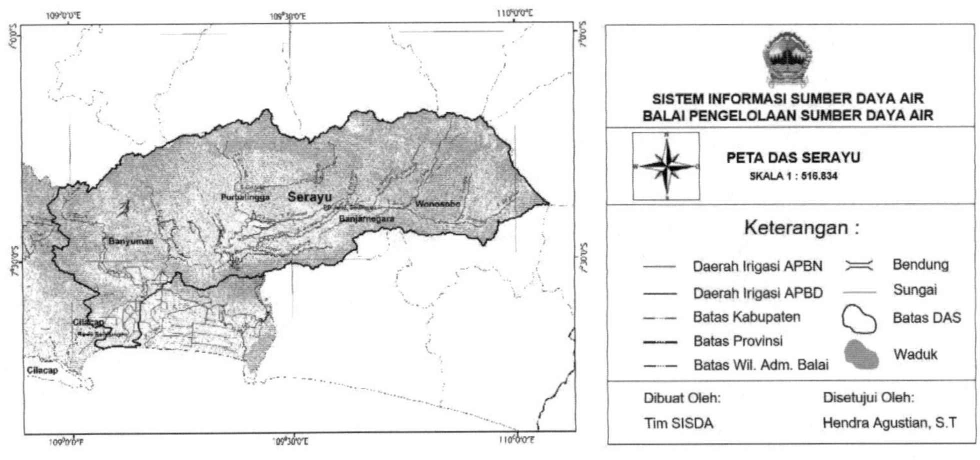

Figure 2. Maps of Serayu Basin [2]

2.2. Polygon Thiessen

Rainfall analyses were performed using the Thiessen Polygon Method. This method was chosen because it has sufficient accuracy, so it is very good if it is used to calculate the average rainfall Serayu Basin, each of which is influenced by the location of the rainfall observation station based on the river network maps and the location of the observation station [3] [10] [12] [13]. Thiessen's coefficient can be determined by the following equation [4] [16] [17]:

$$C_i = \frac{A_i}{A_{total}} \tag{1}$$

$$\overline{R} = \frac{A_1R_1+A_2R_2+A_3R_3+\cdots+A_nR_n}{A_1+A_2+\cdots+A_n} \tag{2}$$

with :
C = Thiessen coeficien
Ai = Influence of areas from observation of station i (km^2)
A = Total Areas Basin (km^2)
$\overline{R}$ = Average Rainfall (mm)
R1, R2, ..., Rn = Rainfall at each measurement point (mm)

2.3. *Rainfall Intencity*

To obtain rainfall intensity data, Mononobe formula is used as follows [5] [7] [9] [17]
:

$$I = \frac{R24}{24} \times [\frac{24}{t}]^{\frac{2}{3}} \tag{3}$$

with :
I = rainfall intensity (mm).
R24 = maximum daily rainfall (mm).
t = rainfall duration or time of concentration (hr).

2.4. *HSS Gama I*

HSS Gama I consists of four main variables namely[5][6]:

2.4.1. *Time of rise (Tr)*

$$T_r = 0{,}43 \times \frac{L}{100 \times SF} + 1{,}0665 \times SIM + 1{,}2775 \tag{4}$$

2.4.2. *Peak fo Discharge (Qp)*

$$Q_p = 0{,}1836 \times A^{0,5886} \times T_r^{-0,4008} \times JN^{0,2381} \tag{5}$$

2.4.3. *Time of base (Tb)*

$$T_b = 27{,}4132 \times T_r^{0,1457} \times S^{-0,0986} \times SN^{0,7344} \times RUA^{0,2574} \tag{6}$$

2.4.4. *Storage Coefficient (K)*

$$K = 0{,}561 \times A^{0,1457} \times S^{-0,1446} \times SF^{-1,0897} \times D^{0,0452} \tag{7}$$

with :
A = Basin Areas
L = length of the main river channel
S = slope length of the riverbed
D = drainage network density
SF = source factor
SN = Source frequency
JN = number of the river junction
RUA = a comparison between the upstream watershed area and the total watershed area
SIM = Symmetry factor

2.5. *HSS Nakayasu*

Nakayasu (1950) has investigated unit hydrographs in Japan and provided a set of equations to form a unit hydrograph as follows [5] [9]:

2.5.1. time lag, (tg)

$$t_g = 0,4 + 0,058 \text{ x } L \qquad \text{for } L > 15 \text{ Km} \qquad (8)$$

$$t_g = 0,21 \text{ x } L^{0,7} \qquad \text{for } L < 15 \text{ Km} \qquad (9)$$

2.5.2. Time to Peak

$$t_p = t_g + 0,8\ T_r \qquad (10)$$

2.5.3. The time when discharge equal to 0,3-time peak of discharge

$$t_{0,3} = \alpha \text{ x } T_g \qquad (11)$$

2.5.4. Peak discharge

$$Q_p = \frac{1}{3,6} \text{ x A x } R_o\text{x } \frac{1}{(0,3\ x\ t_p\ x\ t_{0,3})} \qquad (12)$$

with :

tg = time lag (hr)
L = length of river (Km)
t0,3 = Time when discharge equal to 0,3 time peak of discharge (hr)
α = coefficient, value untill 1,5 – 3,0
tp = time to peak (hr)
Qp = peak discharge (m^3/s)
A = basin areas (Km^2)
Tr = rainfall duration (hr
Ro = rainfall depth unit (mm)

3. Result and Discussion

3.1. Rainfall Calculation with Polygon Thiessen Method

Calculation of rainfall used in the analysis of flood discharges in the Serayu River uses the Polygon Thiessen model. The division of rain areas according to Thiessen polygons can be seen in the figure below.

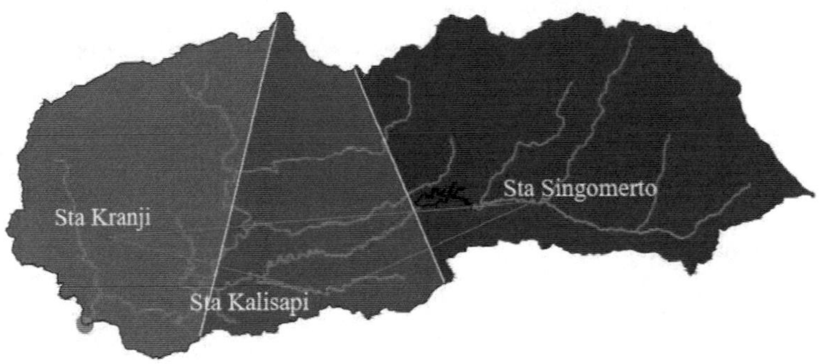

Figure 3. Serayu Basin and Polygon Thiessen

The results of the calculation of the average rainfall of the Thiessen method are as follows:

Table 2. Calculation average daily rainfall

No	year	Kranji		Kalisapi		Singomerto		Average daily (mm)
		0,3106		0,2474		0,4420		
		RI	CRI	RI	CRI	RI	CRI	
1	2009	95	29,367	86	21,279	121	53,302	103,948
2	2010	223	69,262	128	31,672	187	82,649	183,583
3	2011	438	136,038	110	27,218	150	66,296	229,552
4	2012	128	39,755	108	26,723	117	51,711	118,190
5	2013	83	25,779	223	55,178	149	65,633	146,590
6	2014	88	27,332	67	16,578	113	49,722	93,632
7	2015	83	25,779	114	28,084	103	45,523	99,386
8	2016	97	30,127	88	21,830	110	48,750	100,707
9	2017	52	16,151	99	24,496	96	42,341	82,988
10	2018	80	24,847	102	25,238	315	139,001	189,087
								134,766

3.2. Nakayasu Method

3.2.1. Time of concentration

L = 118,276 km > 15 km,
tg = 0,4 + 0,058 x 118,276 = 7,26 hr
tr = was taken 0,5 x 7,26 = 3,63 hr
Tp = tg + 0,8tr
= 10,164 hr
T0,3 = α x tg
= 2 x 7,26 = 14,52 hr

3.2.2. Rainfall Intensity.

The intensity used is the intensity obtained from the average monthly maximum rainfall for ten years. From table 4.3 the average rainfall ($\bar{X}$) value is 134,766 mm.

To obtain rainfall intensity data Mononobe formula is used as follows:

$$I = \frac{R24}{24} \times \left[\frac{24}{t}\right]^{\frac{2}{3}}$$

$$I = 7{,}389 \text{ mm/hr}$$

3.2.3. Peak discharge calculation

$$Q_p = \frac{1}{3{,}6} \times A \times R_0 \times \frac{1}{(0{,}3 \times t_p \times t_{0{,}3})}$$

$$Q_p = 18{,}833 \text{ m}^3/\text{s}$$

3.2.4. Nakayasu hydrograph

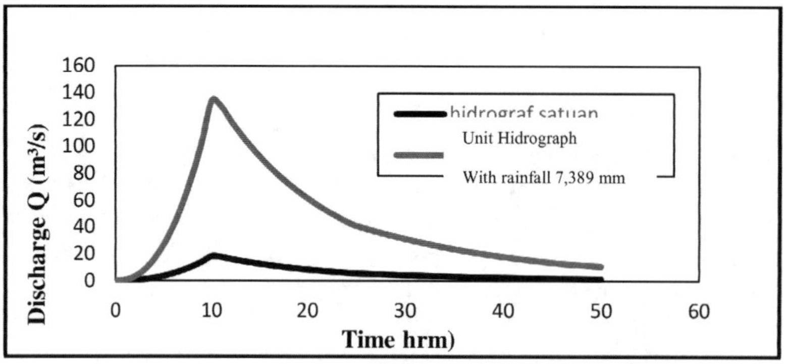

Figure 4. Nakayasu hydrograph of Serayu River

3.3. HSS Gama I Method

3.3.1. HSS Gama I Parameter

The watershed parameters needed in the calculation using the HSS Gama I method can be seen in Table 3 below:

Table 3. HSS Gama I Parameter

No	Parameter	Symbol	unit	value
1	Number of kevel 1 river length	L1	km	118,276
2	Total river length all levels	LN	km	257,591
3	Number of level 1 river junction	P1		7
4	number of all level river junction	PN		8
5	Basin Upper width	W_U	km	30,198
6	Basin bottom width	W_L	km	38,200
7	Elevation of upstream		m	868,03
8	Elevation of downstream		m	10,065
9	Areas of basin	A	km^2	3001,73
10	Areas of upstream basin	AU	km^2	1163,43
11	Length of the main river	L	km	118,276
12	Source factorer	SF	-	0,459
13	Symmetry factor	SIM	-	0,306
14	Width factor	WF	-	0,791
15	Number of river junction	JN	-	8
16	Slope length average of river	S	-	0,007
17	Source Frequency	SN	-	0,875
18	Relative Areas of Upper basin	RUA	-	0,388
19	Drainage network density	D	km/km^2	0,086

3.3.2. Time of rise (Tr)

$$T_r = 0{,}43T \times \frac{L}{100 \times SF}^{3} + 1{,}0665 \times SIM + 1{,}2775$$
$$T_r = 8{,}95 \text{ jam}$$

3.3.3. Time of base (Tb)

$$T_b = 27{,}4132 \times T_r^{0,1457} \times S^{-0,0986} \times SN^{0,7344} \times RUA^{0,2574}$$
$$T_b = 43{,}558 \text{ jam}$$

3.3.4. Storage coefficient (K)

$$K = 0{,}561 \times A^{0,1457} \times S^{-0,1446} \times SF^{-1,0897} \times D^{0,0452}$$
$$K = 10{,}098 \text{ jam}$$

3.3.5. Peak discharge (Qp)

$$Q_p = 0{,}1836 \times A^{0{,}5886} \times T_r^{-0{,}4008} \times JN^{0{,}2381}$$
$$Q_p = 13{,}913 \ m^3/dt$$

3.3.6. HSS Gama I Hydrograph

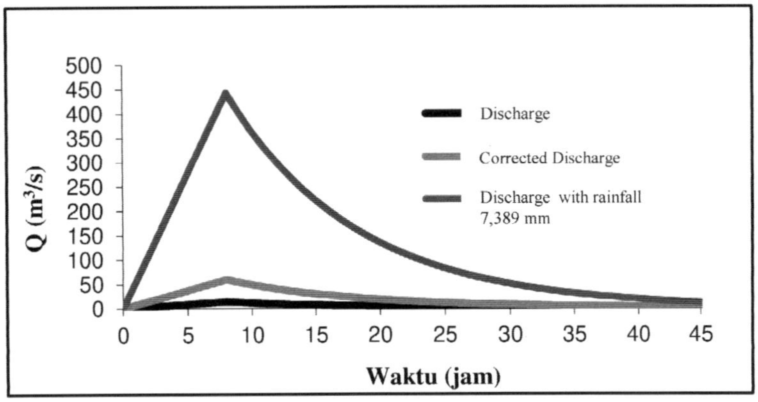

Figure 5. HSS Gama I Hydrograph Serayu River

4. Conclusion

Based on the results of the analyzes that has been done, the results of flood discharge calculations using 2 methods are as follows: Nakayasu Method Qp = 18,833 m³ / sec at 10,164 hours, HSS Gama I Method Qp = 59,953 m³ / sec at 8.95 hours, With a rainfall intensity of 7,431 mm / hour the peak discharge becomes: Nakayasu Method Qp = 139,156 m³ / sec at 10,164 hours, HSS Gama I Method Qp = 445,509 m³ / sec at 8.95 hours.

References

[1] Munir, A. 2009. Karakteristik Daerah Aliran Sungai (Das) Serayu Provinsi Jawa Tengah Berdasarkan Kondisi Fisik, Sosial Serta Ekonomi. Universitas Indonesia. Depok

[2] Anonim, 2019, Balai Pengelolaan Sumber Daya Air Serayu-Citanduy. Purwokerto

[3] Sodikin, W. 2017. Analisis Debit Banjir Kala Ulang Sungai Kali Sapi (Studi Kasus Keruntuhan Jembatan Kali Sapi). Skripsi. Universitas Muhammadiyah Purwokerto. Purwokerto

[4] C.D.,Soemarto. (1999). Hidrologi Teknik, Erlangga, Jakarta

[5] Kamiana, I made. 2011. Teknik Perhitungan Debit Rencana Bangunan Air Graha Ilmu. Yogyakarta

[6] Andiese, Vera Wim. 2012. Pengujian Metode Hidrograf Satuan Sintetik Gama I Dalam Analisis Debit Banjir Rancangan Das Bangga. Mektek tahun XIV No. 1, Januari 2012

[7] Harahap, R., Kemala Jeumpa, Bambang Hadibroto. 2017. *Flood Discharge Analysis with Nakayasu Method Using Combination of HEC-RAS Method on Deli River in Medan City*. ICIESC-2017. IOP Conf. Series: Journal of Physics: Conf. Series 970 (2018) 012011

[8] Halim, Fuad. 2014. *Pengaruh Hubungan Tata Guna Lahan Dengan Debit Banjir Pada Daerah Aliran Sungai Malalayang*. Jurnal Ilmiah Media Engineering Vol.4 No.1, Maret 2014 (45-54) ISSN: 2087-9334

[9] Harun, Lukman 2002. Perbedaan Hitungan Debit Banjir Menggunakan Metode Rasional, Nakayasu dan Synder. Skripsi. Universitas Muhammadiyah Purwokerto. Purwokerto

[10] J, Loebis, 1984. *Banjir Rencana Untuk Bangunan Air,* Departemen Pekerjaan Umum, Badan Penerbit Pekerjaan Umum, Jakarta.

[11] Maryani, Nesti. 2019. *Analisis Debit Andalan Sungai Pedes Hulu Di Kaligiri Sirampog, Brebes*. Skripsi. Universitas Muhammadiyah Purwokerto. Purwokerto

[12] Prastica, Rian Mantasa Salve., Maitri, Caya., dkk. 2017. *Analisis Banjir dan Perencanaan Desain Transportasi Sungai di Kota Bojonegoro*. Media Komunikasi Teknik Sipil, Vol 23, No. 2, 2017, 91-101

[13] Soewarno, 1995. Hidrologi Aplikasi Metode Statistik Untuk Analisa Data, Nova, Bandung.

[14] Sartika, Isri. 2016. *Analisis debit banjir sungai molompar kabupaten minahasa tenggara*. Jurnal Sipil Statik. Vol.4 No.2 Hal. 123-133)

[15] Salim Noor. 2015. Determining The Coefficient Of Tr, A And River Length (L) Of Flood Runoff Model Using Synthetic Unit Hydrograph Nakayasu (Case Study Bedadung Watershed Jember). IOSR Journal of Mechanical and Civil Engineering (IOSR-JMCE). e-ISSN: 2278-1684, p-ISSN : 2320-334X, Volume 12, Issue 6 Ver. VI

[16] Sriyono, edy. 2012. *Analisis Debit Banjir Rancangan Rehabilitasi Situ Sidomukti*. Jurnal Teknik vol. 2 no. 2 / Oktober 2012

[17] SNI 2415:2016. *Tata Cara Perhitungan Debit Banjir Rencana*. Badan Standardisasi Nasional. Jakarta.

IOP Conf. Series: Materials Science and Engineering **771** (2020) 012045 doi:10.1088/1757-899X/771/1/012045

The Assessment of BRT Trans Semarang Service Quality

Amalia Nur Adibah[1,*] and Okto Risdianto Manullang[2]

[1]University of Muhammadiyah Malang, Malang, Indonesia
[2]Diponegoro University, Semarang, Indonesia

[*]Corresponding author e-mail: amalianuradibah@umm.ac.id

Abstract. Semarang city is one of the metropolitan cities that has congestion problems. The existence of BRT Trans Semarang becoming an alternative to reduce city congestion problems. Besides the service performance of BRT Trans Semarang, the service quality is also important in considering the travel modes choice. Tangibles are assessed based on the condition of the facility, the existence of information boards, the presence of officers, and the availability of information. Empathy is assessed based on the existence of accident insurance. Reliability is assessed based on service personnel. Responsiveness is assessed based on officer responsiveness. Assurance is assessed based on the level of knowledge and politeness of the officers. The main purpose of this research is to assess the service quality of BRT Trans Semarang related to tangibles, empathy, reliability, responsiveness, and assurance on weekdays and weekend. This research used the Kano Model to assess the service quality of BRT Trans Semarang based on user preferences. The results of the Kano Model analysis dividing the service quality indicators into several treatments can be repaired or added, can be used as a strategy, and can be prepared for the future. The government must prioritise improvements in corridor five because that corridor has the most deficiencies and need for improvement.

1. Introduction

Bus Rapid Transit (BRT) is one of the popular transportation systems as a more alternative affordable travel against trains. BRT is a medium mode of transportation between conventional buses and trains, so BRT has the speed and flexibility of an affordable train like conventional buses [1].

BRT system is one approach to solve the congestion problems in urban areas. BRT has the privilege from capacity, low cost, and being a priority alternative in big cities, considering its short application [2]. BRT is one of the mass transportation services that fast and efficient service against to conventional buses, and has low fares but still pay attention in terms of comfort, security, safety, time and cost efficiency [3].

SERVQUAL is a combination of several measures that have good reliability and validity and can be used to understand service expectations and user perceptions for improving service quality [4]. Measuring service quality is used to understand the needs and desires of users by analysing based on user experience and satisfaction with the services provided. The purpose of measuring service quality is to increase the productivity of service quality [5]. SERVQUAL is divided into five dimensions, tangibles, empathy, reliability, responsiveness and assurance. Tangibles are assessed by the appearance of physical facilities, equipment used, personnel, and communication materials. Empathy is assessed by the caring from service providers and attention to passengers. Reliability is assessed by the ability to provide reliable and accurate services. Responsiveness are assessed by the willingness to help passengers and provide quickly services. Assurance are assessed by the knowledge and politeness of officers and the ability to provide trust and confidence [4].

In Indonesia, BRT is one solution that is being intensively used in big cities, to reduce the level of congestion, one of which is in Semarang City. Public opinion regarding BRT services is also important to provide an assessment of the service provided. Assessment by involving the community will be used as input for improving the service quality so that it can attract more users. The more public transport users, in this case, BRT, can reduce congestion, especially in Semarang City.

2. Methodology

This research uses a descriptive quantitative method by observing the condition of BRT service quality in the Semarang city based on user preferences on weekdays and weekend. The data that has been collected then compiled based on each indicator of BRT service quality for each line. The methodology of the research study will be described in the **Figure 1**.

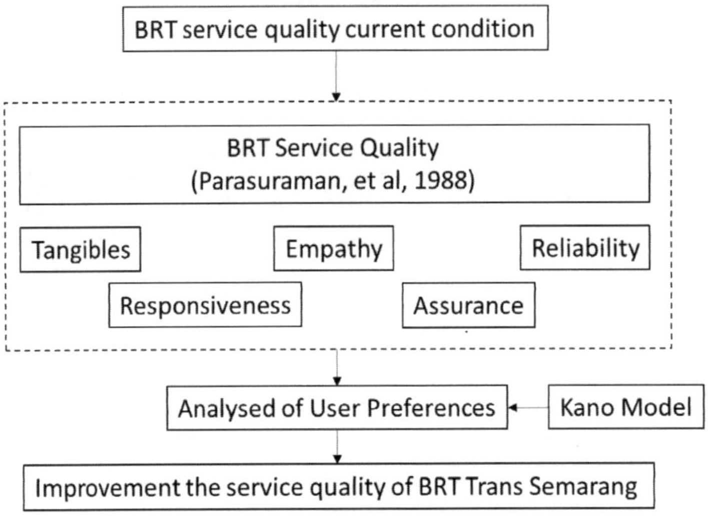

Figure 1. The methodology of the Research Study.

The quadrant in Kano Model analysis divided into 4 (four), i.e. [6]:

a. Must be
   If a product or service doesn't exist, it will make users dissatisfied, but the existence of a product or service doesn't have a significant effect on user satisfaction
b. One dimensional
   The attributes have a parallel correlation with the satisfaction level of a product or service. The higher level of service, the higher the level of user satisfaction.
c. Attractive
   The attributes have a parallel correlation between satisfaction and level of performance. User satisfaction increasing along with the increase in product or service performance, but the decline in product or service performance doesn't reduce the level of user satisfaction.
d. Indifferent
   The existence of attributes will not affect the user satisfaction level

Based on the importance and satisfaction level divided into 4 (four), i.e. [7]:

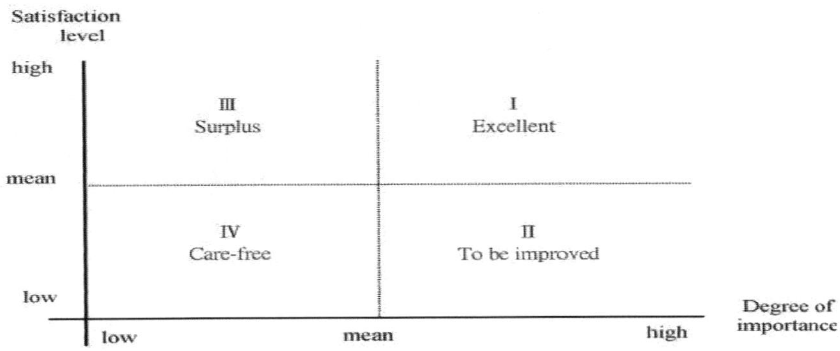

Figure 2. Importance and Satisfaction Model.

a. Quadrant 1 – excellent
Attributes in this area considered to be important by users, so the performance of this attribute be continued.
b. Quadrant 2 – to be improved
Attributes in this area considered important for users, but have bot been satisfactory or as expected. Improvements should be made to the attributes.
c. Quadrant 3 – surplus
The attributes quality in this area are not very important to users, but users feel very satisfied. This attributes can put in additional, depend on the priority cost.
d. Quadrant 4 – care-free
The attributes have a low satisfaction level and importance level. These attributes can be eliminated from the evaluation process.

3. Service Quality of BRT Trans Semarang

3.1. Tangibles
The assessment for BRT Trans Semarang service quality uses several tangibles indicators, such as the condition of bus stop, conditions on the bus, information boards at the bus stop, information boards on the bus, officers at the bus stop, officers on the bus, information availability at the bus stop, and information availability on the bus.

3.2. Empathy
The assessment for BRT Trans Semarang service quality uses one empathy indicator, that is accident insurance.

3.3. Reliability
The assessment for BRT Trans Semarang service quality uses two reliability indicators, service officers at the bus stop and on the bus.

3.4. Responsiveness
The assessment for BRT Trans Semarang service quality uses two responsiveness indicators, the responsiveness of the officers at the bus stop and on the bus.

3.5. Assurance
The assessment for BRT Trans Semarang service quality uses several assurance indicators, knowledge of the officers at the bus stop and on the bus, and also politeness of officers at the bus stop and on the bus.

3.6. TERRA
From the results of the analysis based on several indicators of tangibles, empathy, reliability, responsiveness, and assurance on weekdays and weekend. There are several alternative strategies from

assessing the quality service of BRT Trans Semarang. There are three alternative strategies from TERRA, improvements/addition, strategy, and preparation for the future. The result will be shown in **Table 1** for weekday, and **Table 2** for the weekend.

Table 1. TERRA Indicators Weekday.

Variable	Sub Variable	*Weekday*		
		Improvements/ additions	***Strategy***	***Preparation for the future***
Tangibles	Bus stop conditions	All corridors	-	-
	Conditions on the bus	Corridor 1, 3, and 4	-	Corridor 2 and 5
	Information board at the bus stop	All corridors	-	-
	Information board on the bus	Corridors 1, 2 and 3	Corridor 5	-
	Officers at the bus stop	Corridor 1 and 4	Corridor 3	Corridor 5
	Officers on the bus	Corridor 3 and 4	Corridor 1 and 5	-
	Information availability at the bus stop	Corridor 1, 2, 3, and 4	-	-
	Information availability on the bus	Corridor 1 and 4	-	-
Empathy	Accident insurance	Corridor 2, 3, and 5	-	-
Reliability	Service officers at the bus stop	Corridor 3, 4, and 5	Corridor 1	Corridor 2
	Service officers on the bus	Corridor 2, 3, 4, and 5	Corridor 1	-
Responsiveness	Responsiveness of the officers at the bus stop	Corridor 2, 3, 4, and 5	Corridor 1	-
	Responsiveness of the officers on the bus	Corridor 3 and 5	Corridor 2	Corridor 1 and 4
Assurance	Knowledge of the officers at the bus stop	Corridor 4	Corridor 2, 3 and 5	Corridor 1
	Knowledge of the officers on the bus	Corridor 2	Corridor 1 and 4	Corridor 3 and 5
	Politeness of officers at the bus stop	Corridor 4 and 5	Corridor 2	Corridor 1 and 3
	Politeness of officers on the bus	All corridors	-	-

Table 2. TERRA Indicators Weekend.

Variable	Sub Variable	*Weekend*		
		Improvements/ additions	*Strategy*	*Preparation for the future*
Tangibles	Bus stop conditions	Corridor 2, 3, and 4	-	Corridor 1 and 5
	Conditions on the bus	Corridor 1, 2, 3, and 4	-	Corridor 5
	Information board at the bus stop	All corridors	-	-
	Information board on the bus	All corridors	-	-
	Officers at the bus stop	Corridor 1, 2, and 4	Corridor 3	Corridor 5
	Officers on the bus	Corridor 2	Corridor 3, 4, and 5	Corridor 1
	The information available at the bus stop	Corridor 1, 2, and 4	-	-
	The information available on the bus	Corridor 1	Corridor 2 and 4	Corridor 3
Empathy	Accident insurance	Corridor 1, 2, 4, and 5	-	-
Reliability	Service officers at the bus stop	Corridor 2, 3, and 5	-	-
	Service officers on the bus	Corridor 2, 3, 4, and 5	Corridor 1	-
Responsiveness	Responsiveness of the officers at the bus stop	Corridor 1, 3, 4, and 5	Corridor 2	-
	Responsiveness of the officers on the bus	Corridor 2	Corridor 1 and 5	Corridor 3 and 4
Assurance	Knowledge of the officers at the bus stop	Corridor 4	Corridor 1, 2, and 5	-
	Knowledge of the officers on the bus	Corridor 1 and 3	Corridor 2, 4, and 5	-
	Politeness of officers at the bus stop	Corridor 3 and 5	Corridor 1 and 4	Corridor 2
	Politeness of officers on the bus	Corridor 3	Corridor 1 and 4	Corridor 2 dan 5

4. Results

The service quality of BRT Trans Semarang based on user preference shows that for tangibles, sub-variables that prioritised during the weekday are the conditions on the bus for corridor 3 and 4, while during the weekend is information available on the bus stop for corridor 2. For empathy, no prioritised sub-variables. For reliability, sub-variables that prioritised during the weekday are service officers at bus stops for corridors 3, 4, and 5, and also service officers on the buses for corridor 5, while during the weekend is service officers on the buses for corridor 3. For responsiveness, sub-variables that prioritised during the weekday is the responsiveness of the officers at the bus stop for corridor 3, while during the weekend is the responsiveness of the officers at the bus stop for corridors 1 and 5. For assurance, sub-variables that prioritised during weekday are politeness of officers at the bus stop for corridor 5, and politeness of officers on the buses for corridor 1, 4, and 5, while during weekend are politeness of officers at the bus stop for corridor 5, and politeness of officers on the buses for corridor 3.

5. Discussion

For improving the service quality of public transportation, it should pay attention to user preferences. This is because users can feel directly related to the service quality of public transportation. User preferences are juxtaposed with service evaluations that have been carried out previously, can improve service quality that appropriates in terms of user standards and needs.

6. Recommendation

This research can be used as a recommendation for the Government of Semarang City related to improving the service quality of BRT Trans Semarang. Prioritised improvement in corridor 5, because that corridor has the most deficiencies and needs to improve. To improve the service quality of BRT Trans Semarang, the role of the community is needed.

References

[1] R. Cervero, "Bus Rapid Transit (BRT): An Efficient and Competitive Mode of Public Transport," *IURD Work. Pap. 2013-01*, no. October, pp. 1–36, 2013.

[2] W. Liu and J. Teng, "Analysis and Optimization of the Capacity Bottlenecks of Elevated Bus Rapid Transit System," *Procedia - Soc. Behav. Sci.*, vol. 138, pp. 386–393, 2014.

[3] T. K. A. Putra and A. S. Kurnia, "Analisis Preferensi Masyarakat Terhadap Bus Rapid Transit (BRT) Trans Semarang," *Diponegoro J. Econ.*, vol. 3, 2014.

[4] A. Parasuraman, V. A. Zeithaml, and L. L. Berry, "A Multiple Item Scale for Measuring Consumer Perceptions of Service Quality," *J. Retail.*, vol. 64, no. 1, pp. 12–37, 1988.

[5] A. R. Ghotbabadi, S. Feiz, and R. Baharun, "Service Quality Measurements: A Review," *Int. J. Acad. Res. Bus. Soc. Sci.*, vol. 5, no. 2, pp. 267–286, 2015.

[6] E. Sauerwein, F. Bailom, K. Matzler, and H. H. Hinterhuber, "the Kano Model: How To Delight Your Customers," vol. I, pp. 313–327, 1996.

[7] C. Yang, "The refined Kano 's model and its application," *Total Qual. Manag. Bus. Excell.*, vol. 16, no. 10, pp. 1127–1137, 2005.

Landslide Analysis Study and Cisewu Countermeasures in District, Garut Regency, West Java

A Salimah[1,*], M Ammar[2], D Rahmawati[2], Yelvi[1] and Tri Widya S[1]

[1]Civil Engineering Department, Politenik Negeri Jakarta, Jl. Prof. DR. G.A. Siwabessy, Depok City, Jawa Barat. 16425

[2]Civil Engineering Department, Nusa Putra University, Jl No.21, Cibolang Kaler, Kec. Cisaat, Sukabumi Regency, Jawa Barat 43155

[*]e-mail: aisyah.salimah@sipil.pnj.ac.id

Abstract. Cisewu-Garut is one of the landslide-prone areas in Indonesia which occurs almost every year, particularly in the rainy season. The landslide in this study takes place in the slope where a BTS tower is located. The peripheral fence is hanging due to ground movement, while housings stay beneath the foot of the slope. This problem could harm its surroundings and needs immediate countermeasures to prevent any hazard. At the initial stage, a survey is conducted to investigate slope condition, soil state (including soil test), and drainage. Based on available soil data, the second stage is slope stability analysis by creating a model corresponding to real condition data using the element method to Plaxis2D software. As for the third stage, a landslide disaster prevention is established by strengthening the foot of slope using gabion and retaining walls. The gabion wall cage dimension would be 5m wide and 2.5m tall. Two retaining walls by the size of 2m high to 1m wide and 1m high to 0.6m wide are build. The result of slope stability analysis is an increase of safety factor by 1.729, which existing factor is 1.244. The countermeasures also involve drainage refinement and water flow diversion so that water will not penetrate to landslide zone.

1. Introduction

According to [1] landslide is a prominent natural disaster in Indonesia, notably in intense rainfall season. Parts of Indonesia archipelago have the form of highlands with steep slopes resulting in frequent landslide disasters. Several factors affecting the incidents are topography, soil and sediment characteristics, rainfall intensity, vegetation, and earthquake [2]. Landslide occurs under three conditions: a steep slope, a slipping zone in the impermeable subsurface, and an adequate amount of water from rains which infiltrate to soil voids above the slipping zone generating escalated force of soil to slope [3].

Landslide often happens in the landscape of steep mountain area and substantial amount deforestation [4]. A shifting slope has an effect on the shear stress of the soil. The greater the slope dipping, the greater the shear stress and also more unstable the slope [5, 6]. If the soil shear stress is smaller than the shear stress along the slipping zone (the increase of water saturation effect), soil erosion would accordingly follow [7].

Cisewu is located in Garut regency, West Java. This area is situated in between the boundary of South Bandung regency, South Garut regency, and South Cianjur regency. The regional forecast map indicates that a ground movement has been detected in West Java on March 2017 [8], Cisewu district is including in the potential zone of ground movement middle-high as depicted in Figure 1, which means a high possibility of ground movement would likely happen in intense rainfall. This has resulted mainly in the adjacent area of river valleys, escarpments, cliffs, or disrupted slope and re-activated ground movement.

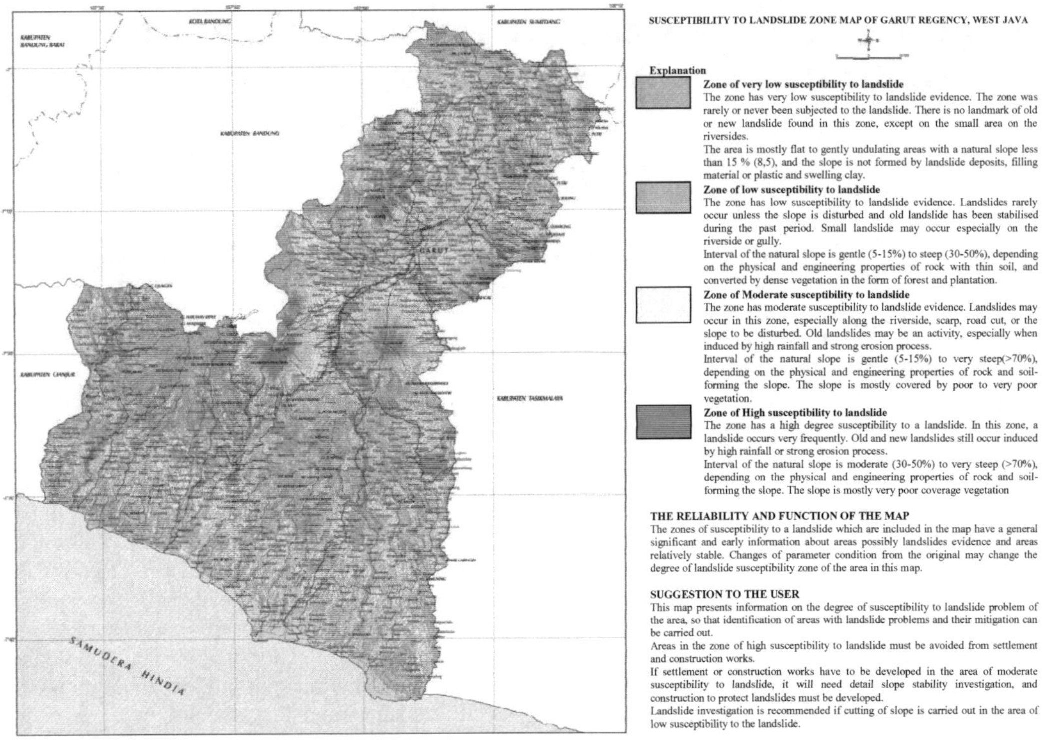

Figure 1. Susceptibility To Landslide Zone Map Of Garut Regency, West Java [8]

The climate and weather in Cisewu district, Garut regency are influenced by three principal factors: monsoonal circulation pattern, mountainous regional topography in the central area of West Java, and topographic elevation in Bandung regency. The average rainfall intensity is normally more than 3500 mm/year, in the El Nino phase the intensity is about 2250-3000 mm/year, and the La Nina phase produces 3000-3500 mm/year [9]. The variety of monthly temperature is the range of 24°C - 27°C.

This research is part of a study on the landslide in a building area Base Transceiver Station (BTS) in Cisewu district. The landslide area is located in the south of Cisewu district, Garut regency, west java where there are extreme contour differences. The research scopes are analysis study and an effort to control the landslide effect by building reinforcement and drainage refinement.

Based on visual observation on Figure 2, the tilt of slope is quite steep where the edge of the fence is hanging as the effect of soil movement under construction. The rainfall condition in a normal year, El Nino phase and La Nina phase are high enough [9], whereas the existing water canal in this tower building could not accommodate the rainwater stream resulting water overflow and leaked water into the soil. This condition would reduce the shear strength of soil and rock that would generate ground movement or landslide as in Figure 3.

Figure 2. Slope failure area in the edge of fence

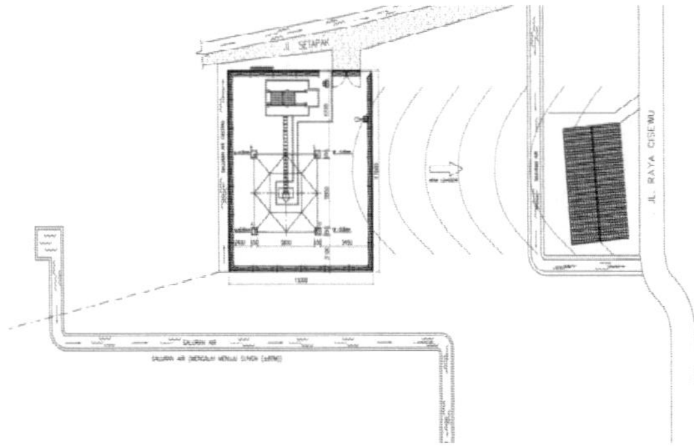

Figure 3. Slope failure direction

Landslide countermeasures include; survey investigation, analysis and slope reinforcement. Appropriate countermeasures shall be selected or designed based on the results obtained from the survey and the analysis [10]. The prevention is established by constructing gabion mesh and retaining walls in the foothills to decrease the slope dipping and to increase safety factor number. Further improvement is to recondition the previous canal and to build the new canal to anticipate rainwater leaked that could lead to the rise of the soil water table and slope failure.

The analysis of site condition is done shortly after the disaster and right after refinement utilising Plaxis2D. The reason for using a finite element method using Plaxis2D approach is that it is being widely used in landslide susceptibility modelling in recent years finite element method [11]. Plaxis is a finite element program, developed for the analysis of deformation, stability and groundwater flow in geotechnical engineering [12]. The finite element method for slope stability analysis has been applied by [13, 14, 15, 16] for further countermeasure.

2. Methods

This study is performed in Cisewu district, Garut regency, West Java. The research location is in the landslide area where a BTS tower building is constructed. The landslide countermeasures are conducted in several stages: survey investigation, slope stability analysis, and an effort to control landslide by applying slope reinforcement and drainage refinement.

2.1 Survey Investigation

This stage includes slope condition survey, soil investigation, and drainage survey. The slope condition survey after soil movement is required to determine the needed test and to formulate the initial hypothesis of landslide cause as well. Furthermore, a description of soil strength could be produced to prevent future landslide disaster. A contour of slope measurement is also depicted in this stage.

The soil test is done using the cone penetration test (CPT) and soil sample collection by hand boring. The penetration test is to recognise the bearing capacity of soil while soil collection by hand boring is conducted to retrieve soil parameters. The soil sample is tested in a laboratory to detect water content, bulk density, particle density, particle analysis, soil consistency limit, and soil shear strength test. These tests also refer to ASTM standards [17, 18, 19, 20].

2.2 Slope Stability Analysis

2.2.1. Model material property. The Mohr-Coulomb model has been used to describe soil material properties. Material shear strength to cohesion, normal stress and angle of internal friction are associated with the criteria of the Mohr-Coulomb model. Table 1 shows soil parameters used as input to Plaxis2D software.

Table 1. Material design parameters in the landslide simulation

Parameter	Name	Layer 1	Layer 2	Gabion dan retaining wall	Units
Material Model	Model	Mohr-Coulomb	Mohr-Coulomb	Elastic Plastis	-
Material Behavior	Type	UnDrained	UnDrained	Non - Porous	-
Unit Weight of Unsaturated Soil	γunsat	10.860	10.110	23.5	kN/m^3
Unit Weight of Saturated Soil	γsat	16.500	16.160	23.5	kN/m^3
Horizontal Direction Permeability	kx	0.000497	0.000916	-	m/hari
Vertical Direction Permeability	ky	0.000497	0.000916	-	m/hari
Young Modulus (contant)	Eref	3428	4000	20.000.000	kN/m^2
Poisson's Ratio	V	0.35	0.35	0.15	-
Cohesion (constant)	Cref	9.270	11.930		kN/m^2
Friction Angle	Φ	23.960	27.920		°
Dilation Angle	Ψ	0	0		°

2.2.2. Slope stability analysis using Plaxis2D.

There are three major aspects that involve in slope stability analysis. The first is about the material properties of the slope forming material. The second is the calculation of factor of safety, and the third is the definitions of the slope failure [21].

The input of material properties is in the form of a graphical geometric model such as soil layering, structure elements, construction stages, burdening, and limit conditions performed using a graphical procedure which is facilitated by Plaxis2D. from the geometrical model, a finite element up to 2D could be easily obtained. Plaxis would automatically generate finite element until random 2D configuration with either global or local choices [12]. Figure 4illustrates initial contour condition after slope failure, and Figure 5 shows slope condition after refinement using gabion mesh and retaining walls as well as soil improvement in the landslide area. The blue area describes Layer 1, whereas the green area shows Layer 2 in the contour design. The slope gains outer loads from both tower weight and building foundation as depicted in these figures. The input for tower load is 52 kN/m^2.

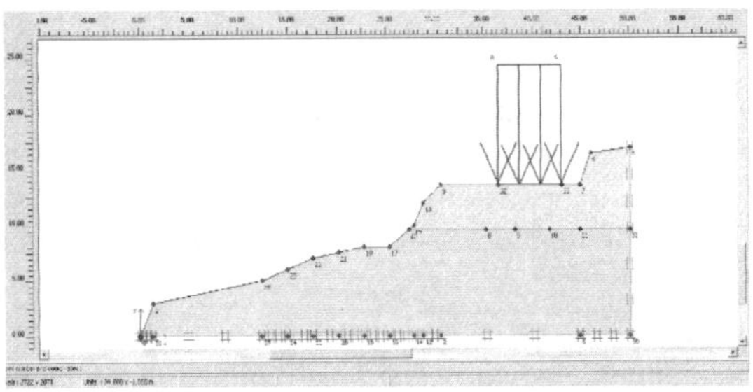

Figure 4. The initial contour design after slope failure.

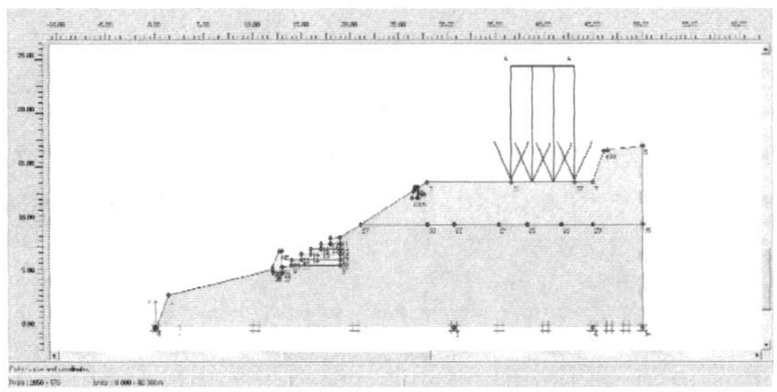

Figure 5. The refined contour design after slope failure.

Calculation of safety factor is obtained by seeking the failure surface (FS) in soil/rocks structure. FS is calculated by subtracting cohesion value to friction angle in the soil gradually until the failure zone produced [14].

Slope failure occurs in case of reduction in shear strength due to saturation and swelling coupled with the condition of seepage, which causes the shear strength equal or less than shear stress [22]. In soil mechanics, it is important to understand the forces acted on slopes, which is the driving force and the resisting force, which prevent from slope failing. The ratio of resisting forces to driving forces is a Factor of Safety (FOS). Slopes with lower FOS means the potential of failing are higher then slopes with higher FOS [23]. Therefore slopes with higher FOS more safer.

2.3 Landslide countermeasures

The countermeasures are performed by applying reinforcement in the foot of slope utilising gabion mesh and retaining walls. These actions also include drainage refinement and constructing new drainage canal to divert the water flow direction heading to the tower building. During heavy rain, the canal should be able to turn the surface water flow from the hill above the tower.

3. Results and discussion

3.1 Survey Investigation

3.1.1. Road shape. The road to the tower area is still accessible enough, however, to reach the residential areas exposed to landslide disaster needs more effort since the road access has been cracked due to slope failure although it can be passed yet as in Figure 6.

Figure 6. The road shape after slope failure

3.1.2. Drainage condition. According to the field survey, the existing drainage condition is not sufficient to accommodate all water flow from the slope as can be seen in Figure 7, this condition triggered seepage to soil and increased the surface water changed to inclining water table and slope failure accordingly.

Figure 7. Drainage condition after landslide

3.1.3. Slope condition. Based on an initial survey result, the slope condition after failure shows muddy, silty clay, with the rise of the water table. The analysis figures that the soil condition is under the safety limit. The safety factor, in this case, is 1.244 while the minimum value is 1.5 in which the failure zone is right under the tower load.

3.2 Analysis and countermeasures

3.2.1. Slope stability analysis using Plaxis2D. The first analysis is done under the condition of after slope failure. The slope receives an outer load in the form of tower weight and building foundation modelled as equal load. In Figure 4, it can be seen that the biggest movement occurs in the tower building area. This movement could be developed since the soil in that area is consists of silty clay or soft soil which could lead to soil displacement in the region. Figure 8 and 9 illustrate the results of numerical analysis by applying the parameters shown in Table 1.

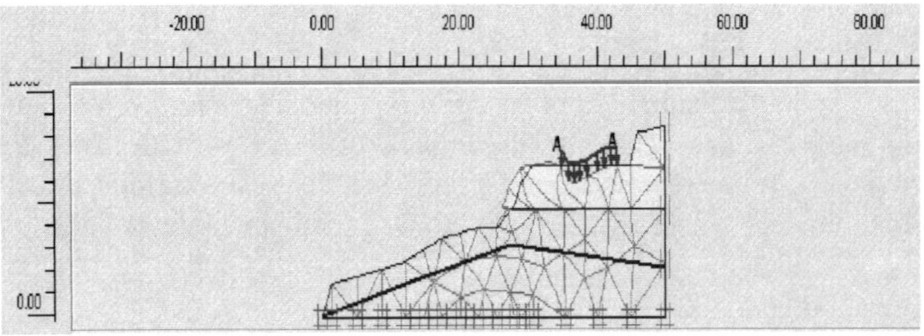

(a)

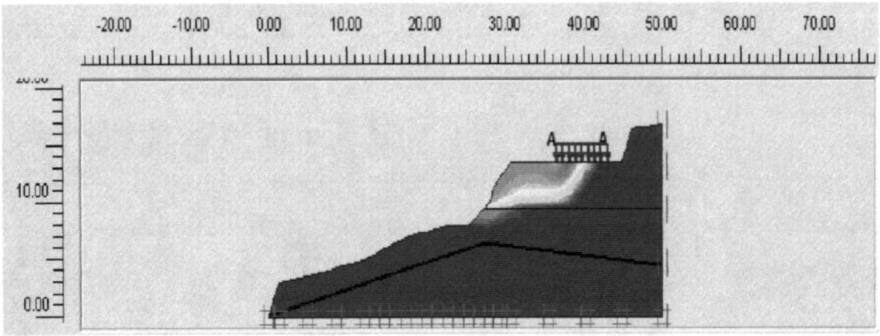

(b)

Figure 8. (a) Deformed mesh after the landslide, (b) Failure plane before refinement

The deformed mesh is shown in Figure 8a. The potential slipping zone is described in Figure 8b, which displacement scaling and total regional displacement are depicted as well. Analysis using The Finite Element Method (FEM) could simulate failure plane reaching toe of slope in the current state. FoS, in this case, is 1.244 as in Figure 10a, which shows that slope is very much likely to fail.

The second analysis is obtained by assuming reinforcement in the foot of slope using gabion mesh and two retaining walls. The mesh dimension is 5m wide and 2.5m tall. The size of the first retaining wall is 2m tall and 1m wide, while the second size is 1m tall and 0.6m wide.

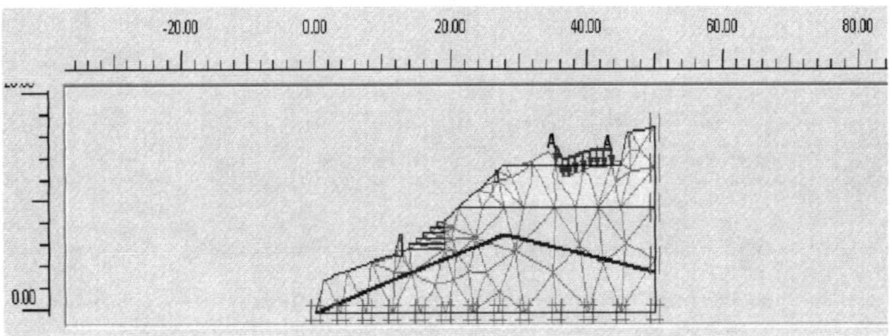

(a)

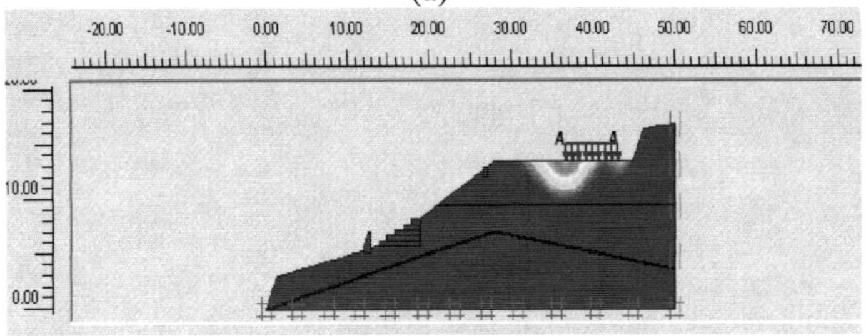

(b)

Figure 9. (a) Deformed mesh after strengthened, (b) Failure plane condition after reinforcement in the foot of the slope

Figure 9a illustrates deformed mesh after strengthened utilising gabion and retaining walls. The potential slipping zone is described in Figure 9b. The reinforcement could improve the value of FoS in which this case is 1.729 as in Figure 10b, greater than the minimum safety value. Therefore it can be seen that slope condition after reinforcement is under control.

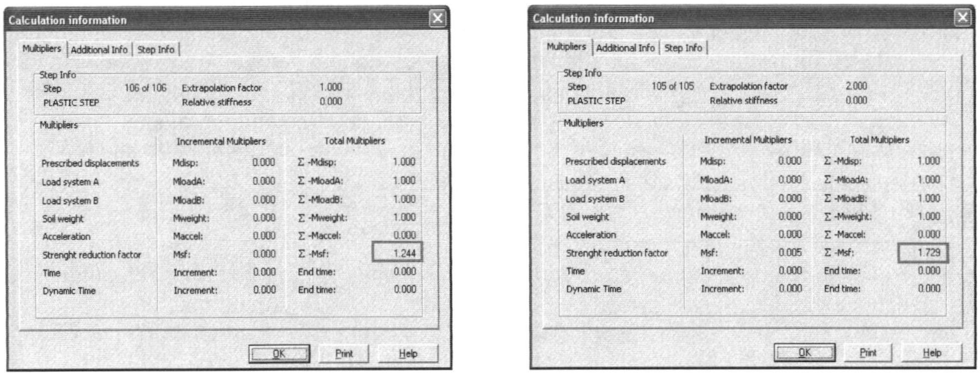

(a) (b)

Figure10. (a) Safety Factor Existing Condition, (b) Safety Factor Strengthening Condition

Further countermeasure is designing drainage canal which is divided into two parts: repairing the existing drainage canal and designing diverted canal dimension that is required to contain rainwater flow from the hill above tower building and to prevent water overflow leaked into the soil (seepage) triggering slope failure. The catchment area and the proposed canal design are described in Figure 11. The diverted canal dimension resulted to accommodate all water flow is 1m wide and 0.9m depth.

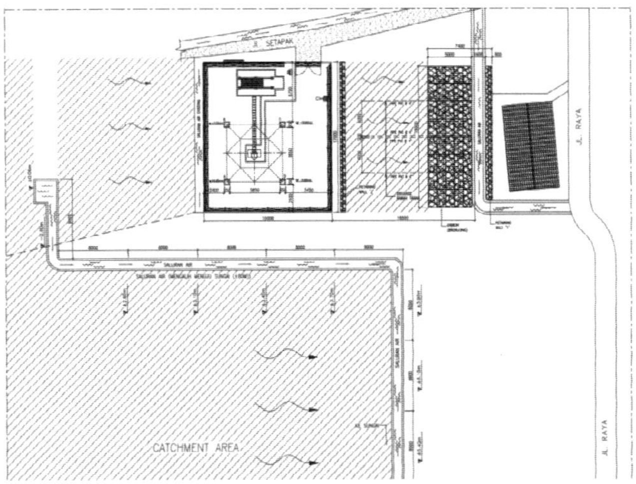

Figure 11. Proposed drainage design

4. Conclusion

According to the data resulted from survey investigation, several tests, and calculations, it can be concluded that landslide countermeasures could be performed in several means: constructing reinforcement using gabion mesh and retaining walls, along with enlarging the existing drainage canal, and building proposed drainage canal. These measures could certainly increase safety factor value to 1.729 greater than the minimum safety value, therefore the safety condition could be reached under the requirement of reinforcement in the foot of the slope.

Acknowledgement

This research is carried out by remarkable assistance from the local community around Cisewu district for the help during the study and thank you to all the teams for their support so that this paper can be completed.

References

[1] Abidin Hasanuddin Z, Andreas H, Gamal M, Surono, and Hendrasto M 2004 *On the Use of GPS Survey Method for Studying Land Displacements on the Landslide Prone Areas* FIG Working Week (Athens Greece) pp 13

[2] Harjadi P, Mezak A Ratag, D Karnawati, S Rizal, Surono, Sutardi, Triwibowo, Hermono Sigit, A Harjadi, P., Mezak A Ratag, D. Karnawati, S.Rizal, Surono, Sutardi, Triwibowo, Hermono Sigit, A. Wasiati, Yusharmen, Pariatmono, S. Triutomo, dan B. W. Widjaja, 2007 *Pengenalan Karakteristik Bencana dan Upaya Mitigasinya di Indonesia* 2nd edition Jakarta.

[3] Brook, K.N., Folliott, P. F., Gregersen, H.M., & Thames, J.K. 1991 *Hydrology and the management of watersheds* Ames, USA: Iowa State University Press.

[4] Paimin, Sukreso, dan Pramono, I.B. 2009 *Teknik Mitigasi Banjir dan Tanah Longsor* Tropenbos International Programme. Bogor.

[5] Guedjeo C. S., Kagou Dongmo A., Ngapgue F., Nkouathio D. G., Zangmo Tefogoum G., Gountié Dedzo

M., Nono A 2013 *Natural hazards along the Bamenda escarpment and its environs: The case of landslide, rock fall and flood risks* (Cameroon volcanic line, North-West Region). Glo. Adv. Res. J. Geol. Min.Vol. 2(1) pp. 015-026

[6] Henriques C 2014 *Landslide Susceptibility Evaluation and Validation at a Regional Scale.* Tese de Doutoramento, Universidade de Lisboa, 563p.

[7] Pakasi, G.A., K. L. Theffie, Z. E. Tamod, dan M. Montolalu 2015 *Identifikasi Proses Pergerakan Massa Tanah Di Desa Rumengkor Kecamatan Tombulu.* Cocos - Jurnal Ilmiah Fakultas Pertanian Universitas Sam Ratulangi. Volume 6 No. 16 pp 1 – 12

[8] Badan Geologi Pusat Vulkanologi dan Mitigasi Bencana Geologi 2017 *Usceptibility To Landslide Zone Map Of Garut Regency, West Java* Kementrian Energi dan Sumber Daya Mineral Badan Geologi

[9] A Pramudia, Y Koesmaryono, I Las, T June, IW Astika, E Runtunuwu 2008 *Pewilayahan hujan dan model prediksi curah hujan untuk mendukung analisis ketersediaan dan kerentanan pangan di sentra produksi padi.* Forum Pascasarjana Vol. 31 No. 2April2008 pp 131-142.

[10] Kokusai Kogyo and JICA 2012 *Landslide Investigation Manual*

[11] Dhananjai Verma, Ashutosh Kainthola, S.S. Guptel, T.N. Singh 2013 *A Finite Element Approach of Stability Analysis of Internal Dump Slope in Wardha Valley Coal Field, India, Maharashtra*, American Journal of Mining and Metallurgy, Vol. 1, No. 1 pp 1-6

[12] Brinkgreve, R B J 2007 PLAXIS 2D Versi 8, Delft University of Technology & PLAXIS b.v., Belanda

[13] Matsui, T, and San, K C 1992 *Finite Element Slope Stability Analysis by Shear Strength Reduction Technique* Soils and Foundations, 32 (1) pp 59-70

[14] Duncan, J M 1996 *State of the Art: Limit Equilibrium and Finite Element Analysis of Slopes* Journal of Geotechnical Engineering, ASCE, 122 (7) pp 577-596

[15] Griffiths, D V, and Lane, P A. 1999 *Slope Stability Analysis by Finite Elements* Geotechnique, 49(3) pp 387-403

[16] Kainthola, A., Verma, D., Gupte, S.S. and Singh, T N 2011a *A Coal mine dump stability analysis – a case study*, Int. Journal of Geomaterial. 1 pp 1-13.

[17] American Standard Testing and Material (ASTM) 2000 D 2487-00: *Standard Practice for Classification of Soils for Engineering Purposes (UnifiedSoil Classification System)* Buku Panduan p 13

[18] American Standard Testing and Material (ASTM) 2000 D 4318-00: *Standard Test Methods for Liquid Limit, Plastic Limit, and Plasticity Index of Soils* Buku Panduan p 14

[19] American Standard Testing and Material (ASTM) 2002 D 5731-02: *Standard Test Method for Determination of the Point Load Strength Index of Rock* Buku Panduan p 9

[20] American Standard Testing and Material (ASTM) 2007 C127-07: *Standard Test Method for Density, Relative Density (Specific Gravity), and Absorption of Fine Aggregate* Buku Panduan p 7

[21] Rocscience Inc 2004 *Application of the Finite Element Method to Slope Stability* Toronto pp 2-6

[22] Neoh, Cheng Aik 2001 *Practical Design Aspects for Slopes in Mountainous Terrains* Proceedings of National Slope Seminar

[23] Hamed, H, Khairul A K, Amin G, Ramli N, Sayyed Y Z F 2012 *Investigation of Slope Failures in Soil Mechanics* Electronic Journal of Geotechnical Engineering Vol. 17. http://www.ejge.com/2012/Abs12.227.htm, accessed on August 13, 2019

Effect of change of channel width in the downstream of the check dam on controlling sedimentation in Mrica Reservoir

D Ulfiana[1,*], D A Wulandari[1], P N Parmantoro[1], Susilowati[1]

[1]Civil Engineering Department, Engineering Faculty, Diponegoro University, Semarang, Indonesia

[*]Corresponding author e-mail: desyta@live.undip.ac.id

Abstract. The increasing sedimentation of Mrica reservoir shows that the sedimentation control has not been effective. The sedimentation control has been made include the disposal of sediments that settle in the reservoir by dredging or flushing. However, the results are not optimal. This is caused by the large volume of sediment entering the reservoir each year. The controlling volume of sediment that goes into the reservoir has also been carried out by conducting a watershed conservation program and building a check dam in the upstream Mrica reservoir. However, the existing check dam is still less effective in controlling sediments, especially suspended load sediments, that are transported in the river. Therefore this study was carried out by adopting the sediment trap concept to deposit sediments that escaped from the check dam. This study simulated changes in the width of the downstream channel of the check dam by three variations using HEC-RAS. There was a width of 20%, 40%, and 60% wider than the initial width. The results showed the effect of changing the channel width on volume sediment deposited. Based on the results of the study, the change in width to wider than the initial width could increase the volume of sediment deposits.

1. Introduction

Mrica Reservoir or Panglima Besar Soedirman Reservoir is one of the strategic reservoirs located in Bawang District, Banjarnegara Regency, Central Java Province. It began operating in 1988 and has the main function as a hydropower plant with an installed capacity of 180.93 MW. Besides, this reservoir is also used as a provider of irrigation water in Banjarcayana, covering an area of 6550 hectares and Penaruban covering an area of 900 hectares. Mrica reservoir has a watershed area of 957 km^2. This reservoir has an area of 8.26 million m^2, with a capacity of 148.287 million m^3. It is planned to have a service life of 60 years with a sedimentation rate of 2.4 million m^3 / year.

The problem that occurs in the management of Mrica Reservoirs is siltation caused by sedimentation. Sediment deposited above the dead reservoir will reduce the effective volume of the reservoir, thereby reducing the economic service life of the reservoir [1]. According to reports from PT Indonesia Power [2] as the manager of the Mrica Reservoir in 2016, the effective volume of this reservoir decreased by 48.7%. The total sediment volume of the Mrica Reservoir up to October 2016 was 114.254 million m^3, meaning 22.40% of the reservoir volume has been filled with sediment.

Based on PT Indonesia Power's research from year to year, the sediment deposition in Mrica Reservoir has increased [2]. Mrica Reservoir sedimentation, which continues to grow, shows that sedimentation control is less effective. There have been some studies on sedimentation control of the Mrica Reservoir. A study conducted by Alfianto and Soewarno [3] developed a micro sabo dam on agricultural land in the upstream watershed of the Mrica Reservoir. Suroso and Widiyanto [4] also researched controlling sedimentation using a combination of flushing and fluidisation.

Several sedimentation control has been done, including the removal of sediments that deposited in reservoirs by dredging and flushing [5]. However, sediment dredging requires a very high cost. While sediment flushing has been done regularly, but the results are less than optimal. It is caused by the large volume of sediment entering the reservoir each year. Efforts to reduce the rate of sedimentation entering reservoirs have also been carried out by watershed conservation programs and building sediment control structures (check dam) in the upper reaches of the Mrica Reservoir. However, the existing check dam is still less effective in depositing sediments, especially suspended sediments.

Based on the problems above, a study was carried out by adopting a sediment trap concept to deposit suspended sediments that escaped from the check dam. Some research on the sediment trap has also been carried out. One of them is a study conducted by Guillaume and Alain [6] that combines sediment traps with an open check dam. In this study, the sediment trap was designed at the upstream of the check dam. Schwindt et al. [7] also conducted a study on modifying the design of a sediment trap with an open check dam by adding a guiding channel. A modification of a sediment trap was also carried out by Thaxton and Mclaughlin [8] by adding baffle technology.

The design approach of sediment trap structures used in this study is irrigation sediment traps. In irrigation channel, flow discharge passing through sediment trap is the amount of water used for irrigating plants, so that the passing water is planned and tends to be stable. While on the river, flow discharges are more fluctuating. Therefore, research is needed to determine the effectiveness of the sediment trap concept in the river channel, mainly to deposit sediment suspension. The adaptation of the sediment trap concept is made by varying the width of the river.

This study simulated changes in the width of the downstream channel of the check dam by three variations using HEC-RAS. There was a width of 20%, 40%, and 60% wider than the initial width. This research is expected to be able to reduce the volume of sediment effectively to maximize the management of sedimentation in Mrica Reservoir.

2. Methods

2.1. Analysis of study area

Based on research conducted by PT Indonesia Power, three main rivers become sediment entry media into Mrica Reservoir, namely Serayu, Merawu, and Lumajang rivers. In the Lumajang river channel, there is a check dam that serves to reduce sediment entering the reservoir. The check dam, which was built in 2007, is located in Linggasari Village, Wanadadi, Banjarnegara, or geographically located at coordinates 109^0 40' 34" E, 07^0 21' 40" S.

The existing check dam has a design capacity of 1009 m^3 of sediment storage with a flood discharge of 41.0 m^3/s. Based on PT Indonesia Power's data, the volume of suspended load sediment entering the Mrica Reservoir from November 2015 until October 2016 is 16544 m^3 while the bedload sediment is 9416 m^3. With a sediment storage capacity of 1009 m^3, the Lumajang River check dam in 2016 could only reduce sediment entering the reservoir by 3.88%. Besides, based on the research of Soewarno in 2007 until 2008, the volume of suspended load sediment entering the reservoir was 6821.33 m^3, and the bedload sediment was 5634.41 m^3. Then the effectiveness of the check dam in 2008 was 8.1%. It could be calculated from the data that there was an increase in sediment volume by 108.3% in 8 years. It could be said that the check dam performance is no longer effective in controlling sedimentation in Mrica Reservoir compared to when it was first built.

Another problem that exists is based on research conducted by Soewrno, showing that the check dam storage has been filled in only one rainy season. This is very ineffective, considering the high maintenance costs. Besides, most sediment that deposited in the check dam is bedload sediment, while suspended load sediment still escapes, even though around 63.73% of the sediment entering the Mrica Reservoir is suspended sediment. Therefore it is necessary to analyze the effectiveness of the sediment trap at the downstream of the check dam in overcoming the problem of suspended load sediment.

2.2. Data collection methods and analysis

Data used in this study are primary data, secondary data, and data from previous studies. Primary data used are sediment samples taken at the site. The sample is taken on upstream and downstream of the check dam to obtain concentration and gradation of bedload sediment. Secondary data used are river flow discharge, topographic maps, and design drawings of existing check dams. Other data needed are taken from previous studies that have been conducted by Soewarno and Sukatja [9] and PT Indonesia Power [2].

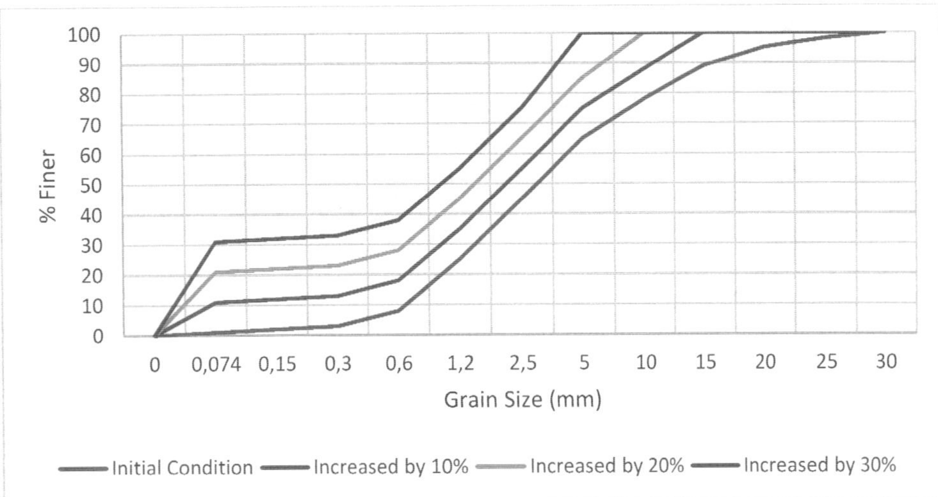

Figure 1. Variations of Sediment Gradations.

Suspended load sediment concentration and gradation data used are data from Soewarno's research. Considering that check dams are only effective for depositing sediment bedload sediment and the increasing of suspended load sediment volume from year to year, a variation of the percentage of suspended load sediment gradation is carried out to see the effectiveness of the sediment trap design. Three gradation variations are an increase in the percentage of suspended load sediment by 10%, 20%, and 30%, as seen in Figure 1.

2.3. Preliminary design

The next step is to make a preliminary design of a sediment trap as a basis for modelling in HEC-RAS software. This modelling is carried out to determine the effect of widening the channel cross-section on the volume of sediment deposited, especially on river flow conditions that have fluctuating discharge.

2.3.1. Sediment trap location. The location of the sediment trap is planned to adjust according to the downstream condition of the existing check dam. There is a bridge located on the downstream of the check dam. Therefore the sediment trap is planned to be made in the downstream of the bridge so that it does not disturb the bridge structure. Based on the topographic map, the length of the Lumajang River that could be used as a sediment trap is 145 m.

2.3.2. Sediment trap dimension. The cross-section of the river in the downstream of the check dam has a width of 10 m and a depth of 5.8 m. The channel cross-section dimension is then used as an initial condition of the sediment trap. Variations are then made to the cross-section width of the river by widening the initial width by 20%, 40%, and 60%. Then, four designs of river cross-section are obtained, which is cross-section with widths (B) of 10 m; 12 m; 14 m; and 16 m.

2.3.3. Inflow. There were three types of discharge data inputted in modeling in the HEC-RAS software, that is check dam flood discharge, daily discharge of Lumajang River, and dependable flow discharge 50% of the river. Based on check dam technical data, the flood discharge is 41.0 m^3/s. The dependable flow discharge 50%v of the river is obtained from the flow duration curve, which is 5 m^3/s.

2.3.4. Sediment inflow. Sediment inflow discharge is calculated using equation (1).

$$Q_s = 17.94\, Q_w^{\ 2.342} \tag{1}$$

Where Q_s is suspended sediment discharge and Q_w is flow discharge

2.3.5. Model scenario. Modelling the sediment trap is done by simulating four variations of sediment trap width with three types of discharge and four variations of the percentage of sediment gradations. Simulations are carried out to determine the volume of sediment that is deposited for each scenario.

3. Results and Discussion

The analysis of the modelling result is determined by measuring the total volume of sediment deposited in each variation of the sediment trap. The result of analysis for each variation of sediment traps include:

1. Modelling on Daily discharge of Lumajang River
   Based on the result of modelling by HEC-RAS software, the graph comparison between the width of the channel and the volume of sediment deposited was generate, as shown in Figure 2. The graph shows the percentage of suspended sediment gradation increased by 10% shows that the wider the channel, the greater the volume of sediment deposited. While the percentage increased by 20% and 30%, there is no sediment deposited.

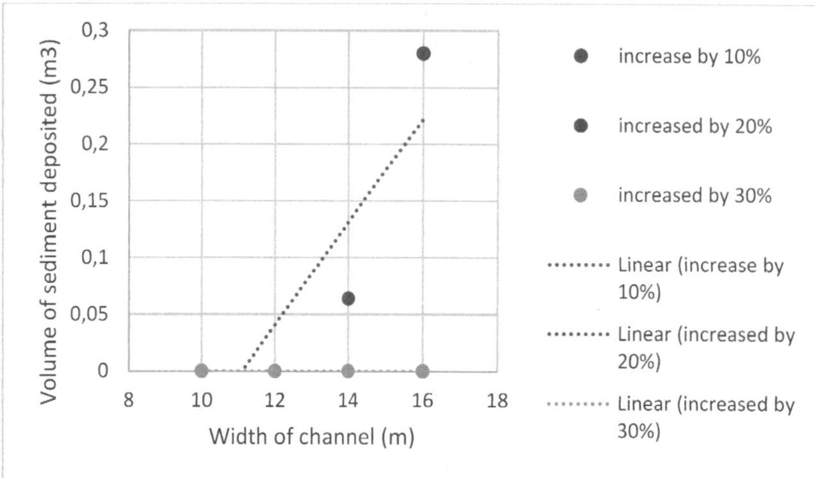

Figure 2. Width of Channel vs. Volume of Sediment Deposited Graph on Daily Discharge.

2. Modelling on Dependable Flow Discharge 50%
   The results of sediment trap modelling on dependable flow discharge, 50% are generated on the graph, as shown in Figure 3. The graph shows that the greater the width of the channel, the greater the volume of sediment that could be deposited. Based on the results of the analysis, the application of sediment traps with an increase in channel width of 20% could increase sediment volume deposited by 19.22%.
3. Modelling on Designed Flood Discharge
   The results of sediment traps modelling on designed flood discharge are generated on the graph, as shown in Figure 4. The graph shows that the greater the width of the channel, the greater the volume of sediment that could be deposited. Based on the results of the analysis, the application of sediment traps with an increase in channel width of 20% could increase sediment volume deposited by 19.40%.

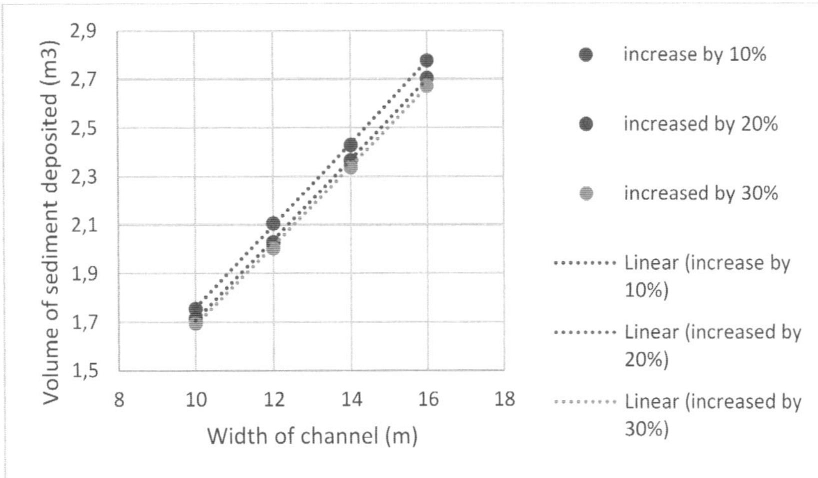

Figure 3. Width of Channel vs. Volume of Sediment Deposited Graph on Designed Flood Discharge.

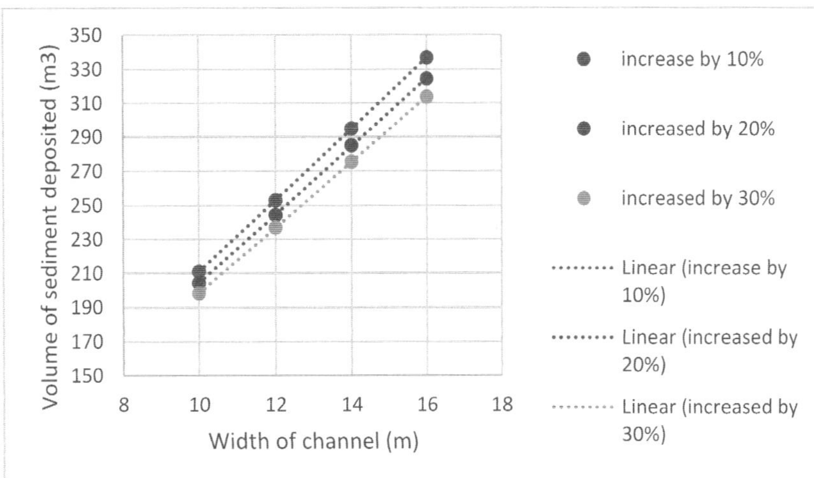

Figure 4. Width of Channel vs. Volume of Sediment Deposited Graph on Dependable Flow Discharge 50%.

Based on the results of the analysis on each discharge variations, it can be concluded that the wider the channel, there is an increase in sediment volume deposited. However, the ability of sediment traps to deposit suspended sediment is limited to this value. This is proven even though the percentage of suspended sediment gradation is increased, the volume of sediment that has settled tends to be constant.

4. Conclusion

Based on the analysis of the result, the application of the concept of sediment trap in a river channel by widening the cross-section gives effective results because it could increase the volume of sediment that deposited compared to the initial condition of the river cross-section. These results indicate that making sediment traps in the downstream of check dams could support the effectiveness of check dams in reducing the volume of sediment entering the Mrica Reservoir, especially in dealing with suspended sediment that cannot be deposited at check dams. However, the widening of the cross-section of the river has not been able to accommodate the condition of the sedimentation rate, which continues to increase from year to year because based on the results of the analysis of the increasing percentage of suspended sediment gradation that tends to remain constant.

Acknowledgements
The authors thank Diponegoro University for providing funding support to the author through the "Penelitian Strategis 2019" scheme.

References

[1] D. A. Wulandari, "Penanganan sedimentasi waduk mrica," *Berk. Ilm. Tek. Keair.*, vol. 13, no. 4, pp. 264–271, 2007.
[2] PT Indonesia Power, "Laporan Pelaksanaan Penyelidikan Sedimentasi Waduk PLTA PB. Soedirman," 2016.
[3] A. Alfianto and Soewarno, "Techno Sabo for Overcoming Reservoir Watershed Sedimentation (Case Reservoir Mrica)," *J. Tek. Hidraul.*, vol. 5, no. 1, pp. 83–98, 2014.
[4] S. Suroso and W. Widiyanto, "Model Pengendalian Sedimentasi Waduk Mrica dengan Fluidisasi (Sedimentation Control Model Of Mrica Reservoir Using Fluidization)," *Din. Rekayasa*, vol. 5, Aug. 2009.
[5] P. Utomo, "Mrica Reservoir Sedimentation: Current Situation and Future Necessary Management," *J. Civ. Eng. Forum*, vol. 3, p. 365, Sep. 2017.
[6] P. Guillaume and R. Alain, "Design of Sediment Traps with Open Check Dams. I: Hydraulic and Deposition Processes," *J. Hydraul. Eng.*, vol. 142, no. 2, p. 4015045, Feb. 2016.
[7] S. Schwindt, M. J. Franca, A. Reffo, and A. J. Schleiss, "Sediment traps with guiding channel and hybrid check dams improve controlled sediment retention," *Nat. Hazards earth Syst. Sci.*, vol. 18, pp. 647–668, 2018.
[8] C. Thaxton and R. Mclaughlin, "Sediment capture effectiveness of various baffle types in a sediment retention pond," *Trans. ASAE*, vol. 48, Sep. 2005.
[9] Soewarno; and C. B. Sukatja, "Kinerja dam sabo k. lumajang untuk pengendalian sedimentasi waduk mrica," *J. Sumber Daya Air*, vol. 6, no. 1, pp. 17–32, 2010.

IOP Conf. Series: Materials Science and Engineering **771** (2020) 012048 doi:10.1088/1757-899X/771/1/012048

Effect of Combination of Steel Fibre and Silica Fume on the Mechanical Strength of Concrete

M Sofyan[1,*], B Wicaksono[2], A O Irlan[3]

[1]Sekolah Tinggi Teknik PLN Jakarta, Puri Permata, Tangerang, Indonesia
[2]Cilimus, Cirebon, Indonesia
[3]Puri Permata, Tangerang, Indonesia

[*]e-mail : liorente1688@gmail.com

Abstract. Concrete currently consists of several types. Among them are Fibrous Concrete, Lightweight Concrete, SCC Concrete, Hollow Concrete etc. All have advantages and disadvantages to each. The use of various types of concrete is adjusted to the needs of construction. In this research, fibrous steel concrete with a combination of silica fume will be the main object to be studied. So far, fibrous concrete has been actively studied in various countries. From the results of previous studies, fibrous concrete can improve the quality of concrete, especially reducing the brittle nature and initial cracking in concrete. While Silica Fume can increase the compressive strength of concrete. The main objective of this study to determine and understand the mechanical properties of concrete with a combination of steel fibre and silica fume such as compressive strength and split tensile strength. The percentage of steel fibre percentage used in this study is from 0 to 3 per cent, while the percentage of silica fume used is 10 per cent by weight of cement. The results of a study are the optimum compressive strength is in the DF1 variation with the percentage of silica fume 10 per cent and 0 per cent steel fibre. The use of steel fibre combined with silica fume does not significantly affect the compressive strength. Although it tends to decrease the compressive strength. The strength of tensile strength obtained ranges from 11 to 14 per cent of the compressive strength.

1. Introduction

Concrete is a material commonly used in various infrastructure buildings. The development of concrete technology from time to time is very rapid. The high volume of utilisation of concrete materials in construction encourages the Engineers and Scientists to continue to innovate to produce concrete material products that have good quality in terms of strength and durability. Concrete with good strength and durability will reduce maintenance costs to achieve economic material criteria. Combining several materials that have different physical and chemical characteristics is an innovative step in producing a quality construction material product such as concrete. Material developed in this study is concrete material that uses silica fume as filler and steel fibres. These two types of material will be combined in a concrete mixture to obtain optimal mechanical strength.

There have been many studies conducted regarding the use of silica fume in concrete. Various combinations of a local natural pozzolan and silica fume were used to produce workable high to very high strength mortars and concrete with a compressive strength in the range of 69±110 MPa [1] Shannag, 2000. Almusallam et al,2004 [2] investigated the effects of silica fume (15%) on the compressive strength of concrete made with low-quality coarse aggregates (calcareous, dolomitic, and quartzitic limestone and steel slag). It was observed that compressive strength increased with age in all the concrete specimens. After 180 days of curing, the highest compressive strength was noted in the 15% silica fume cement concrete specimens (54 MPa) followed by those prepared with 10% silica fume (52 MPa). Silica fume (SF) is a by-product of the silicon and ferrosilicon industry. The reduction of high-purity quartz to silicon at temperatures up to 2000 ◦C produces SiO2 vapours, which oxidises and condense in the low-temperature zone to tiny particles consisting of non-crystalline silica. When silica fume is added to concrete, it results in a significant change in the compressive strength of the mix. This is mainly due to the aggregate-paste bond improvement and enhanced microstructure [3] Siddique et al., 2011.

Steel fibre is widely used in concrete mixtures. Steel fibres have a considerably larger length and higher Young's modulus as compared to the polypropylene fibre. This leads to an improved potential for crack control [4] Qian et al.,2000. In 2007 Sivakumar et al. has been made to study the synergistic behaviour of hybrid combinations of steel and non-metallic fibres in controlling plastic shrinkage cracks. It revealed Plastic shrinkage cracks were reduced significantly by fibre addition (by 50–99% compared to plain concrete without fibres); hybrid fibres were more effective in crack reduction compared to individual steel fibres [5]. Because of the cases above, the combination of two material components with different characteristics can have a good effect on the mechanical properties of concrete.

2. Materials

The main concrete forming material to be used is prepared in advance, namely, Cement, Split, Sand, Silica Fume and Steel Fiber. The cement used is PCC cement. Silica Fume used is silica fume produced by PT Sika. Whereas steel fibre that will be used as a substitute material is steel fibre produced by Bekaert.

2.1. Aggregates

An aggregate test is performed first to find out the physical and mechanical characteristics of the aggregate. Aggregate testing will later become a reference in the concrete mix design. The aggregates to be tested are coarse and fine aggregates. The test that will be carried out is the Filter analysis test to find out the gradation/size of aggregate grains, sludge content testing, which aims to determine the percentage of sludge content in the aggregate. Sludge levels exceeding the 5% limit must be carried out by aggregate washing. Besides, water absorption and specific gravity testing will be carried out so that the amount of cement water factor can be determined as a factor in determining the compressive strength of concrete.

All of the aggregate testings follow ASTM standards. Physical characteristics of fine aggregate is a reference in determining the water-cement ratio in a concrete mix design. Fine aggregate test results can be seen in Table 1.

Table 1. Fine Aggregates Properties

Physical Properties	Value
Fine modulus	2.73
Specific Gravity (SSD)	2.17
Water Absorption (%)	2.04
Water Content (%)	4.16

Coarse aggregates are gravel as a result of natural disintegration of rocks or in the form of broken stones produced by the stone-breaking industry and have grain sizes between 5 mm - 40 mm. Physical Coarse aggregate test results can be seen in Table 2. Aggregate testing is carried out in surface saturated dry conditions (SSD). The total dry aggregate condition will affect the value of the concrete slump. Slump values that are too small will reduce concrete workability.

Table 2. Coarse Aggregates Properties

Physical Properties	Value
Fine Modulus	7.11
Specific Gravity (SSD)	2.65

Water Absorption (%)	1
Dry volume weight (kg/m^3)	1575
Size (mm)	38.1
Water Content (%)	2.46

2.2. *Silica Fume and Steel Fibre*

Compared to cement granules. With a very fine particle size, silica fume can fill the space so that the concrete mixture has a saturation process (tighter) which can increase its compressive strength and impermeability. More than 95% of silica fume particles are finer than 1 μm. Its typical physical properties are given in Table 3 [3].

Table 3. Silica Fume Properties (Silica Fume and Association,2005 [6])

Physical Properties	Value
Particle Size	< 1 μm
Bulk Density	
(as-produced)	130–430 kg/m^3
(slurry)	1320–1440 kg/m^3
(densified)	480-720 kg/m^3
Specific Gravity	2.22
Surface Area	13000-30000 m^2/kg

Steel fibre Dramix is used in this study. The steel fibres Dramix 3D have 60 mm in length (l), 0.90 mm in diameter (d), 65 length/diameter ratio (l/d), with 1160 MPa of tensile strength and 210 GPa of modulus of elasticity [7]. Silica fume and steel fibre used in this study can be seen in Figure 1.

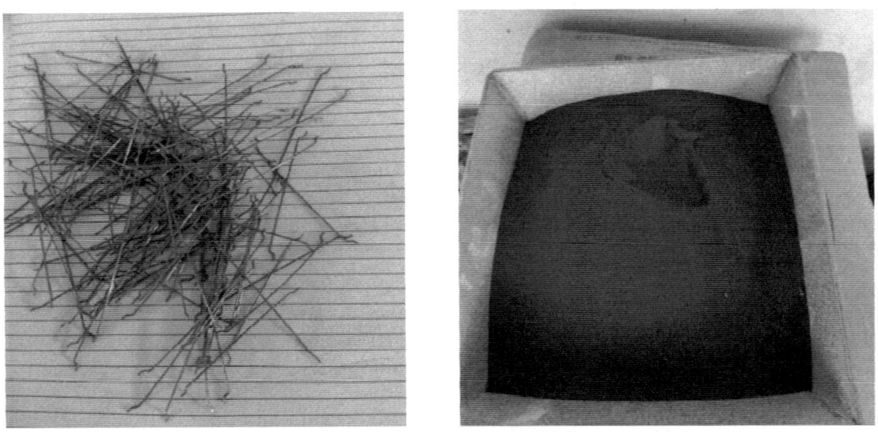

Figure 1. Silica Fume And Dramix Steel Fibre

3. Experimental Programs

Concrete will reach optimal strength after 28 days of immersion. To test the strength of concrete steps taken is to test the compressive strength of concrete using a compression testing machine that aims to determine the strength of concrete in bearing axial compressive load. In addition, a tensile strength test will be conducted / Split Tensile Test to determine the concrete split tensile capacity. The Unconfined Compression Strength tests were conducted according to ASTM C873 (ASTM International 2010a) and ASTM C496 (ASTM International 2010b), respectively. Test Set up of the specimen can be seen in Figure 2.

Figure 2. Compression Test And Split Tensile Test Set Up

In this research, eight variations of concrete cylinder specimens will be made. Each specimen uses 10% silica fume, which will be combined with steel fibre. The composition of Steel Fibre varies from 0 to 3 per cent by weight of cement. The compressive strength of the normal concrete plan used is 21 MPa with a water-cement ratio of 0.674. The variation of the percentage of silica fume and steel fibre can be seen in Table 4.

Table 4. Mix proportions of Silica Fume and Steel Fibre by weight of cement

Specimen Label	Cement (%)	Silica Fume (%)	Steel Fibre (%)
DF0	100	0	0
DF1	90	10	0
DF2	89.5	10	0.5
DF3	89	10	1
DF4	88.5	10	1.5
DF5	88	10	2
DF6	87.5	10	2.5
DF7	87	10	3

4. Results and Discussion

Concrete compressive strength is the maximum compressive strength that can be carried concrete per unit area. Normal concrete compressive strength between 20 - 40 MPa. Concrete compressive strength is influenced by water-cement ratio (w / c), the nature and type of aggregate, type of mixture, workability, curing and concrete age. Water Cement ratio (w / c) greatly affects the compressive strength of concrete. The smaller the w / c value, the less amount of water will produce a large concrete compressive strength. Concrete compressive strength can be calculated by equation 1.

$$f'c = \frac{P}{A} \tag{1}$$

Where f'c is the compressive strength of concrete. P is the compressive force and A is the area of the compressive plane. Concrete tensile strength is a measure of the tensile strength of concrete. concrete has a low tensile strength. The tensile strength of concrete can be measured by conducting a concrete split tensile test. the tensile strength of concrete slabs can be calculated by Equation 2.

$$T = \frac{2P}{\pi L d} \tag{2}$$

Where T is the tensile strength of concrete, P is the Maximum applied load, d is the cross-section diameter, and L is the length of the cylinder. After the concrete compressive strength test and split, tensile test results are obtained as shown in Table 5.

Table 5. Mechanical Test results of Specimen

Mixture	Compressive Strength 28 Day (MPa)	Split Tensile Strength 28 Day (MPa)
DF0	21.40	2.866
DF1	26.32	2.87
DF2	24.57	3.014
DF3	24.33	3.027
DF4	24.20	2.979
DF5	23.52	2.974
DF6	22.18	2.931
DF7	22.11	2.939

Table 5 shows the highest compressive strength obtained by the variation of DF1 specimens of 26.32 MPa. In the DF1 specimen, the percentage of silica fume used was 10 per cent. While the percentage of steel fibre use in DF1 is 0 per cent. In the variation of the DF1 mixture, there has been no contribution to the increase in strength provided by steel fibre. Based on the compressive strength data of DF0 and DF1 specimens, there was an increase in strength of 18.7 per cent with the addition of silica fume. Meanwhile, when there is a contribution from the addition of 5% steel fibre to the mixture, the compressive strength decreases in the DF2 specimen. The decrease in strength that occurs is 6.7%.

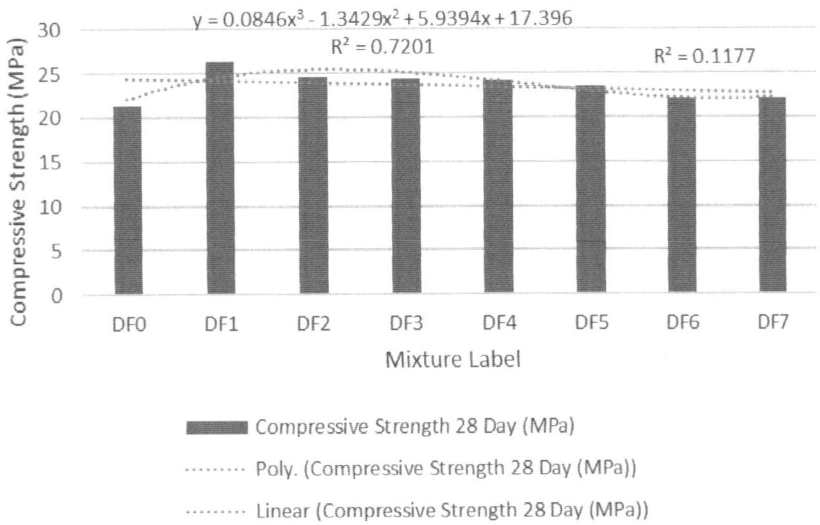

Figure 3. Compressive Strength of Specimen

Graphically the relationship of the variation of the composition of the test specimen to the compressive strength can be seen in Figure 3. Figure 3 shows that the strength of DF2 to DF7 has a difference below 6%. The smallest strength occurs between DF6 and DF7 by 0.2%. The tendency of the curve formed is ideally a 3rd order polynomial curve where the value of R square is 0.72. Linear curves are not very relevant to be used to see the number R squared below 0.5 that is equal to 0.117.

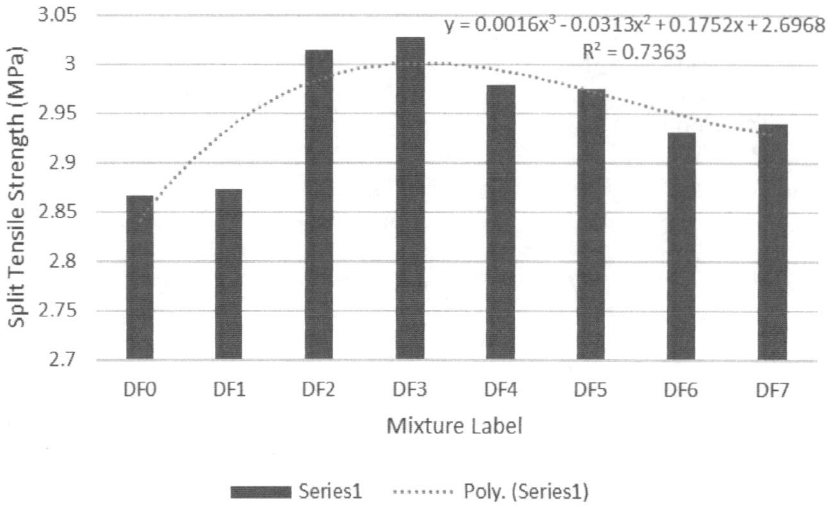

Figure 4. Split Tensile Strength of Specimen

Based on the data in Table 3 and Figure 4, it can be seen that the highest split tensile strength is in the DF3 specimen with a percentage of Steel fibre of 1%. The difference between the Split tensile strength of concrete to the compressive strength of concrete ranges from 86 to 89 per cent. This shows that the tensile strength of concrete slides ranges from 11 to 14 per cent of the compressive strength. the ideal curve that can be used is the polynomial 3rd order curve with the number R square 0.736.

5. Conclusion

From the results of a study conducted obtained several conclusions. The optimum compressive strength is in the DF1 variation with the percentage of silica fume 10 per cent and 0 per cent steel fibre. The use of steel fibre combined with silica fume does not significantly affect the compressive strength. Although it tends to decrease the compressive strength, the decline is no more than 10 per cent. The highest split tensile strength was obtained in the DF3 variation with the percentage of silica fume 10 per cent and 1 per cent steel fibre. The strength of tensile strength obtained ranges from 11 to 14 per cent of the compressive strength.

References

[1] Shannag, M. J. (2000). High strength concrete containing natural pozzolan and silica fume. *Cement and Concrete Composites*, *22*(6), 399–406. https://doi.org/10.1016/S0958-9465(00)00037-8

[2] Almusallam AA et al, (2004). Effect of silica fume on the mechanical properties of low-quality coarse aggregate concrete. Cement & Concrete Composites 26(7):891–900.

[3] Siddique, R. (2011). Utilization of silica fume in concrete: Review of hardened properties. Resources, Conservation and Recycling, 55(11), 923–932. https://doi.org/10.1016/j.resconrec.2011.06.012

[4] Qian, C. X., & Stroeven, P. (2000). Development of hybrid polypropylene-steel fibre-reinforced concrete. Cement and Concrete Research, 30(1), 63–69. https://doi.org/10.1016/S0008 8846(99)00202-1

[5] Sivakumar, A., & Santhanam, M. (2007). A quantitative study on the plastic shrinkage cracking in high strength hybrid fibre reinforced concrete. Cement and Concrete Composites, 29(7), 575–581. https://doi.org/10.1016/j.cemconcomp.2007.03.005

[6] Silica Fume Association. (2005). Silica fume manual. Lovettsville, VA: Silica Fume Association

[7] Caetano, H., Ferreira, G., Rodrigues, J. P. C., & Pimienta, P. (2019). Effect of the high temperatures on the microstructure and compressive strength of high strength fibre concretes.

Construction and Building Materials, *199*, 717–736. https://doi.org/10.1016/j.conbuildmat.2018.12.074

[8] ASTM International (2010a). Standard test method for compressive strength of concrete cylinders cast in place in cylindrical molds. Designation: C872/873M. West Conshohocken, PA:ASTM International. doi: 10.1520/C0873_C0873M-10A. www.astm.org.

[9] ASTM International (2011). Standard test method for splitting tensile strength of cylindrical concrete specimens. Designation: C 496/496M. West Conshohocken, PA:ASTM International. doi: 10.1520/C0496_C0496M-11. www.astm.org.

Community Perception of The Use of Shelter as Tsunami Mitigation In The Coastal Areas of Padang City

Y Hesna[1,*], J Sunaryati[1], A Suandi[1], Supani[2]

[1]Civil Engineering Department, Andalas University, Padang, Indonesia
[2]Civil Engineering Department, Sepuluh November Institute of Technology, Surabaya, Indonesia

[*]Corresponding Author: yervi@eng.unand.ac.id

Abstract. After the 2009 earthquake occurred, found liquefaction at several points of the Padang coast. To respond to this, the government built several shelters in Padang City to minimise the impact of tsunami or earthquakes in the future. However, it was found that there was still a minimum of people to evacuate themself to the shelter. Based on these conditions, this research intends to conduct on community's perception of Shelters as a tsunami mitigation effort in the coastal areas of Padang City. This research was carried out with direct studies in the field, questionnaire interviews with the community, and those with experience in this field, namely BPBD (Regional Disaster Management Agency) and KOGAMI (Tsunami Alert Community). The objects of this research are four main shelters of Padang City, namely Ulak Karang shelter, Air Tawar shelter, Darussalam Mosque shelter, Nurul Haq mosque shelter. Fifteen respondents for each shelter will be interviewed. Based on the results of the study obtained important things that affect people's perceptions of the existence of Shelters and the use of shelter as an evacuation point.

1. Introduction

Besides tsunami, liquefaction is a disaster that also threatens the coast of Padang [1]. The potential liquefaction on the coast of West Sumatra is at 4-8 m and also will occur after the earthquake. In 2009 West Sumatra earthquake found that water seepage emerged from the fractures around the coast, and the many buildings sloped as an indication of liquefaction. The vulnerability of the West Sumatra region to the potential for earthquake and tsunami threatens 273.755 people who live in the inundation areas [2]. Tsunami disaster can destroy community social structure, not only the loss of humans and material. Based on the devastating impact of tsunami needs disaster resiliency of coastal community [3].

According to Singh [4], it takes about 20 - 30 minutes a tsunami hit the coast of Padang. While residents have to walk 3-5 km to the safe area, it can be said that the time for tsunami evacuation in Padang city is short. Therefore the choice of vertical evacuation is urgent rather than walking along to the safe area. So the development of shelter that minimises the impact of disasters is one of disaster resiliency strategies.

Besides provide a barrier for tsunami waves, the shelter also functions as a socioeconomic building that is formed by cultivating tsunami evacuation culture through symbols on the streets and buildings used in social activities [3]. But there are still many people who not become aware of the existence of the shelter and their functions. Disaster management planning should include community planning and participation because it can contribute to community safety [5]. People rarely use the shelters either when there is a moderate earthquake or during a disaster simulation. From these conditions, this research is intended to get people's perceptions of the use of vertical shelters as tsunami mitigation in the coastal city of Padang.

2. Research Method

The methodology used for this research is a semi-structured interview. A semi-structured interview is an open interview process, but the interviewer must follow the theme and flow of the conversation. The surveys were conducted at four shelters, which are Ulak Karang shelter, Air Tawar shelter, Darussalam Mosque shelter, and Nurul Haq Mosque shelter. The interviewer asked 15 people who live and or have activity near the shelter as a respondent for each shelter. The total respondents are 60 people. The in-depth interview also took to the Regional Disaster Management Agency (BPBD) and the Tsunami Alert Community (Kogami).

The survey was conducted 4 hours a day (14.00-18.00) for two weeks. The questionnaire form related to this research were filled with four sections. The question regards to the aim of the interview are explained in Table 1.

Table 1. The List of Interview Topics

No	Topic
1	Respondent Profile
2	Respondents' Perception of the Existence of Shelter Buildings
3	Respondents' Perception of the Use of Shelter as Vertical Evacuation Buildings
4	Management of Shelter Operation

3. Result and Discussion

3.1. Respondent Profile

The gender survey results at shelters are 56% female respondents and 44% male respondents. The occupation of respondents in the research location is relatively varied, as seen in the following figure.

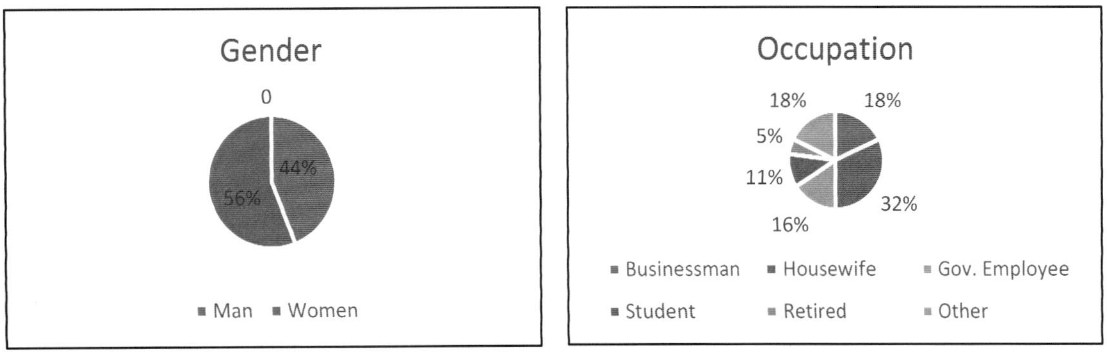

Figure 1. Respondent Profiles

3.2. Shelter Building Profile

Ulak Karang shelter located 500 meters from the beach. This shelter completely builds in 2015 with a total of 4 floors and a building height of 20.46 m. This shelter can accommodate ± 4000 people. The shelter built by the National Disaster Management Agency (BNPB), and it not been handed over to the local government. Ulak Karang shelter is also rarely used by the community due to inadequate and poorly maintained shelter conditions except on the 1st floor, which is often used by children to play football.

Air Tawar shelter is on Polonia Street, Padang Utara district. Air Tawar Shelter was built in 2012 and consisted of 3 floors, which 35 m height. This shelter can accommodate ± 4000 people. According to the local community, this shelter is often used by wild parties for immoral acts because the shelter is not well maintained. Some of the existing facilities at Air Tawar shelter also not function.

Darussalam Mosque shelter is located in Parupuk-Tabing, Koto Tangah, Padang, and is located ± 1 km from the beach. Five floors of this shelter have an area of 3500 m² with a building height of 25 m. This shelter can accommodate ± 5000 people. Built with complete facilities, but due to lack of maintenance and the act of wild parties, this shelter is often locked.

Nurul Haq Mosque shelter is located in the Jondul Parupuak Tabiang complex. This shelter was inaugurated in 2015, which has spent a total of Rp. 17,564,343,000. This shelter has 22 m height. This shelter can accommodate up to ± 4000 inhabitants around the shelter. Each floor of the shelter is equipped with toilets and clean water facilities. The management of this shelter is quite good because the community also often uses the shelter for community activity.

(a) Nurul Haq Shelter (b) Darussalam Mosque Shelter

(c) Air Tawar Shelter (d) Ulak Karang Shelter

Figure 2 All of Shelter Buildings in Padang City

3.3. Identify the perception of coastal communities on the existence of shelter buildings as tsunami evacuation building

Only 3% of 59 respondents did not know information about shelters, and the remaining 97% of respondents are aware of the existence of shelters or information about shelters. The community is given counselling by the local government every April 26 so that the community should be more vigilant facing disaster. The public should easy to know where the shelter is. Evacuation signs should indicate the location and direction of the evacuation to the shelter. The adequacy of evacuation signs at the research location confirms this result.

76% of respondents who live on the coast say that the community will use the shelter if there is a warning for a potential tsunami disaster. However, not all of the people choose this, as evidenced by the fact that 24% of other respondents will stay at home or go to a safer area like a higher place, the bypass street, and other alternatives.

According to some respondents who choose to stay at home because they had another place that is higher and closer to their home like a neighbour's house. While some others said, access to the nearest shelter is not easy to reach. This was conveyed by the people who lived a few meters away from the Ulak Karang shelter and Darussalam Mosque shelter. It was challenging to get to the shelter because of the narrow and hampered roads.

Other respondents also said the shelter building was quite low regarding tsunami wave height. So that respondents felt it was still unsafe to escape to the shelter. As for other reasons such as going to a higher place, for example, go to hills, especially for respondents whose residence is quite far from the beach.

One reason for the lack of public distrust to use shelter buildings is because of the physical condition of the shelter due to the lack of proper management in the operational and maintenance. It can be seen from the completeness or not of the available instruments and facilities at the shelter, for example, lighting, electrical installations, and bathrooms. There is a leak in the room and poor in the cleanliness of the room.

Figure 3. Inadequate Maintenance of Shelter Building

Also, the observations show the lack of a security system for shelters. The shelter does not have special security guards. In the absence of guarding, people choose to lock in access to the shelter, which aims to reduce thievery and other unwanted action. But it also brings anxiety from the community because if the shelter is often locked will cause unclear response when a disaster occurs.

Figure 4. Access to Ulak Karang shelters and Darussalam mosque shelter is locked.

Another example that happened at the Ulak Karang shelter is the missing of the lamp and the electrical cable in the room. Improperly use of shelter by local people found in Ulak Karang shelter as an indication of lack of security system. It is dating on the top floor of shelters and the presence of teenagers who sniff glue in one of the shelters' rooms. Lack of security system will make people uncomfortable and unwilling to use the shelter.

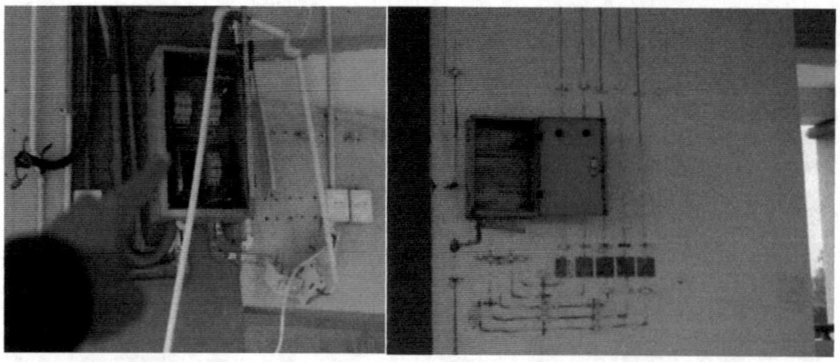

Figures 5. Theft of electrical installations at shelters

The operational management of shelter building should involve the participation of the community. The prime component of disaster preparedness is to involve the participation of the community [6]. At Nurul Haq Mosque shelter, government institution has transferred the management of the shelter to the community surrounding. Therefore, the security problem did not exist at the Nurul Haq Mosque shelter.

3.4. Identify the perception of coastal communities of the potential liquefaction in the area of shelter buildings

Based on the questionnaire, 68% of the respondent felt safe for the shelter position, which means that the respondent knew about the threat of liquidation and would continue to use the shelter while disaster happened. Another 32% said it was unsafe, and they see the danger of liquefaction disaster. When the earthquake occurs, the shelter building is expected to have an immediate performance [7]. For instance, no structural damage, the building is still functioned without any rehabilitation, and the area is not followed by liquefaction.

The lack of knowledge about liquefaction and the absence of appeal from the government for threats of the liquefaction make the respondent feel not safe from the liquidation. The position of the Darussalam Mosque shelters that are close to the swamp is one of the most substantial reasons for the respondent. It can be seen in figure 6 below.

According to Minister of Public Works Regulation No. 06/ PRT/M/2009, states that infrastructure development in disaster-prone areas requires the placement of building floors above puddles. The infrastructure should provide barrier walls for inundation, or adjust the distance and orientation of buildings to avoid the tsunami force on the surrounding buildings and buildings.

Figure 6. The Position of Darussalam Mosque Shelter that Close to The Swamp

The lack of information about liquefaction makes community doubt the safety of shelters from liquefaction disasters. The liquefaction map can be the source of information for the public. The government should publish it to the public to make people well informed about the location or areas of the potential liquidation. Developing communication strategies to the community regarding disaster will influence the social and psychological factor upon recommendation escape strategy [8].

Besides this, the government should consider the vulnerability of the building location against liquefaction. FEMA P-646, the guideline for vertical evacuation, reveals that if it is possible, the building should far from the hazard that can cause additional damage to the structure and endanger the safety of the building users like liquefaction. Zulfiar [7] revealed that the vulnerability of building is technically caused by topography or location. If the location of the shelter has a risk of liquefaction, the government should do some practical action to improve the ground structure [9, 10].

4. Conclusion

Padang City has four main shelters that can be used as temporary evacuation sites, but this number is still insufficient for all residents on the coast of Padang. Some people do not know the use of the shelter. This condition is as an indication of the community's lack of knowledge about the danger of a tsunami. The results of the research found that several things influenced the community's perception of the existence of shelters on the coast of Padang. Namely, providing information about shelters as an option in disaster evacuation, evacuation signs located in disaster-prone areas and areas toward shelters, and disaster evacuation maps.

This research also found factors that affect the community's perception of the use of shelters as evacuation places from the tsunami disaster are the location of shelters, the direction to get to shelters, physical shelter conditions, and management in shelter maintenance. And factors of community's reliability of the use of shelters that will not be affected by liquefaction disasters are shelter locations, dissemination from the government, and publication of liquefaction maps. It is hoped that disaster planning and mitigation will be more success by defining this factor.

Acknowledgment

Publication of this article is supported by the Andalas University Grant Number: T/54/UN.16.17/PT.01.03/RD-Inovasi Sains/2019Tahun Anggaran 2019.

References

[1] Abdul H 2013 *J.Teknik Sipil 20*

[2] BPBD Padang 2017 *Rencana Kontinjensi Menghadapi Bencana Tsunami Kota Padang [Mitigation Plan towards tsunami Hazard in Padang City]* BPBD (Regional Disaster Management Agency) .

[3] Yu T.J.K, Marcin P.J, Fatma D, Ricardo S.C, Jianping G, Miguel E, Rafael A, and Tomohoro A 2015 *Inter'l J. of Disaster Risk Reduction* 13

[4] Faisal A, Dilanthi A, Richard H 2014 *Procedia Economics and Finance* 18

[5] Laurie P 2004 *Natural Hazard* 28

[6] Jeyanth K.N, Godfrey G.P.J 2003 *Disaster Prevention and Management* 12, 1

[7] Billy R 2018 *J. of the Civil Engineering Forum* 4, 3

[8] Douglas P, David J 2001 *Disaster Prevention and Management : An Int'l J.* 10, 4

[9] Adrin T 2013 *Prosiding pemaparan hasil penelitian puslit geoteknologi*

[10] Hamzah L 2012 *37th HAGI Annual Convention & Exhibition Palembang*

Sediment Trap in the Downstream of Check Dam: The Effect of Changes in Channel Depth on Sediment Deposition

D A Wulandari[1,*], Susilowati[1], D Ulfiana[1], P N Parmantoro[1]

[1] Civil Engineering Department, Engineering Faculty, Diponegoro University, Semarang, Indonesia

*dyahariwulandari@yahoo.co.id

Abstract. Reservoir sedimentation is a major problem experienced by reservoirs in Indonesia. Reservoir sedimentation can be overcome by reducing the amount of sediment that enters the reservoir, among others, by building a Check dam. Check dam deposits more sediment loads with coarse grains, while sediment loads with fine grains will escape and enter the reservoir. So it is necessary to have construction to hold and settle fine sediment loads downstream of the Check dam. Construction is planned like a sediment trap in an irrigation system with variations in the cross-sectional area. In this study, the cross-section was expanded by changing its depth. The purpose of this research is to determine the effect of changes in the channel depth with discharge and grain size distribution variation on sediment deposition that occurs. The calculation is done using the HEC-RAS software. Simulation results show the deeper the sediment trap, the smaller the amount of sediment deposit. The large discharge, the smaller of deposit percentage. The greater of fine sediment percentage, the less the deposit.

1. Introduction

A Reservoir is a pool of water formed by the construction of dams in rivers. In addition to storing water, the reservoir also holds sediment that is carried by the flow. Most of this sediment will then settle in the reservoir, and some will come out through the release building. The amount of sediment that settles is used to predict reservoir life. Changes in land use in the catchment area of the reservoir cause higher erosion rates so that the reservoir sedimentation rate is also higher. Many reservoirs in Indonesia experience this problem. What can be done to overcome this problem is to reduce the amount of sediment that enters the reservoir, such as by building a Check dam. However, not all sediments can settle at the check dam. Some sediments with coarse grains will settle, while sediment loads with fine grains will escape and enter the reservoir. According to Fauzi (2006), in [6], sediments with a diameter < 0.1 mm do not settle even though the water velocity is close to zero. Whereas sediments with a diameter > 0.1 mm will settle if the water velocity is zero or close to zero. Research on the size of sediment grains deposited in the Lumajang river check dam has been conducted by [6], from this study it was found that the gradation of sediment deposits deposited on check dams in the form of rocks, gravel, sand and a small portion of fine sand or silt. Small sediments only settle at the time of initial filling; this is because upstream of the check dam forms a puddle, which results in a decrease in flow velocity at the start of submergence. When 20 months after submergence, sediments measuring 0.10 - 1.0 mm only settle by 1% while sand, gravel until the rock rises to 99%. The efficiency of sediment capture itself depends on the size of the inflow and its storage capacity. The greater the capacity of the check dam, the greater the efficiency of sediment capture. Therefore it is necessary to have a building to hold and deposit sediment loads that still pass through a check dam. While research that has been done on average is research to improve the effectiveness of deposition at check dams ([5], [7]).

This research is part of the research on Sediment Trap in the Downstream of Check dam. The construction is planned as a mud bag system that is placed on a river body in the downstream part of the check dam. After determining the dimensions requirements, it is necessary to know the effect of changes in the depth of the sediment trap on sediment deposition to obtain an effective depth of sediment deposition. Because the discharge in the river always experiences fluctuations where the difference between the maximum discharge and the minimum discharge is very large, and the gradation of the suspended sediment grains will also fluctuate depending on the size of the sediment that settles at check dam. There are several studies have been conducted to determine the sedimentation behavior of sediments that occur in sediment traps in irrigation channels with some modifications and conditions ([1], [3], [4], [8]). This needs to be adjusted to the hydraulic conditions and sediment transport in the river.

2. Methods

The research location is in the downstream section of a check dam in Lumajang River, Linggasari, Banjarnegara. The purpose of this study was to determine the effect of changes in channel depth with variations in discharge and suspended sediment grain size distribution on sediment deposition. Changes in depth are done to get variations in the cross-sectional area because the sedimentation velocity of sediment particles is inversely proportional to the channel cross-sectional area. The wider the channel cross-section, the faster the sediment deposition. In this research, the modelling scenario is as follows:

a. The basic dimensions of the sediment trap are 145 m long, 10 m wide, 5.8 m deep, and the channel slope is 0.034.
b. The cross-section is expanded by changing the existing depth with three depth variations, namely depths of 6.4 m, 7.0 m, and 7.6 m.
c. The discharge used is an average discharge of 5 m^3/s, and the maximum flood discharge at 41 m^3/s
d. This sediment trap is planned to deposition the sediment load that escapes from the check dam. So that the fine sediment granules are varied to increase with a maximum grain size of 15 mm. Gradations of sediment grains are made in three scenarios, as in Table 1.
e. The relationship between water discharge and sediment is expressed by equation [1].

Table 1. Grain Size distribution Scenario

Scenario	Diameter of sediment grains (mm)	Percentage of Grains (%)
1	≤ 0.074	11
	0.075 – 15	89
2	≤ 0.074	21
	0.075 – 10	79
3	≤ 0.074	31
	0.075 – 5	69

The calculation is done using the HEC-RAS software. There are 24 models combination of scenarios, as mentioned above.

3. Results and Discussion

Based on the results of calculations with HEC-RAS, in each scenario, a cross-section and a long section will be obtained, as in Figure 1 and Figure 2 respectively. So that in each scenario the elevation of the river bed can be determined. If the elevation of the initial riverbed is greater than the end condition of the river bed, it means that the section has been scoured. Meanwhile, if the elevation of the initial river bed is lower than the end condition of the river bed, it means that the deposition occurred in the section (Figure 3). Based on the results of changes in river bed elevation, the volume of sediments that settles and the percentage of sediments that settles is calculated, as in Table 2 and Table 3 respectively.

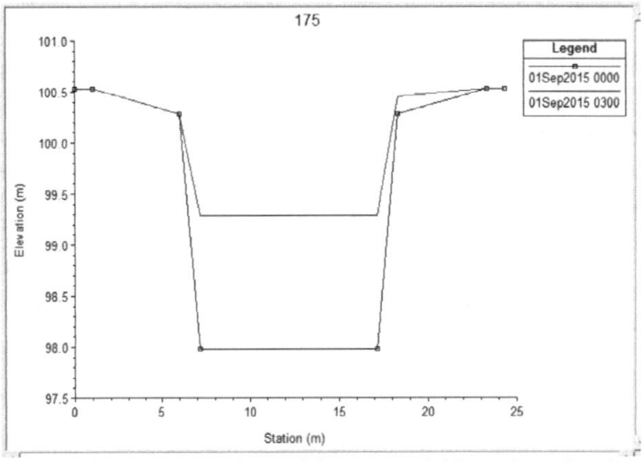

Figure 1. STA 175 cross section in scenarios with 16 m width, 5.8 m depth, 41 m^3/s discharge and grain size distribution 2.

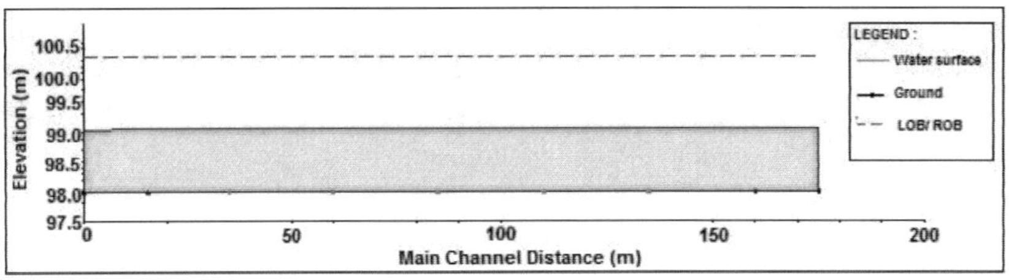

a. Initial condition

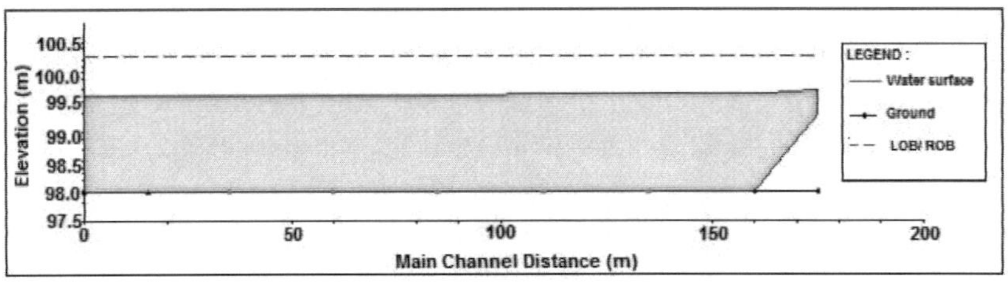

b. Running result

Figure 2. Long section STA 175 in scenarios with 16 m width, 5.8 m depth, 41 m^3/s discharge and grain size distribution 2.

	01Sep2015 0000	Before	01Sep2015 0300	After
	Station	Elevation	Station	Elevation
1	0.00	100.5301	0.00	100.5301
2	1.00	100.5301	1.00	100.5301
3	6.00	100.2801	6.00	100.2801
4	7.15	97.9801	7.15	99.2939
5	17.15	97.9801	17.15	99.2939
6	[illegible]	[illegible]	[illegible]	[illegible]
7	23.30	100.5301	23.30	100.5301
8	24.30	100.5301	24.30	100.5301

Figure 3. Bed Elevation STA 175 in scenarios with 16 m width, 5.8 m depth, 41 m^3/s discharge and grain size distribution 2.

Based on Table 2 and Table 3, it can be seen that the discharge of 5 m^3/s and 7.0 m depth that settles sediment is the least, while the flow of 41 m^3/s is deeper than the sediment trap depth then the sediment that settles is also getting smaller. According to the [2], [3] and [4] the particles entering the pool at A, with the particle deposition velocity w and the velocity of water v must reach the bottom at C (Figure 4). Particles take time for H/w to reach the bottom and will move horizontally along the L distance in L/v time. If the depth of the sediment trap gets deeper than before, the cross-sectional area will be greater, and the flow velocity will be smaller. So that the horizontal sediment transfer time slows.

However, the deeper the sediment trap is, the distance traveled by the sediment grains to reach the bottom is also getting farther and takes longer. In this case, it means slowing down the horizontal displacement time due to decreased flow velocity is still faster than the time required for sediment to reach the bottom, so the deeper the sediment trap, the smaller the amount of sediment that settles. In this case, the length of the sediment trap needs to be longer. Reducing the flow velocity by increasing the cross-sectional area by increasing the depth of the sediment trap proved ineffective. Consideration should be given to enlarging the cross-sectional area by increasing the width of the sediment trap.

Based on Table 3, it can be seen that in all depth variations, the percentage of sediment that settles at 41 m^3/s is smaller than the percentage of sediment that settles at 5 m^3/s. This is because a large discharge will increase the flow velocity, thereby reducing the sedimentation rate of sediment particles. In all variations of sediment trap depth, the difference in grain gradation also results in different sediment deposition. The greater the percentage of fine sediment, the less sediment settles. Fine sediment grains have slower deposition speeds so that the time needed to reach the bottom is longer, so the size of the sediment trap must be longer so that the fine sediment settles optimally.

Table 2. Sediment deposited

Length (m)	Wide (m)	Depth (m)	Sediment deposited (m^3)					
			Discharge (m^3/s)					
			5			41		
			Grain size Distribution			Grain size Distribution		
			1	2	3	1	2	3
145	10	5.8	1.755	1.715	1.695	211.090	204.190	198.200
		6.4	1.760	1.715	1.695	211.080	203.875	197.665
		7.0	1.710	1.690	1.645	211.050	203.765	197.510
		7.6	1.760	1.715	1.695	211.048	203.785	197.510

Table 3. Percentage of sediment that settles

Length (m)	Width (m)	Depth (m)	Percentage of sediment that settles (%)					
			Discharge(m^3/dt)					
			5			41		
			Grain size Distribution			Grain size Distribution		
			1	2	3	1	2	3
145	10	5.8	1.48	1.45	1.43	1.29	1.25	1.21
		6.4	1.49	1.45	1.43	1.29	1.25	1.21
		7.0	1.44	1.43	1.39	1.29	1.25	1.21
		7.6	1.49	1.45	1.43	1.29	1.25	1.21

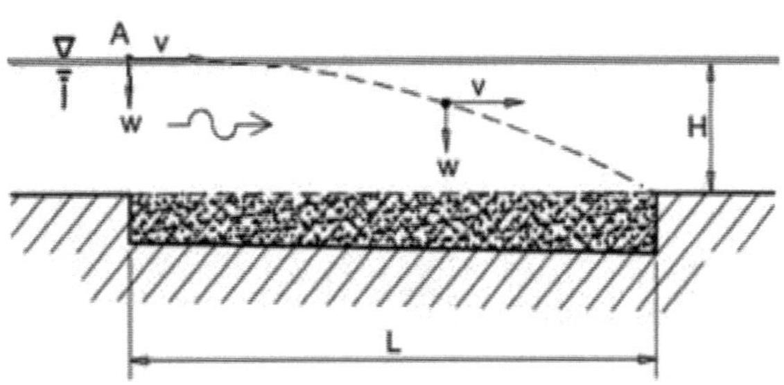

Figure 4. Schematic sediment deposition [2]

4. Conclusion

Based on the results of the analysis that has been done, it can be concluded that:

a. The deeper the sediment trap the smaller the amount of sediment that settles because the deeper, the longer time needed for the sediment grains to settle
b. In all depth variations, large discharges produce smaller sediment percentages than those generated in small discharges because large discharges increase the flow velocity, thereby reducing the rate of deposition of sediment particles.
c. In all depth variations, the greater the percentage of fine sediment the less the sediment settles because fine sediment has slower deposition velocity, so the time needed to reach the bottom is longer

Acknowledgements
The authors are grateful to UNDIP, who funded this research through the Penelitian Strategies scheme in 2019.

References
[1] J. Baade and R. Hesse, "An Overlooked Sediment Trap In Arid Environments: Ancient Irrigation Agriculture In The Coastal Desert Of Peru", *Sediment Dynamics In Changing Environments* (Proceedings Of A Symposium Held In Christchurch, New Zealand, December 2008), IAHS Publ. 325, 2008.
[2] Departemen Pekerjaan Umum, Standar Perencanaan Irigasi KP-02. 1986. Jakarta: Badan Penerbit Pekerjaan Umum
[3] Hanwar, Suhendrik and Revalin Herdianto, "Desain Bangunan Penangkap Sedimen dengan Teknologi Baffle", *Jurnal Teknik Sipil dan Perencanaan*, vol. 9, no. 2, pp. 145-154, 2007.
[4] R. Herdianto, and R. Hidayati, "Kombinasi Sekat Dan Tanaman Air Untuk Optimalisasi Bangunan Penangkap Sedimen", *Rekayasa Sipil,* vol. VI, no. 1, 2010.
[5] G. Piton and A. Recking, "Design of Sediment Traps with Open Check Dams. I: Hydraulic and Deposition Processes", *J. of Hydraulic Engineering*, 142(2): 4015045, 2016, https://doi.org/10.1061/(ASCE)HY.1943-7900.0001048
[6] Soewarno; and C. B. Sukatja, "Kinerja dam sabo k. lumajang untuk pengendalian sedimentasi waduk mrica," *J. Sumber Daya Air*, vol. 6, no. 1, pp. 17–32, 2010.
[7] S. Schwindt, M. J. Franca, A. Reffo, and A. J. Schleiss, "Sediment traps with guiding channel and hybrid check dams improve controlled sediment retention," *Nat. Hazards earth Syst. Sci.*, vol. 18, pp. 647–668, 2018.
[8] J.I.F. Slaets, P. Schmitter, T. Hilger, T.D. Vien, and G. Cadisch, "Sediment Trap Efficiency Of Paddy Fields AtThe Watershed Scale In A Mountainous Catchment In Northwest Vietnam", Biogeosciences Discuss., 12, pp. 20437–20473, 2015, Doi:10.5194/Bgd-12-20437-2015.

Setting Time of the Geopolymer Binder with White Soil Substitution

Y A Priastiwi [1,*], Muhrozi[1], R Y Adi[1], D Daryanto[1], and Z Salamsyah[1]

[1]Department of Civil Engineering, Diponegoro University, Semarang, Indonesia.

[*]yulita_tiwi@ymail.com

Abstract. This study analyzes the effect of white soil as a substitution of fly ash on geopolymer binders. Fly ash from Tanjung Jati Jepara, white soil from Kupang, water and NaOH as alkaline activators are mixed into binders and tested for setting time with a Vicat tool. The water factor of the binder was used 0.5. The white soil substitution in fly ash using a percentage of 0 to 30 of the weight of fly ash. Fly ash and white soil used had the particle size smaller than 0.074 mm or passed the number 200 sieve and had a moisture content of 0%. The results show that the setting time of geopolymer binder with substitution white soil in the range of 52.5 to 105 minutes for the initial setting time and is between 180 to 225 minutes for the final setting time. This value approaches the same as the setting time of Portland cement in SNI 15-2049-2004. From the 7 variations tested, the binder with 15% of white soil has a faster setting time compared to other variations, i.e 52.5 minutes for initial and 180 minutes for final setting time. The temperature when testing affects the setting time of the geopolymer binder

1. Introduction

White soil is natural ingredients from Kupang, East Nusa Tenggara. The white soil is a rock that has undergone changes in chemical composition, caused by weathering and the influence of underground water conditions. The white soil has brownish-white features; it is compact and rather difficult to dig with simple tools. It has chemical constituents, namely SiO_2, P_2O_5, Al_2O_3, Fe_2O_3, FeO, MnO, MgO, CaO, Na_2O, K_2O, TiO_2 and SO_3 [1]. The content of Si and Al in white soil is one source of material that can be used for geopolymer materials [2].

More recently, fly ash has been used as a component in geopolymers, where the reactivity of the fly ash can be used to create a binder similar to a hydrated Portland cement in appearance, but with potentially superior properties, including reduced CO_2 emissions [3]. Research on geopolymers made from fly ash and white soil still needs to be developed. This study will examine the setting time of geopolymer binders made from fly ash with white soil substitution.

2. Research Methods

2.1. Binder Geopolymer Composite Material

This study intends to obtain the setting time of geopolymer binder with fly ash and NaOH as an activator and compare it with geopolymer binder with the white soil substitution on part of fly ash. The amount of white soil substitution given varies and each variation will be measured when initial and final setting time. The composition of the geopolymer binder is fly ash, white soil, water and alkaline activator (NaOH) with molarity level 8 M. These materials mixed formed geopolymer binders. The material used must meet certain criteria so that it can act as a binder such as size and moisture content.

In this study, white soil is a natural product from Kupang East Nusa Tenggara was used. The white soil has several chemical compounds, including calcium, magnesium, silica, alumina and iron, expressed in the form of oxides (CaO, MgO, SiO_2, Al_2O_3, Fe_2O_3. The white soil used a size smaller than 0.075 mm or passed the number 200 sieve, as shown in Figure 1(a), while the fly ash used was class F from PLTU Tanjung Jati B Jepara. The fly ash is a mineral residue produced from the combustion of refined coal at a power plant center. It used had the particle size smaller than 0.075 mm or passed the number 200 sieve, as shown in Figure 1(b). The alkaline activator used was sodium hydroxide (NaOH) 8M[4]. NaOH used to facilitate the condensation polymerisation reaction on the geopolymer, as shown in Figure 1(c)

(a) White soil **(b)** Fly Ash **(c)** NaOH

Figure 1. The constituent material of geopolymer binder

2.2. Making Binder Test Objects

In conducting this research, several tools were needed to support the work process and testing, which are as follows a set of Vicat tools with needles Ø 1 mm, scales, ebonite ring, glass plate, stopwatch, porcelain bowl and stirrer. The Vicat tool shows in Figure 2.

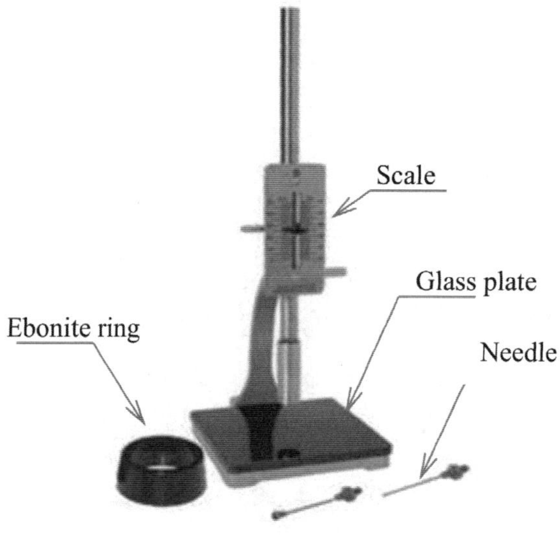

Figure 2. Vicat tools

2.3 Composition of Binder Geopolymer

There are 7 variations of the test object with different compositions according to the percentage of white soil substitution, which varies 0, 5, 10, 15, 20, 25, and 30% from fly ash proportion. The composition of binder geopolymer mixtures with NaOH activator as shown in Table 1.

Table 1. Composition of binder geopolymer mixtures with NaOH activator

Test Object	White soil substitution (%)	Fly Ash (%)	White Soil (gr)	Fly Ash (gr)	Water (ml)	NaOH (gram)	(% White soil to fly ash)
1	0	100	0	200	80	25.6	White soil substitution 0%
2	5	95	10	190	80	25.6	White soil substitution 5%
3	10	90	20	180	80	25.6	White soil substitution 10%
4	15	85	30	170	80	25.6	White soil substitution 15%
5	20	80	40	160	80	25.6	White soil substitution 20%
6	25	75	50	150	80	25.6	White soil substitution 25%
7	30	70	60	140	80	25.6	White soil substitution 30%

Usage of 8 M NaOH
Calculation of NaOH molarity with 8 M NaOH
M = 8 M
Water = 80 ml
Mr NaOH = 40

$$M = \frac{NaOH\ weight}{Mr} \text{ x } \frac{1000}{Water\ Volume}$$

$$8 = \frac{NaOH\ Weight}{40} \text{ x } \frac{1000}{80}$$

NaOH Weight = 25.6 gram

2.4. Method for Making Test Objects

When cement is mixed with water, it hydrates and makes cement paste. This paste can be moulded into any desired shape due to its plasticity. Within this time cement continues with reacting water and slowly cement starts losing its plasticity and set harden. This complete cycle is called setting time of cement.

The initial setting time is the time needed for the cement from the time it starts reacting with the water to cement paste until it loses its plasticizing properties. The initial setting test method uses ASTM C 191[5] standard and uses a Vicat tool with a 1 mm diameter needle. The initial setting time of the cement is obtained when the reduction reaches 25 mm. The time of setting to cement between 45-120 mm, while the final setting time is the time where needle penetration is not visually visible within a maximum of 360 minutes[6]. The setting time test on this geopolymer binder also uses the same method as follows:

1. The white soil and fly ash used were the particles that pass sieve no. 200. The white soil and fly ash had water content of 0%.
2) The white soil, fly ash, water and NaOH were weighed as planned.
3) The water and NaOH were mixed. Because of the heat produced during the mixing process, some waiting time was needed for the sample to be cold.
4) Insert the white soil and fly ash into the bowl and stir until evenly distributed.
5) Check and adjust the equipment needed, adjust the Vicat tool so that it shows the position of 0 mm.
6) The surface of the ebonite ring smeared with the oil and then placed the ebonite ring on the glass plate with a small diameter of the ebonite ring at the top.
7) The mixture of white soil and fly ash is then mixed with a solution of NaOH and water, then stir for 3 minutes to form a paste.
8) Insert the paste mixture into the ebonite ring and tap the ebonite ring with a spoon so that the air cavity in the paste disappears, then flatten the surface of the paste.
9) Release the Vicat needle freely from the starting point 0.
10) The decrease in binder at 30 seconds after the needle was removed was noted.

IOP Conf. Series: Materials Science and Engineering **771** (2020) 012051 doi:10.1088/1757-899X/771/1/012051

11) Furthermore, every 15 minutes tested and a decrease that occurs is recorded. Created the drop graph. The initial time starts to the binder at a decrease of 25 mm. Time starts to harden at a decrease of 0 mm.

During testing the equipment must be free of vibrations. Keep the needle 1mm straight and clean because the lumps of the binder attached to the side of the needle will slow penetration and if the binder sticks to the tip of the needle will accelerate penetration.

3. Result and Analysis

The results of the setting time test on the geopolymer binder could be illustrated in the graph of penetration (mm) – time (minute). For the setting time of the geopolymer binder without white soil, substitution is shown in Figure.3, while the geopolymer binder with white soil substitution in Figure 4.

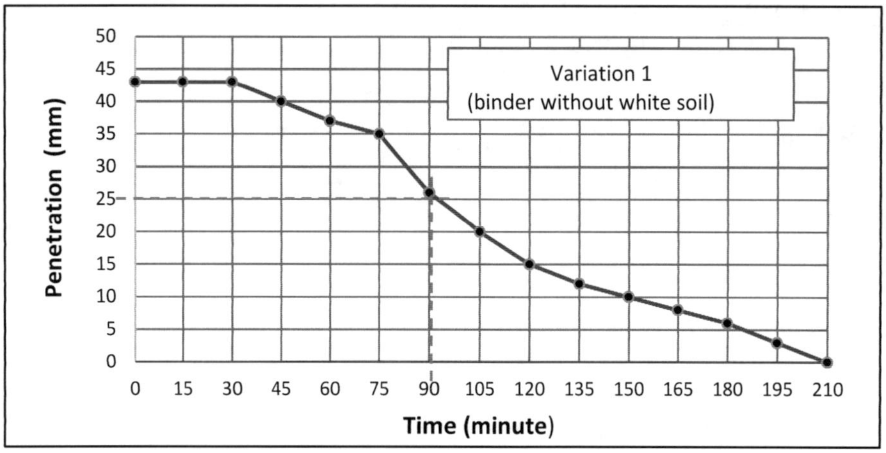

Figure 3. Penetration- time relationship in geopolymer binders (without white soil substitution)

From Figure 3 it is shown that the binder which is composed of a mixture of fly ash, water and NaOH activator and tested at a temperature of 29 °C has an initial setting time of 90 minutes and a final setting time of 210 minutes. The chemical composition of fly ash which contains Silica and Alumina almost resembles the chemical composition of Portland cement, making the setting time that occurs is still within the time span given to Portland cement.

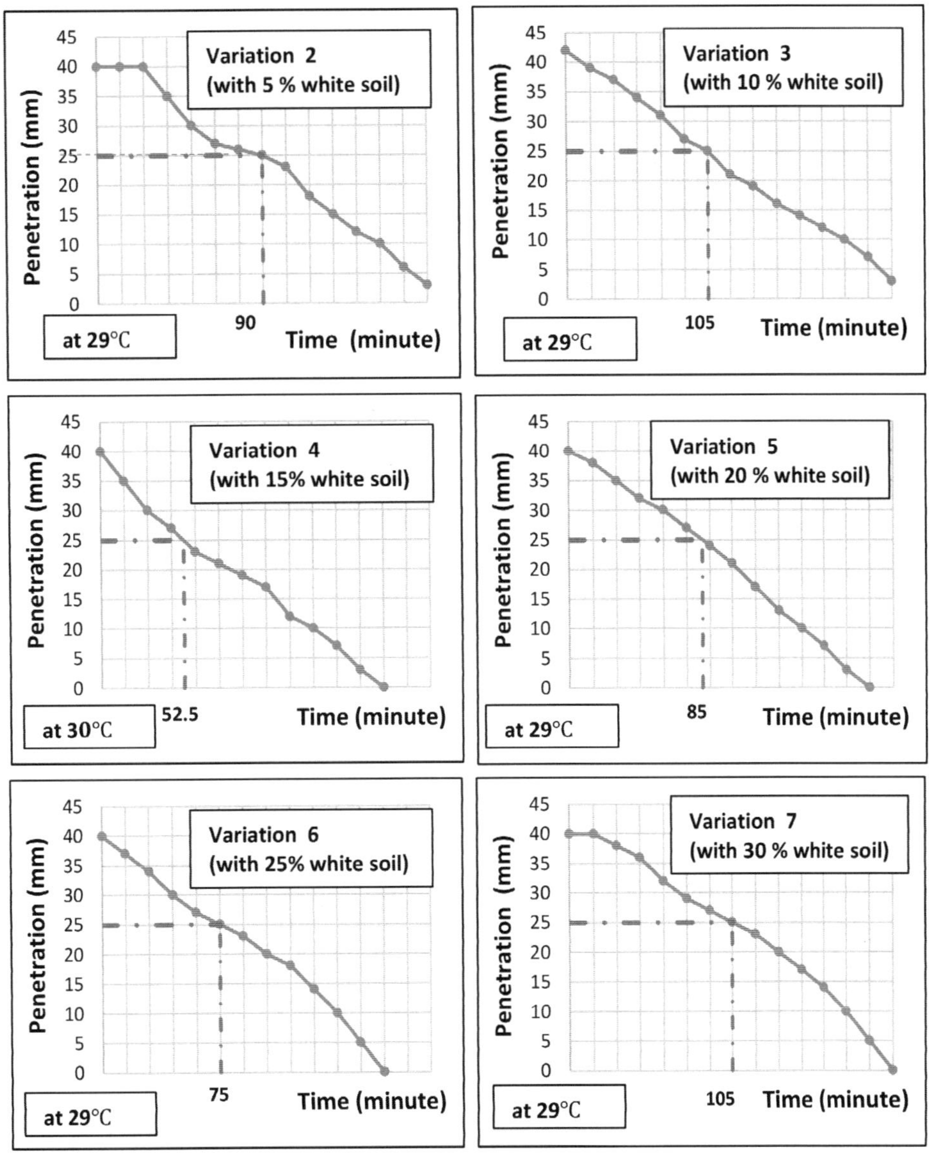

Figure 4 . Penetration- time relationship in geopolymer binders
(with percentage white soil substitution)

The Figure 4 shows the graph of the setting time that occurs in the binder given substitution in the proportion of fly ash with white soil, with a percentage of 5 to 30. White soil which has several chemical constituents including Calcium, Magnesium, Silica, Alumina, and Ferron affects the setting time of geopolymer binders. From the six variations of binder with white soil substitution, it can be seen that variation 4 has a faster binding time compared to other variations, that is 52,5 minute. The influence of higher temperatures during testing (30°C) is thought to be the cause of faster setting time while the other binders tested at 29 °C have an initial setting time of 75 to 105 minutes, so it can be said that the test temperature also affects the setting time of the geopolymer binder.

The experiment on 7 variations showed that the longest initial setting time occurred in binder variations with white soil substitution of 5 and 30% i.e. during 105 minutes, which means longer than the binder without white soil as shown in Figure 5.

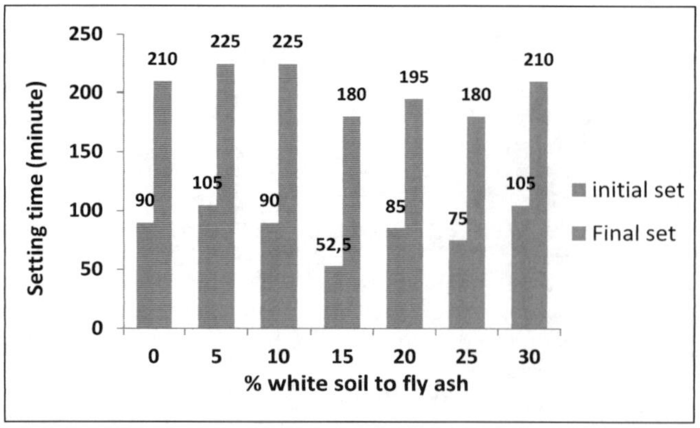

Figure 5. Setting time vs % white soil in binder geopolymer.

. Whereas the final setting time was seen that variations 4 and 6 which are the binders with white soil substitutions 15 and 25% have the shortest time, which means the binder hardens the fastest. The time range of the binder begins to lose its plasticizing properties to a solid mass for all variations ranging from 105 to 135 minutes. The initial and the final setting time of the geopolymer binder, both with and without white soil substitution are in the same time span as portland cement, which is a minimum of 45 minutes and a maximum of 360 minutes.

4. Conclusion

Based on the results of the analysis in this study, it can be concluded that :

1. The initial setting time of geopolymer binders with or without white soil substitution is about 52.5 to 105 minutes, while the final setting time is between 180 to 225 minutes. This range approaches the same as the setting time of Portland cement in SNI 15-2049-2004.
2. The fastest initial setting time in binder geopolymer with NaOH activator occurs in binder 4, namely by adding 15% white soil substitution, the initial setting time occurs at 52.5 minutes and the final setting time occurs at the 180 minutes.
3. The temperature when testing affects the setting time of the geopolymer binder.

Acknowledgements

This research was financially supported by The Faculty of Engineering Diponegoro University, Indonesia trough Strategic Research Grant 2019.

References

[1] Agus 2013 Pemanfaatan Tanah Putih sebagai Bahan Tambahan dalam Pembuatan Paving *Skripsi* Universitas Nusa Cendana Kupang

[2] Hunggurami E, Simatupang PH and Lori AL 2015 Studi Kelayakan Penggunaan Tanah Putih Sebagai Pengganti Agregat Halus (Pasir) Terhadap Kualitas Beton *Jurnal Teknik Sipil* **V4(1)** p29-38

[3] Hardjito D, Wallah S E, Sumajouw D M J, and Rangan B V 2004 On the Development of Fly Ash-Based Geopolymer Concrete *ACI Materials Journal* V.101(6) pp. 467-472

[4] Hardjito D and Rangan B V 2005 Development and Properties of Low-Calcium Fly Ash Based Geopolymer Concrete Perth Australia

[5] Duxson P, Provis JL, Lukey GC, van Deventer JSJ 2007 The role of inorganic polymer technology in the development of Green Concrete *Cement and Concrete Research* 37(12) p159M 266-99

[6] ASTM Standart C 191-08a, 2008, Standart Test Method for Time of Setting of Hydraulic Cement by Vicat Needles.

[7] SNI 15-2049-2004, Semen Portland, Badan Standarisasi Nasional (BSN).

The planning of campus bus in Islamic University of Indonesia with geographic system information

P J Romadhona[1,*] and A I Hapsari[1]

[1]Department of Civil Engineering, Islamic University of Indonesia, Yogyakarta, Indonesia

*e-mail : prima_dhona@uii.ac.id

Abstract. Islamic University of Indonesia (UII) has more than 23.000 active students during 2018/2019 period. With a large number of students, there is high mobility on campus. Traffic jam often occurs in UII during rush hour since students have classes on campus. Additionally, the parking lot always reach the maximum index. Therefore, one of many solutions can be done to reduce traffic jam and lack of parking lot is by planning a campus bus in UII so that students do not need to ride a vehicle to campus. The planning of campus bus in UII consists of demand analysis, route analysis, bus stop analysis, and calculate the number of campus bus needed. Demand analysis was by distributed 500 questionnaires to respondents. The route analysis considered demands, potential areas, and road width. Bus stop analysis based on Decree of Directorate General of Transportation number 271 of 1996, catchment areas, and demands. Analysis for calculating the number of bus campus based on Decree of Directorate General of Transportation number 271 of 1996. The result of this research were three routes for campus bus in UII. Route 1 had 20 bus stops, route 2 had 19 stops, and route 3 had 17. The type of bus used were 16 seats of Isuzu Elf. There were six units for route 1, 7 units for route 2, and 5 units for route 3.

1. Introduction

Islamic University of Indonesian (UII) is a university in Yogyakarta that located on Jalan Kaliurang Km. 14,5. It was noted on the official website of the Islamic University of Indonesia that UII received approximately 4.800 new students and had more than 23.000 active students in 2018/2019 period. For this reason, many students cause high mobility on this campus. That is because access to the campus is very limited due to the limited public transportation that passes and can only be reached by private vehicles and online transportation. Thus, many students use private vehicles like motorcycles or cars to campus and causing a traffic jam. Congestion on campus occurs during certain hours caused by a long queue for a motorcycle when entering the parking lot, as well as limited parking space for cars, so students must looking for several times to find an empty parking space. Limited parking space on campus has hampered student activities. The reason students prefer to use private vehicles to get to campus is because of the limited public transportation that serves the route to campus.

Based on the problems above, the UII campus bus transportation planning will be able to support and improve all activities carried out by students, lecturers, and employees in the area around the UII campus. Because compared to private modes, public transport provides higher capacity and safer environment [1]. The planned bus lane will have a special route network that connects the UII region from the south with a radius of 10 km and to the west with a radius of 5 km. Research conducted by Hashim R et al. and Apriyudha R et al. regarding the procurement of campus buses for environmentally friendly, not only for the green campus to minimise gas emissions but also for handling the increasing population of students [2,3]. The reasonable design of campus bus routes can meet the travel needs of campus personnel, and at the same time, it can realise the green, safe and harmonious campus traffic [4]. Reducing greenhouse gas emissions being caused by motor vehicles also important. Therefore, the sustainable transport system is an alternative to save energy by utilising mass transportation system [5]. This research is using a geographic information system as a model. A model can be defined as a form of simplification of reality (or the real world), including physical model (architect model, civil engineering model, etc.), maps and diagrams (graphics), statistical and mathematical models [6]. Geographic information systems are digital computer design applications that are used to retrieve, store, manipulate, analyse, and display geographic information [7]. For more details, geographic information systems are computer-based systems that have four capabilities to handle geographic reference data, namely data entry (store), data storage management and retrieve, analysis and manipulation, and produce data (output) [8].

2. Research Methods

The method used in this research is descriptive quantitative analysis. Two types of data will be taken. There are primary data and secondary data. Primary data is taken directly from the original field data conducted by interviews and questionnaires to students, lecturers, and UII employees. The stages in planning the UII Campus Bus are as follows.

2.1. Demand Analysis

Islamic University of Indonesia (UII) has 16.764 students and 1.057 lecturers and employees from 2018/2019 period. Demand analysis was obtained from questionnaires and distributed to respondents that consist of students, lecturers, and employees of UII. The questionnaires mainly investigate about the intention of campus bus, address, and type of campus bus., some respondents determined in questionnaires sampling were calculated based on the following Slovin (1960) formula [9] to get a valid sample questionnaire which represents the entire campus

$$\text{n} = \frac{N}{1+Ne^2} \tag{1}$$

In the formula, n is the number of samples; N is the number of population; E is limit error tolerance.

Questionnaires that had been distributed to students, lecturers, and employees are total of 500 questionnaires. Data from the questionnaires were processed and recapitulated. Respondent's addresses are plotted on Google Earth to find out travel demands.

2.2. Route Analysis

Route analysis in this study based on certain considerations as follows.

1. Respondent's travel demands results. Higher demands, the greater the potential for the area to be passed by campus buses.
2. Areas that have potential public facilities such as health centres, sub-district offices, shopping areas and traditional markets.
3. Road width.

Next, modelling on the ArcGIS Network Analyst is done by entering road network data of Yogyakarta and potential locations in the form of shapefiles. This platform solves the tasks of data pre-processing, processing and publishing very efficiently because it consists of several components [10]. ArcGIS has functions such as to model the interaction between traffic information data, route selection, and the appearance of route results and the addition of other parameters in its analysis with the help of a network analyst toolbox. Network Analyst tool helps the decision-makers determine the best routes among all of the existing road networks for transportation [11]. Display the Network Analyst toolbar to do network analysis. After matching map information, the operating track could be divided into a highway, arterial road, and subsidiary road [12]. Continue to solve the analysis. It will produce a route that can be seen on the map, can be chosen based on the shortest route or the route with the fastest travel time.

2.3. Location of Bus Stop Analysis

The bus route network is a system of bus lanes, stops, terminals, and infrastructure needed for the safe and efficient movement of buses and passengers [13]. In this study, there are 3 bus routes for Islamic

University of Indonesia. The route uses a radial network pattern, which is from the Central Business District (CBD) to the suburbs. Bus stop location must be guided by three guidelines as follows.

1. The maximum distance of bus stops to pedestrian crossing facilities is 100 meters, the minimum distance of bus stops from intersections is 50 meters, and the minimum distance of bus stops from buildings that requires calm (for example hospitals) is 100 meters away [14].
2. A number of travel demands.
3. Catchment area. The catchment area in this study is an area that has great potential demand for UII's Campus Bus planning. The catchment area is made to find out which locations are included in that area. In this study, the selection of a large catchment area states that the convenience for pedestrians to reach public transportation stops is 500-1000 m [15]. Therefore, a radius of the catchment area of 500 m is used for locations that have a lot of demands and 1000 m for a particular point that has small demands.

2.4. Number of Bus Needed

The number of campus bus to be used is based on the number of passengers, vehicle capacity, circulation time, and headway on each route [16]. Demands on route 1 are 444 passengers, route 2 are 531 passengers, and route 3 is 477 passengers.

2.4.1 Cycle Time

The deviation is 5 per cent of travel time and stops time is 10% of travel time [17]. Then, the circulation time can be calculated by the following formula [17,18].

$$CT_{ABA} = (T_{AB} + T_{BA}) + (\sigma_{AB}{}^{2} + \sigma_{BA}{}^{2}) + (T_{TA} + T_{TB}) \tag{2}$$

In the formula, CT_{ABA} is circulation time from A to B, back to A again;
T_{AB} is the average cycle time from A to B;
T_{BA} is the average travel time from A to B;
σ_{AB} is the deviation of travel time from A to B;
σ_{BA} is the deviation of travel time from B to A;
T_{TA} is vehicle downtime in A
T_{BA} is vehicle downtime in B

2.4.2 Headway

Headway is the difference between the arrival time of one vehicle and the next vehicle. When the passenger demands increases, the optimal headway must be inversely proportionally shortened to ensure the satisfied service of operation[19]. The ideal headway is 5-10 minutes[17]. In this study, a 10-minute headway design was used to determine the number of passengers per hour. Load factor calculation using load factor 0.7 with dynamic conditions.

2.4.3 Number of Campus Bus

The number of campus bus per cycle time calculated using the formula below [17,18].

$$K = \frac{CT}{H\,x\,FA} \tag{3}$$

In the formula, K is bus needed per cycle time;
CT is cycle time (minutes);
H is headway (minutes);
FA is a vehicle availability factor (assumed 1).

3. Result

The result of this research were 3 routes for campus bus in UII. Route 1 serves the trip from UII along Kaliurang st. Before the ring road and back to UII. Route 2 serves trips from UII - Kaliurang st. - ring road - Palagan st. - Kapten Haryadi st. - Kaliurang st. - Sunan Pandanaran st. - UII. Route 3 serves trips starting from UII – Kaliurang st. – Damai st. – Palagan st. – Rejodani st. - Sunan Pandanaran st. - Kaliurang st. - UII. Routes in this study use ArcGIS as modelling seen in Figure 1.

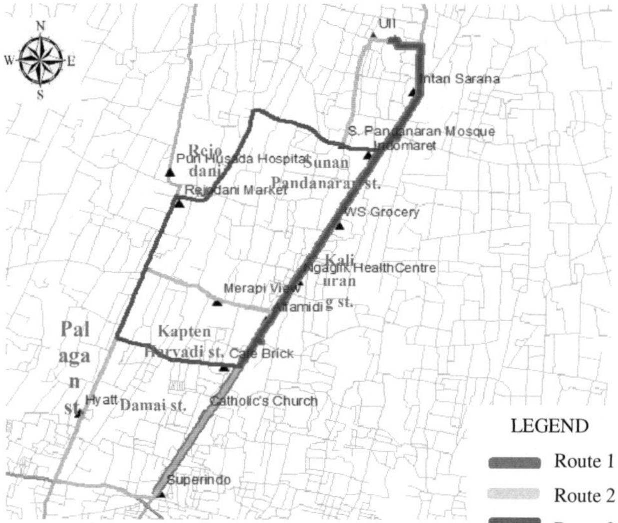

Figure 1. Routes for Campus Bus in

According to Figure 2, there was the location of bus stops below.

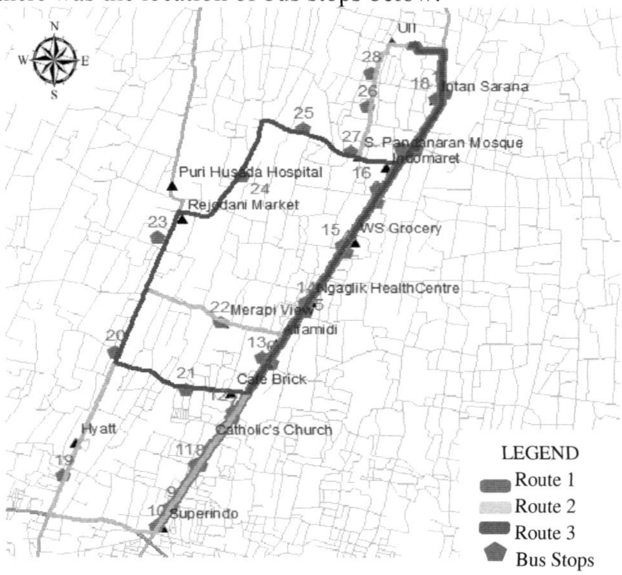

Figure 2. Bus Stops Location

Bus stop 1 had 53 demands, bus stop 2 had 112 demands, bus stop 3 had three demands, bus stop 4 had 15 demands, bus stop 5 had two demands, bus stop 6 had ten demands, bus stop 7 had 13 demands, bus stop 8, 9, 10 had two demands, bus stop 11 had three demands, bus stop 12 had 14 demands, bus stop 13 had ten demands, bus stop 14 had ten demands, bus stop 15 had seven demands, bus stop 16 had 12 demands, bus stop 17 had 108 demands, bus stop 18 had 66 demands, bus stop 19 had 11 demands, bus stop 20 had eight demands, bus stop 21 had nine demands, bus stop 22 had four demands, bus stop 23 had nine demands, bus stop 24 had 17 demands, bus stop 25 had 16 demands, bus stop 26 had 82 demands, bus stop 27 had 49 demands, and bus stop 28 had 131 demands. This is a stop and demand diagram to illustrate the data above.

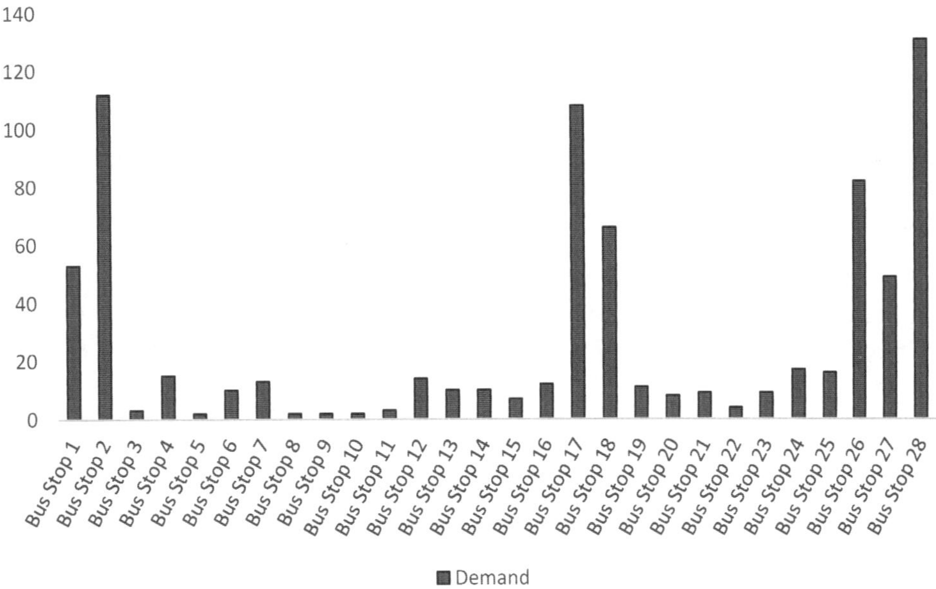

Figure 3. Diagram of Bus Stops and Demands

The final calculation for the number of campus bus needed are :

Table 1. Number of Bus Campus Needed

	Load Factor	Cycle Time (minutes)	Headway (minutes)	Number of buses needed (unit)
Route 1		50,643	10	6
Route 2	0,7	62,1	10	7
Route 3		49,224	10	5

4. Conclusion

Based on the analysis that had been done in Islamic University of Indonesia, the following conclusions can be obtained.

1. There are 3 routes for UII Campus Bus planning. Route 1 has a length of 18,349 km with 20 bus stops, starting at UII - Kaliurang st. KM 6 - UII. Route 2 has a length of 22,5 km with 19 bus stops, starting at UII - Kaliurang st. - ring road – Palagan st. - Kapten Haryadi st. – Kaliurang st. - Sunan Pandanaran st. - UII. Route 3 has a length of 17,835 km with 17 bus stops, starting at UII – Kaliurang st. - Damai st. - Palagan st. – Rejodani st. - Sunan Pandanaran st. – Kaliurang st. - UII.
2. There are a total of 28 bus stops in this study, with details of 18 bus stops on Kaliurang st., three stops on Palagan st., one-stop on Damai st., one-stop on Captain Haryadi st., one-stop on Rejodani st., and four bus stops on Sunan Pandanaran st.
3. The total number of campus bus needed is 18 units with details of 6 units for route 1, 7 units for route 2, and 5 units for route 3, with the type of vehicle 16 seats Isuzu Elf.

References

[1] Abousaeidi M Fauzi R and Muhamad R 2016 Geographic Information System (GIS) Modeling Approach to Determine The Fastest Delivery Routes *Saudi Journal of Biological Sciences* Vol 23 pp 555-564
[2] Hashim R, Haron S, Mohamad S, and Hassan F 2013 Assessment of Campus Bus Service Efficacy: An application towards green environment *Procedia – Social and Behavioral Science* pp 294-303
[3] Apriyudha R, Handayani D, and Djumari 2015 Analysis of Amount Bus Needed and Operational

Bus Schedule In Order to Support Green Campus UNS Program *e-Jurnal Matriks Teknik Sipil* pp 268-276
[4] Shu Z and Wu X 2018 The Route Planning on Campus Bus H University *American Journal of Industrial and Bussiness Management* 8 pp 473-486
[5] Sulviawan A P and Susantono B 2015 Modeling the Tembalang UNDIP Campus Bus Route with the Application of Geographic Information Systems (GIS) *Journal of PWK Engineering* Vol 3 pp 841-855
[6] Tamin O Z 1997 *Transportation Planning and Modeling* 1st ed (Bandung: ITB Publisher)
[7] Muslim M A 2005 Best Route Determination Application Based on Geographic Information Systems *DINAMIC Scientific Information Technology Journal* Vol X pp 76-83
[8] Aronoff 1989 *Geographic Information System: A Management Perspective* (Ottawa: WDL Publication)
[9] Akdon and Riduwan 2005 *Formulas and Data in Statistics Applications* (Bandung: Alfabeta)
[10] Wilkening J, Kapaj A, and Cron J 2019 Creating a 3D Campus Routing Information System with ArcGIS Indoors *Publications of the DGPF* Vol 28
[11] Hung P, Doi K, and Inoi H 2019 User Retention Tendency of Bus Routes Based on User Behaviour Transition in an Area with Low Mode Share of Public Transport *IATSS Research* 00222
[12] Xiru T, Yueyan Z, and Liping X 2016 The Analysis of Space-time Characteristics of Bus Operation and Energy Consumption based on ArcGIS *Science Direct Applied Energy Symposium and Forum* pp 456-461
[13] Giannopoulos G A 1989 *Bus Planning and Operation in Urban Areas: A Practical Guide* (Aldershot: Gower Publishing Company)
[14] Directorate General of Land Transportation 1996 Technical Guidelines for the Engineer of Public Passenger Stop Stations *Decree of the Director-General of Land Transportation Number 271*
[15] National Standardization Agency of Indonesia 2004 *Procedure for Planning A Residential Environment in Urban Areas SNI 1733*
[16] Winaya A 2017 Analysis of Amount Bus of the Public Transportation Route in the Benowo-Kalimas Barat Terminal (Lyn BJ) of Surabaya *Jurnal Rekayasa Teknik Spil Universitas Madura* Vol 2
[17] Directorate General of Land Transportation 2001 Technical Guidelines for the Engineer of Public Passenger Stop Stations *Decree of the Director-General of Land Transportation Number 687*
[18] Munawar A 2005 *Fundamentals of Transportation Engineering* (Yogyakarta: Beta Offset)
[19] Ding J, Feng S, Li L, and Zhang Y 2017 Campus Bus Network Design and Evaluation Based on the Route Property *Tsinghua Science and Technology* Vol 22 pp 539-550

IOP Conf. Series: Materials Science and Engineering **771** (2020) 012053 doi:10.1088/1757-899X/771/1/012053

Shear Strength of Reinforced Concrete Beams Strengthened with Near Surface Mounted Steel Plates

R Thamrin[1,*], S Haris[1], and Zaidir[1]

[1]Civil Engineering Department, Engineering Faculty, Universitas Andalas, Padang, Indonesia

[*]Corresponding author: rendythamrin@eng.unand.ac.id

Abstract. Near surface mounted horizontal steel plates was applied to study the contribution of this strengthening method on the shear capacity of reinforced concrete beams. Specimens were nine reinforced concrete beams without stirrups consisting of three control beams and six beams strengthened with steel plates. The test variables were the ratio of the longitudinal reinforcement and the position of steel plates in the vertical direction of the web side of the beam. The specimens were loaded up to failure using a four-point bending test. The contribution of steel plates on the shear capacity was observed by monitoring the ultimate load and deflection of the beams. The test results show that the shear capacity of beams strengthened with steel plates 17% to 50% higher than the control beams depending on the position of steel plates and the ratio of longitudinal reinforcement. It was also observed that the ratio of longitudinal reinforcement influences the failure mode of the beams.

1. Introduction

It is widely known that one of the simplest methods to strengthen an existing reinforced concrete structure is to increase the quantity of reinforcement by gluing steel plates, steel bars, FRP plates, FRP bars or FRP sheets to the surface of the structural elements [1]. However, inadequate protection from environmental attacks may decrease the service life of the strengthened structure. In a decade, near surface mounted (NSM) reinforcement was introduced as a technique for strengthening existing reinforced concrete structures [2,3].

Some reports have been published related to the application of the NSM technique [4-10]. It is reported that structural elements strengthened with NSM reinforcement show higher load-bearing capacity [4]. It has also been proven that this method is an effective way to increase the shear capacity and ductility of reinforced concrete elements [5,6]. This method also provides better fire resistance than lamination with externally bonded reinforcement (EBR) [6]. In another report [7] the position of NSM steel plates significantly affects the stiffness and flexural capacity of the beam. Comparison between the NSM and EBR methods is presented in another report [9] where the report states that the adhesion of the NSM method is better than the EBR.

From the reports in the above literature, further studies on the contribution of steel plates to the shear strength of reinforced concrete beams are still needed. The objective of this study is to determine the effect of steel plate installation on the shear strength of reinforced concrete beams. In addition, the position of the plate was also investigated by installing steel plates in two different positions. Variations in shear capacity requirements are carried out using three types of longitudinal reinforcement ratios as test variables. Furthermore, the theoretical flexural capacity of each specimen needs to be known to determine the ultimate flexural capacity of the beam. In this study, the calculation of the complete response of the beam flexural capacity is carried out with a computer program that has been developed by the author.

2. Experimental study

The effect of steel plate installation on the shear strength of reinforced concrete beams was experimentally investigated by means of four point bending tests. Test set-up and loading arrangement for each beam are shown schematically in Figure 1. Applied load and resulting deflections at each beam were recorded continuously using a load cell and LVDT's. Load cell and LVDT's were connected to a data logger and the data was stored on data media.

Two different positions of steel plates and three different longitudinal reinforcement ratios were chosen as the test variables to observe the effect of steel plate installation on the shear strength as shown in Figure 2. The beam cross-section was 125x250mm and the diameter of longitudinal reinforcing bars was 13 mm with a yield strength of 394 MPa. The concrete compressive strength on the basis of 28 days cylinder crushing strength was 23.7 MPa.

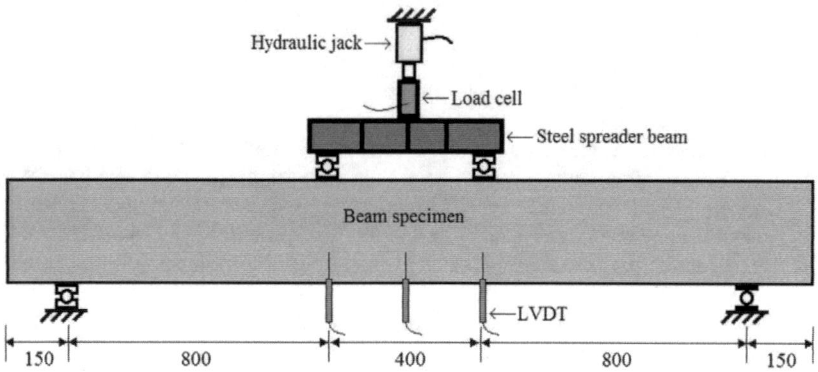

Figure 1. Beam dimension and four point bending tests mechanism.

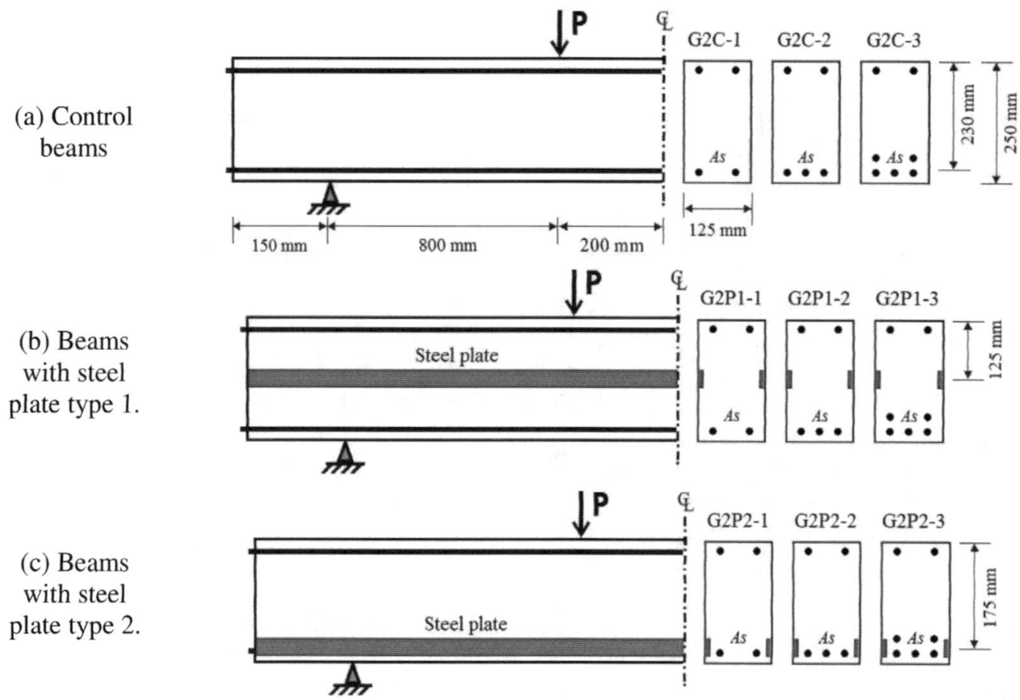

Figure 2. Beam cross-section and positions of steel plates.

For each of the specimens, the groove cuts were initially prepared on the vertical side of the formwork. The groove dimension was 50 mm wide and 5 mm deep. The formwork was released after 28 days of curing time. Then, the groove surfaces were cleaned to remove out fine particles and dust.

The epoxy adhesive paste (Sikadur 31) is filled halfway into the grooves using a palette knife. Then the steel plates (3 mm thick and 50 mm width and yield strength of 304 MPa) were placed on the top of the adhesive layer and pressed lightly along the plates. The grooves were finally filled with epoxy adhesive paste and the surface was levelled. Table 1 shows the beams data, details of the reinforcement and test results.

Table 1. Details of test beams and test results.

Beams data							Longitudinal reinforcement (Tensile)			Calculated Flexural Capacity	Experimental	
											First Crack	Max. Load
Specimens		fc' (MPa)	b (mm)	H (mm)	d (mm)	a (mm)	ϕ_b (mm)	N	ρ (%)	V_f (kN)	(kN)	V_m (kN)
Control beams	G2C-1	23.7	125	250	220	800	13	2	1.0	26.2	4.5	17.8
	G2C-2				220			3	1.4	38.4	8.4	19.0
	G2C-3				205			5	2.6	57.5	5.0	28.6
Type I	G2P1-1				220			2	1.0	37.0	5.3	35.2
	G2P1-2				220			3	1.4	48.3	6.1	40.3
	G2P1-3				205			5	2.6	61.5	5.2	34.3
Type II	G2P2-1				220			2	1.0	42.7	7.5	29.4
	G2P2-2				220			3	1.5	54.0	6.2	25.0
	G2P2-3				205			5	2.5	68.9	5.9	34.5

3. Analytical study

The moment-curvature analysis was carried out in order to obtain the theoretical flexural capacity of the beam. This calculation requires the formulation of reinforced concrete sectional response [11]. It is assumed that the stress-strain relationship of concrete and steel are known. The stress-strain model of concrete adopted from the literature [12] was applied while the bilinear stress-strain model for steel reinforcement and steel plates was used in this study. The procedure is started by dividing the beam cross-section into several thin layers. The next step is assuming the value of curvature and axial strain. Then the stresses and internal forces are calculated by using assumed stress-strain laws of concrete and steel. The equilibrium of internal forces is achieved by the iteration process. Once the equilibrium is satisfied the moment at corresponding curvature can be calculated. Then the calculation process can be continued to the next value of curvature. The maximum concrete compression strain was taken as 0.003. Figure 3 shows an analytical model for the strengthened beam cross-section and schematic algorithm of an analytical procedure is shown in Figure 4.

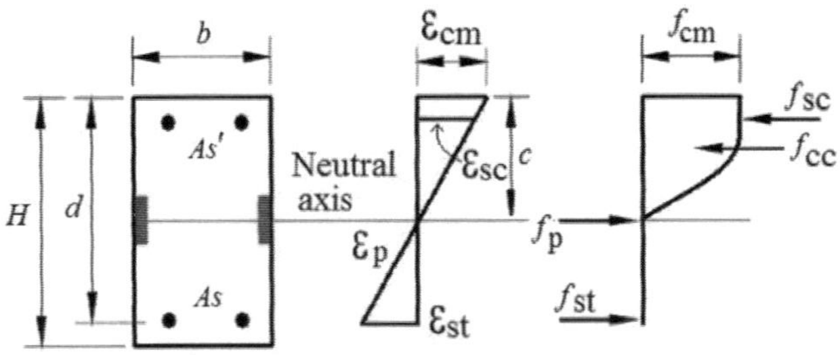

Figure 3. Analytical model for the strengthened beam cross-section [8].

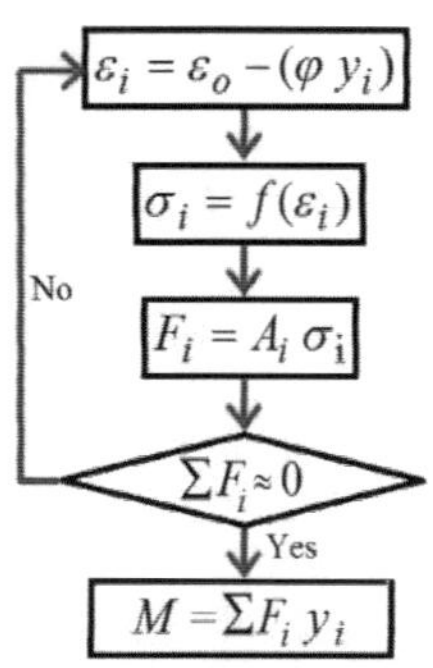

Figure 4. Schematic algorithm of the analytical procedure [8,9].

4. Test results and discussion

At the early steps of loading, the flexural crack was first developed in the constant moment zone at an average load level of 6 kN. As the load increased, the flexural crack developed to the shear span zone which then propagated to become diagonal shear cracks.

G2C-1

G2C-2

G2C-3

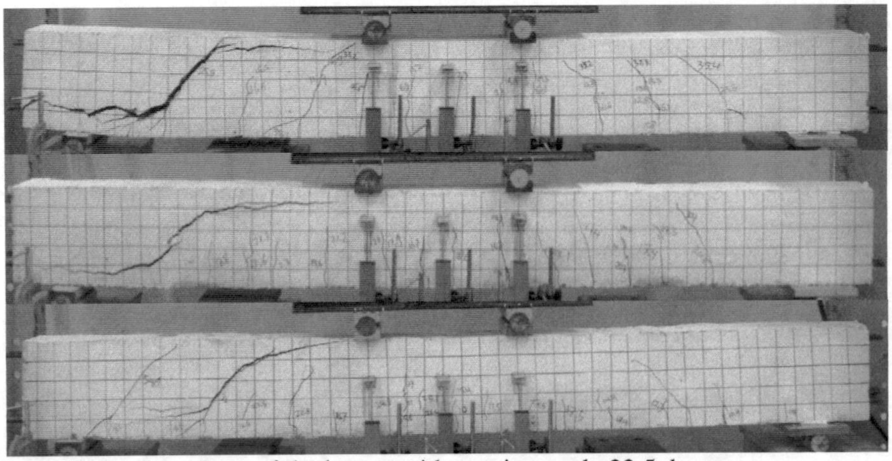

Figure 5. Crack patterns of the beams with rotation angle 22.5 degrees.

G2P1-1

G2P1-2

G2P1-3

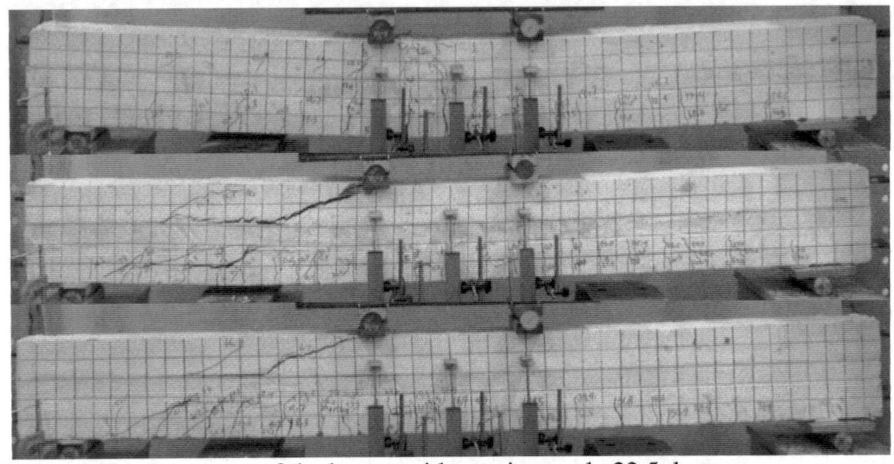

Figure 6. Crack patterns of the beams with rotation angle 22.5 degrees.

G2P2-1

G2P2-2

G2P2-3

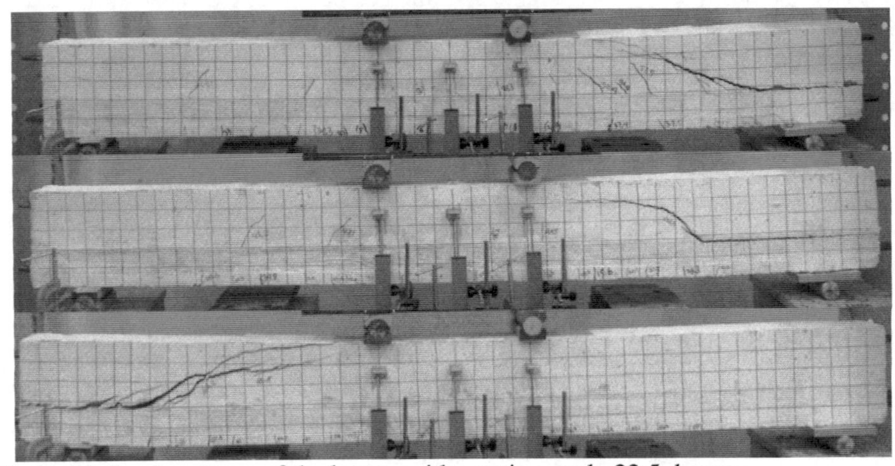

Figure 7. Crack patterns of the beams with rotation angle 22.5 degrees.

In all tested beams diagonal shear crack were observed in the shear span zone as shown in Figure 5 to Figure 7. Soon after the occurrence of diagonal shear cracks, the beams failed in shear except for beam G2P1-1. Beam G2P1-1 was failed in the flexural type of failure indicated by crushing of compression zone between two point loads. Typical crack patterns of the tested beams at failure are shown in Figure 5, Figure 6 and Figure 7.

Figure 8 shows the load-deflection curve of the tested beams. It is shown that the curves suddenly drop due to shear failure except for beam G2P1-1. Analytical predictions of flexural capacity for both beams with and without steel plates are shown in Figure 9 and Figure 10. It can be seen that only the G2P1-1 beam reaches an ultimate bending capacity which is also confirmed by the analytical results.

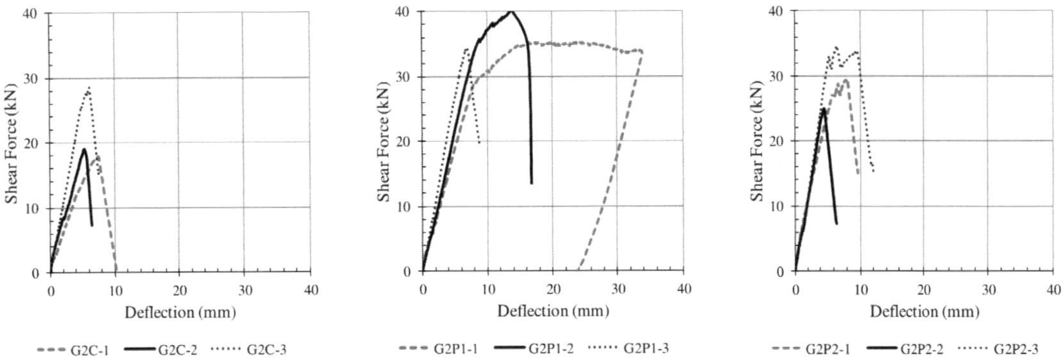

(a) Control beams. (b) Strengthened beams type 1. (c) Strengthened beams type 2.

Figure 8. Test results based on beams categories.

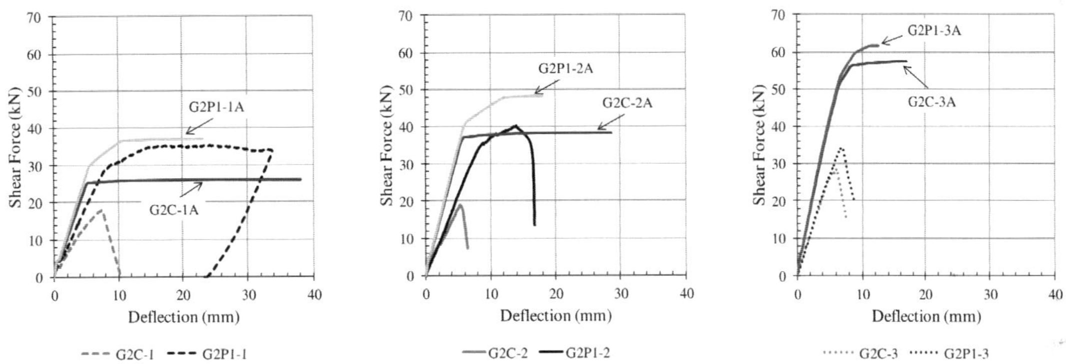

Figure 9. Comparison between control beams, strengthened beams type 1 and analytical results.

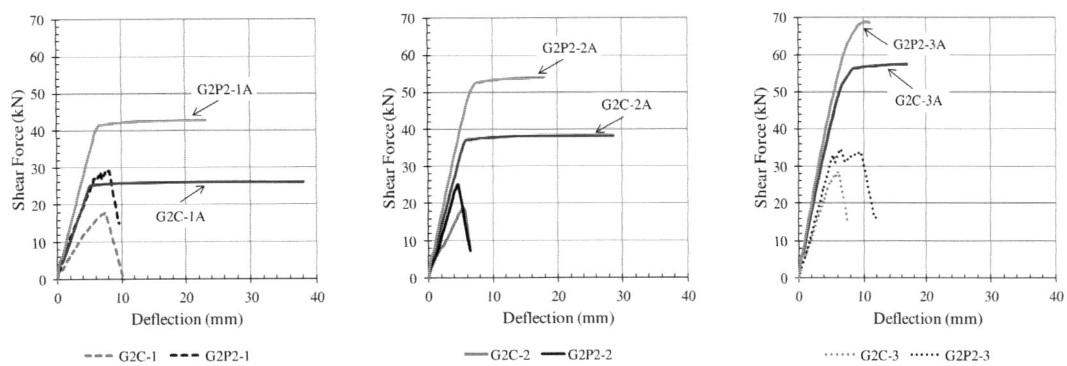

Figure 10. Comparison between control beams, strengthened beams type 2 and analytical results.

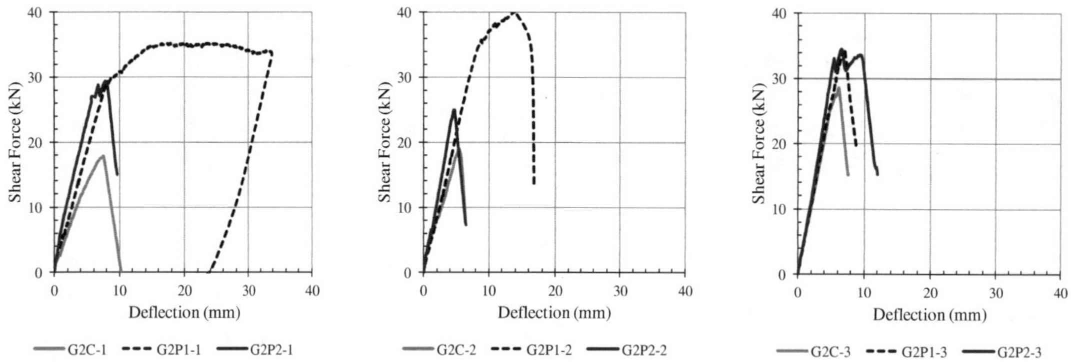

Figure 11. Comparison between strengthened beams type 1 and strengthened beams type 2.

Even though majority of the beams failed in shear but the shear capacity of the strengthened beams higher than control beams for about 17% to 50% as shown in Figure 9, Figure 10 and Figure 11. These results also prove that steel plates contribute significantly to the shear capacity of the beam. It was found from the test results that the plate in the centre of the beam height contributed to the greater shear strength although analytical results show the opposite results. It was also observed that the contribution of steel plates to resist shear forces decreases with increasing ratio of longitudinal reinforcement.

5. Conclusions

Nine reinforced concrete beams were tested. The comparison between two positions of steel plates are discussed and based on the test results the following conclusions are drawn:

1. Shear capacity of the strengthened beams higher than control beams for about 17% to 50%.
2. The plate position in the centre of the beam height contributed to the greater shear strength.
3. The contribution of steel plates to resist shear forces decreases with increasing ratio of longitudinal reinforcement.

Acknowledgements

This study was financially supported by Penelitian Dasar Unggulan Klaster Riset-Publikasi ke Guru Besar Universitas Andalas (PDU-KRP2GB-Unand) financial year 2019, with contract number T/15/UN.16.17/PP.IS-KRP2GB/LPPM/2019.

References

[1] Adhikary B B, Mutsuyosh H and Sano M 2000 Construction and Building materials 14 237-244.
[2] Lorenzis L D and Nanni A 2001 ACI Structural Journal 98 1 60–68
[3] El-Hacha R and Rizkalla S H 2004 ACI Structural Journal 101 5 717–726
[4] Szabó Z K and Balázs G L 2007 Periodica Polytechnica Civil Engineering 51 1 33–38
[5] Sabol P and Priganc S 2013 Procedia Engineering 65 364–369
[6] Parvin A and Shah T S 2016 Journal of Polymers 8 298
[7] Thamrin R and Sari R P 2017 Procedia Engineering 171 957-964
[8] Thamrin R 2017 MATEC Web of Conferences 103 02012
[9] Thamrin R 2018 Journal of Engineering Science and Technology 13 11 3781 - 3794
[10] Thamrin R, Haris S and Zaidir 2019 MATEC Web of Conferences 276 01004
[11] Park R and Paulay T 1975 *Reinforced Concrete Structures* (New York: John Wiley)
[12] Mander J B, Priestley M J N, and Park R 1988 Journal of Structural Engineering 114 8 1804-1826

Landslide Vulnerability Mapping for Landslide Victim Relocation

R Hidayat[1]

[1]SABO Research and Development Center, Research Center for Water Resources, Ministry of Publics works, Sopalan, Maguwoharjo, Sleman, D.I. Yogyakarta, Indonesia.

rokhmathidayat33@gmail.com

Abstract, some victims of landslides must be relocated if the location affected by a landslide impossible to be occupied again. The relocation site of the Sijeruk-Banjarnegara landslide victims had experienced a landslide again, due to the inappropriate location selection. It was necessary to mapping the landslide vulnerability for the selection of relocation sites so that the new location must safe from landslides. Settlements that were still located in the high-risk landslide were as the danger the occupants. The selection of relocation sites must be appropriate and the risk of landslides was low. This study aims to produce a vulnerability landslide map so that land suitability can be obtained for the development of settlements by referring to the threat of landslide hazards. Slope stability modelling uses the TRIGRS (Transient Rainfall Infiltration and Grid-Based Regional Slope-Stability) program version 2.0. The principles are modelling the effect of rainfall on regional slope stability. The data needed in modelling was digital elevation model (DEM), soil data index properties, soil engineering properties, and rainfall. The modelling results show that the safety factor values in the study were ranged from 0.8 to 10. Safe numbers below 1.2 means are prone to landslides. Slope stability map can be used to determine the location of the relocation of landslide victims.

1. Introduction

Some area in Indonesia had a medium to high scale landslide vulnerability. Landslide disasters had a high frequency of occurrence by causing victims of property, lives and infrastructure damage. The landslide disaster event disrupted the Indonesian economy sector, so disaster mitigation must be done to reduce the risk. This high landslide threat was caused by hilly topography and relatively steep slope and also other factors such as rainfall. Geological factors also affect the risk of landslides. To reduce the number of victims in landslide-prone, relocation was needed. The choice of location for relocation must be appropriate so that it was safe from landslide hazards, so it was necessary to map the landslide vulnerability level for the location to be used for relocation. Like the incident in Sijeruk Banjarnegara, the relocation site of the landslide victims returned to a landslide. Micro-landslide modelling was carried out with the TRIGRS program.

Based on landforms in Banjarnegara Regency, especially Karangkobar District, had steep slope topography. Karangkobar District was a sub-district in Banjarnegara Regency which was dominated by the threat of high and medium scale landslides. There were still settlements in the threat of high landslides. Settlement-based landslide mitigation planning needed to be done, so no major losses and fatalities in the event of a landslide.

Disasters were events or series of events that threaten and disrupt the lives and livelihoods of people caused by natural factors or non-natural factors and human factors resulting in human casualties, environmental damage, property losses, and psychological impacts [1]. The disaster was a combination of three elements that disaster threat, vulnerabilities, and abilities triggered by an event.

Concerning Housing and Settlement as article 19 settlements were the implementation of houses and housing that were carried out to fulfil the needs of homes as a basic human need for the improvement and equitable distribution of people's welfare. The types and shapes of houses in residential were as in Article 21 include commercial, self-help, special, public, and statehouses with single, series and collated houses [2].

The results of planning houses and settlements must meet technical, ecological, administrative and spatial requirements. Technical requirements relating to the safety and comfort of the building, and the reliability of its environmental facilities and infrastructure. Ecological requirements were related to harmony and balance, both between the artificial environment and the natural environment and the socio-cultural environment, including the cultural values of the nation that need to be preserved. Administrative requirements relate to the granting of business licenses, location permits, and building permits and the granting of land rights. The physical characteristics that must be considered in a residential were so that it was habitable was free from the danger of topography. Chen et al. conduct research on the relocation of landslide disaster victims in Mrarakot typhoon, Taiwan. However, very little financial effort has been made by moving people from landslides prone to safety zones [3]. One of the main causes of relocation failures was considered physical structure, then addressing the emotional and psychological requirements of the community resettlement process [4]. The way to address the relocation issues and challenges best solutions, community-based strategies are most welcome solutions [5].

Mitigation was a series of efforts to reduce disaster risk, both through physical development and awareness and enhancement of the ability of the community to deal with danger threats (Law Number 24 Year 2007 concerning Disaster Management Article 1 paragraph 9). An early warning was a series of activities to give warning as soon as possible to the community about the possibility of a disaster at a place by an authorized institution. Whereas an early warning system was an effort carried out by the government together with the community in order to reduce the impact that might occur if a natural disaster landslides.

Rainwater infiltration will fill the air pores in the soil so that the soil mass increases and affects the saturation of the soil. Increased saturation of land affects changes in the ability of the soil to absorb water suction. Changes in groundwater level result in changes in parameters on soil properties, both property and technical parameters [6].

Mapping of land movement vulnerability was one of the efforts to predict the occurrence of ground movements for preparedness against the threat of land movement. The land movement vulnerability model must be able to provide information on when and where land movements will occur in a hilly were a. The Center for Volcanology and Geological Disaster Mitigation [7], made a Vulnerability Map for the Soil Movement of the Sukabumi region using a geographic information system based weighting approach. Mapping the vulnerability of regional land movements due to rainfall using deterministic and empirical ground motion modelling had been carried out by many previous researchers. TRIGRS was one model of slope stability that was widely used to map the vulnerability of shallow ground movements. The study of soil movements triggered by rain in an was a was interesting as an effort to estimate the potential of ground movements triggered by rain. To determine the causes of landslides, especially seen from topographic and rainfall factors, numerical modelling using TRIGRS was one method that can be applied to spatial were as [8, 9].

The ability of the TRIGRS model to predict shallow ground motion due to rain was influenced by the resolution of temporal rainfall variations [10, 11], resolution of spatially geotechnical parameters [12] and resolution of digital elevation models. This software was developed for landslide prediction using the Matlab program and the programming language used was FORTRAN. Rain in the ground causes infiltration of water to the surface, which was distinguished into a partially saturated zone and a saturation zone. Part of the saturation zone was generally above the groundwater level with a partially filled pore with water. This zone was also called the aeration zone and water that spreads from the ground surface goes down through the main root zone. The thickness varies depending on the type of soil and vegetation. In this zone, the space between the particles was partly filled with water and partly filled with air [13]. The data needed to be the nature of the geotextile, slope (DEM), and rainfall. [14] examined the limitations of TRIGRS, development, examples of modelling the effect of variations in rainfall on landslide susceptibility. Modelling the slope stability with TRIGRS based on rainfall infiltration with the intensity and duration of rainfall that occurred [15].

2. Research Methodology

The research location was in Tangkil Village, Pulung District, Ponorogo. This location on April 1, 2017, experienced a landslide. This landslide incident happened to the settlement, causing a large amount of material losses and fatalities.

Modelling of regional slope stability was carried out with the TRIGRS program. The primary data of the study were soil geotechnical data including soil type weight, cohesion value, deep friction angle, soil permeability, soil depth, and groundwater depth. Primary data were obtained by field investigations, field testing, and laboratory testing carried out by the Sabo Research and Development Laboratory. While secondary data in the form of topographic maps were obtained from BIG (Geospatial Information Agency).

Map results from modelling slope stability can be used to determine the location of the relocation of landslide victims. Landslide victims need to be relocated if it was not possible to be occupied again. A concept that was in accordance with the regional development plan as a settlement based on landslide disaster mitigation was needed. Landslide risk analysis was carried out to find out were as that had the threat of landslides, both from low to high scale. If there was a threat of landslides, it can be avoided for relocation sites, so that the area used for the relocation site must be safe from landslides.

3. Results and Discussion

Sites for the relocation of landslide victims need to be analyzed for land suitability for settlement development by referring to the threat of landslide hazards. Suitability analysis can be from aspects of the slope, elevation, land use, hydrology, soil, and the threat of landslides. Analysis of the suitability of the area for disaster evacuation was also carried out to determine to be used as an evacuation site for landslides. The mitigation concept used for settlement planning which developed into spatial plans, activities, evacuations, facilities, facilities and infrastructure, circulation, and vegetation plans. The spatial plan was supported by plans for facilities and infrastructure for evacuation activities such as early warning systems (EWS) for landslides.

One example of a failed relocation was the location of the victims of the Sijeruk landslide, Banjarnegara. The relocation position of the Sijeruk landslide victims in 2004, experienced a landslide again in 2008 (Figure 1). That means the location selection for relocation was not safe from the landslide. Land suitability analysis for relocation locations was not appropriate. From the results of the location plots on the landslide vulnerability level map, it turns out that the relocation point was in a zone prone to landslides.

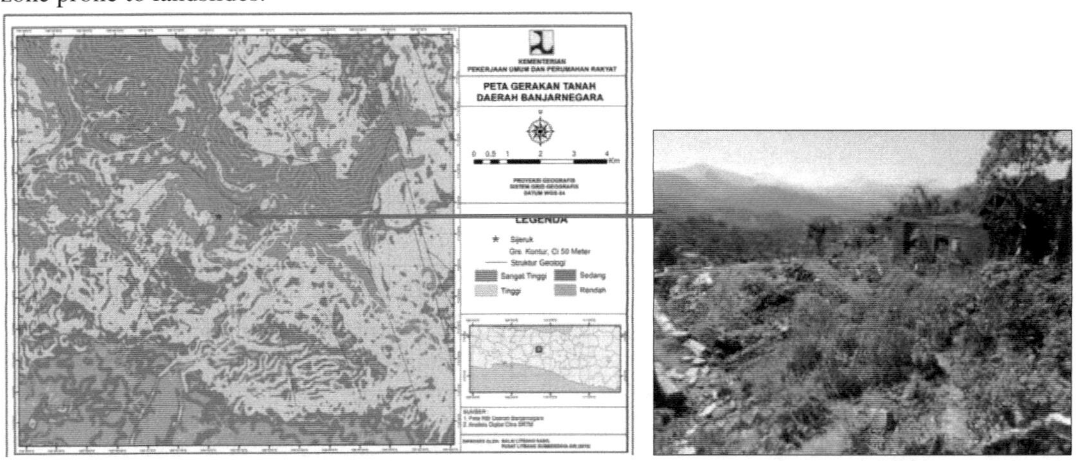

Figure 1. Relocation of Sijeruk-Banjarnegara landslide victims that had returned landslides, and plots of relocation points on landslide hazard level maps

3.1. Tangkil Landslide, Banaran, Ponorogo

Landslides in Tangkil Hamlet, Banaran Village, Pulung Sub-District, Ponorogo District occurs on April 1, 2017, at 07.30 WIB. The incident claimed about 30 lives with the number of houses affected by 32 houses. Subsequent landslides occurred on April 9, 2017, at 11:30 WIB, the landslide occurred in Krajan Hamlet, Banaran Village, with the widening of the landslide affected were resulting in around 37 houses damaged [16]. The landslide in Tangkil Hamlet resulted in huge casualties and material losses. From the picture can be seen the condition of the slope and type of soil. The slope of the landslide location after the landslide was around 50° (Figure 2). Tangkil Hamlet landslide was included in the type C volcano morphology, Mount Ngliman. The area of the landslide had morphology in the form of hills and valleys that become rivers. The lithology is volcanic breccia that further weathering condition. Dark brown color with matrix grain size from clay domination clays, fragment size from 10-50cm, fragment condition had also experienced high weathering until the initial rock minerals were no longer visible. The weathered rock conditions make the soil thicker than more than 10 meters.

From figure 3, it can be seen that the victims were not only were as of that experienced landslides, but also the area buried by landslides. From the picture, it can be seen that the houses did not in experience landslides but were buried by landslides. In the landslide hazard map, the area was likely not prone to landslides but was buried by landslides. Thick soil makes it easily saturated with water, so if high rainfall will easily experience landslides. If it had high slope levels, it will easily to landslides.

Figure 2. Landslide conditions in Tangkil Village, Ponorogo

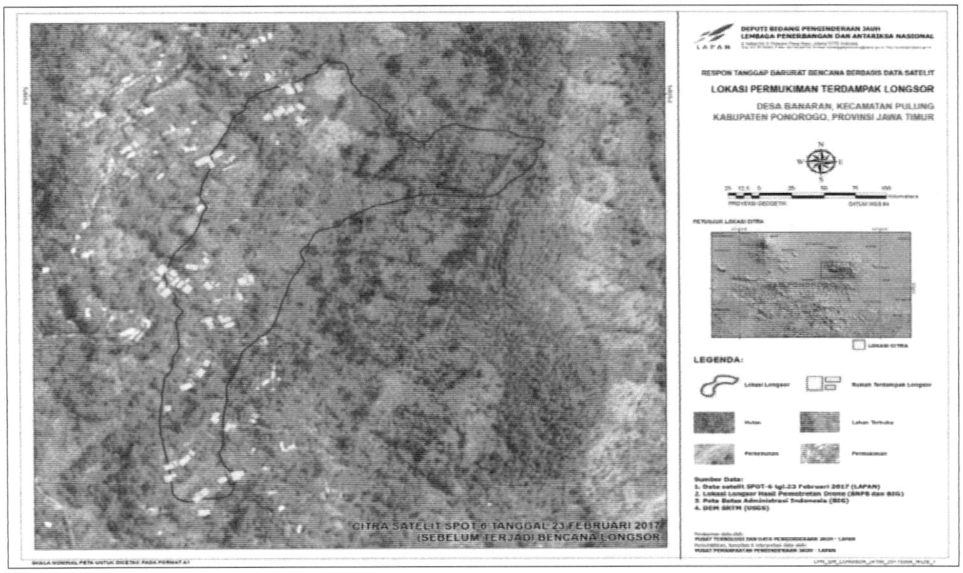

Figure 3. Landslide affected some house damage in Tangkil Village, Ponorogo (source: LAPAN)

3.2. Modelling of Regional Slope Stability

Some input variables in TRIGRS were taken based on field observations. For example the value of the depth of the groundwater and the soil depth. The surface of the groundwater in Tangkil Village was at depth of -3.5m, while the depth of the soil layer was -5m based on the field of landslides on the landslide crown. For parameter values of shear angle and friction angle (c, ϕ) taken from the lab test results. The value of volumetric water content was approached by SWCC (Soil Water Characteristic Curve) with fitting obtained by simulating soil samples at SEEP / W software. The variable geotechnical value of land for TRIGRS input can be seen in table 1. Rainfall intensity values were taken based on rainfall data at several rain stations around Pulung District 4 days before the landslide. The value of the diffusivity variable (D1) was obtained by doing calculations.

Table 1. Input variable in TRIGRS

No	Variable	Value unit
1	Saturated permeability coefficient (Ks) m/s	3.867e-07
2	Diffusivity Saturated (Do)	6,805e-06
3	Volumetric water content (θs, θr)	0.58, 0.13
4	Volume weight (γb) N / m	1.66e + 04
5	Cohesion (c), friction angle (ϕ) N / m	0.200e + 03, 1.8^{o}
6	The slope of the curve (α)	-2,017
7	Rainfall (ri) m / s	2.9e-7
8	Permeable (Zmax) soil depth m	5
9	Groundwater level, m	4

Source: Lab test results

The topographic aspect was an important thing for the selection of settlement locations. In addition, the condition of oxygen levels. The higher location, the available oxygen levels will be lower. Their height place affects the comfort level, especially for residential. The suitability of the development of settlements in Tangkil Village showed if this area dominated by marginal suitability for the development of settlements. This was because an altitude of 750 to 1000 msl dominating this was a. Elevation less than <750m was comfort for settlements [17].

Based on the limit equilibrium approach slope stability analysis, the slope will landslide if it had an FS value <1. According Ward et al. (1979), category of SF value was high if < 1,2, middle class if 1,2 - 1,7, and low if > 1,7 (Table 2). However, [18] mentioned that erroneous errors in determining FS in spatial were as ranged from 20% to 30%, so the determination of landslide potential can be assessed from FS value.

The results of the stability modelling simulation with TRIGRS showed FS values in the study area had ranged from 0.83 to 10 (Figure 4). Using the classification in Table 2, the TRIGRS simulation results show that there were parts of the region experiencing high landslide potential. From Figure 4, the area that experienced landslides was mainly the upper part of the steep slope. The top of the slope usually part of the landslide crown which a source, and the movement from the top of the slope.

Some previous studies [11] explain that the results of modelling will be more accurate by using higher spatial resolution for rainfall and soil data in zones that had high geomorphological variation. Topography and DEM influence on TRIGRS prediction results. The use of smaller sized grids of 10 - 15m will provide better predictive results [19].

The greatly factor influences the threat level of landslides was topographic conditions, although there were still other factors causing landslides besides topography. The spaces used for evacuation spaces were open spaces in were as those were predicted to be safe from landslide hazards. The topographic component consists of land slope and altitude. The slope of the land was a major factor in the occurrence of landslides due to the influence of gravity, so that compared with other factors in calculating the likelihood of landslide threats; the slope of the slope produces the highest weight value.

Tabel 2. Classification model for potential and probability of landslide [18]

Classification	Potential for landslides	Landslide Probability P
High	< 1,2	> 60%
Middle class	1,2 - 1,7	30 - 60%
Low	> 1,7	< 30%

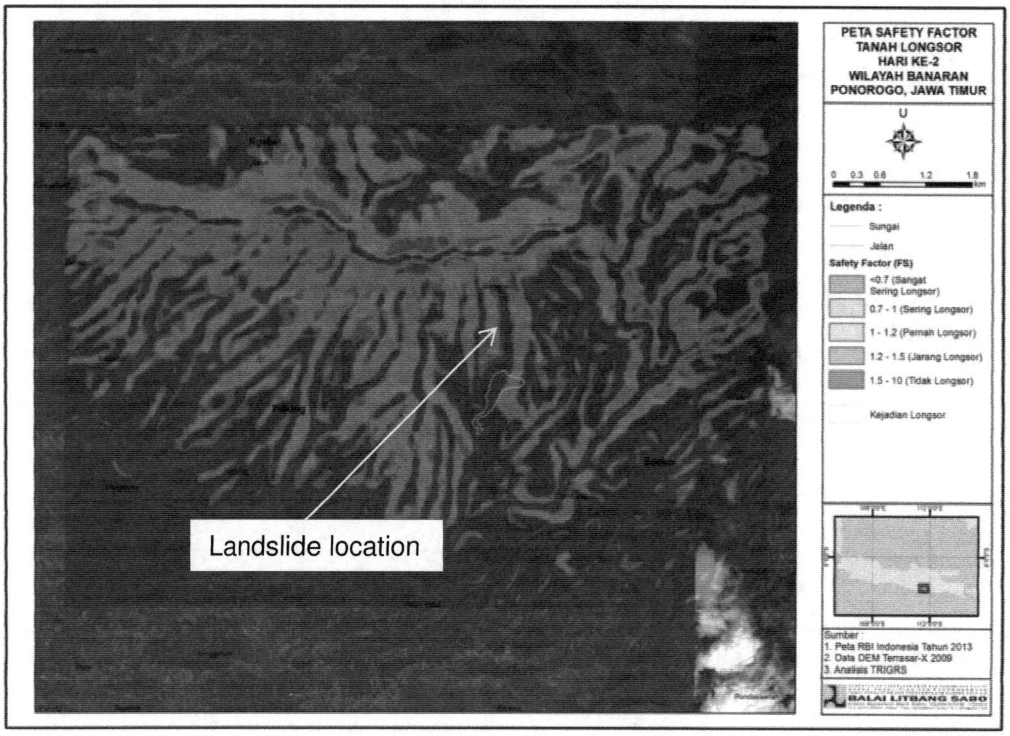

Figure 4. Slope stability map of Tangkil Village, Ponorogo

Community activities cannot be separated from economic activities. The economic support in the location of the Ponorogo landslide was very dependent on agriculture, especially horticulture plants. The relocation were a was certainly not far from their lading, because they make a living from the farm. The level of marginal and inappropriate conformity was not permitted for settlement development or as a cultivation room in accordance with Minister of Public Works Regulation No.22 / PRT / M / 2007 because it had the threat of moderate and very high landslides. Development of settlements was only allowed in were as with a low threat of landslides.

The extent of the area that had the threat of medium and high scale landslides makes the suitability of the area for settlements to be limited. There were also existing settlements which were in the threat of high landslides. The threat of a disaster must be avoided or engineered if the settlement will be carried out. This disaster threat analysis greatly determines the level of land suitability for settlement development. Landslide threats were used as a limiting factor in the analysis process because of the influence that was very difficult to remove and greatly affects land suitability.

4. Conclusion

Many landslides area were not suitable to be used as residential so that residents need to be relocated. Before relocation, it was necessary to map the landslide vulnerability level for the selection of relocation sites, so that the new location was safe from landslides. So in the relocation site, the landslides will not occur again. Settlements that were still located in the high-risk landslide were as endanger the occupants. The relocation must be in a safe location from landslides. The relocation of Sijeruk landslide victim did not in a safe area. The location for relocation experience landslide again. Modelling slope stability using the TRIGRS program in principle was measuring the effect of rainfall on regional slope stability. The data needed in modelling was digital elevation model (DEM), soil data index properties, soil engineering properties, and rainfall. Field data collection activities, in the undisturbed soil sampling and geotechnical laboratory testing on soil samples, to obtain the values of physical properties and soil engineering. The modelling results show that the safety factor values in the study were ranged from 0.8 to 10. Safe numbers below 1.2 means are prone to landslides. The relocation was a must had a safe number above 1.2. so we suspect the relocation site was the safe area from the landslide.

Acknowledgement

The author wishes to acknowledge to the Head of the Sabo Research and Development Center for the opportunity in this study. Acknowledge to the head of BPBD of Banjarnegara and BPBD of Ponorogo. So thank you for all my friends in Balai Sabo for all cooperation.

References

[1] [DPR-RI] Indonesian Parlianment. 2011. Constitution Number 1/2011 about housing and residential areas
[2] [KemenPU] Public work ministry. 2007. Minister of Public Works Regulation No.22 / PRT / M / 2007 concerning Spatial Planning Guidelines for Landslide Disaster Areas
[3] Chen H., Liu Y., Maki N., Hayashi H., Sawada M. (2012). Disaster recovery collective relocation and life recovery issues: fundamental analysis of permanent houses sites of Typhoon Morakot, Taiwan. Papers on city planning, 47(3), 919-924
[4] Perera T. Weerasoori I, Karunarathne H., (2012). An evaluation of success and failures in Hambantota, Siribopura resettlement housing program: lessons learned, Sri Lanka Journal of Real Estate, No. 6, 1-15 pp.
[5] Sangasumana R. P., 2018, Post Disaster Relocation Issues: A Case Study of Samasarakanda Landslide in Sri Lanka, European Scientific Journal November 2018 edition Vol.14, No.32 ISSN: 1857 – 7881 (Print) e - ISSN 1857- 7431
[6] Fredlund D. G., & Rahardjo R., 1993, Soil Mechanics for Unsaturated Soils, John Wiley & Sons, Inc.
[7] PVMBG 2009, Landslide and Planning Hazard Assessment, Geology Agency, Department of Energy and Mineral Resources, Bandung1
[8] Baum R. L., Savage W. Z. dan Godt J. W., "TRIGRS—A FORTRAN program for transient rainfallinfiltration and grid-based regional slope-stability analysis, version 2.0", U.S. Geological. Survey, 2008
[9] Chen C.Y., Chen T.C., Yu F. C., and Lin S.C., 2005. Analysis of time-varying rainfall infiltration induced landslide, *Engineering Geology*, Vol. 48, 466–479.

[10] Liao Z., Hong Y., Kirschbaum D., Adler R.F., Gourley J.J., and Wooten R., 2011. Evaluation of TRIGRS (Transient Rainfall Infiltration and Grid-Based Regional Slope-Stability Analysis)'s Predictive Skill for Hurricane-Triggered Landslides: A Case Study in Macon County, North Carolina, *Natural Hazards*, Vol. 58, 325–339.

[11] Park D.W., Nikhil N.V., and Lee S.R., 2013. Landslide and Debris Flow Susceptibility Zonation using TRIGRS for the 2011 Seoul Landslide Event, *Natural Hazards and Earth System Sciences*, Vol. 13, 2833–2849.

[13] Alveoli M., Guzzetti F., Rossia M., 2014, Scaling properties of rainfall induced landslides predicted by a physically based model, Geomorphology 213, Elseiver

[14] Abramson, Lee W., 1995, Slope Stability and Stabilization Methods, John Willey & Sons, Inc, Newyork.

[15] Sugiarti K. and Sukristiyanti S., 2017, TRIGRS Application for landslide susceptibility mapping, Global Colloquium on GeoSciences and Engineering, IOP Publishing

[16] https://regional.kompas.com/read/2017/04/02/07585811/longsor.di.ponorogo.27.orang. belum. ditemukan.21.rumah.tertimbun (diakses 5 April 2017)

[17] Mulyana R. 2009. The Concept of Healthy and Environmentally Friendly Settlements in the Cianjur River Basin, Cianjur Regency, West Java Province [dissertation]. Bogor (ID): Bogor Institute of Agricultural.

[18] Ward T. J.; Li R. M.; Simons D.B., 1979. Mathematical Modeling Approach for Delineating Landslide Hazards in Watersheds. In: C.D. Humphrey (ed.), Proceedings of the 17th Annual Engineering Geology and Soils Engineering; 1979 April 4-6; Moscow, 109-142.

[19] Viet T.T., Lee G., Thu T.M., and An H.K., 2016, Effect of Digital Elevation Model Resolution on Shallow Landslide Modeling Using TRIGRS, *Natural Hazards Review*, 04016011: 1 – 12

Effective Performance Evaluation to Estimate Cost and Time Using Earned Value

F Susilowati[1,*] and W M Kurniaji[2]

[1] Civil Engineering, Universitas Tidar.
Jl. Kapten Suparman 39 Potrobangsan, Magelang. Telp (0293) 364113
[2] Civil Engineering, Politeknik Negeri Jakarta.
Jl.Prof. Dr. G.A. Siwabessy, Campus of UI Depok. Telp (021) 7863534

[*]Corresponding author e-mail: fajar.susilowati@untidar.ac.id

Abstract. The data from Badan Pusat Statistik (BPS) Indonesia, show that the development sector in the Yogyakarta region is changing rapidly, especially with the construction of new airports in the region. That demands the effectiveness and efficiency of cost and time in every construction. It requires supervision, evaluation, and performance control in the construction process. In this effort, this research discusses more detail about the calculation of construction project performance using the earned value method. This research conducted on an integrated development project consisting of malls, hotels, and apartments in Sleman, Yogyakarta. The research data collection was carried out by direct observation and analysis of supporting project data. The data obtained were processed and analysed using the accepted value method. The results of this study indicate that at the end of the review, the Schedule Performance Index (SPI) obtained a value of 0.58. Project progress in the field shows that the project has a 41.48% delay. It can see from the initially planned project of 99.06%, but at the time of review, the realisation was only 57.22%. In terms of costs, the Cost Performance Index (CPI) is 1.25. It shows that the costs incurred during the review are still less than the existing budget plan. At the end of the review period, the estimated final project costs are obtained, both direct and indirect costs. With the tendency of project performance at the time of review, the value is still less than the project budget, while the estimated time of project completion will be longer or late at around 56.52% of the planned schedule.

1. Introduction

The data from BPS Indonesia shows that the development sector in transportation, buildings, factories, and other infrastructures that support the economic sustainability of the present has increased rapidly. One of them can see from almost every corner of the city in the Yogyakarta region is undergoing construction, starting from the construction of malls or shopping centres, hotels, apartments, and other tourism support facilities.

This development process cannot separate from the management of construction projects. In construction projects, planning and control are the most basic functions in realising the success of activity in a construction project [1]. Cost and time control of the project over issues of effectiveness and efficiency can see from actual progress in the field. Effectiveness and efficiency of the cost and time of construction at the time of implementation need to be monitored, evaluated, and controlled. The uncontrol of cost and time can cause construction projects are different from the planned [2][3].

In recent years, various new approaches to management have been carried out to accurately predict project duration based on the Earned Value Management (EVM) method [4] [5]. Management of this project is an effective tool for managing project performance [6]. The earned value method is one of the tools that can be used in project management to integrate costs and time. The concept of earned value presents three dimensions e.i., the physical completion of the project (the per cent complete), the planned absorption of costs (budgeted cost), and the actual costs as well as what obtained from the costs that have incurred (earned value). Of the three dimensions, the concept of earned value can connect between cost performance and time derived from the calculation of variants of cost and time [7] [8].

Based on cost and time performance, a project manager can identify the overall performance of the project, and the work packages then predict the cost performance and project completion time, assuming that the trends occurred during reporting until the project completed. The results of the project performance evaluation can be used as preliminary information if there is an inefficient performance in project completion. Based on this initial information, management policies or changes in implementation methods can take, so that cost overruns and delays in project completion can prevent and handled adequately. The application of the concept of earned value to the project can provide a comprehensive described of the status of project performance [9].

Some previous studies explain that using Earned Value can lead to improved performance and practical excellence [10] [11]. The Earned Value technique is also able to monitor the potential for non-linear accruals from the number of incentives throughout the project [12]. In some project profiles, the earned value method can describe the planned value that will provide benefits, reliability, and accuracy of the duration shown by some concrete examples on several large projects [13].

Success in implementing Earned Value Analysis is also a combination of the design and operational aspects of the system. This framework can be used to develop Earned Value-based systems in the project environment [14]. In the Earned Value system, the prediction of Plan Value before the project starts will inform management about the performance of Earned Value and Actual Cost, thus enabling more effective proactive actions to ensure favourable performance results [15]. The results of the Earned Value analysis are complete and objective reports to evaluate the value management process that will be obtained later [16].

One area of potential for future research is an integrated model that focuses on time and cost as well as other factors such as quality of sustainability [17]. Based on information from various previous studies, the next will discuss in more detail on the direct application of Earned Value in one of the projects in Yogyakarta. It can provide real information about Earned Value calculations, especially in the construction sector.

2. Earned Value Concept

Three essential elements from some references in analysing the performance of the project based on the concept of earned value [18].

The first element is the Budgeted Cost of Work Scheduled (BCWS), or Planned Value (PV) is a budget that is allocated based on a work plan that has prepared against time. BCWS calculated from the accumulated planned budget for work in a certain period. BCWS, at the end of the project (100% completion), is called the Budget at Completion (BAC). BCWS also measures the time performance of project implementation. BCWS reflects the cumulative absorption of plan costs for each work package based on the order according to the planned schedule.

The second element is the Actual Cost of Work Performed (ACWP), or Actual Cost (AC) is a representation of the total expenditure incurred to complete the work within a specified period. ACWP can be cumulative up to the period of performance calculation or the number of expenses in a certain period

The third element is the Budgeted Cost of Work Performed (BCWP) or Earned Value (EV), which is the value received from the completion of work during a specified period. This BCWP is called earned value (EV). This BCWP is calculated based on the accumulation of completed works. Project performance appraisal can be known based on analysis of variance, analysis of performance indexes, and analysis of estimated project completion costs.

2.1. Analysis of Variance

Cost Variance (CV) is the difference between the value obtained after completing work packages with the actual costs incurred during project implementation. Positive cost variance indicates that the value of the work packages obtained is greater than the cost incurred to work on the work packages, on the opposite, the negative value indicates that the value of the completed work packages is lower than the cost already spent [19].

$$CV = BCWP - ACWP$$

Schedule Variance (SV) is used to calculate deviations between BCWS and BCWP. Positive values indicate that the project work packages that implemented are more than planned, on the opposite, negative values indicate poor work performance because the work packages that implemented are less than the planned schedule [19].

$$SV = BCWP - BCWS$$

Table 1 explains the project description about the variance analysis that consists of Schedule Variance and Cost Variance values.

Table 1. Integrated Variance Analysis.

SV	CV	Description
Positive	Positive	Work is carried out faster than Schedule Planned and lower costs
Zero	Positive	Work is carried out on schedule and lower cost
Positive	Zero	Work is carried out faster than Schedule Planned and on budget
Zero	Zero	Work is carried out on schedule and budget
Negative	Negative	Work is carried out late than schedule planned and over budget
Zero	Negative	Work is carried out on schedule but over budget
Negative	Zero	Work is carried out late than schedule planned but on budget
Positive	Negative	Work is carried out faster than Schedule Planned but over budget

Based on Table 1, it can see that SV value positive means that work is carried out faster than Schedule Planned and SV value negative means that work is carried out late than Schedule Planned. SV value zero means that work carried out on Schedule. CV value positive means that work is carried out lower than budget and CV value negative means that work is carried out over budget. The CV value zero means that work carried out on Budget.

2.2. *Performance Index Analysis*

Cost Performance Index (CPI) is a cost-efficiency factor that has incurred. It is shown by comparing the value of physically completed work (BCWP) with costs that have incurred in the same period (ACWP). This CPI value shows the value obtained (relative to the overall project value) against the costs incurred. CPI less than one shows a low-cost performance because the costs incurred (ACWP) are greater than the value obtained (BCWP), or in other words, waste occurs [20].

$$CPI = BCWP / ACWP$$

Schedule Performance Index (SPI) is a factor of performance efficiency in completing work. It is shown by comparing the value of physically completed work (BCWP) with the planned expenditure of costs incurred based on the work plan (BCWS). The SPI value indicates how much work can be completed (relative to the whole project) to the planned work unit. SPI value less than one indicates that job performance is not as expected because it is not able to achieve the planned work targets [20]

$$SPI = BCWP/BCWS$$

Table 2 explains the project description about the Performance Index Analysis that consists of the Schedule Performance Index and Cost Performance Index values.

Table 2. Performance Index Analysis.

Index	Score	Description
SPI	>1	Project performance is faster than Schedule Planned
	<1	Project performance is late than Schedule Planned
	$=1$	Project performance is on schedule
CPI	>1	ACWP is less than the value of the work obtained (BCWP)
	<1	ACWP is more than the value of the work obtained (BCWP)
	$=1$	ACWP is same as the value of the work obtained (BCWP)

Based on Table 2, it can see that SPI value more than one means that Project Performance is carried out faster than Schedule Planned and SPI value less than one means that Project Performance is carried out late than Schedule Planned. SPI value is one, it means that Project Performance carried out on Schedule. CPI value more than one means that Actual Cost is carried out less than Budget Cost and CPI value less than one means that Actual Cost is carried out more than Budget Cost. CPI value is one, it means that work carried out on Budget.

Estimate to Complete (ETC) is an estimate of the cost of remaining work, assuming that the trend of project performance is the same until the end of the project. The estimate can explain with some assumptions. The first is to assume that the remaining work costs a budget. The assumption used is the cost of remaining work within the budget, and it does not depend on performance at the time of review. The second assumption is that performance is equal to the end of the project. The assumption used is that performance at the time of review will remain until the end of the project [20].

Some approaches combine the two methods. If the percentage of work below 50% uses the formula:

$$ETC = (\text{Budget Cost} - BCWP)$$

If the percentage of work above 50% uses the formula:

$$ETC = (\text{Budget Cost} - BCWP) / CPI$$

Estimate at Completion (EAC) is the estimated total cost at the end of the project obtained from the actual costs plus ETC [20].

$$EAC = ACWP + ETC$$

The importance of CPI and SPI is to predict the costs needed to complete a project statistically. There are many methods for predicting project completion costs (EAC). But calculations with SPI and CPI are more comfortable and faster to use.

EAC calculation is the sum of the actual costs that have incurred and the remaining costs that will be needed to complete the project. The remaining costs to be required are predicted statistically by calculating cost performance (CPI) and schedule performance (SPI). From the EAC value, it can estimate the difference between the project completion plan cost (BAC) and the cost of project completion based on the work performance that has been achieved (EAC) or variance at completion (VAC).

CPI and SPI indicators are more often used to assess project performance compared to SV and CV. CPI and SPI values are weight values that do not have dimensions so that they compared between project performance with one another. Besides, the SPI and CPI values provide a relative comparison to BCWS, which is the basis for assessing project status in terms of cost and time.

Time Estimated (TE) is the estimated time of project completion. The assumption used to estimate the completion time is the tendency for project performance to remain the same as at the time of review.

$$TE = \textit{Actual Time Expended} + (\textit{Original Duration} - (\textit{Actual Time Expended} \times SPI)) / SPI$$

3. Methodology

This research was conducted by direct observation in the field for 11 months (month 2nd until month 12th) to evaluate the performance of the project using the Earned Value method. From the observations in the field obtained data as in Table 3 below.

Table 3. Recapitulation Data.

Month period	% Planned	% Realization	% Actual Cost	BCWS (IDR millions)	BCWP (IDR millions)	ACWP (IDR millions)
2	3.21	1.34	1.06	8,346	3,484	2,752
3	13.36	6.32	4.49	34,736	16,432	11,686
4	33.05	18.00	13.49	85,930	46,800	35,070
5	45.56	24.41	18.28	118,456	62,764	47,522
6	66.31	32.03	24.35	169,806	83,278	63,318
7	75.48	39.01	29.87	196,248	101,426	77,654
8	81.35	46.80	36.64	211,510	121,680	95,275
9	88.75	51.55	40.78	230,750	134,030	106,020

Month period	% Planned	% Realization	% Actual Cost	BCWS (IDR millions)	BCWP (IDR millions)	ACWP (IDR millions)
10	93.48	53.52	42.51	242,996	139,152	110,527
11	98.06	55.10	43.87	254,956	143,260	114,060
12	99.06	57.22	45.69	257,556	148,772	118,800

From Table 3 can see that the SV, CV, SPI, and CPI coefficients calculated to determine the time and cost performance of the project. The results are estimated that the final value of the project calculated on the performance of time and cost. Based on Table 3 can see that the per cent of realisation is lower than the per cent of planned, so budget cost work performance and actual cost work performance also lower than budget cost work schedules.

4. Results dan Discussion

The total duration in this project is 13th months, but project performance in12th months tends to be negative. It can see in Table 4 that shows the analysis of schedule variance (SV) was negative. It means that the project late than scheduled planned. The analysis of cost variance (CV) was positive. It means that the costs incurred are below from the budget planned. Project performance in the 12th months was negative. It means that the progress of work increased slightly due to changes in the design of the structure that had done. It results in demolition and reworks according to the new design.

Variant Analysis, based on Table 4, in the 2nd to 12th month, shows that SV is negative and CV is positive. The SV is negative means that the project is behind schedule. The CV is positive indicates that the budget spent is below the budget planned.

Performance Index Analysis based on Table 4, in the 2nd to 12th month, the SPI value less than one means the project late from the schedule planned, and the CPI value more than one means that the Actual Cost is less than Budget Cost.

Table 4. Calculation of Variants and Performance Index.

Month Period	SV (IDR millions)	CV (IDR millions)	SPI	CPI
2	(4,862)	732	0.42	1.27
3	(18,304)	4,746	0.47	1.41
4	(39,130)	11,730	0.54	1.33
5	(55,692)	15,242	0.53	1.32
6	(86,528)	19,960	0.49	1.32
7	(94,822)	23,772	0.52	1.31
8	(89,830)	26,405	0.58	1.28
9	(96,720)	28,010	0.58	1.26
10	(103,844)	28,624	0.57	1.26
11	(111,696)	29,200	0.56	1.26
12	(108,784)	29,972	0.58	1.25

From Table 3 and Table 4 can describe using the graph in Figure 1 as below.

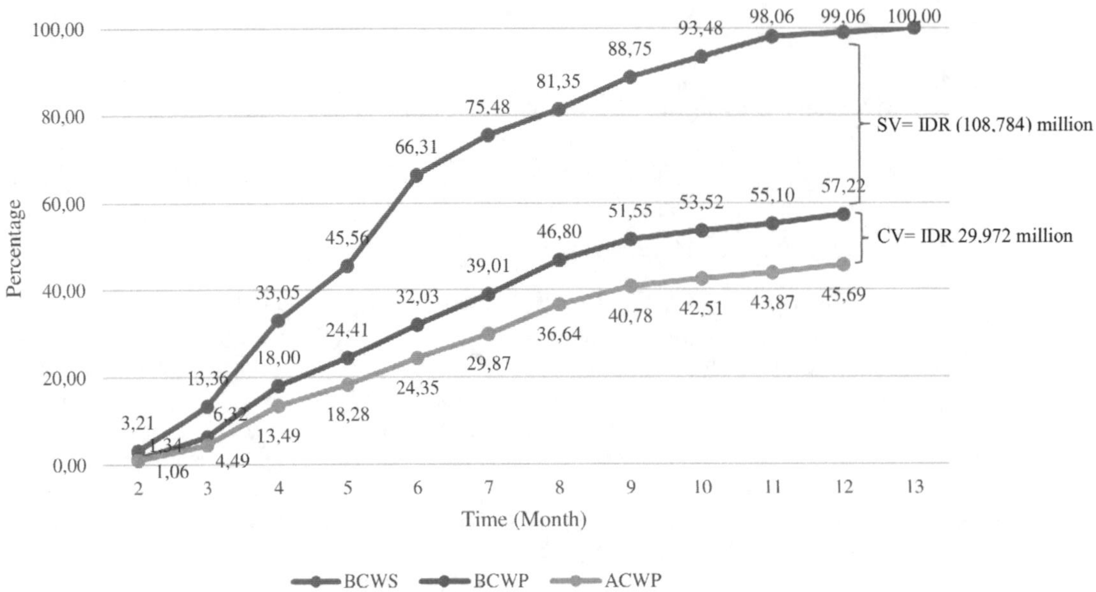

Figure 1. The relationship among BCWS, BCWP, and ACWP in 12th Month.

Based on figure 1 shows that the percentage of Budget Cost Work Performance and Actual Cost Work Performance is below of the Budget Cost Work Schedule. From Tabel 4 and Figure 1 can describe that the Schedule Variance in 12th month is negative, which is IDR 108,784 million, dan Cost Variance in 12th month is positive, which is IDR 29,972 million. The following is a graph of the SPI and CPI movement values during the review period based on Table 4.

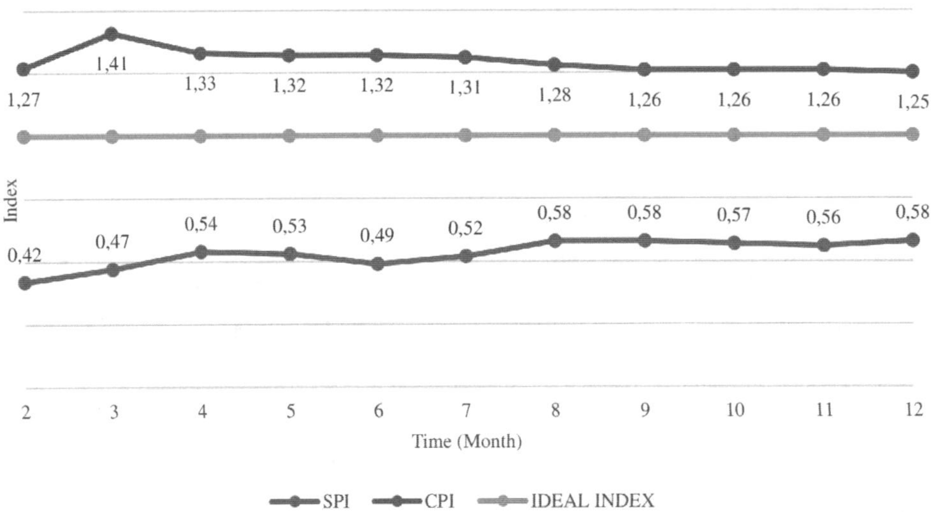

Figure 2. SPI and CPI Graph in 12th Month.

From Figure 2 can describe that the CPI is above the ideal index dan the SPI is below the Ideal Index. Then the estimated value of the final cost and the total cost of project completion calculated as well as the estimated time of completion of the project as shown in Table 5.

Table 5. Analysis of Time and Cost Completion.

Month Period	ETC (IDR millions)	EAC (IDR millions)	TE (month)
2	256,516	259,268	31

Month Period	ETC (IDR millions)	EAC (IDR millions)	TE (month)
3	243,568	255,254	27
4	213,200	248,270	24
5	197,236	244,758	25
6	176,722	240,039	27
7	158,574	236,228	25
8	138,320	233,595	23
9	99,644	205,664	22
10	95,989	206,516	23
11	92,946	207,006	23
12	88,820	207,621	23

Cost and Time Estimation of Project Completion in the 12th Month based on a review conducted every month estimate that cost obtained which is an estimate for the remaining costs (ETC) of IDR 88,820 (millions), and an estimated total project completion cost (EAC) of IDR 207,621 (millions). The time estimate of project performance (TE) for 23 months, it means longer or will be late compared to the schedule planned that has a duration of 13 months. It can interpret that the project implementation delayed by 56.52%.

5. Conclusion

Based on the results of the analysis can be explained that at the end of the review was obtained The Schedule Performance Index (SPI) of 0.58. Project progress in the field shows that the project is experiencing 41.48% delays. It can see from the initially planned project of 99.06%, but at the time of review, the realisation was only 57.22%. In terms of cost was obtained The Cost Performance Index (CPI) of 1.25. It shows that the costs incurred are still below from the budget planned. At the end of the review period, direct and indirect costs estimated at IDR 207,620 (millions). If the tendency of project performance as at the end of the review, this cost is below the project budget or lower than the budget planned. While the estimated completion time of the project is 23 months means that it is longer or will be late than the schedule planned, which has a duration of 13 months. It can interpret that the project implementation delayed by 56.52% of the schedule planned.

References

[1] F. Rochman and H. C. Wahyuni, "Analisa Pengaruh Pengendalian Kinerja Proyek Terhadap Mutu Proyek Konstruksi Dengan Menggunakan Uji Statistika," *J. Tek. Ind.*, 2018.

[2] A. Mishakova, A. Vakhrushkina, V. Murgul, and T. Sazonova, "Project Control Based on a Mutual Application of Pert and Earned Value Management Methods," *Procedia Eng.*, vol. 165, pp. 1812–1817, 2016.

[3] P. A. de Andrade, A. Martens, and M. Vanhoucke, "Using real project schedule data to compare earned schedule and earned duration management project time forecasting capabilities," *Autom. Constr.*, vol. 99, no. October 2018, pp. 68–78, 2019.

[4] J. Batselier and M. Vanhoucke, "Evaluation of deterministic state-of-the-art forecasting approaches for project duration based on earned value management," *Int. J. Proj. Manag.*, vol. 33, no. 7, pp. 1588–1596, 2015.

[5] P. Ballesteros-Pérez, E. Sanz-Ablanedo, D. Mora-Melià, M. C. González-Cruz, J. L. Fuentes-Bargues, and E. Pellicer, "Earned Schedule min-max: Two new EVM metrics for monitoring and controlling projects," *Autom. Constr.*, vol. 103, no. March, pp. 279–290, 2019.

[6] H. L. Chen, W. T. Chen, and Y. L. Lin, "Earned value project management: Improving the predictive power of planned value," *Int. J. Proj. Manag.*, vol. 34, no. 1, pp. 22–29, 2016.

[7] R. D. H. Warburton and D. F. Cioffi, "Estimating a project's earned and final duration," *Int. J. Proj. Manag.*, vol. 34, no. 8, pp. 1493–1504, 2016.

[8] D. A. Wood, "High-level integrated deterministic, stochastic and fuzzy cost-duration analysis aids project planning and monitoring, focusing on uncertainties and earned value metrics," *J.*

Nat. Gas Sci. Eng., vol. 37, pp. 303–326, 2017.
[9] B. Susanti, M. Melisah, and I. Juliantina, "Penerapan Konsep Earned Value Pada Proyek Konstruksi Jalan Tol (Studi Kasus Ruas Jalan Tol Kayuagung - Palembang -Betung)," *J. Rekayasa Sipil*, 2019.
[10] J. Colin, A. Martens, M. Vanhoucke, and M. Wauters, "A multivariate approach for top-down project control using earned value management," *Decis. Support Syst.*, 2015.
[11] G. Espinosa-Garza and I. Loera-Hernández, "Proposed model to improve the forecast of the planned value in the estimation of the final cost of the construction projects," *Procedia Manuf.*, vol. 13, pp. 1011–1018, 2017.
[12] L. P. Kerkhove and M. Vanhoucke, "Extensions of earned value management: Using the earned incentive metric to improve signal quality," *Int. J. Proj. Manag.*, 2017.
[13] R. D. H. Warburton and D. F. Cioffi, "Estimating a project's earned and final duration," *Int. J. Proj. Manag.*, 2016.
[14] D. Bryde, C. Unterhitzenberger, and R. Joby, "Conditions of success for earned value analysis in projects," *Int. J. Proj. Manag.*, 2018.
[15] H. L. Chen, W. T. Chen, and Y. L. Lin, "Earned value project management: Improving the predictive power of planned value," *Int. J. Proj. Manag.*, 2016.
[16] D. S. Christensen, "The Costs and Benefits of the Earned Value Management Process," *J. Parametr.*, 2018.
[17] L. L. Willems and M. Vanhoucke, "Classification of articles and journals on project control and earned value management," *Int. J. Proj. Manag.*, 2015.
[18] B. Wahab, "Penialaian Pengendalaian Biaya dan Waktu Pada Proyek Peningkatan Jalan Menggunakan Earned Value," *Teras J.*, 2019.
[19] R. A. N, D. R. S, and F. Kistiani, "Pengendalian Biaya Dan Waktu Proyek Dengan Metode Konsep Nilai Hasil (Earned Value)," *J. Tek.*, 2015.
[20] K. A. B. Bombana, A. Efendi, and L. O. Sumarsiddin, "Pengendalian Waktu Dan Biaya Pada Proyek Dengan Metode Earned Value (Studi Kasus Pembangunan Dermaga Kasipute," *J. Ilm. Tek. Mesin*, 2015.

Macroscopic Flow Models of State Highway Midblocks under Mixed Traffic Flow Conditions: Ramon Magsaysay Boulevard Case Study

M S Delos Santos[1,*], A R Alzona[1], M A Taguba[1], C M Vahdanipour[1]

[1]Department of Civil Engineering, National University-Manila, 551 M F Jhocson St., Manila, Philippines.

[*]msdelossantos@national-u.edu.ph, aralzona@national-u.edu.ph

Abstract. Macroscopic flow modelling has been fundamentally used to characterize and predict the traffic flow parameters of a continuous road utilizing the analogy for one-dimensional fluid flow. In the Philippines, Ramon Magsaysay Boulevard is a major arterial road that links Manila City to various important cities in the Metropolis. Due to its noticeable deteriorating performance in addressing the increasing traffic flow demand, this work investigated the corridor capacity and characterized the key traffic elements along the major mid blocks using Greenshields model. It is assumed that road users behave uniformly and continuously distributed over the highway to choose a path to minimize their total travel time based on instantaneous traffic information. Investigation shows volume demands exceeding the design traffic capacities equivalent to 0.7429, 1.7959, 0.7226, and 1.4726 VCR ratios.

1. Introduction

The macroscopic flow of models that are currently used and further utilized today originated from the fundamental assumption that there is a degree of relationship among vehicle positions and their velocities [1]. Elementary principles of fluid motion for one-dimensional flow such as the equation of continuity and the equation of motion have been adapted to traffic modelling, arriving with the general equation

$$\frac{du}{dk} = -ck^{\frac{n-1}{2}} \tag{1}$$

where u = speed
k = density
c, n = arbitrary constants.

On the other hand, Greenshields [2], after assessment of recorded vehicle speeds and densities along a continuous road, conjectured that there is a linear relationship between the two traffic variables, shown as

$$u = u_f\left[1 - \frac{k}{k_{jam}}\right] \tag{2}$$

where u_f = free-flow speed
k_{kam} = jam density.

This clearly suggests that the density-flow relationship would in turn be parabolic. The speeds were obtained by locating the position of vehicles from two or more pictures taken at short time intervals. It can also be perceived that this formula can be directly obtained from (1) by substituting n=1 and applying the boundary conditions for the case of free-flow velocity and jam density. Frequently, the density-flow scatter plots are widely scattered due to non-equilibrium traffic condition [1]. Zhang proposed that there should be a distinction in acceleration and equilibrium flow with the speed-density relation; over the entire road length speed and density remains constant [3]. Plots not satisfying this criterion must be discarded from the graph.

Meanwhile, Ramon Magsaysay Boulevard is an eight-lane divided arterial road that connects Nagtahan interchange, A.H. Lacson Avenue, and Legarda Street, Manila City in the west wing up to G. Araneta Avenue, Quezon City and Aurora Boulevard, San Juan City in the east wing. In its entire length, the route is traversed by Light Rail Transit Line 2, with which two of its significant stations are located. Total travel demand along this route is high because it is a highly urbanized area, and various institutions are located nearby the corridor. Thus, the road is noticeably deteriorating in addressing traffic flow performance as time progresses.

This paper aims to evaluate the current traffic flow capacity of Ramon Magsaysay Boulevard utilizing the fundamental macroscopic model (Greenshields model). Finally, the Levels of Service of the selected mid blocks will be determined according to the guidelines stipulated in Highway Capacity Manual 2000.

2. Methodologies

This study aims to evaluate the traffic flow performance of Ramon Magsaysay Boulevard using macroscopic flow modelling. Fundamental traffic flow parameters utilized in this study are detailed below.

2.1 Volume studies

Traffic volume is the most fundamental data needed for the study of traffic flow. Volume studies correspond to the number of vehicles passing thru a point within the area of study in relation to time; hence, the data shall be expressed in the number of vehicles with respect to time.

The volumes to be gathered by the researchers utilizing Cordon counts are as follows: 1) the average daily traffic (ADT), which corresponds to the actual demand for service by the road, 2) hourly traffic and short term (15-minute) count to estimate the maximum flow rate and to identify the peak hour volumes. In terms of the time of the study, this work shall perform the count during 7 AM to 9 AM, and 5 PM to 7 PM to represent the very high level of traffic volume, and short counts of one hour with 15 minute-interval to analyze peak hour volumes. Figure 1 illustrates the data collection performed during the study. Situations that would cause abnormal conditions, such as holidays or fiestas, bad weather conditions, or demonstrations shall not be allowed in the study [4]. It is important to consider the passenger car equivalent factors (PCEF) depending on the type of vehicles entering the study section as stipulated in the Department of Public Works and Highways (DPWH) Highway Capacity Manual [5].

2.2 Spot speed studies

Like volume studies, situations that would cause abnormal conditions with the traffic flow shall also not be avoided during spot speed studies. The ideal duration for this study shall be around one hour. This paper shall use the manual method, also known as "trap length method," because it is the most efficient to execute among all methods. In this technique, two lines, 30 meters apart, shall be drawn transversely on the pavement. A stopwatch is then used by the field researcher to measure the "elapsed time" of the vehicle within the trap. This procedure shall be done 30 times for the analysis and presentation of the spot speeds. Furthermore, to avoid bias in sampling systematic random sampling shall be employed, wherein every nth vehicle represents the whole spot speeds within that closed area for that specific time range alone.

Figure 1. Volume counting along Pureza St. midblock-outbound

2.3 Analysis and presentation of traffic data

Using the data obtained from the quantitative studies (volume studies, spot speeds, and density), fundamental traffic flow parameters can be readily fitted in graphs. For speed-density relationship along mid blocks, simple linear regression modelling can be used using the equations.

$$u = \frac{\sum k_i u_i - n\bar{k}\bar{u}}{\sum k_i{}^2 - n\bar{k}^2} \quad \text{and} \tag{3}$$

$$a = \bar{u} - b\bar{k} \tag{4}$$

Moreover, the Levels of service can also be determined at this point base on the design capacity assigned for this type of road.

3. Results and discussion

Figure 2 shows the histogram of vehicle speeds observed at the first eastbound midblock. The arithmetic means speed is 25.0 kph with a standard deviation of 9.33 kph, showing a widely scattered speed data. In addition, Figure 3 shows the cumulative distribution of frequencies at midblock 1. It can be depicted that the 50th percentile speed representing median is 21.64 kph, while the modal speed is 29.0 kph. Finally, the speed limit that can be imposed given these traffic conditions is 31.89 kph, which is the 85th percentile speed.

Applying the principles used and developed in Greenshields model, the speed-density relationship of at midblock 1 is

$$u = -0.1066x + 31.2855 \tag{5}$$

With an R squared of 0.698749, depicting low correlation. Plots not satisfying equilibrium should not be considered in the relationship. Figure 4 shows the abovementioned relationship.

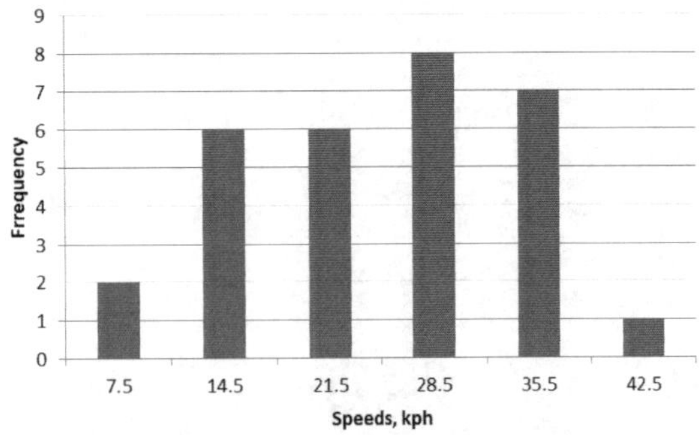

Figure 2. Spot speed frequency histogram at midblock 1: Legarda (eastbound).

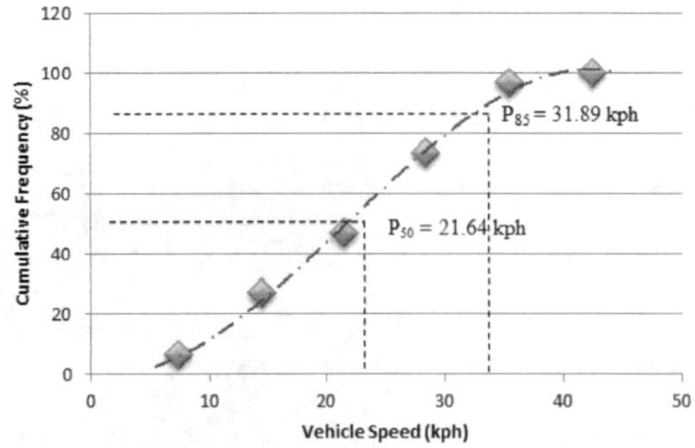

Figure 3. Cumulative frequency distribution at midblock 1: Legarda (eastbound).

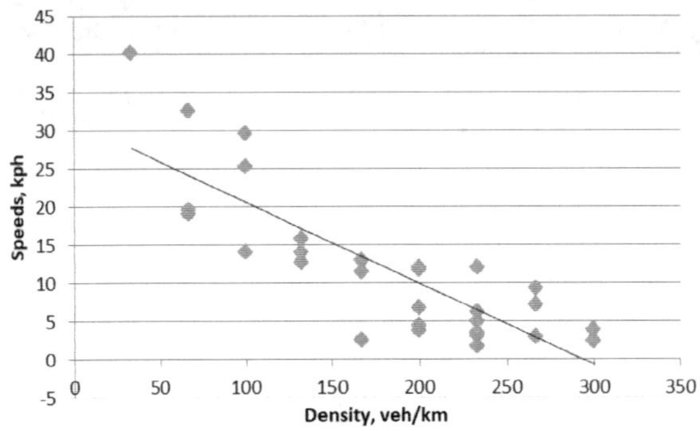

Figure 4. Speed-density scatterplot at midblock 1: Legarda (eastbound).

On the other hand, Figure 5 shows the histogram of vehicle speeds observed at the second eastbound midblock. The time-mean speed for this road section is 20.67 kph with a standard deviation of 9.75 kph. The figure shows irregularities in spikes within the region of lower speeds, suggesting that vehicle arrivals using stochastic methods cannot be applied here. Furthermore, Figure 6 displays the cumulative distribution of frequencies at midblock 2. It can be depicted that the 50th percentile speed representing median is 18.41 kph, while the modal speed is 24.0 kph. Finally, the speed limit that can be imposed given these traffic conditions is 28.04 kph.

The relationship of density and speed when applying Greenshields model again is

$$u = -0.097k + 34.249 \tag{6}$$

with an R squared value of 0.716, indicating a low correlation between the parameters. Figure 7 illustrates this relationship.

For the westbound direction of flow, Figure 8 shows the histogram of vehicle speeds observed at the first midblock. The arithmetic mean speed in this section is 12.17 kph with a standard deviation of 9.54 kph. It can be observed that there should have been a normal curve plot with its peak approximately at 14.5 kph, but there is an odd frequency for middle-class speed of 4.5 kph. In addition, Figure 9 displays the cumulative distribution of frequencies. It can be depicted that the 50th percentile speed representing median is 6.26 kph, while the modal speed is 4.5 kph. The speed limit that can be imposed given these traffic conditions is 19.85 kph; however, this value is so small that imposing this speed limit can be quite unimportant.

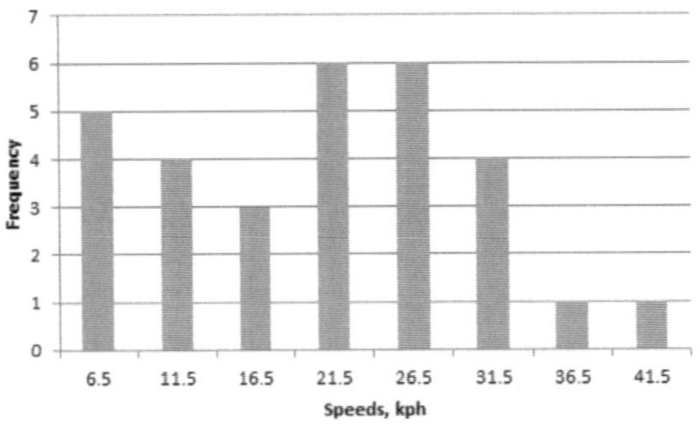

Figure 5. Spot speed frequency histogram at midblock 2: Pureza (eastbound).

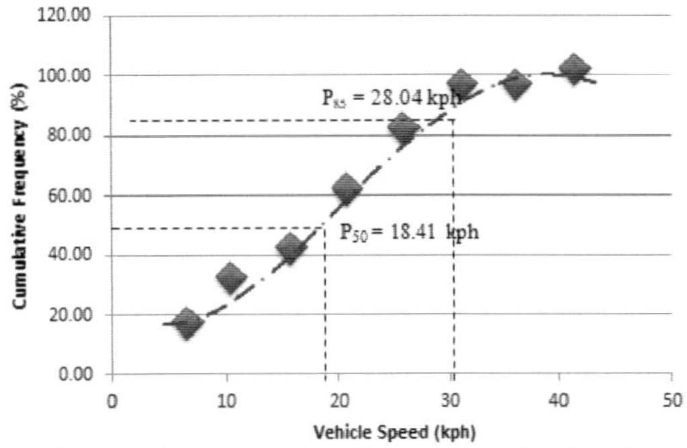

Figure 6. Cumulative frequency distribution at midblock 2: Pureza (eastbound).

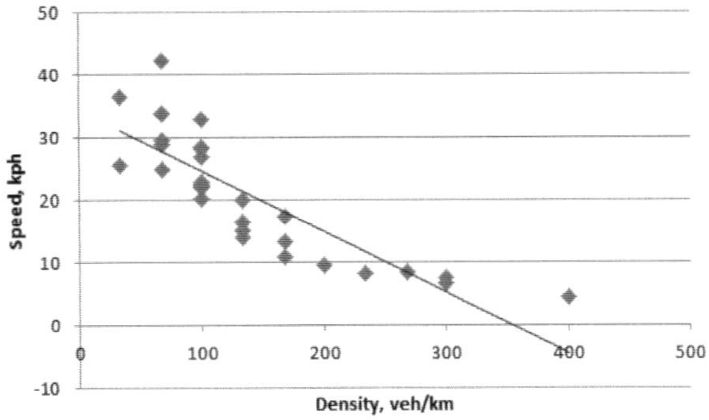

Figure 7. Spot density scatterplot at midblock 2: Pureza (eastbound).

Greenshields model of speed-density relationship for this midblock is

$$u = -0.107k + 31.278 \tag{7}$$

yielding a correlation coefficient of 0.607082, denoting a low correlation of plots. Figure 10 shows the relationship.

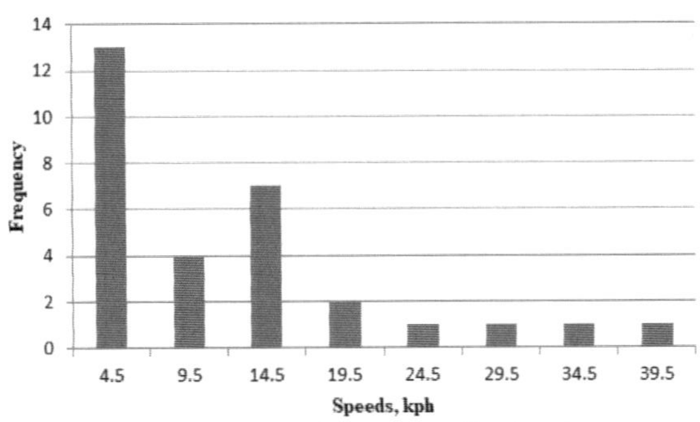

Figure 8. Cumulative frequency distribution at midblock 1: Legarda (westbound).

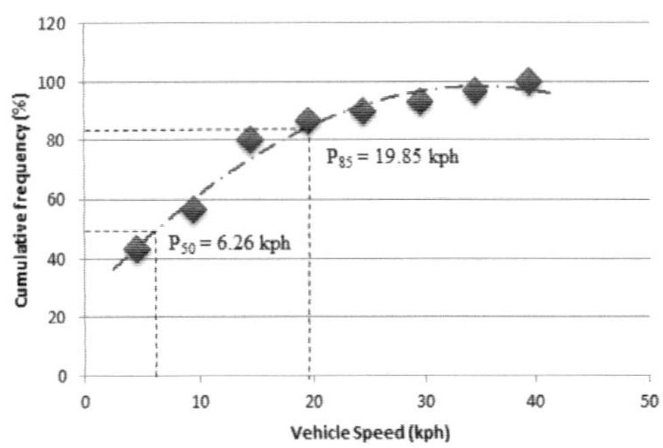

Figure 9. Cumulative frequency distribution at midblock 1: Legarda (westbound).

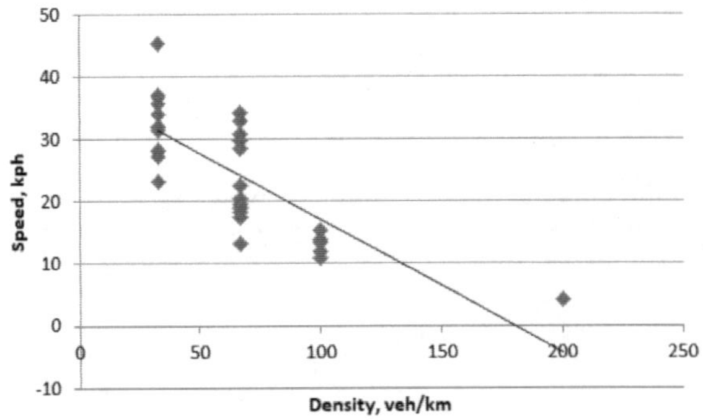

Figure 10. Speed-density scatterplot at midblock 1: Legarda (westbound).

Finally, Figure 11 illustrates the histogram of vehicle speeds observed at the second midblock westbound direction. The arithmetic mean speed corresponding in this section is 25.63 kph with a standard deviation of 9.85 kph. It can be observed too that the frequency shows a stochastic distribution of vehicle speeds with its modal speed at 27.5 kph. Figure 12 displays the cumulative distribution of frequencies with the 50th percentile speed of 22.83 kph. The speed limit that can be imposed given these traffic conditions is 32.57 kph.

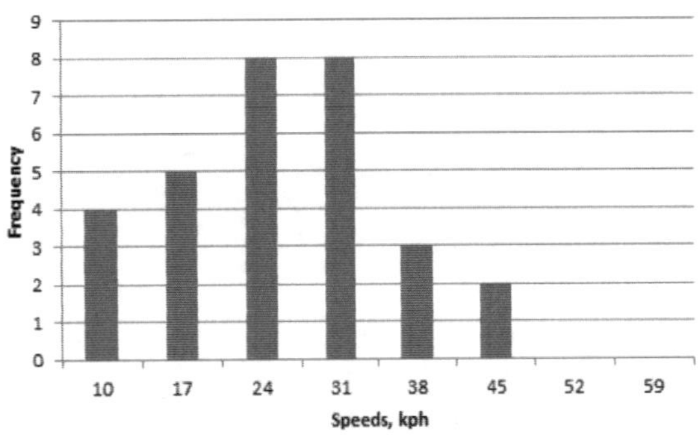

Figure 11. Spot speed frequency histogram at midblock 2: Pureza (westbound).

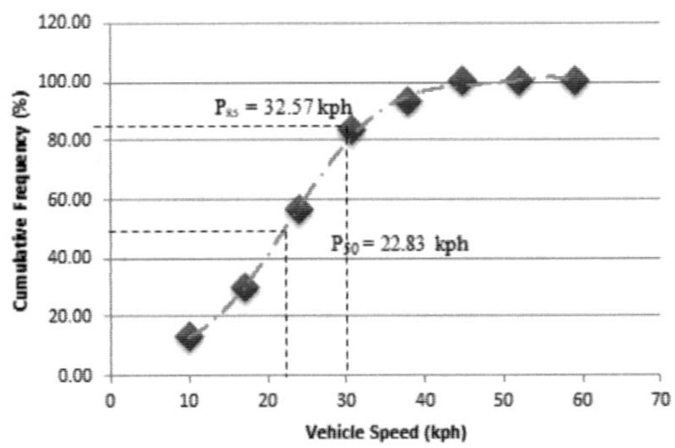

Figure 12. Spot speed frequency histogram at midblock 2: Pureza (westbound).

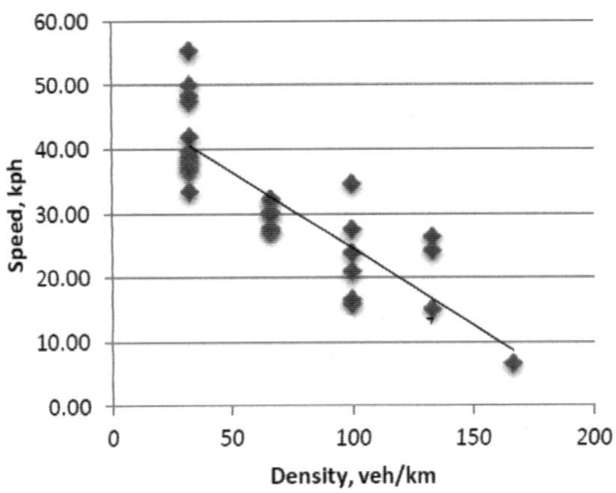

Figure 13. Speed-density scatterplot at midblock 2: Pureza (westbound).

Applying the principles used and developed in Greenshields model, the speed-density relationship of at midblock 1 is

$$u = -0.239k + 48.438 \tag{8}$$

with an R squared of 0.712, also depicting low correlation.

The maximum sustainable flow rate at which vehicles can be expected to traverse a point during a certain period of time for four-lane two-way divided highway in the Philippines is 3,400 veh/hr. Based on the recorded flow rates at 15-minute time intervals, the flow rate at midblock 1 for eastbound and westbound are 2,526 veh/hr and 2,457 veh/hr. On the other hand, the recorded flow rates for midblock 2 for eastbound and westbound are 6,106 veh/hr and 5,007 veh/hr, respectively. Table 1 shows the levels of service at the mid blocks.

Table 1. Levels of Service at Midblock.

Midblock	Volume	Capacity	LOS
1 (Eastbound)	2,526	3,400	D
2 (Eastbound)	6,106	3,400	F
1 (Westbound)	2,457	3,400	D
2 (Westbound)	5,007	3,400	F

4. Conclusions

Ramon Magsaysay Boulevard's important mid blocks traffic flow capacities were investigated using Greenshields model. The following conclusions can be drawn from the results:

- The actual traffic flow variables and capacity of Ramon Magsaysay Boulevard is not efficiently capable in sustaining the vehicular flow demand passing through it. Very low levels of service directly affect the ridership, transportation cost, and comfortability of the road users. Additional traffic measures must be implemented to decongest the thoroughfare.
- High traffic congestion in the mid blocks directly affected the consistency of the principles used in macroscopic flow modelling. Equilibrium flows and accelerations must be distinguished evidently with the corresponding speed and density relationships in order to acquire more correct models.
- Most of the diagrams constituted a widely scattered set of points. Standard deviations of speeds ranged around ± 9 kph showing a need for the more accurate methodology of recording data. Moreover, several factors were neglected in the analysis such as road friction factor, obstructions of traffic regulations, lane changing of some drivers, to name a few. Accuracy would increase significantly if these would be addressed in the future researches.

5. References

[1] van Wageningen-Kessels, F., van Lint, H., Vuik, K., & Hoogendoorn, S. (2015). Genealogy of traffic flow models. *EURO Journal on Transportation and Logistics*, *4*(4), 445–473. https://doi.org/10.1007/s13676-014-0045-5
[2] Greenshields, B. D. (1934). The Photographic Method of Studying Traffic Behavior.
[3] Zhang, H. M. (1999). *A mathematical theory of tra • c hysteresis. 33.*
[4] Highways, D. P. (2013). *Department Order No. 22; Clarification on the Definition and Difference Between Widening and Paving of Shoulders.* Manila.
[5] Sigua, R. (2008). *Fundamentals of Traffic Engineering.* The University of the Philippines Press.

IOP Conf. Series: Materials Science and Engineering **771** (2020) 012057 doi:10.1088/1757-899X/771/1/012057

Fresh and Hardened Properties of Self Fiber Compacting Concrete (SFCC) Incorporated with Zeolite and Nylon

Besty Afriandini[1,*], Fanny Monika[2], Fadillawaty Saleh[2], Hakas Prayuda[2], Martyana Dwi Cahyati[2], Siti Isnaini Kurniawati Djaha[3]

[1]Department of Civil Engineering, Faculty of Engineering, Universitas Muhammadiyah Purwokerto, 53182 Kembaran, Banyumas, Central Java, Indonesia.
[2]Department of Civil Engineering, Faculty of Engineering, Universitas Muhammadiyah Yogyakarta, 55183 Kasihan, Bantul, Yogyakarta, Indonesia.
[3]Department of Civil Engineering, Faculty of Engineering, Akademi Teknik Kupang, 85118 Kota Raja, Kupang, Nusa Tenggara Timur, Indonesia.

[*]e-mail: bestyafriandini@ump.ac.id, hakasprayuda@umy.ac.id

Abstract. This study discusses the use of zeolite waste and nylon fiber which are used as added material in making concrete. The type of concrete used is self fiber compacting concrete (SFCC). Self-Compacting Concrete (SCC) is a type of concrete that emphasizes a high enough workability value so that the concrete can flow and solidify itself during the casting process. In this study, fresh and hardened properties will be tested. The fresh properties tested consisted of slump flow, T50, V-Funnel and L-Box while the hardened test of concrete consisted of compressive strength, splitting tensile strength and flexural strength. Nylon fiber is used as much as 1% of the weight of cement with length of 50 mm, while zeolite ash is used as a substitute for cement with variations of 5%, 10% and 15%. The purpose of this study was to determine the effect of nylon and zeolite ash variations on fresh and hardened properties of self fiber compacting concrete. The results show that using these two additional materials can be applied to self-compacting concrete because it meets predetermined standards for the fresh properties of concrete. In the hardened properties of the mixture of 5% zeolite produces higher compressive strength, splitting tensile strength and flexural strength than other variations. It can be concluded that the content of 1% fiber and 5% zeolite ash instead of cement is the optimum level and suitable for self-fiber compacting concrete.

1. Introduction

Self-compacting concrete is a type of concrete developed from conventional concrete to increase the value of its workability. Self-compacting concrete was first introduced in Japan in the mid-1980s to produce concrete with fairly good performance [1, 2]. Self-compacting concrete can be defined as a type of concrete that can flow and compact by itself due to gravity without the help of vibrators and workers, so this concrete requires a high level of fresh performance to avoid segregation [3, 4]. In its development, self-compacting concrete has been applied in various infrastructures in several countries [5], implementation of this type of concrete is also used for structural pre-stressed beam girder [6], Dams [7], and in buildings [8]. This concrete is used to facilitate its specific work in areas that are difficult to be compacted manually during the casting process.

Self-compacting concrete is very suitable for building structures that have a high level of difficulty, this type of concrete is very helpful in maintaining the quality of concrete to remain in accordance with standards and reduce the number of pores produced by concrete due to not being perfectly solid. In the manufacturing process, self-compacting concrete tends to require a fairly high number of binders, especially cement, so that a lot of research is done to reduce the amount of cement used but still produce a pretty good performance. Previous researchers have used fly ash as a substitute for cement [9, 10], Palm oil ash waste ash [11, 12], baggase ash [13, 14], and also recycled concrete aggregates [15, 16]. Utilization of this waste is based on its chemical properties which contain pozzolan properties making it suitable to be used as a cement substitute partially.

The nature of concrete which has a high level of resistance to the compressive force, but is weak in holding the tensile force. If the tensile strain produced exceeds the tensile strain capacity, cracking in the concrete will occur. One method for increasing the tensile strength of concrete is to add fibrous material. Research on fiber concrete has been carried out such as by adding nylon to normal concrete [17, 18], Polypropylene fiber [19, 20], plastic waste and recycled fiber [21, 22], and with various other types of fibers both organic and inorganic fibers. While the use of fiber in self-compacting concrete has not been done much so it needs to be tested further on the effect of fiber on the workability value of self-compacting concrete.

In this study fresh and hardened properties will be tested on self-fiber compacting concrete with fiber composition of 1% to the weight of cement and added with zeolite variations of 5%, 10% and 15% as a substitute for cement. The purpose of this study is to determine the effect from fiber and zeolite as a substitute for cement to the characteristics of self-compacting concrete. Fresh properties testing is done to determine the ability of concrete in maintaining performance so that it can still be categorized as self-compacting concrete. The fresh properties testing is done in the form of slump flow, T50, V-Funnel and L-Box. Whereas the testing of hardened properties is carried out to determine the mechanical properties of concrete, the tests are in the form of compressive strength, spitting tensile strength and flexural strength. The testing period starts from 7, 14 and 28 days. Through this research it is hoped that it can improve the performance of self-fiber compacting concrete by utilizing waste materials so as to reduce the production price of concrete.

2. Experimental Program

2.1. Materials and Mix Design

In this study using materials composed of cement, water, fine aggregate, coarse aggregate, zeolite, superplasticizer and fiber. Before being used as a constituent, coarse and fine aggregates are first tested for properties as shown in Table 1. Tests are carried out in the form of examining specific gravity, moisture content, absorption, roughness values up to mud content. The test results of each aggregate indicate that this type of aggregate is suitable for use as a material making up concrete.

Table 1. Fines and coarse aggregate properties

Test Item	Unit	Fines Aggregate	Coarse Aggregate
Water Content	%	3.71	1.97
Specific Gravity	-	2.58	2.65
Absorption	%	0.82	0.75
Mass Density	gr/cm^3	1.53	1.55
Mud Content	%	4.01	0.91
Roughness content	%	-	32.87

The use of cement in the manufacture of self-compacting concrete is generally required in high enough quantities so that it will increase the price of this type of concrete production. In addition, the large number of waste materials that have pozzolanic characteristics in Indonesia is certainly a good enough method to produce concrete that is both efficient and environmentally friendly. In this study using zeolite waste as a substitute for cement. The percentage used varies from 5%, 10% to 15% of the weight of normal cement used. In Table 2 a comparison of the chemical content in the zeolite ash used and the chemical content in cement. Table 2 shows that Portland cement has high CaO, SiO_2 and Al_2O_3 contents, as well as the results obtained by zeolite ash.

Table 2. Chemical Properties of Portland cement and Zeolite Ash [23]

Materials	SiO_2	K_2O	Fe_2O_3	CaO	MgO	Al_2O_3	Na_2O	SO_3
PC	18.68	0.73	3.53	64.56	0.98	4.67	0.14	3.00

ZA	68.85	2.19	1.29	3.97	1.06	11.71	0.29	0.18

One of the important constituents in this research is nylon fiber. Figure 1 (a) shows the shape of Zeolite ash, while Figure 1 (b) shows the shape of the nylon fiber used. Nylon fiber is a composite added material used to increase the tensile strength of concrete and flexural strength. The nylon fiber used in this study has a length of 50 mm aimed at controlling early cracking and increasing durability. Nylon fiber is used as much as 1% of the weight of cement.

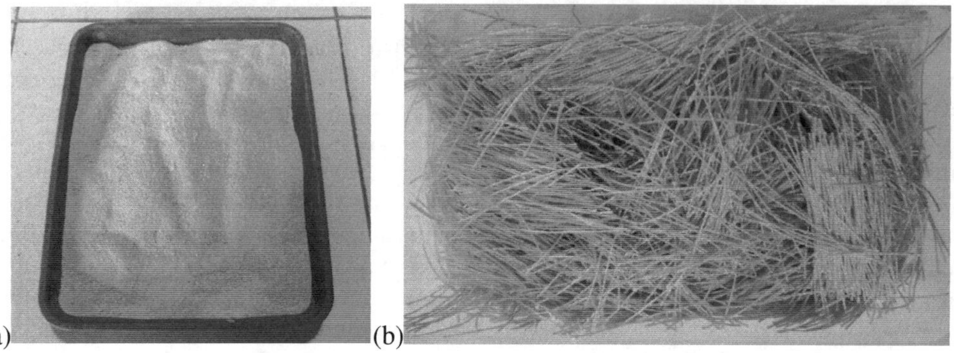

Figure 1. (a) Zeolite Ash; (b) Nylon fiber

In this study, using 4 variations of the mixture that there are differences in the content of zeolite as a substitute for cement. In Table 3 the mix design results for each variation used for 1m3. The mix design used in this study adopts Agrawal, et al in 2008 [24].

Table 3. Mix design for each variation in kg/m^3

Materials (Kg)	0% ZA	5% ZA	10% ZA	15% ZA
Cement	485.00	460.75	436.50	412.25
Zeolite Ash	0.00	24.25	48.50	72.75
Fines Aggregate	600	600	600	600
Coarse Aggregate	561	561	561	561
Superplasticizer	7.275	7.275	7.275	7.275
Nylon Fiber	0.00	4.85	4.85	4.85
Water	135	135	135	135

In this study, using the water to binder ratio with the same conditions for all variations so that it can be known whether the increasing amount of zeolite used will affect the workability of each mixture. The superplasticizer used in this study is a special SP for self-compacting concrete that is widely used in Indonesia, namely the Sikament NN type. The superlastisizer content used is 0.8% with the expectation that it can improve workability even though the water to binder ratio used is only 0.28.

2.2. Fresh Properties Test

Fresh properties testing conducted in this test consists of four parts, namely slump flow, T50, V-Funnel and L-Box. All testing criteria follow the standards set by the European Federation of National Trade Associations Representing Producers and Applicators of Specialist Building Products (EFNARC) [25]. In the testing of slump flow and T50 using the same equipment as shown in Figure 2 (a) where the purpose of this test aims to determine the ability of flow ability of fresh concrete mix. Whereas in Figure 2 (b) is a form of L-Box testing tool that aims to determine the ability of passing ability that is the ability to pass obstacles that have been provided in such a way according to established standards. V-funnel testing is carried out to determine the value of viscosity and filling ability using tools as shown in Figure 2 (c).

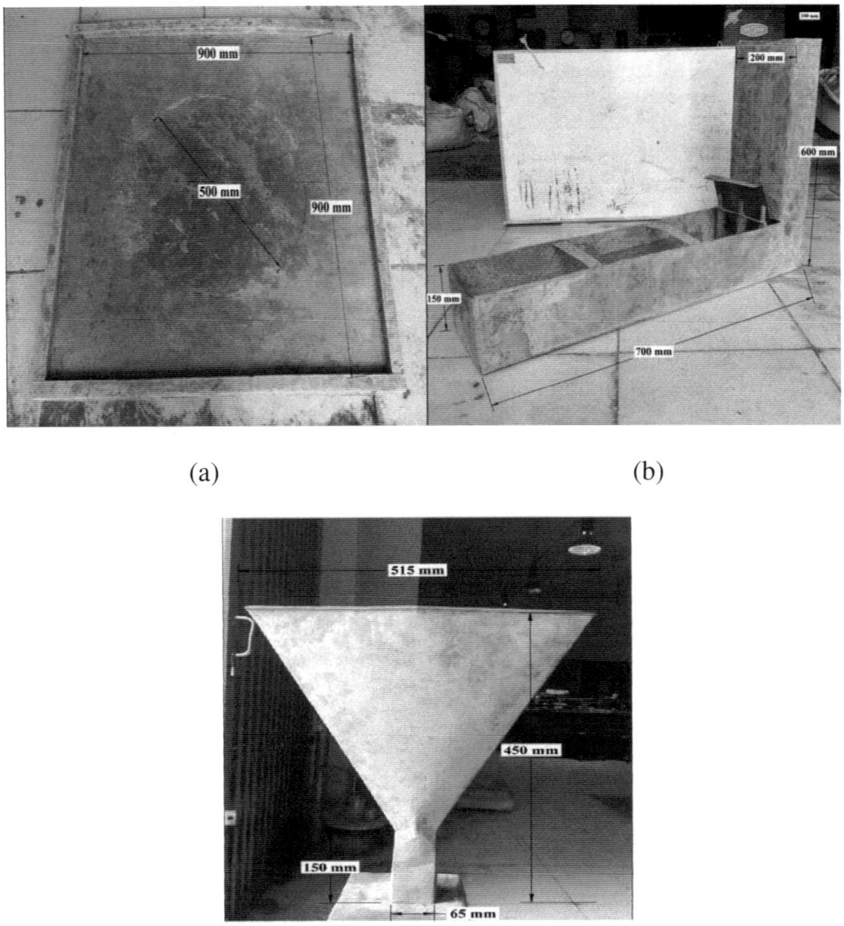

(a) (b)

(c)

Figure 2. Fresh properties check tools (a) Slump flow; (b) L-Box; (c) V-funnel

2.3. Hardened Properties Test

Tests carried out in the form of compressive strength, tensile strength and flexural strength in each test follow different standards. Each test is done at the age of 7, 14 and 28 days by curing using water curing. The compressive strength test uses dimensions of 150 mm diameter with 300 mm height [26], while the tensile strength test uses specimens with the same dimensions as the compressive strength, only my test position is different [27]. Flexural strength testing uses point loading in the middle of the test specimen with a 150x150x600 mm specimen size [28].

3. Result and Discussion

3.1. Fresh Properties

Fresh properties testing conducted in this study consisted of checking the slump flow, T50, V-Funnel and L-Box with the results that can be seen in Figure 3. In the slump test it was seen that overall it met the specified specifications. The results show that the higher the zeolite content in the concrete mixture the slump flow value will decrease so this causes the flow ability value of fresh concrete to be reduced. The higher levels of zeolite used cause more water is up so that the spread value in the concrete becomes smaller.

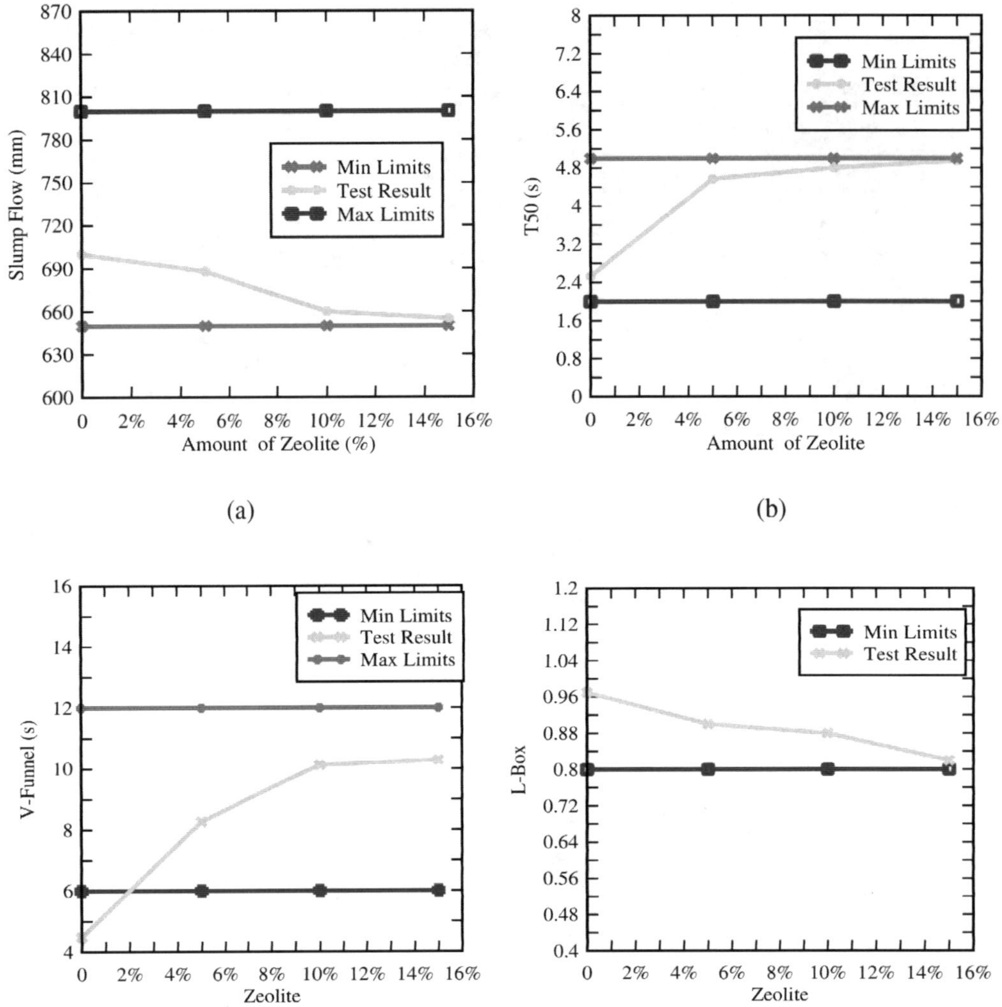

Figure 3. Fresh properties result (a) Slump Flow; (b) T50; (c) V-Funnel; (d) L-Box

With the reduced value of the slump flow it will increase the T50 value because the thicker the concrete produced takes longer the concrete to spread during the slump flow test. The level of plasticity of fresh concrete is very decisive sooner or later the concrete has spread, in Figure 3 (c) is the result of V-Funnel test. In self-fiber compacting concrete without using zeolite added ingredients, it can be seen that fresh concrete does not meet the specified standards. However, concrete with the addition of zeolite shows an increase in value so that it becomes better and meets specified specifications.

The L-Box test shows that the resulting value decreases with increasing levels of zeolite which is used as an added ingredient in place of cement. Overall shows that the maximum level of zeolite that can be used as self-fiber compacting concrete is 15%, when using more than 15% it is very likely that fresh concrete does not meet the specifications specified for self-compacting concrete.

3.2. Hardened Properties

Concrete compressive strength test carried out at the age of 7, 14 and 28 days in Figure 4 is the result of the compressive strength relationship with curing time. Curing conducted in this study is water curing for all test specimens. In Figure 4 (a) explains that concrete with a percentage of 5% zeolite produces a higher compressive strength than normal concrete without cement replacement. In normal self-concrete, the compressive strength is 27.4 MPa when the concrete is 7 days, 29.2 MPa when the concrete is 14 days and 32.9 MPa when the concrete is 28 days. Whereas concrete with 5% zeolite content produces compressive strength of 29.8 MPa when it is 7 days, 33 MPa when concrete is 14 days and 35.7 MPa for concrete with 28 days age. Concrete with variation of 10% and 15% zeolite produces a compressive strength that is smaller than normal concrete. Through this compressive strength results it can be concluded that the optimum optimal level in the use of zeolite in concrete is 5%.

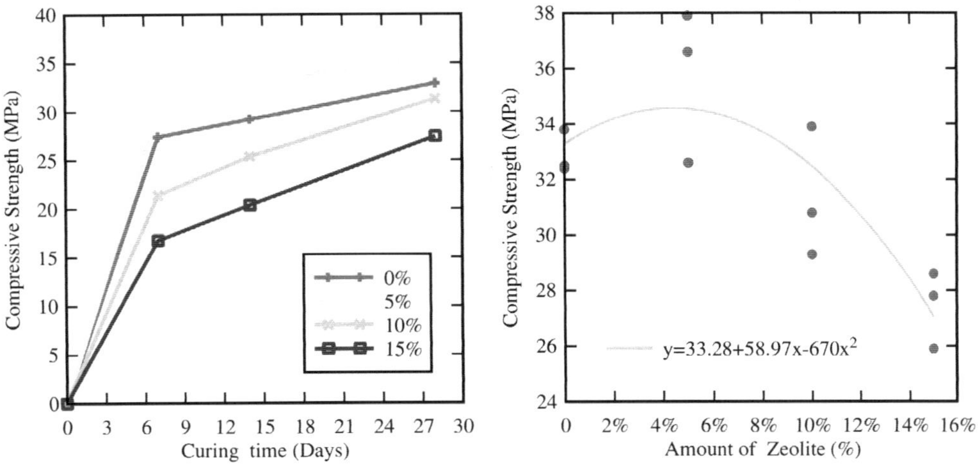

(a) (b)

Figure 4. (a) Compressive strength; (b) relationship between compressive strength with amount of zeolite

Spitting tensile strength testing is done at the same age as compressive strength. The results of splitting tensile strength testing on self-fiber compacting concrete can be seen in Figure 5. The results show that 5% zeolite levels produce better tensile strength compared with 10% and 15% zeolite levels. This has similarities with the compressive strength described earlier. So it can be concluded that the tensile strength will produce optimum value when using zeolite levels of 5%. In Figure 5 (b) explains the relationship between splitting tensile strength and the amount of zeolite. The results show that there is an optimum level of cement replacement conditions as much as 5%, but as the amount of zeolite increases, the resulting tensile strength decreases.

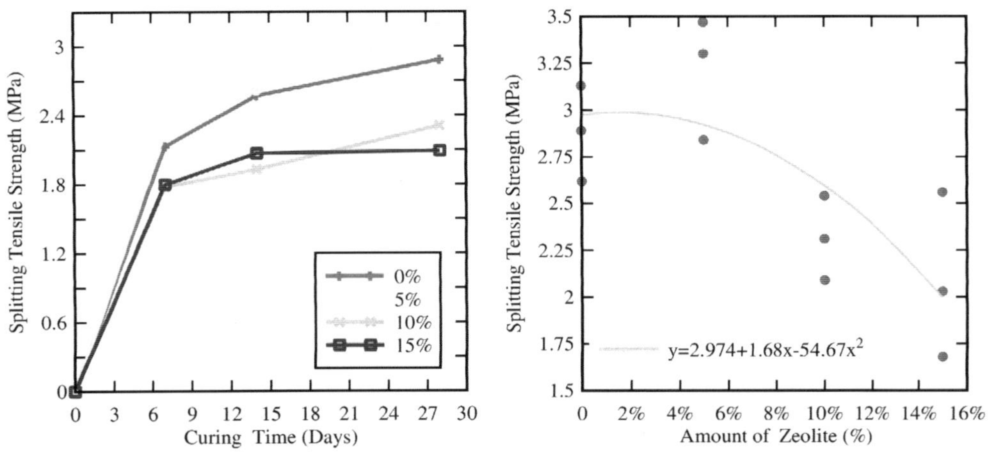

(a) (b)

Figure 5. (a) Splitting tensile strength; (b) relationship between splitting tensile strength with amount of zeolite

Flexural strength testing is performed at the age of 7, 14 and 28 days of concrete. The results showed that the flexural strength produced at 5% zeolite levels produced the highest value compared to the flexural strength in other variations. In Figure 6 shows the results of flexural strength testing and displacement generated in each mixture. In concrete using 5% zeolite, the displacement value is greater, this also corresponds to the value of the flexural strength produced. Meanwhile, with the increase in the amount of zeolite used, the resulting displacement value decreases, but the smallest displacement value is produced on the test object without using zeolite added ingredients. Overall on the hardened properties test it can be concluded that the 5% zeolite content is optimum for cement substitute added ingredients.

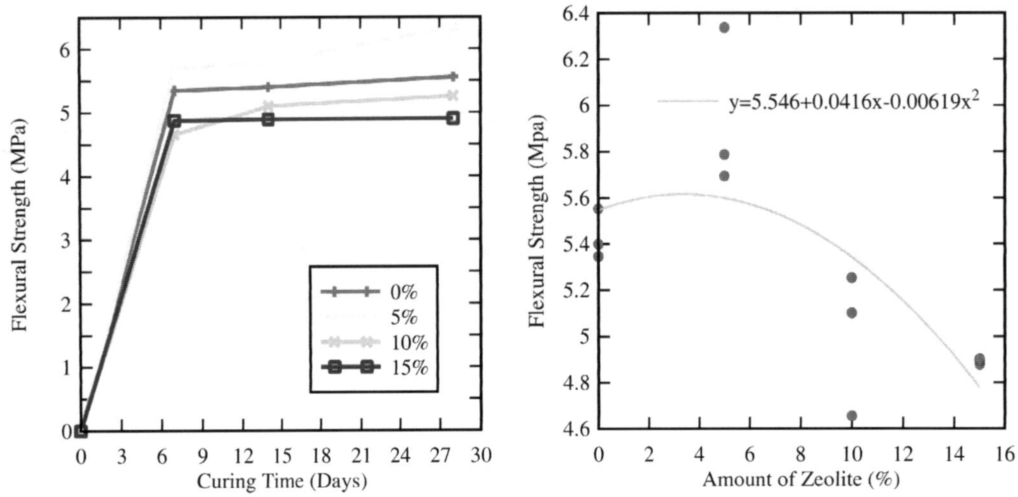

(a) (b)

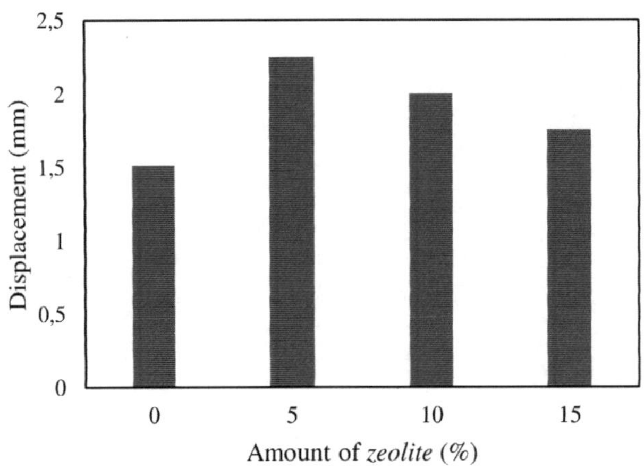

(c)

Figure 6. (a) Flexural strength; (b) relationship between flexural strength with amount of zeolite; (c) maximum displacement from flexural strength

4. Conclusion

Based on the results and discussion can be summarized as follows:

1. Fresh properties test results show that the increasing value of zeolite added ingredients instead of cement will reduce the workability value. The 15% mixture level is the maximum level that can be used so that it meets the fresh properties requirements required by the standard.

2. The results of hardened properties show that 5% zeolite levels produce higher compressive strength, splitting tensile strength and flexural strength than in other mixes. So it can be concluded that the optimum level of zeolite is 5%.

References

[1] V. Anusuya, C. V. Saranya, and S. N. K. Pravin, "Comparison of the performance of self-compacting concrete", 6th National Conference on Innovative Practices in Construction and Waste Management, pp. 181-185, April 2017.
[2] M. M. Kamal, M. A. Safan, A. A. Bashandy, and A. M. Khalil, "Experimental Investigation on the behavior of normal strength and high strength self-curing self-compacting concrete", Journal of Building Engineering, Vol. 16, pp. 79-93, 2018.
[3] H. Prayuda, F. Saleh, T. I. Maulana, and F. Monika, "Fresh and Mechanical Properties of Self-Compacting Concrete with Coarse Aggregate Replacement using Waste of Oil Palm Shell", The 7th AIC-ICMR on Science and Engineering, Vol. 352. February 2018.
[4] H. Okumura, and M. Ouchi, "Self-Compacting Concrete", Journal of Advance Concrete Technology, Vol. 1, pp. 5-15, 2003.
[5] M. Ouchi, S. Nakamura, T. Osterberg, and S. E. Hallberg, "Application of self-compacting concrete in Japan, Europe, and The United States", Proceeding ISHPC, pp. 1-20, 2003.
[6] A. Pozolo, and B. Andrawes, "). Analytical prediction of transfer length in prestressed self-consolidating concrete girders using pull-out test result", Construction and Building Materials, Vol. 25, pp. 1026-1036, 2011.
[7] M. H. Beigi, J. Berenjian, O. L. Omran, A. S. Nik, and I. M. Nikbin, "An experimental survey on combined effects of fibers and nanosilica on the mechanical, rheological, and durability properties of self-compacting concrete", Materials and Design, Vol. 50, pp. 1019-1029, 2013.
[8] A. Khaloo, E. Molaei, R. P. Hosseini, and H. Tahsiri, "Mechanical performance of self-compacting concrete reinforced with steel fibers", Construction and Building Materials, Vol. 51, pp. 179-186, 2014.
[9] G. Sua-Iam, and N. Makul, "Utilization of high volumes of unprocessed lignite-coal fly ash and rice husk ash in self-consolidating concrete", Journal Clean Production, Vol. 78, pp. 184-194, 2014.
[10] V. Kannan, "Strength and durability performance of self-compacting concrete containing solf-combusted rice husk ash and metakaolin", Construction and Building Materials, Vol. 160, pp. 169-179, 2018.
[11] B. S. Thomas, S. Kumar, and H. S. Ariel, "Sustainable concrete containing palm oil fuel ash as a supplementary cementitious material a review", Renewable and Sustainable Energy Reviews, Vol. 80, pp. 550-561, 2017.
[12] H. Mohammadhosseini, A. Awal, and A. Ekhsan, "Influence of palm oil fuel ash on fresh and mechanical properties of self-compacting concrete", Sadhana, Vol. 40, pp. 1989-1999, 2015.
[13] J. P. Moretti, S. Nunes, and A. Sales, "Self-compacting concrete incorporating sugarcane bagasse ash", Construction and Building Materials, Vol. 172, pp. 635-649, 2018.
[14] T. Akram, S. A. Memon, and H. Obaid, "Production of low-cost self-compacting concrete using bagasse ash", Construction and Building Materials, Vol. 23, pp. 703-712, 2008.
[15] M. Omrane, S. Kenai, E. Kadri, A. Ait-Mokhtar, "Performance and Durability of Self Compacting Concrete using Recycled Concrete Aggregates and Natural Pozzolan", Journal of Clean Production, Vol. 165, pp. 415-430, 2017.
[16] S. Manzi, C. Mazzotti, and M. C. Bignozzi, "Self-compacting concrete with recycled concrete aggregate: study of the long-term properties", Construction and Building Materials, Vol. 157, pp. 582-590, 2017.
[17] M. Khan, and M. Ali, "Use of glass and nylon fibers in concrete for controlling early age micro cracking in bridge decks", Construction and Building Materials", Vol. 125. Pp. 800-808, 2016.
[18] S. Lee, "Effect of Nylon Fiber Addition on the Performance of Recycled Aggregate Concrete", Applied Sciences, Vol. 9. Pp. 1-14. 2019.
[19] Y. Qin, X. Zhang, J. Chai, Z. Xu, and S. Li, "Experimental study of compressive behavior of polypropylene fiber reinforced and polypropylene fiber fabric reinforced concrete", Cnstructin and Building Materials, Vol. 194, pp. 216-225, 2019.

[20] F. M. Z. Hossain, M. Shahjalal. K. Islam, M. Tiznobaik, and M. S. Alam, "Mechanical properties of recycled aggregate concrete containing crumb rubber and polypropylene fiber", Construction and Building Materials, Vol. 225, pp. 983-996, 2019.

[21] F. S. Khalid, J. M. Irwan, M. H. W. Ibrahim, N. Othman, and S. Shahidan, "Performance of plastic wastes in fiber-reinforced concrete beams", Construction and Building Materials, Vol. 183, pp. 451-464, 2018.

[22] F. Grzymski, M. Musial, and T. Trapko, "Mechanical properties of fibre reinforced concrete with recycled fibres", Cnstruction and Building Materials, Vol. 198, pp. 323-331, 2019.

[23] E. Gulbandilar, and Y. Kocak, "Prediction of Splitting Tensile Strength of Concrete Containing Zeolite and Diatomite by ANN", International Journal of Economic and Environmental Geology, Vol. 8, Pp. 32-40, 2016.

[24] P. Aggarwal, R. Siddique, Y. Aggarwal, and S. M. Gupta, "Self-Compacting Concrete Procedure for Mix Design", Leonardo Electronic Journal of Practices and Technologies, Vol. 12, pp. 15-24, 2008.

[25] European Federation of National Trade Associations Representing Producers and Applicator of Specialist Building Products (EFNARC), "Specification and Guidelines for Self-Compacting Concrete", Hampshire, U.K, 2002.

[26] ASTM International, "ASTM C 39/C 39M: Standard test method for compressive strength of cylindrical concrete specimens", USA, 2018.

[27] ASTM International, "ASTM C 46/C 46M: Standard test method for splitting tensile strength of cylindrical concrete specimens", USA, 2004.

[28] ASTM International, "Standard test method for flexural strength of concrete (using simple beam with center-point loading)", USA, 2016..

Analysis of concrete mixtures with marble waste material as cement substitution and glass waste as sand substitution on concrete compressive strength

F Zulkarnain[1,*], J Hadipramana[1] and Krisnianda[2]
[1]Lecturer, Civil Engineering, University of Muhammadiyah Sumatera Utara, Medan, Indonesia.

[2]Freshgraduate, Civil Engineering, University of Muhammadiyah Sumatera Utara, Medan, Indonesia.

[*]Email: fahrizalzulkarnain@umsu.ac.id

Abstract. Concrete technology that continues to develop produces high quality concrete which will be a problem of increasing concrete needs. Along with the soaring price of cement as the main ingredient in making concrete, the cost of making concrete is a problem that needs to be sought as a solution for technological development in the field of building construction. Therefore making concrete mixes by suppressing the amount of cement used can be one of the solutions to the problem. The many needs of the Indonesian community for buildings cannot be separated from the increasing demand for building materials. One of them is marble and glass which in the process there is trash or waste from the results of the remains of the pieces or discarded without any prior management. Therefore the need for efforts that can be made to handle the amount of marble or glass waste by utilizing it as a filler in making concrete. This research is to find out the quality of compressive strength of concrete mixed with marble waste. This study used a compressive strength of 20 MPa plan after the study was carried out obtained the average compressive strength of 10% marble concrete waste at 14 days at 22,35 MPa and at 28 days at 23,20 MPa. The average compressive strength value of glass waste concrete 8% at 14 days is 22,64 MPa and at 28 days is 23,48 MPa. Meanwhile, the average compressive strength value of the concrete combination of 10% marble waste + 8% glass waste at 14 days is 22,92 MPa and at 28 days is 23,77 MPa it can be concluded that all concrete variations in marble waste of 10% and glass waste of 8% have increased. This proves that marble and glass waste can be used in certain variations.

1. Introduction

Concrete is a composite material consisting of cement, coarse aggregate and fine aggregate which has a certain ratio and mixed with water as a hardening aid which when stirred with water and without any additives will have a certain solid mass and quality. Concrete is also a mixture of cement, coarse aggregate, fine aggregate, water and with or with additional materials with a certain ratio that will from fresh concrete. The use of concrete is more desirable because it is relatively strong, easily formed, and more economical when compared to construction using steel, wood or using other materials. The development of the construction world in Indonesia at this time is very impacting on the increasing use of concrete as a material in strengthening structures, in addition to that technology in concrete also always experiences more dynamic development. Fine and coarse aggregates are referred to as mixed coarse materials which are the main components of concrete [1].

Along with the soaring price of cement as the main material for making concrete, the cost of making concrete is a problem that needs to be solved in order to develop technology in the field of construction, especially in the cost of making a building structure. For the reason, it is necessary to have a substitute for cement in the manufacture of concrete or just a filler to reduce the amount of cement needed in the manufacture of concrete, but it does not reduce the quality of concrete so that it still meets the requirements in construction work. The use of marble waste is based on research [2]. The study was conducted by utilizing marble as a partial cement filler with a composition of 10% of the volume of cement. In general, the most marble content is Calcium Oxide (CaO) and a little Magnesium Oxide (MgO), Alumunium Dioxide (Al_2O_3) and Silicon Dioxide/Silicate (SiO_2) marble chemical content is mostly also found in cement. Whereas for the use of glass waste is based on [3]. The study was conducted utilizing glass waste as partial sand filler with a percentage variation of 8% of the volume of fine aggregate. Chemical content that is in the glass in the form of Silicate (SiO_2), Calcium Oxide (CaO), Magnesium Oxide (MgO), Aluminium Dioxide (Al_2O_3) dan Soda Fluks (Na_2O). For that the authors conducted research using marble waste and glass waste in a concrete mixture, to determine the compressive strength of concrete.

2. Review the library Concrete Understanding

In construction, is a composite building made of a material consisting of a mixture of cement, water, aggregates (coarse and fine) and additives when needed. Normal concrete in concrete which has a weight of 2200-2500 kg/m³ by using natural aggregates which are broken down or without being broken down [4].

Concrete itself is a function of the preparation material consisting of cement or portland cement, coarse aggregate, fine aggregate, water and additives or admixture. Concrete also has several advantages compared to other construction materials, including having a large compressive strength, resistant to fire, easily formed, no special expertise is required in its manufacture, and raw materials are easy to obtain so that concrete contains about 1%-2% air cavity, cement paste (cement and water) around 25%-40%, and aggregate (fine aggregate and coarse aggregate) around 60%-75% [5].

Concrete preparation materials include water, portland cement, coarse aggregates and fine aggregates as well as added materials, where each constituent material has different functions and influences. Important properties, in general, are also good. The factors that influence the compressive strength of concrete consist of the quality of the constituent material, the value of the cement water factor, aggregate gradation, maximum aggregate size, method of working (mixing, transportation, compaction, and maintenance) and the age of concrete [6].

2.1. Marble

Marble or marble is the result of metamorphic rocks or metamorphic rock. The influence of temperature and pressure produced by endogenous forces causes recrystallization of these rocks to form various foliation and non-foliation. Due to the recrystallization of the structure of origin of the rock to form a new texture and grain order.

Figure 1. Before and after the marble is processed.

Marble waste used in this study is the result of the rest of the pieces of whole marble carried out in building work. The pieces of marble are then pounded to get the ashes, then the ashes are sieved using sieve number 200.

Chemical content in marble fractions containing 55,07% Calcium Oxide (CaO) and other chemical elements [7]. The results of chemical analysis of marble shards can be seen in Table 1.

Table 1. Chemical contents of marble fragments [7].

No.	Chemical Elements	Ingredients (%)
1.	Silica Dioxide (SiO_2)	0,13
2.	Aluminium Dioxide (AlO_3)	0,31
3.	Oxide Ferries (FeO_3)	0,04
4.	Calcium Oxide (CaO)	55,07
5.	Magnesium Oxide (MgO)	0,36
6.	Potash (K_2O)	0,01
7.	Sulfur Trioxide (SO_3)	0,08
8.	(LoI)	44

Calcium Oxide (CaO) is the biggest chemical element in the marble content, which is 55,07% the same as the basic material for making portland cement, so that marble can serve to increase the binding distribution in the concrete mixture.

2.2. Glass

Glass bottles are products of the chemical industry that are most familiar with our daily lives. In terms of physical glass is a very cold liquid. So called because the structure of the particles are arranged far apart from each other as in liquid but they themselves are solid. This occurs due to the rapid cooling process, so Silica particles do not have time to arrange themselves regularly. In terms of chemistry, glass is a combination of various non-volatile inorganic oxides, which are produced from the decomposition and melting of alkaline and alkaline earth compounds, sand and various other preparations. Typical glass properties are influenced by the unique Silica (SiO_2) and formation process.

Glass is a substance made by cooling down melted materials. Not in the form of crystals but remains in a hollow condition. Waste glass is usually separated based on the color of the glass, in general, the glass is divided into 3 colors:

a. Clear/colorless, usually used as a household appliance.
b. Green, usually used as a beer or wine bottle.
c. Chocolate, usually used as a soft drink bottle.

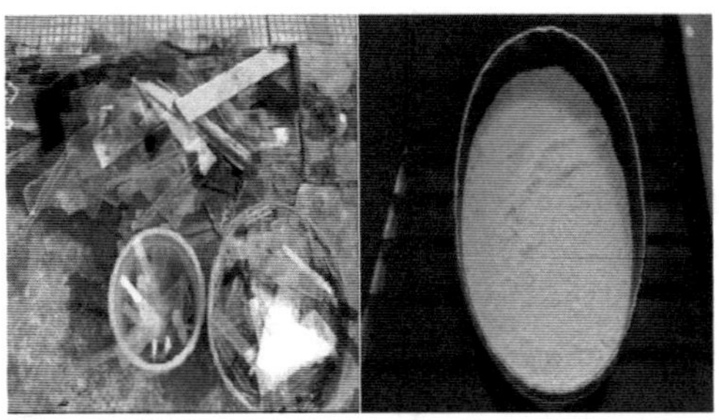

Figure 2. Before and after the glass is processed.

Glass waste can be obtained from the remaining pieces of glass in building work such as window glass or others. In this study, the glass used was clear/colorless. The broken glass pieces are then crushed which will later become powder, then sifted using a filter number 30 held in filter number 50.

Chemical content in broken glass contains 73,5% Silica Oxide (SiO_2) and other chemical elements [8]. The results of chemical analysis of broken glass can be seen in Table 2.

IOP Conf. Series: Materials Science and Engineering **771** (2020) 012058 doi:10.1088/1757-899X/771/1/012058

Table 2. Chemical content of broken glass (Setiawan, 2006).

No	Chemical Elements	Ingredients (%) Type of Glass (Clear Glass)
1	Silica Dioxide (SiO_2)	73,2-73,5
2	Aluminium Dioxide (AlO_3)	1,7-1,9
3	Sodium Oxide + Potash (Na_2O + K_2O)	13,6-14,1
4	Calcium Oxide + Magnesium Oxide (CaO + MgO)	10,7-10,8
5	Sulfur Trioxide (SO_3)	0,2-0,24
6	Oxide Ferries (Fe_2O_3)	0,04-0,05
7	Chromium Oxide (Cr_2O_3)	-

Many potentially beneficial things from using glass as a concrete aggregate, between others:

a) Has a high degree of durability, considering that glass is a material that does not absorb water.

b) Glass has a high resistance to abrasion and this characteristic is a rare characteristic found in other natural aggregates, as for the use of additives for natural aggregates to achieve the same strength.

3. Data Analysis And Discussion

3.1. Concrete Mix Data

In this case, the author will analyze the data that has been obtained during the study in order to obtain the values of the aggregate basic examination results listed in Table 3 below. These data can be used for planning concrete mix (Mix Design) with a compressive strength required of 20 MPa using the method (SNI 03-2834-2000).

Table 3. Aggregate base inspection data.

No	Basic Inspection Data	Values
1	Fine aggregate specific gravity	2,57 gr/cm^3
2	Coarse aggregate specific gravity	2,70 gr/cm^3
3	Fine aggregate mud content	3 %
4	Coarse aggregate mud content	0,83 %
5	Weight of fine aggregate content	1,15 gr/cm^3
6	Weight of coarse aggregate content	1,69 gr/cm^3
7	Fine aggregate modulus of refinement	2,73 %
8	Coarse aggregate modulus of refinement	7,28 %
9	Fine aggregate water content	2,22 %
10	Coarse aggregate water content	0,86 %
11	Absorption of fine aggregates	1,73 %
12	Absorption of coarse aggregates	0,72 %
13	Rough aggregate wear	16,36 %
14	The value of the plan slump	30-60 mm
15	Maximum aggregate size	40 mm

3.2. Concrete Compressive Strength Testing

Concrete compressive strength testing is performed when the concrete is 14 days old, and 28 days old, using a press machine with a capacity of 1500 KN, the test object to be tested is cylindrical with a diameter of 15 cm and a height of 30 cm with a total of 24 specimens, with grouping test specimens according to the variation of the mixture. Concrete compressive strength testing is done to get a picture of the quality of concrete, and also proof of the Mix Design that is made based on the quality of the concrete that has been planned.

IOP Conf. Series: Materials Science and Engineering **771** (2020) 012058 doi:10.1088/1757-899X/771/1/012058

Mix ratio for 1 cylinder test specimen,

Cement	: Sand	: Broken stone	: Water
2,09 kg	: 3,39 kg	: 6,56 kg	: 0,88 kg

Based on the comparison above explains the amount of cement weight for 1 cylindrical specimen is 2,09 kg, the weight of marble waste is 10% of the cement weight for 1 cylindrical specimen which is 0,209 kg. Hence, the weight of cement for 1 cylinder test object is 1,881 kg.

Based on the comparison above explains the amount of cement weight for 1 cylindrical specimen 3,39 kg – the weight of sand weight 8% of the sand weight for 1 cylinder test object that is 0,271 kg. Hence, the weight of sand for 1 cylinder test object is 3,12 kg.

3.3. Normal Compressive Strength of Concrete

Table 4. Normal concrete compressive strength test results.

Concrete age 14 days			
Test object	Compressive load (P) (kg)	A = 176.71 cm² $f'_c = (P/A)$	Estimated 28 days $f'_c/0,88$ (MPa)
I	43500	24,62	27,97
II	33000	18,67	21,22
II	34500	19,52	22,19
	f'_c average (MPa)	20,94	23,79
Concrete age 28 days			
Test object	Compressive load (P) (kg)	A = 176.71 cm² $f'_c = (P/A)$	Estimated 28 days $f'_c/1,00$ (MPa)
I	42000	23,77	23,77
II	45000	25,47	25,47
II	30000	16,98	16,98
	f'_c average (MPa)	22,07	22,07

Based on the results of normal concrete compressive strength test 14 days and 28 days, the average value of 14 days of normal concrete compressive strength is 20,94 MPa, 28 days is 22,07 MPa.

3.4. Compressive Strength of Concrete Mixture of 10% Marble Waste

Table 5. Concrete compressive strength test results of mixed marble waste 10%.

Concrete age 14 days			
Test object	Compressive load (P) (kg)	A = 176.71 cm² $f'_c = (P/A)$	Estimated 28 days $f'_c/0,88$ (MPa)
I	36000	20,37	23,15
II	40500	22,92	26,04
II	42000	23,77	27,01
	f'_c average (MPa)	22,35	25,40
Concrete age 28 days			

Test object	Compressive load (P) (kg)	A = 176.71 cm² f'_c = (P/A)	Estimated 28 days f'_c/1,00 (MPa)
I	40000	22,92	22,92
II	39000	22,07	22,07
II	43500	24,62	24,62
	f'_c average (MPa)	23,20	23,20

Based on the results of the compressive strength test of concrete with a mixture of marble waste 10% age 14 days and 28 days, the average value of compressive strength of concrete with marble waste 10% age 14 days is 22,35 MPa, age 28 days is 23,20 MPa.

3.5. Compressive Strength of Concrete Mixture of 8% Glass Waste

Table 6. Concrete compressive strength test results of mixed glass waste 8%.

Concrete age 14 days			
Test object	Compressive load (P) (kg)	A = 176.71 cm² f'_c = (P/A)	Estimated 28 days f'_c/0,88 (MPa)
I	45000	25,47	28,94
II	42000	23,77	27,01
II	33000	18,67	21,22
	f'_c average (MPa)	22,64	25,72
Concrete age 28 days			
Test object	Compressive load (P) (kg)	A = 176.71 cm² f'_c = (P/A)	Estimated 28 days f'_c/1,00 (MPa)
I	49500	28,01	28,01
II	39000	22,07	22,07
II	36000	20,37	20,37
	f'_c average (MPa)	23,48	23,48

Based on the results of the compressive strength test of concrete with a mixture of glass waste 8% age 14 days and 28 days, the average value of compressive strength of concrete with glass waste 8% age 14 days is 22,64 MPa, age 28 days is 23,48 MPa.

3.6. Compressive Strength of Concrete Mixture Marble Waste 10% + Glass Waste 8%

Table 7. Compressive strength test results of concrete mixture marble waste 10% + glass waste 8%.

Concrete age 14 days			
Test object	Compressive load (P) (kg)	A = 176.71 cm² f'_c = (P/A)	Estimated 28 days f'_c/0,88 (MPa)
I	33000	18,67	21,22
II	42000	23,77	27,01
II	46500	26,31	29,90
	f'_c average	22,92	26,04

IOP Conf. Series: Materials Science and Engineering **771** (2020) 012058 doi:10.1088/1757-899X/771/1/012058

(MPa)

Concrete age 28 days			
Test object	Compressive load (P) (kg)	A = 176.71 cm² f'_c = (P/A)	Estimated 28 days f'_c/1,00 (MPa)
I	51000	28,86	28,86
II	39000	22,07	22,07
II	36000	20,37	20,37
	f'_c average (MPa)	23,77	23,77

Based on the results of the compressive strength test of concrete with a mixture of marble waste 10% + glass waste 8% age 14 days and 28 days, the average value of concrete compressive strength with a mixture of marble waste 10% + glass waste 8% age of 14 days is 22,92 MPa, the age 28 days is 23,77 MPa.

From the compressive strength test results obtained compressive strength value for each variation of mixtures with different concrete ages, as attached in Figure 3, as follows:

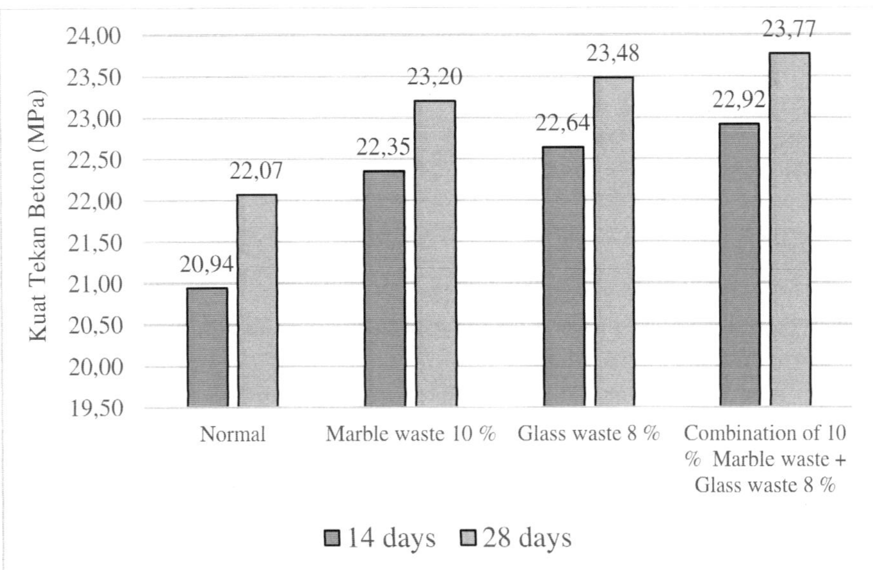

Figure 3. Graph of average compressive strength at 14 days and 28 days.

From the picture above we can see an increase in the compressive strength of concrete from the age of 14 days and 28 days on each variation of concrete mixes, all have a compressive strength value of concrete according to the quality of the concrete plan of 20 MPa.

4. Discussion

When compared between the compressive strength values of normal concrete with concrete that uses a mixture of marble waste 10%, glass waste 8%, and marble waste mixture 10% + glass waste 8% for concrete age 14 days and 28 days, it can be seen an increase or not on the compressive strength value of concrete that uses a mixture of materials in each variation.

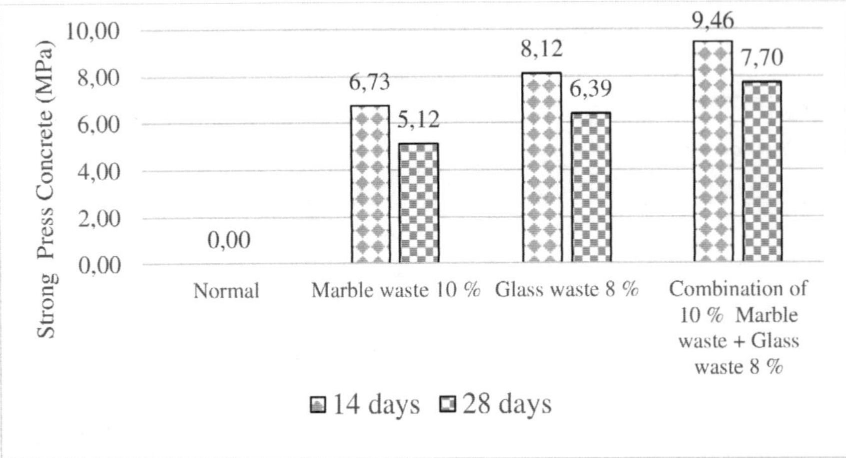

Figure 4. Graph of the ratio of the percentage increase in compressive strength of normal concrete at the age of 14 days and 28 days.

From the picture above it can be seen that the percentage of compressive strength of concrete combined with 10% marble waste + 8% glass waste has increased, with a percentage of 9,46% of normal concrete at 14 days concrete age and at 28 days concrete age has increased, with a percentage of 7,70% of normal concrete. The increase in compressive strength of glass waste concrete has increased, with a percentage of 8,12% of normal concrete at 14 days concrete age and at 28 days concrete age has increased, with a percentage of 6,39% of normal concrete. The increase in compressive strength of marble waste concrete has increased, with a percentage of 6,73% of normal concrete at 14 days concrete age and at 28 days concrete age has increased, with a percentage of 5,12% of normal concrete. In this case, it can prove that marble waste as a partial filler of cement and glass waste as a partial filler of sand can increase the compressive strength of a concrete plan.

5. Conclusion

From the results of research and testing of concrete compressive strength age 14 days and 28 days, several conclusions can be drawn including:

1. As for the results of the compressive strength testing of 10% marble waste as a partial filler of cement experienced a percentage increase of 6,73% at 14 days and 5,12% at 28 days of normal concrete. While the glass waste of 8% as a partial filler of sand also increased by 8,12% at the age of 14 days and 6,39% at 28 days of normal concrete.
2. As for the results of the compressive strength test of the combination of marble waste 10% + glass waste 8% as a partial filler of cement and sand experienced the most optimum percentage increase of 9,46% at 14 days and 7,70% at 28 days of concrete normal. Based on the results of research on the influence of the use of marble waste 10% and glass waste 8% as filler material for concrete mix is the reduced amount of cement and sand in the concrete mixture.

Acknowledgements

Thank you to Dr Fahrizal Zulkarnain as a Supervisor I and thank you to Dr Josef Hadiprmana as a Supervisor II who has provided direction and input to this study. Thank you to Munawar Alfansury Siregar, S.T, M.T as the Dean of the Faculty of Engineering, University of Muhammadiyah Sumatera Utara.

References

[1] Mulyono, T. (2004). "*Teknologi Beton*". Penerbit: Andi, Yogyakarta.

[2] Istiqomah & Kurnia, S. (2013). "*Pengaruh Limbah Marmer Sebagai Bahan Pengisi Pada Beton (179S)*". *Jurnal Teknik Sipil* Istiqomah & Kurnia, S. Fakultas Pendidikan Teknik Sipil Universitas Pendidikan Indonesia, Bandung, Jawa Barat, 2013.

[3] Suhartini, A., Gunarti, A. S. S. & Hasan, A. (2014). "*Pengaruh Penambahan Tumbukan Limbah Botol Kaca Sebagai Bahan Subtitusi Agregat Halus Terhadap Kuat Tekan Dan*

Kuat Lentur Beton". *Jurnal Teknik Sipil* Suhartini, dkk. Fakultas Teknik Sipil Universitas Islam 45 Bekasi, Jawa Barat, *Vol. 2* No. 1 Januari 2014.

[4] Sugiyanto. (2000). "*Pengaruh Abu Terbang Sebagai Bahan Pengganti Semen Pada Beton Beragregat Halus Bottom Ash*". *Skripsi* Lincolen, K. Fakultas Teknik Universitas Lampung, Bandar Lampung, 2017.

[5] Mulyono, T. (2005). "*Teknologi Beton*". Penerbit: Andi, Yogyakarta.

[6] Tjokrodimuljo, K. (1996). "*Teknologi Beton*". Penerbit: Nafiri, Yogyakarta.

[7] Wihardi dkk. (2006). "*Penggunaan Limbah Batu Marmer Dari Gunung Batu Naitapan Kabupaten Timor Tengah Selatan Sebagai Alternatif Pengganti Agregat Pada Campuran Beton*". *Jurnal Teknik Sipil* Nauk, S. S. Fakultas Teknik Sipil Universitas Nusa Cendana, Nusa Tenggara Timur, *Vol. 1* No. 4 September 2012.

[8] Setiawan. (2006). "*Analisa Kuat Tekan Beton Dan Penyerapan Air Kombinasi Filler Abu Ampas Tebu Dan Botol Kaca Subtitusi Pasir* (*Studi Penelitian*)". *Skripsi* Harahap, R. S. Fakultas Teknik Sipil Universitas Muhammadiyah Sumatera Utara, Medan, 2018.

Optimization of the compressive strength of hair fiber reinforced concrete using central composite design

F D Santos[1,*], M A M Taguba[1,2], A R Alzona[1,2], R C S Basada[1], J M A Elbit[1], G L Nones[1], K P M Olido[1], E U G Ramos[1]

[1] Department of Civil Engineering, National University, Sampaloc, Manila 1008 Philippines
[2] Department of Environmental and Sanitary Engineering, National University, Sampaloc, Manila 1008 Philippines

Corresponding author: fdsantos@national-u.edu.ph

Abstract. In society today, improper waste disposal brings about numerous problems. One of the most common residual wastes found in the municipal solid waste is human hair. This waste, which is evidently continuing to accumulate in disposal streams poses a threat if not disposed properly. Several studies have been done in the search for possibilities of incorporating hair to construction materials. In this study, the compressive strength of Hair Fiber Reinforced Concrete (HFRC) was studied aiming to find the combination that would yield the optimum compressive strength. A parametric study on the effect of amount of hair (1%, 2% and 3%) as well as length of hair strand (0.5-inch, 1 inch and 1.5 inch) on the compressive strength of HFRC was done. The results of the parametric study were used to find the optimum compressive strength. Utilizing the Central Composite Design (CCD) of Response Surface Methodology (RSM), a mathematical model was produced relating the parameters and the combination of parameters that would yield the maximum compressive strength. Based on the results, the mathematical model obtained a Coefficient of Regression of 0.9807 showing that the probability plot of the residuals fits the regression line. As generated by the mathematical model, a combination of 3.2% amount of hair by volume of concrete and a hair strand length of 0.752 inch will yield an optimum compressive strength of 38.15 MPa.

1. Introduction

One of the most widely used construction materials in the construction industry is concrete due to its high compressive strength, long service life and low cost [1]. However, concrete has low crack resistance and has limited tensile strength. The use of concrete ranges from its utilization in the construction of buildings to its versatility in highway construction particularly in the construction of roadways [2]. Moreover, the development of infrastructures (residential, commercial and industrial) depends on the performance of concrete [3]. Researchers have devoted their time in finding materials which can be added to concrete in order to improve its qualities. Due to the advances of concrete technology, fiber reinforcement was introduced. The reliability of Fiber reinforced concrete has already been established and is widely implemented in various range of applications in the industry. It has also been observed in previous studies that fiber reinforced concrete has a better performance as compared to normal concrete [4].

One of society's major problems at present time is improper waste disposal. Often, inappropriate disposal of refuse (i.e. indiscriminate disposal of litter to streets) leads to clogging of drainages which is one of the culprits of street flooding. In all the residual wastes that are causing problems, human hair is one of the most overlooked. A common constituent found in municipal waste streams in most parts of the world is human hair [5]. Hair is characterized by its high tensile strength, thermal insulation and elastic recovery [6]. Studies have been done in efforts of managing our waste by means of incorporating fibers as composite in concrete [7]. Some of these studies are about the physical and mechanical properties of human hair [8-9]. There are also studies about the characterization of human hair and its applications but using it as reinforcement in concrete has not been adequately explored.

Response surface methodology (RSM) is a collection of mathematical and statistical techniques for empirical model building. By careful design of experiments, the objective is to optimize a response (output variable) which is influenced by several independent variables (input variables). RSM was developed to model experimental responses and then migrated into the modeling of numerical experiments [10]. The application of RSM to design optimization is aimed at reducing the cost of expensive analysis methods (e.g. finite element method or CFD analysis) and their associated numerical noise. The response can be represented graphically, either in the three-dimensional space or as contour plots that help visualize the shape of the response surface

Central Composite Design (CCD) is a five-level factorial design that is used in the calibration of the quadratic response model [11]. The design model is utilized when generating a model that would yield the relationship between different variable inputs and the optimum condition of the output of the experiment. This study aimed to generate the relationship between input variables such as the amount of hair fiber (1%, 3% and 5%) by volume and hair strand length (0.5 in, 1 in and 1.5 in) using Central Composite Design of Response Surface Methodology (RSM) and determine the optimum compressive strength of Hair Fiber Reinforced Concrete (HFRC).

2. Materials

The supply of hair used in the preparation of concrete cylinders was collected from hairdresser shops, parlors, and salons along the city of Manila. The hair samples were washed and air-dried before including it in the design mix. Ordinary Portland Cement of a local company, fine and coarse aggregates was used for preparing the concrete samples. The materials used in this research were no less than the materials used in actual practice. The samples consist of cement, coarse and fine aggregates and collected hair. The design mix is the process of determining the amount of cement, fine aggregate, coarse aggregate as well as the water/cement ratio and the target strength to be attained by the samples. The Class B concrete mix of 1:2.5:5 (cement: sand: aggregate) by volume with a water cement ratio of 0.55 was the basis for the proportions of the concrete. Hair fiber was incorporated to the concrete mix and was distributed well in order to prevent agglomeration as shown in Figure 1.

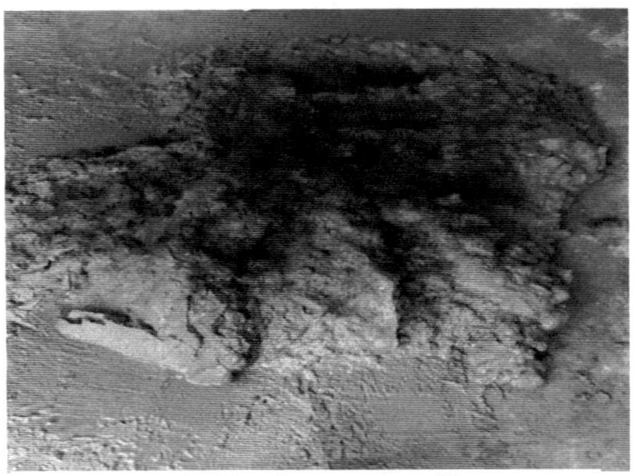

Figure 1. Hair fiber incorporated in concrete.

3. Experimental Methodology

3.1 Methodology on Parametric Samples

The concentration hair samples were varied as 1%, 3% and 5% per volume while the length of hair fibers was maintained and cut by 1-inch length for the initial samples, followed by 1.50-inch and 0.5-inch. The mix was poured into the molds in three layers as per the requirement of ASTM C-39 where each layer was compacted by 25 blows from the tamping rod. The sample was then lifted and dropped freely so that the air voids would be removed from the samples. The concrete samples were removed from the mold after 24 hours and were placed in a curing tank as shown in Figure 2 for curing periods of 7, 14, 21 and 28 days. The normal concrete samples were prepared in the same manner but without the incorporation of the hair fiber. ASTM C-39 was used in testing the compressive strength of the concrete cylinders.

Figure 2. The concrete specimens placed in the curing tank

3.2 Optimization using Central Composite Design

After the parametric study, the process of optimization was carried out. For this process, the Response Surface Methodology (RSM) Central Composite Design (CCD) which is a powerful tool in observing the interaction between the different parameters was utilized. RSM consist of a group of mathematical and statistical techniques for fitting linear response surface models in addition to a description of methods for the determination of optimum operating conditions [12]. Through this, a combination of parameters i.e. amounts of hair by volume and length of hair strand that would yield the highest compressive strength was produced (See Table 1). As suggested by the software, ten runs of dissimilar combinations of the different parameters were generated based on the numerical limits of the parameters presented. The software based the suggested values on the lower limit and higher limit of the parameters. The number of runs corresponded to the number of samples needed for the optimization process. After preparation and curing of these samples at different curing times (7, 14, 21 and 28 days). They were tested again to see the required compressive strength for optimization.

Table 1. Suggested combination of parameters for optimization

Runs	Amount of Hair (% by volume)	Hair Strand Length (in)
1	1	1.50
2	3	1.7071
3	1	0.50
4	3	1
5	0.1716	1
6	5.8284	1
7	3	0.2929
8	5	1.50
9	5	0.50
10	3	1

An additional function of RSM aside from generating the optimum condition for a certain factor, also predicts the equation of the model as a multiple regression analysis. Equation (1) is the multiple regression observation.

$$f(CS) = \beta_0 + \beta_1 A + \beta_2 B \qquad \text{(Equation 1)}$$

where β_0, β_1, and β_2 are the regression coefficients, while the factors of amount of hair by percentage of weight and hair strand length is denoted by A and B. A geometric illustration of the Central Composite Design is shown in Figure 3.

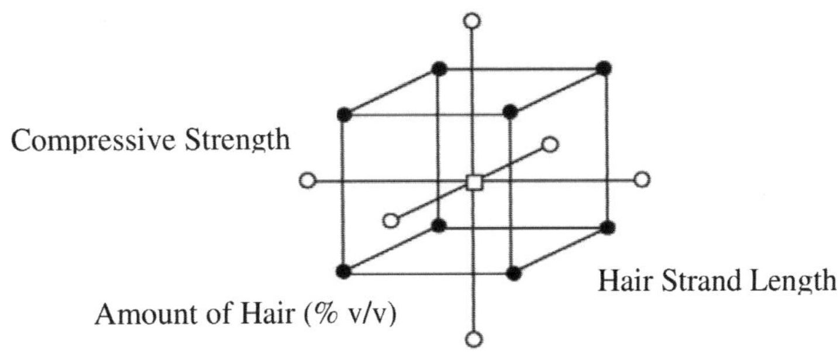

Figure 3. Geometric illustration of CCD

4. Results and Discussion

4.1 Parametric Study of HFRC

The compressive strength of the hair fiber reinforced concrete samples with varying amount of hair as obtained from the test results are shown in Figure 4. The highest compressive strength recorded at 38.10 MPa was obtained by the sample containing 3% hair by volume of concrete composite at its 28th day of curing as compared to 29.25 MPa of the sample without hair fiber. The addition of hair fiber led to a 30.26% increase in compressive strength compared to normal concrete.

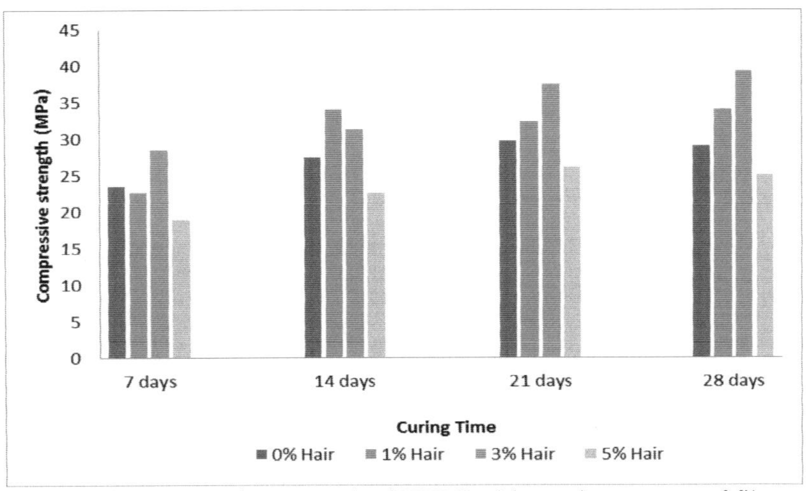

Figure 4. Compressive strength of HFRC with varying amount of fiber.

Meanwhile, the results of the compressive strength test of concrete samples with varying lengths of fiber reinforcement is shown in Figure 5.

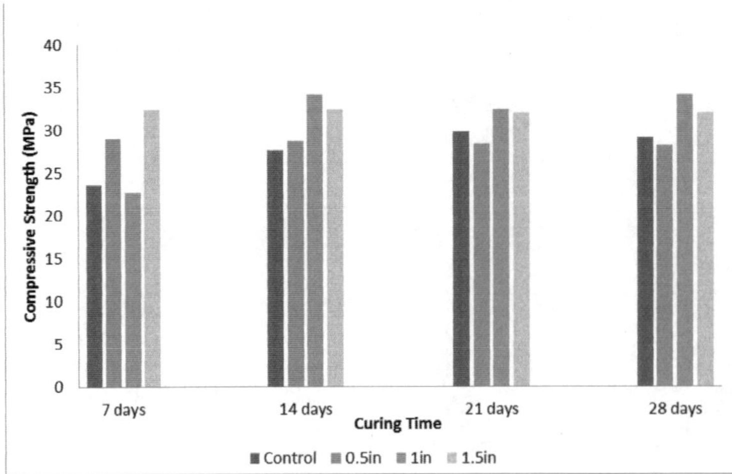

Figure 5. Compressive strength of HFRC with varying hair strand lengths.

Based on testing, results show that the highest compressive strength was recorded at 34.07 MPa during the 28th day of curing with the sample containing 1in hair strand length. Comparing this to the control sample or normal concrete, which was recorded at 31.04 MPa, there is a 9.75% increase in the compressive strength with the addition of hair fiber in the samples.

Furthermore, for both cases of amount of hair fiber and hair strand length in the samples, it can be observed that the compressive strength of the cylinders starts to decrease after reaching a certain peak at the 21st day of curing. The further increase in concentration of fiber greater than 3% possibly led to excessive fibers which may also increase the void spaces in the concrete that reduces the compaction of the mix thereby decreasing the compressive strength.

4.2 Optimization by Response Surface Methodology Central Composite Design

Ten runs were generated by the CCD corresponding to the samples that needed to be produced and tested. After undergoing the compressive strength as per ASTM C-39, the results of the ten samples are shown in Table 2.

Table 2. Compressive strength of samples generated by CCD

Runs	Amount of Hair (% by volume)	Hair Strand Length (in)	Compressive Strength (MPa)
1	1	1.50	30.18
2	3	1.7071	33.79
3	1	0.50	29.02
4	3	1	39.25
5	0.1716	1	24.36
6	5.8284	1	28.99
7	3	0.2929	31.38
8	5	1.50	31.67
9	5	0.50	28.70
10	3	1	39.25

By applying multiple regression analysis to the experimental data, the following mathematical model was found to represent the relationship between the responses indicated with the help of Sigma XL then obtained the generated model equation below:

RSM Regression Model: Compressive Strength = (39.25) + (0.964787789) * A: Amount of Hair (% by volume) + (0.941644994) * B: Hair Strand Length + (0.454894) * AB + (-6.2207545) * AA + (-3.2684865) * BB

Aside from the regression model, an estimate of parameters by Central Composite Design for Compressive Strength is shown on Table 3. As shown in the table, the p value of the parameter AB exceeds 0.05 (>0.05); thus, the significant effect on parameters on output response.

Table 3. Estimate of Different Statistical Parameters

Term	Coefficient	SE Coefficient	T	P
Constant	39.25	0.69095734	56.805	0.0000
A: Amount of Hair (% by volume)	0.964787789	0.34547867	2.793	0.0492
B: Hair Strand Length	0.941644994	0.34547867	2.726	0.0527
AB	0.454894	0.48858062	0.931052	0.4045
AA	-6.2207545	0.45702532	-13.611	0.0002
BB	-3.2684865	0.45702532	-7.152	0.0020

Shown in Table 4 is the Standard Error (SE) Coefficient and Variance Inflation Factor (VIF). The SE Coefficients range from 0.3 to 0.7 indicating that the predictions were precise and significant. SE Coefficient represents the average distance that the observed values fall from the regression line. Conveniently, it tells how wrong the regression model is on average using the units of the response variable. Smaller values are better because it indicates that the observations are closer to the fitted line. With the values of the SE Coefficient shown, the mathematical model is said to be adequate in this parameter.

Table 4. Statistical Performance Criteria

Term	SE Coefficient	VIF
Constant	0.690957345	
A: Amount of Hair (% by volume)	0.345478672	1
B: Hair strand Length	0.345478672	1
AB	0.488580624	1
AA	0.457025325	1.225
BB	0.457025325	1.225

Variance Inflation Factor (VIF) measure how much variance of the estimated regression coefficients are inflated as compared to when the predictor variables are not linearly related. It is used to describe the multicollinear property of the equation or the correlation between the predictors or input variables. Parallel to this, the variables being tested should not be highly correlated so as not to be subject to a domino effect of errors.

Based on the mathematical model, the coefficient of regression (R^2) is 0.9807 showing that the plot of residuals fits the regression line well. The results were used to produce a surface contour plot showing the interaction between the parameters and the compressive strength as shown in Figure 6 and Figure 7. The optimum compressive strength lies between 38 MPa to 40 MPa as shown in the figures.

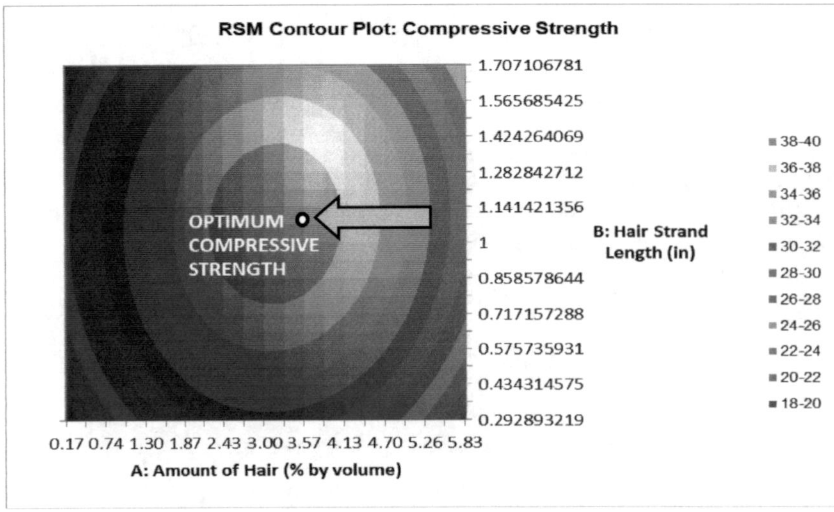

Figure 6. RSM Contour plot for compressive strength

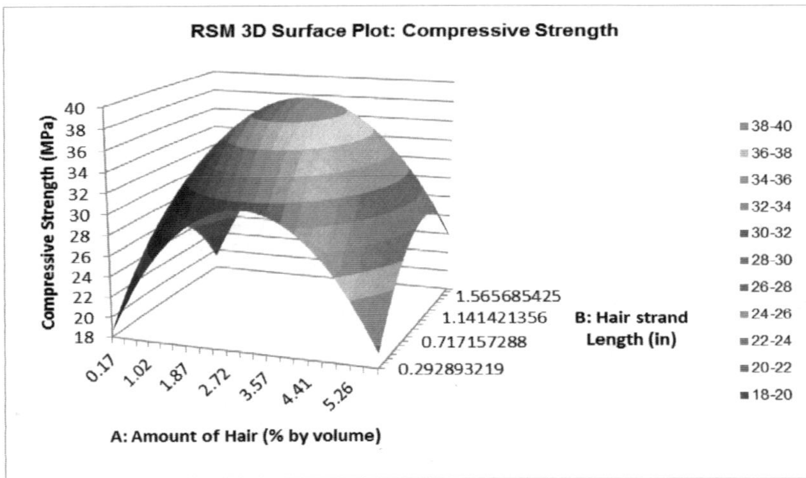

Figure 7. RSM 3D plot for compressive strength

From Figure 6, it can be seen that the parameters of amount of hair (% by volume) and the length of hair strand prove to be significant to the compressive strength because the contour plot yields an increasing trend in the compressive strength with the optimum output can be interpreted as the peak of the contour plot. The value of the concentration of hair which yields the maximum compressive strength is 3.2% and the corresponding length of hair strand is 0.752 inches and obtaining the optimum compressive strength at 38.15 MPa.

5. Conclusion

This paper aimed to evaluate the compressive strength of Hair Fiber Reinforced Concrete and compare it to normal concrete as a potential way of hair disposal and study the influence of concentration of hair and hair strand length on the compressive strength of concrete. The results of the parametric study were used to get the interaction between the parameters to get the combination that would yield the optimum compressive strength using Response Surface Methodology Central Composite Design. From the results, the following conclusions are made:

- Due to the addition of hair fiber, there is an evident increase in the compressive strength of the concrete samples compared to normal (conventional) concrete.

- With respect to the amount of hair added to the samples, the HFRC with 3% amount of hair by volume of concrete composite yielded the highest compressive strength at 38.10 MPa compared to other concentrations.
- Based on hair strand length added to the samples, the HFRC with 1-inch hair strand obtained the highest compressive strength at 34.07 MPa compared to the other hair strand lengths.
- As generated by the Central Composite Design, a combination of 3.2% amount of hair by volume of concrete and a hair strand length of 0.752 inch will produce an optimum compressive strength of 38.15 MPa.

The use of RSM to develop a regression model for the optimization of the compressive strength of concrete proves that it is a powerful tool to determine the relationship between independent variables. Also, the different statistical performance criteria were used to check the validity and adequacy of the model and it showed that the mathematical model was a good fit for the data. The results obtained from the study shows the feasibility of using hair fiber as a form of reinforcement in concrete. The findings also signify that hair fiber can be used in the production of an innovative and low-cost construction material while serving as a waste diversion for hair in order to address the waste disposal problem of the community.

Acknowledgments
The authors wish to acknowledge the efforts of the core group that helped in the success of this research, particularly Engr. Romel Carl Basada, Engr. Jan Michael Elbit, Engr. Gabriel Nones, Engr. Keanu Patric Olido and Engr. Ephraim Uriel Ramos for their kind help during the experimentation. The authors also would like to acknowledge National University Manila for the continuous support to in the research undertakings of its faculty members.

References
[1] J Ilya and C Cheow Chea (2017) *IOP Conf. Ser.: Mater. Sci. Eng.* **271** 012079
[2] Ishaya A, Oyemogum I M, Arinze A, Abah J. C, Polytechnic F and Namoda, K. (2016) Properties of Concrete Produced with Waste Bottle Caps (WBC) as a Partial Replacement of Coarse Aggregate and Orange Leaves Powder as Plasticizer, *The Journal of Civil and Environmental Research* **8**(7), 91–95.
[3] Dumne S. M. (2014) Effect of Superplasticizer on Fresh and Hardened Properties of Self-Compacting Concrete Containing Fly Ash, *American Journal of Engineering Research (AJER),* **Vol. 3** Issue-03, Pages: 205-211.
[4] Johnston C D (1994) Fiber-Reinforced Cement and Concrete *Advances in Concrete Technology* CANMET, (Ottawa: National Resources Canada) pp 603-673
[5] Gupta, A., & Gupta, A. (2014) Human Hair "Waste" and Its Utilization: Gaps and Possibilities. *Journal of Waste Management*, 2014, 1–17.
[6] Bishnu Prasad Nanda and Alok Satapathy 2017 *IOP Conf. Ser.: Mater. Sci. Eng.* **178** 012012
[7] Siddique R 2008 *Waste Material and By-Product in Concrete* (Berlin: Springer-Verlag Heidelberg)
[8] Velasco M V R, Dias T C S, Freitas A Z, Junior N D, Pinto C A S O, Kaneko T M, Baby A R 2009 BJPS **45** 153-161
[9] Seshadri I P, Bhushan B 2008 Acta Mater. **56** 774–781.
[10] Uca et al 2018 Daily Suspended Sediment Discharge Prediction Using Multiple Linear Regression and Artificial Neural Network J. Phys.: Conf. Ser. **954** 012030
[11] Rakesh Potluri et al 2016 IOP Conf. Ser.: Mater. Sci. Eng. **149** 012148
[12] Box, G. E. P., and Draper, N. R., (1987), Empirical Model-Building and Response Surfaces, John Wiley & Sons, New York, NY.

IOP Conf. Series: Materials Science and Engineering **771** (2020) 012060 doi:10.1088/1757-899X/771/1/012060

Evaluation of the implementation of the motorcycle lane based on the riding experience

T Kurniati[1,*], Purnawan[1], Yosritzal[1] and E E Putri[1]

[1]Civil Engineering Department, Andalas University, Padang, Indonesia

[*]Corresponding author e-mail: titi@eng.unand.ac.id

Abstract. Currently, two-wheeled vehicles or motorcycle dominate more than 80% of motorised vehicles in Indonesia. A large number of motorcycles result in complicated traffic operations and reduced levels of safety. Lanes separation is the best engineering solution to overcome this problem. Several cities in Indonesia have implemented special motorcycle lanes. The results of the study stated that this program was not successful. This paper tries to evaluate the application of motorcycle lanes in terms of the user's riding experience. The Sleman Outer Ring Road in Yogyakarta which has implemented a special motorcycle lane is the location of this study. Data collection was carried out through online questionnaire interviews. Respondents provide answers on a choice of four (4) scales, namely, strongly agree, agree, disagree, and strongly disagree. The method of analysis used for evaluating the use of a motorcycle lane is the descriptive statistical method and the chi-square test. Based on the experience of riding on an exclusive motorcycle lane, assessment of the special motorbike lane geometric and traffic flow is good.

1. Introduction

Motorcycles are the most desirable mode of transportation in developing countries, including Indonesia. The limited availability of mass public transport encourages people to look for alternative modes that can meet their needs. The choice of a motorcycle as a mode of transportation is due to its relatively high mobility and relatively affordable price. Data in 2017 in Indonesia, the number of motorcycles reached 113 million units, which is equivalent to 81.6% of the total number of motor vehicles in Indonesia.[1] The growth in the proportion of motorcycle in traffic flow in Indonesia increased from 25-45% in 1997 to 50-85% during 2011.[2]

Various harmful effects arise due to changes in the increase in the number of motorcycles. Among the decrease in space mean speed and accidents. In Indonesia, traffic accident growth of 11.64% per year is dominated by accidents involving motorcycles [3]. The best engineering solution to reduce the negative effect of increasing the volume of the motorcycle in traffic flow and saving motorcyclists from disasters is lane separation [3-5]. Some countries that have implemented special motorcycle lanes include Malaysia, Taiwan, etc. Neighboring Malaysia, which implemented this rule, succeeded in reducing the number of accidents involving motorcycles to 39% per year. [6]

Since 2006 motorcycle lane facilities have been implemented in several cities in Indonesia, including Jakarta, Sragen, Surabaya, Bandung, and Yogyakarta. Evaluation results show that motorcycle users have not optimally utilised the implementation of this facility.

The unsuccessful implementation of a special motorcycle lane in Indonesia attracted researchers to identify the characteristics of users of this facility. The purpose of this study was to determine the factors causing the failure of implementing a special motorcycle lane at the study location, explore the

IOP Conf. Series: Materials Science and Engineering **771** (2020) 012060 doi:10.1088/1757-899X/771/1/012060

desires and experience of motorcycle users on the geometric and traffic conditions on a special motorcycles lanes.

2. Literature review

2.1. Definition and design of motorcycle lanes

According to Act number 22, the Year 2009 regarding traffic and transportation in Indonesia, "motorcycle is a two-wheeled motor vehicle with or without accessories or a three-wheeled vehicle without accessories" [7]. Ministry of transportation ordinance of the Republic of Indonesia number 34 the year 2014 regarding road signs defines that "Lane is a long part of the road with or without road signs which has enough width so that one vehicle can run, except motorcycle"[8]. Motorcycle lane sign is another sign which is marked by a picture of a white motorcycle that is placed on the left of traffic. A motorcycle lane sign is placed on the same lane of other vehicles.

In Malaysia, there are two types of motorcycle lanes; they are exclusive motorcycle lanes and inclusive motorcycle lanes [9]. Exclusive motorcycle lanes are constructed specially to separate motorcycle and other vehicles, which is separated by fence or median. Meanwhile, an inclusive motorcycle lane or shared lane [10] separates motorcycle with other vehicles, which is bordered by the road sign.

Applying inclusive motorcycle lane in Indonesia has been done on some highways in Jakarta, and applying exclusive motorcycle lane Darmo street in Surabaya; Lingkar Utara Road, Sleman, Yogyakarta; Bypass Sukarno-Hatta arterial street, Bandung.

The application of motorcycle lanes in Indonesia was carried out in Jakarta in December 2006. The lane is designed to the left of the traffic lane and must turn on the lights even during the day. This program did not last long, in mid-2007, most motorbikes passed in the middle lane, even some on the right lane [11]. Putranto et al. (2011) evaluated the rate of effectiveness and the rate of exclusiveness in the implementation of motorcycle lanes in Jakarta (3 locations) and Sragen (2 places). Each site with a different geometric design so that the operating characteristics are different. Statistical test results on the correlation between the rate of effectiveness and rate of exclusiveness with V/C show unclear patterns [2].

Jalan Raya Darmo, in the city of Surabaya, has implemented a particular motorbike lane since November 2011. Before and after the implementation of the specific lane, road performance was stable with Level of Service (LOS) C. This performance condition remained until 2014 [12].

The results of the pilot survey on the implementation of special lanes and the sharing of motorcycle lanes implemented in Thailand succeeded in reducing speed. The study recommends a special line width of 3-4 m for exclusive motorcycle lanes and 3 m for the shared lane. With the decline in speed, motorcyclist safety is further improved, and road performance can be maintained at the level of service B [10].

2.2. Likert Scale

The Likert scale is commonly used in survey research. It is often used to measure respondents' attitudes by asking the extent to which they agree or disagree with a particular question or statement. Likert scale is a bipolar scale method that measures both positive and negative responses to a statement. Four choice scales are also sometimes used for Likert scale questionnaires that force people to choose one pole because "neutral" choices are not available. A typical scale might be "strongly agree, agree, disagree, strongly disagree." On the surface, survey data using the Likert scale may seem easy to analyse, but there are important issues for a data analyst to consider [13]

When responding to a question on a Likert scale, respondents determine their level of agreement with a statement by choosing one of the available choices. Four scale options with formats such as:

a. Positive Questions (+)
   Score 1. Very (disagree / bad / less)
   Score 2. No (agree /good) or less
   Score 3. (Agree /Good/Like)
   Score 4. Very (agree / good / like)
b. Negative Questions (-)
   Score 1. Very (agree / good / like)

IOP Conf. Series: Materials Science and Engineering **771** (2020) 012060 doi:10.1088/1757-899X/771/1/012060

Score 2. (Agree / Good / Like)
Score 3. No (agree / good /) or less
Score 4. Very (disagree / bad / less)

3. Methodology

3.1. Questionnaire Design

Data collection was carried out through online questionnaire interviews. Questionnaire questions are designed into three (3) sections. The characteristic of the questions for the three parts of the questionnaire is closed questions. Open-ended questions are facilitated in section 2 regarding user desires for the geometric of motorcycle lanes. The question in part 1 is about the personal characteristics of the respondent (gender, age, last education, car ownership, vehicles that are used more often). Part 2 of this question is intended to determine or evaluate the riding experience on the geometric road of motorbikes lane. The question in section 3 to evaluate the journey of motorcycle users is viewed from the aspect of traffic flow.

There were 26 questions given, consisting of 25 closed questions and 1 open question. Open-ended questions related to improvements that must be made on a special motorcycle lane. In part 1, 7 questions related to the personal characteristics of the respondent and five questions related to the characteristics of the trip on a special motorcycle lane. Part 2, In part 2, 4 questions evaluate the geometric (width, length, signs, and markings) of a motorcycle lane. In part 3, 9 questions regarding evaluating the response experience in terms of traffic flow when using exclusive motorcycle lanes. Answers to questions are provided on a choice of four (4) scales, namely strongly agree, agree, disagree, and strongly disagree. Questionnaires are created in google docs forms. The survey can be accessed via the link: https://intip.in/JalurSepedaMotor.

3.2. Data Collections and processing

The location of the study was conducted on the Sleman Outer Ring Road in Yogyakarta. The method of analysis used for evaluating the use of a motorcycle lane is the descriptive statistical method. The chi-square test is one approach for analysing the data in this way.

Evaluation of respondents' interpretations of riding experience on a motorcycle lane is the result of the value generated using the Index% formula.

$$\text{Index\%} = \text{Total Score} / Y \times 100 \quad (1)$$

$$I = 100 / \text{total Likert score} \quad (2)$$

Y = highest Likert score x number of respondents
I = score interval

Criteria for interpretation of scores based on intervals:

Value 0% - 24.99%	= strongly disagree
Value 25% - 49.99%	= disagree
Value 50% - 74.99%	= agree
Value 75% - 100%	= strongly agree

4. Results and Discussion

The survey results collected 80 responses. The personal characteristics of respondents, as shown in Table 1, the respondent is dominated by the age of 17-30 years with occupation as a student. Even though car ownership of the respondent is 36%, the type of vehicle that is used more frequently is motorcycle by 87.5%. Only 4% of respondents ride without a license.

Table 1. Personal characteristics of the respondent.

Characteristics		Percentage (%)
Gender	Male	55
	Female	45
Age	Age < 17	1.3
	17 ≤ age < 30	80
	30 ≤ age < 45	12.5
	45 ≤ age < 60	6.3
	Age ≥ 60	0
Last education	Primary school	0
	Junior high school	0
	Senior high school	31
	College	69
Occupation	employees	20
	Student	71
	Businessman	4
	housewife	0
	others	5
Car ownership	Owning	36
	Not owning	64
The type of vehicle that is more often used	Motorcycle	87.5
	Car, sometimes motorcycle	12.5
Driving license ownership	Yes	96
	No	4

Table 2 shows the respondents' experiences when riding at the motorcycle lane. From the total sample obtained, there are 60% of respondents using the motorcycle lane every day.

Table 2. Experiences in using motorcycle lanes

Experiences		Percentage(%)
Riding at the motorcycle lane	yes	95
	no	5
Frequency riding at motorcycle lanes	frequently	60
	sometimes	26
	infrequently	14
Time riding at motorcycle lane	peak hour in the morning	26
	off-peak hour in the morning	19
	peak hour in the noon	10
	off peak hour in the noon	10
	peak hour in the afternoon	19
	off peak hour in the afternoon	0
	in the night	16
Time-frequency riding at motorcycle lane	everyday	47.5
	once a week	17.5
	every two weeks	5
	infrequently	30
Improvements that must be done on a motorcycles lane at this time	- Uneven pavement construction due to many utility holes and damaged - need to be widened - street lighting - the car disturbs the motorcycle lanes	

The experience of the respondent riding on a motorcycle lane as shown in Figure 1. Most respondents (>50%) strongly agreed and agreed that the width of the motorcycle lane was sufficient to reach optimal speed, comfort, and safety. The same thing for the length of the motorcycle lane and the availability of signs and markers.

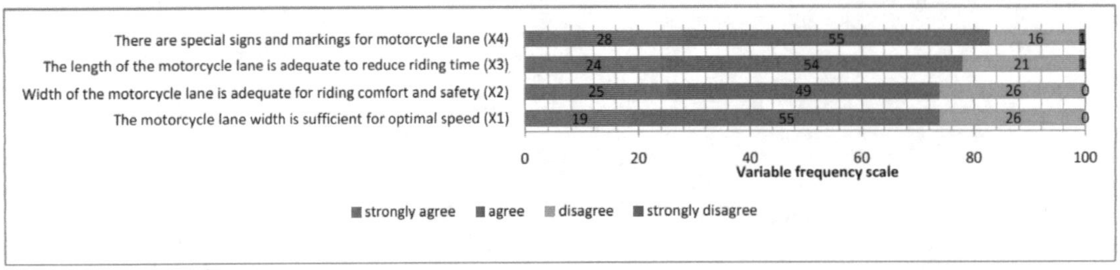

Figure 1. Respondent's experience riding on the geometry of a motorcycle lane.

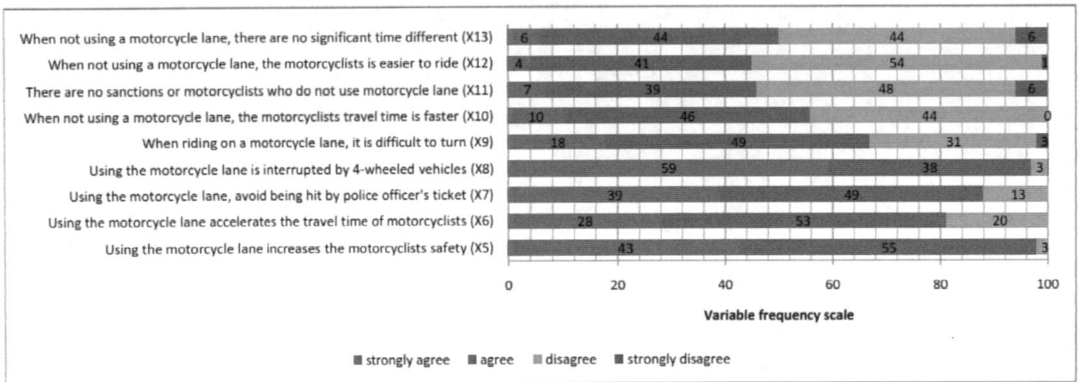

Figure 2. The respondent's riding experience on a motorcycle lane traffic flow.

Respondents' driving experience in terms of traffic flow parameters as shown in Figure 2. Questions X1-X9 are positive questions; the answers given are also positive. Respondents agree with the statement that using the motorcycle lane increases the safety/security of motorcyclists (X5), using a motorcycle lane accelerates the travel time of motorcycle riders (X6), using a motorcycle lane. You obey the rules and avoid being hit by a police officer's ticket (X7), When riding on a motorcycle lane, vehicles that are not allowed to pass through the motorcycle lane disturb the motorcyclists (X8). When riding on a motorcycle lane, it is difficult to turn (X9).

But on questions X10-X13, which are negative questions, only half the respondents expressed disagreement.

The results of determining the index% and its interpretation are recapitulated in Table 3. Questions 1-8 are positive questions, response X3-X4 strongly agrees, but response X1 and X2 agree. Questions 9-13 are negative, the response agrees. Questions 9-13 are negative; the answers agree. The hypothesis expects the results to disagree.

Table 3. Index% and interpretation.

Variable	Index%	Interpretation
X1	73.13	Agree
X2	74.70	Agree
X3	75.00	Strongly agree
X4	77.19	Strongly agree
X5	85.00	Strongly agree
X6	76.90	Strongly agree
X7	81.60	Strongly agree

IOP Conf. Series: Materials Science and Engineering **771** (2020) 012060 doi:10.1088/1757-899X/771/1/012060

X8	88.40	Strongly agree
X9	54.70	Agree
X10	58.40	agree
X11	63.13	agree
X12	63.13	agree
X13	62.50	agree

Chi-square statistical test results on the parameters of personal characteristics and variables X1-X13 are independent. Except, car ownership affects the X5 (using the motorcycle lane increases the safety/security of motorcyclists) and X6 (using a motorcycle lane accelerates the travel time of motorcycle riders) responses. The type of vehicle that is used more frequently affects the X11(when riding without using a motorcycle lane, there are no sanctions from the officers) response.

5. Conclusion

Based on the experience of riding on a motorcycle lane, users feel the benefits of this facility. Geometric conditions and traffic flow on the motorcycle lane are good, but the surface condition of the road pavement decreases the respondents' ratings. Car ownership affects the X5 (using the motorcycle lane increases the safety/security of motorcyclists), and X6 (using a motorcycle lane accelerates the travel time of motorcycle riders) responses. The type of vehicle that is used more frequently affects the X11(when riding without using a motorcycle lane, there are no sanctions from the officers) response.

References

[1] Central Bureau of Statistics 2018 Statistik Transportasi Darat, https://www.bps.go.id/publication/2018

[2] Putranto L S, Suardika G P, Sunggiardi R, Munandar A S and Lutfi I 2011 *Proc. of the Eastern Asia Soc. for Transportation Studies* vol 8

[3] Sulistio H (2018) The Open Transportation Journal, 12, 1-7.

[4] Hamid H , Sohadi R U R, Sadullah A F M and Ma'soem D M 2005 *IATSS Research* 29 Num.1 50-56.

[5] Le T Q and Nurhidayati Z A 2016 *Procedia Engineering* 142 292-298

[6] Khaidir N M, Manan M M A and Johari N M 2015 *Proc. of Conf. of ASEAN Road Safety 2015*, 80-85

[7] Anonymous 2009 Act of the Republic of Indonesia number 22 Year 2009 *Traffic and transportation*

[8] Anonymous 2014 Ministry of transportation ordinance of the Republic of Indonesia number 34 year 2014 *Road markings*

[9] Law T H and Sohadi R U R 2005 *Journal of the Eastern Asia Society for Transportation Studies*, **6**, 3372-3385

[10] Mama S and Taneerananom P 2016 *Engineering Journal* Vol 20 Issue 3 113-121

[11] Sunggiardi R and Putranto L S 2009 *Jurnal Transportasi* Vol.9 Num.2 117-126

[12] Hikmana A A, Djakfar L and Suharyanto A 2014 *Jurnal Rekayasa Sipil* Vol. 8 Num.3.

[13] Shane H 2010 How to Use the Likert Scale in Statistical Analysis

Acknowledgements

We are grateful for financial support from Riset Disertasi Doktor Dana BOPTN Tahun Anggaran 2019 under contract no.T/1/UN.16.17 /PT.01.03/RDD-Inovasi Sains/2019 Andalas University

Fluorescent concrete with strontium phosphorus powder and plastic waste

D D Purnama[1,*], T Iduwin[1], and R Hidayawanti[1]

[1]Sekolah Tinggi Teknik PLN, Lingkar Luar Barat Sreet, Duri Kosambi, Cengkareng, West Jakarta

[*]Corresponding author e-mail: dicki@sttpln.ac.id

Abstract. Concrete as a structural material can also function in the aesthetic field. The use of fluorescent concrete can beautify buildings, roads and other construction buildings with concrete material. The use of plastic waste has added value to this concrete because it can reduce environmental damage. So that beautiful, environmentally friendly concrete is expected to be created. This research was carried out in the hope of utilising strontium phosphorus powder and plastic waste to increase the value of concrete benefits. The research that will be carried out consists of several stages. First, the problem identification phase and the materials were making up the concrete, so that the characteristics of the material can be known. The second stage is the concrete mix design using SNI 03-2834-2000 guidelines. The third stage is making the compressive strength and concrete panel specimens carried out in the laboratory. The fourth step is testing the compressive strength and concrete panel specimens and analyzing the test results. Concrete test results show that the more the percentage of strontium phosphorus powder instead of cement, the lower the compressive strength of the concrete. For the percentage of concrete plus 4% plastic waste and 5%, strontium phosphorus powder has a compressive strength of 11.69 MPa compared to normal concrete 26.03 MPa. Then the longest time the fluorescent concrete produced by panel concrete plus cement paste 10% strontium phosphorus powder is for 3 hours.

1. Introduction

Concrete is generally only intended for construction materials in terms of strength only. However, this changed because of the discovery of fluorescent cement by Mexican scientist named Jose Carlos Rubio Avalos, who discovered cement that could glow at night. This cement is made from energy-saving material that can absorb sunlight and start emitting light at nightfall. Light-emitting cement products can last for 100 years and will shine for 12 hours at night. Every one square meter that maybe 3 mm thick, will cost around 60 to 70 US dollars (Rp. 780,000-Rp. 910,000) [1].

Strontium phosphorus powder material has a function that can be glow in the dark so that it can be used to make fluorescent concrete. However, for a review of strontium phosphorus content in mix design, compressive strength, and how long the light produced is still not much to discuss.

Another thing that is the focus of research to be carried out is the use of plastic waste that is increasing day by day. Jambeck data (2015) recorded that 187.2 million tons of plastic were produced in all regions of Indonesia and ranked second in the world [2]. The characteristic of the plastic that reflects light is a reason for the use of plastic in this study. It is expected that the light produced will be greater than with conventional concrete.

Based on the formulation of the problem, this study has the objective of how to use strontium phosphorus powder and plastic waste as a fluorescent concrete manufacturing material. So that it is known how long the fluorescent concrete can be glowing, the compressive strength of concrete using strontium phosphorus powder and plastic waste, and the strength comparison of concrete using strontium phosphorus powder and plastic waste with conventional concrete strength.

2. Literature review

Rismayasari (2012) examined the manufacture of concrete with a mixture of plastic waste and its characteristics using the percentage of plastic waste of 0%, 2%, 4%, 6%, 8% and 10% obtained the result that the highest value of compressive strength of 21.8 + 0, 2x106 N / m2 at the percentage of plastic waste 4% and the lowest 16 + 0.1x106 N / m2 at the percentage of plastic waste 10% [3].

Malek, et al. (2017) in their research on Post-event damage assessment of concrete using the fluorescent microscopy technique, explains that fluorescent is used to detect the level of damage caused by external loads. Fluorescent itself serves as a source of light energy to illuminate images of damage such as cracks seen using a microscope [4].

The test results showed that the FMWG has pozzolanic characteristics and using it as a mineral admixture in concrete, had a bad effect on workability, but improved the mechanical properties of concrete considerably at later ages. The optimum percentage of FMWG that gives the maximum values of compressive, splitting tensile and bond strengths is 10%. Results also showed that expansion due to the alkali-silica reaction was minimised obviously by the increase in the FMWG content [5].

Investigated mix compositions are based on silica fume and its combination with glass powder. Testing results indicate that silica fume replacement by additionally ground fluorescent glass (up to 50%), slightly increases consumption of water and delays the setting time of cement paste. All concrete mixes with glass powder show decreased early-age strength and considerable strength gain after long-term hardening. It is concluded, that the best way of glass application in High-Performance Concrete is to use fine ground glass powder together with silica fume as complex admixture [6].

Based on consideration of the references used in this test, it is planned that concrete has more properties than conventional concrete. Things to be achieved are concrete that can glow by utilising strontium phosphor powder material that can glow in the dark and environmentally friendly concrete by utilising the remaining plastic bottle waste to reduce environmental damage.

In this study, the percentage of plastic waste that will be used is 4%, and for the percentage of strontium-phosphorus powder is taken at 1%, 3% and 5% for concrete. The hope is the result of compressive strength and light levels produced under optimal conditions.

3. Methods

The research process is divided into several stages:

1. Preparation stage, collecting data and basic theories about the material to be used as well as data collection and preparation of tools and materials.
2. Stage of material testing
    a. Fine Aggregate
    b. Coarse Aggregate
    c. Cement
3. Stage of manufacture specimens (Mix Design)
    a. Determination of Concrete Mix Design (mix proportions)
    b. Slump test
    c. Manufacture of specimens for compression test.
    d. Manufacture of specimens for fluorescent concrete.
4. Curing Stage
   The specimens that have been made are left in a cylindrical mould for ± 24 hours after they have dried enough, the specimens are put into the sink for the planned age of 7, 14, and 28 days. For specimens, concrete panels are allowed to stand and watered periodically.
5. The testing of concrete cylindrical specimens and concrete panels
   This stage is carried out testing the compressive strength of concrete, aged 7, 14, and 28 days for cylindrical concrete specimens. For concrete panel test specimens, the measurement

conditions for long glowing concrete are reviewed in conditions illuminated by the sun, room lights, and by a flashlight.

6. Data Analysis Stage
   At this stage the data obtained from the compressive strength test results are analyzed with the help of a calculation program to get the relationship between the variables examined in the study.
7. Conclusion Stage
   This stage makes conclusions related to the research objectives from data that has been analyzed in the previous stage.

Table 1. Concrete Cylinder Specimens.

Percentage of plastic waste (%)	*Strontium phosphorus powder* (%)	Testing day -			Number of Test Objects	Variation Type
		7	14	28		
4	5	3	3	3	9	Variation 4
4	3	3	3	3	9	Variation 3
4	1	3	3	3	9	Variation 2
Normal Concrete		3	3	3	9	Variation 1
TOTAL		12	12	12	36	

Table 2. Concrete Panel Specimens.

Cement (%)	***Strontium phosphorus powder (%)***	**Number of Test Objects**
90	10	1
95	5	1
97	3	1
	Total	**3**

Concrete cylindrical specimens measuring 15 cm in diameter and 30 cm in height were used as many as 36 pieces with each variation of 9 test specimens. While the concrete panel test specimens measuring 1x1x0.025 m. Strontium phosphorus powder is used as a substitute for cement in the cement paste.

4. Results and Discussion

From the material data obtained, a mix design mix for 4 variations has been made which is explained in Table 1. The mix design procedure follows SNI 03-2834-2000. The mix design results obtained by concrete mix with material requirements as follow :

Table 3 Compressive Strength Test Results.

Type	Age of Concrete Cylinders (Days)	Compressive Strength (MPa)			Average Compressive Strength
	7	18.11	18.11	17.54	17.92
Variation 1	14	22.64	19.24	23.77	21.88
	28	27.16	26.03	24.90	26.03
Variation 2	7	7.36	11.32	11.32	10.00

	14	14.71	14.15	13.58	14.15
	28	14.71	16.98	14.71	15.47
	7	9.05	9.05	9.05	9.05
Variation 3	14	12.45	11.32	11.88	11.88
	28	13.58	14.71	13.58	13.96
	7	9.05	7.92	11.32	9.43
Variation 4	14	9.62	11.32	11.88	10.94
	28	12.45	11.32	11.32	11.69

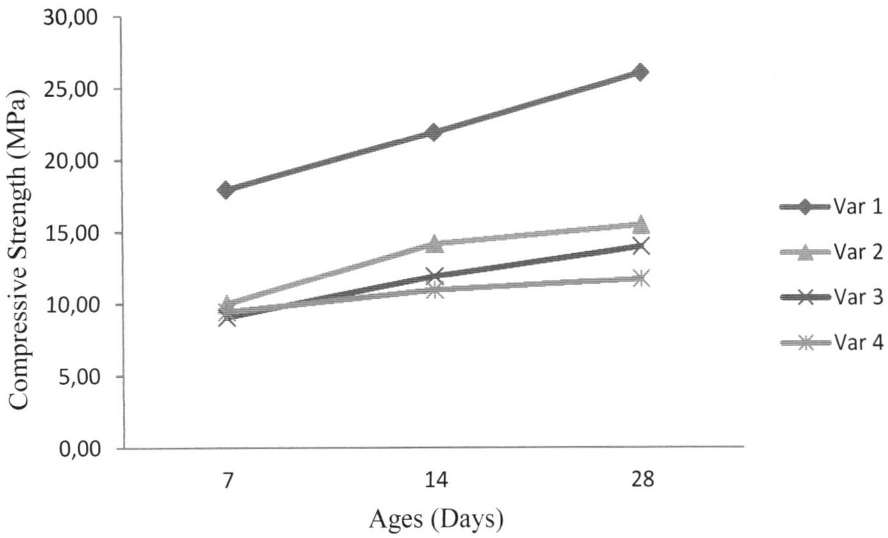

Figure 1. Comparison of Compressive Strength Variations.

The lowest compressive strength values obtained for variation 4 using concrete plus 4% plastic waste and 5% strontium phosphorus powder which has a compressive strength of 11.69 MPa while normal concrete is 26.03 MPa. The compressive strength value of the plan is 25 MPa.

Figure 2. Concrete Panel Specimens.

Concrete panels are made the same as cylindrical concrete variations 2, 3 and 4. The size of the concrete panels is 1x1 m and 2.5 cm thick. Preliminary test results show that concrete panels are almost invisible at all glowing. So it was decided to make cement paste with the percentage replacement of strontium phosphorus powder to cement at 3%, 5% and 10%. Figure 3 shows the condition of the concrete panel after being illuminated with a flashlight with a high lux level.

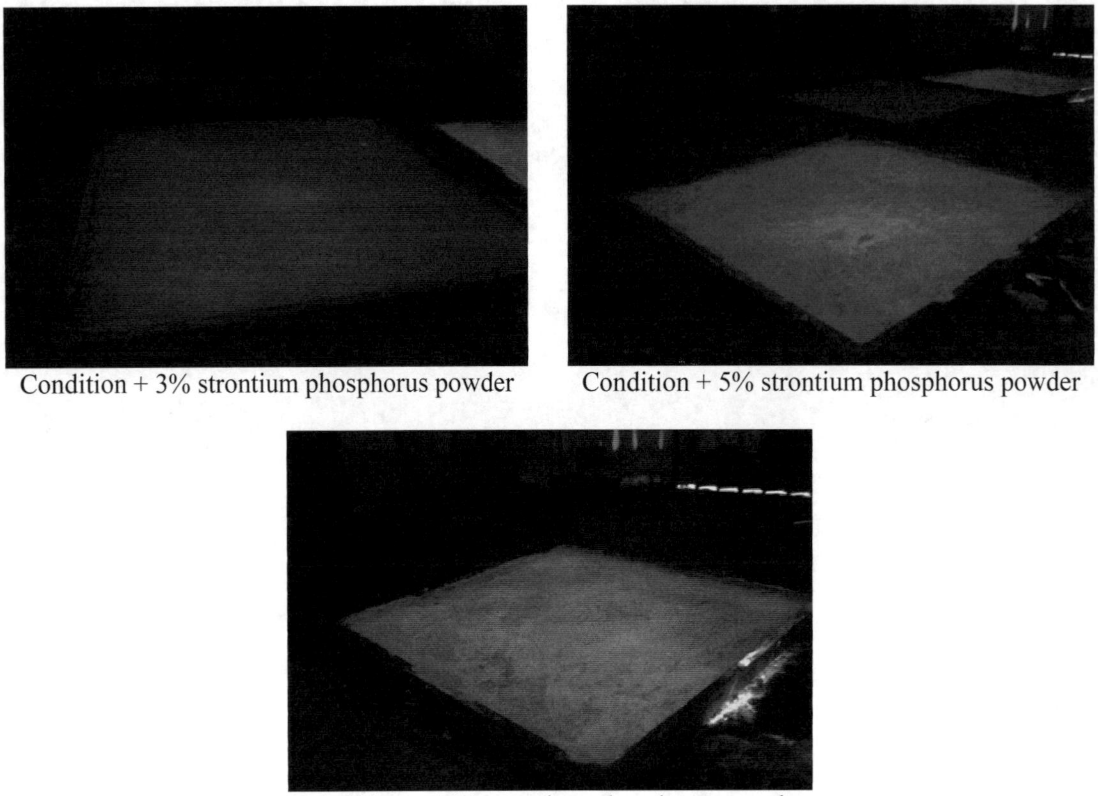

Condition + 3% strontium phosphorus powder

Condition + 5% strontium phosphorus powder

Condition + 10% strontium phosphorus powder

Figure 3. Conditions of Concrete Panels after a Flashlight.

Fluorescent conditions were reviewed in several concrete panel irradiation conditions. It aims to determine the effect of the irradiation process of concrete panels that have been coated by cement paste + strontium phosphor powder with the most visible time and fluorescent level. The best results are obtained by irradiation with a flashlight with a high lux level with a glowing period of 3 hours until the concrete panel is completely invisible to the eye.

Table 4. Test Results for Time and Level of Concrete Panel.

No	Condition	Length of fluorescent (hours)	Fluorescent level
1	By the sun	1	Almost Invisible
2	By the lamp	1	Almost Invisible
3	By Flashlight	3	Visible

5. Conclusion

The test results show that the more levels of strontium phosphorus powder as a substitute for cement, the lower the quality of concrete. This shows that strontium phosphorus powder does not have properties that can replace cement, so it is better to use it for cement paste. The lowest compressive strength values obtained for variation 4 using concrete plus 4% plastic waste and 5% strontium phosphorus powder which has a compressive strength of 11.69 MPa while normal concrete is 26.03 MPa. The compressive strength value of the plan is 25 MPa. For the long glowing time is best obtained by irradiation with a flashlight with a high lux level with a glowing period of 3 hours. The more levels of strontium phosphorus powder, the brighter the level of light produced.

References

[1] Teknik Material dan Metalurgi. Semen Glow In The Dark Bakal Ubah Dunia Konstruksi Global [Internet]. July, 19 2016 07:04 UTC+07:00 [quoted August, 16 2018]. Available from : https://materialmetalurgiku.blogspot.com/2016/07/semen-glow-in-dark-bakal-ubah-dunia/

[2] CNN Indonesia. Indonesia Penyumbang Sampah Plastik Terbesar Ke Dua Dunia. [Internet]. February, 23 2016 07:01 UTC+07:00 [quoted August, 16 2018]. Available from : https://www.cnnindonesia.com/gaya-hidup/20160222182308-277-112685/indonesia-penyumbang-sampah-plastik-terbesar-ke-dua-dunia/

[3] Rismayasari, Y., U, U., & Santosa, U. (2016). Pembuatan Beton dengan Campuran Limbah Plastik dan Karakterisasinya. *Indonesian Journal of Applied Physics*, *2*(02), 24. https://doi.org/10.13057/ijap.v2i02.1284

[4] Malek, A., Scott, A., Pampanin, S., & MacRae, G. (2017). Post-event damage assessment of concrete using the bubuk fosfor strontium microscopy technique. *Cement and Concrete Research*, *102*(October), 203–211. https://doi.org/10.1016/j.cemconres.2017.09.015

[5] Metwally, I. M. (2007). *Investigations on the Performance of Concrete Made with Blended Finely Milled Waste Glass*. *10*(1), 47–54.

[6] Sahmenko, G., Toropovs, N., Sutinis, M., & Justs, J. (2014). Properties of high performance concrete containing waste glass micro-filler. *Key Engineering Materials*, *604*, 161–164. https://doi.org/10.4028/www.scientific.net/KEM.604.161

Concrete Performance with Ground Granulated Blast Furnace Slag as Supplementary Cementitious Materials

R A T Cahyani[1,*], Y Rusdianto[1]

[1]University of Muhammadiyah Malang, Malang, Jawa Timur, Indonesia.

*rizkiatcahyani@umm.ac.id

Abstract. Supplementary Cementitious Materials (SCMs) are materials that contribute to the hardened concrete properties through hydraulic or pozzolanic activities. The use of Ground Granulated Blast Furnace Slag (GGBFS) as SMCs or partial replacement of cementitious components in concrete provides benefits both in terms of economy and performance improvement. However, GGBFS is not widely known in Indonesia, hence the available research for this product is still limited. In this study, the performance of concrete with GGBFS as a partial replacement for cement was investigated, with the percentage of GGBFS varying by 0%, 30%, 50% and 70% of cement weight. The mechanical properties of concrete being observed are compressive strength and elastic modulus. Porosity testing is carried out to assess the durability aspect of concrete. Compressive tests are carried out when the concrete age reached 7, 28, 56 and 90 days to assess the strength development of GGBFS based concrete. From the study, it was found that GGBFS can replace cement up to a percentage of 50%, with its compressive strength comparable to conventional concrete. However, the addition of GGBFS in concrete mixes slowed the development of concrete compressive strength at early age. Concrete modulus of elasticity decreases with increasing GGBFS, but not significantly. The increasing percentage of GGBFS causes the pore volume to decrease, indicating the greater durability of GGBFS based concrete compared with the conventional one.

1. Introduction

Concrete is one of the most widely used construction materials in the world, including in Indonesia. The widely use of concrete is due to its high strength, versatility, durability, and other benefits in terms of low production, transportation and installation costs. However, the production of cement as binding material in concrete has a considerable environmental impact. Until 2014, world production of cement has reached 4 billion tons, which accounted for 8% of total CO_2 gas emissions. Half of the gas emissions are the result of combustion of fossil fuels, while the other half is due to calcination of limestone [1].

Cement as binding material in concrete still cannot be completely replaced by other materials. However, there have been many studies on the partial replacement of cement with pozzolanic materials, such as: fly ash, silica fume, rice hull ash and, the recently developed material in Indonesia, granulated blast furnace slag or GGBFS. Ground granulated blast furnace slag (GGBFS) is obtained from an iron slag, a by-product from the manufacture of iron and steel. The liquid iron slag from the blast furnace is rapidly quenched using water to obtain granular products, which are then dried and mashed to form a powder (very fine grain). With advanced processing technology, the iron and steel industry waste is transformed into an economical cementitious material that has the potential as a sustainable material for concrete mixes.

Many studies of the effect of using GGBFS as cement partial replacement on both concrete properties and characteristics has been conducted. Previous studies shows that by adding GGBFS on concrete mixes, the overall concrete performance was improved, includes: strength, workability, permeability and durability. Research conducted by Wang, et al. [2] and Turu'allo [3] found that concrete with GGBFS had better workability than concrete with ordinary Portland cement or OPC, where the fluidity of the paste is increased. Further studies by Turu'allo [3], Rughooputh, et al. [4] and Raman and Krishnan [5] proves that GGBFS can be used as a sustainable cement replacement material, shown by concrete compressive strength that continues to increase until replacement level of 40% - 50% GGBFS to the weight of cement.

Despite its high potential sustainable partial cement replacement, GGBFS itself is not widely known in Indonesia. The fact is, GGBFS was recently introduced in 2018 with the launch of the product by PT. Krakatau Semen Indonesia. The available research for this new product is still limited, either of GGBFS effect as partial cement replacement on hardened concrete properties, or how high the percentage of cement can be replaced with. In this study, the change in mechanical properties of concrete with the various replacement levels of GGBFS was investigated, which includes compressive strength and its development relative with concrete ages and the stress-strain curve up to peak strength. The workability of the fresh concrete with GGBFS was also studied. The porosity of concrete with GGBFS was investigated to know its durability aspect.

2. Experimental Procedure

2.1. Experiment Materials and Specimen Preparations

The materials used in the concrete mixes was consist of Portland Cement, GGBFS, coarse aggregate and fine aggregate. The coarse aggregate was crushed stone with a diameter within 5 – 30 mm; the fine aggregate was natural sand river with a diameter of 0 – 5mm. Ordinary Portland Cement and ground granulated blast furnace slag from PT. Krakatau Semen Indonesia was used as binder. The cement was replaced with four different replacement levels of GGBFS, which was 0%, 30%, 50% and 70% by weight of cement. The water – binder ration was at a constant value of 0.45. The material proportion for the concrete mixes was shown in table 1. There was no addition of chemical admixture in the concrete mixes.

The size for concrete cylinder specimens were 150 x 300 mm, which was used for determines the concrete compressive strength as well as the stress-strain curve up to the peak strength. Cubic concrete specimens were 100 x 100 x 100 mm, and designated to test the porosity of the concrete. All of the specimens were cast in steel molds. After 24 hours, the specimens were demolded and cured for 7, 28, 56 and 90 days.

Table 1. Mix proportion of concrete specimens.

Specimen	GGBFS%	Water (kg/m^3)	Binder (kg/m^3) Cement	Binder (kg/m^3) GGBFS	Aggregate (kg/m^3)	Sand (kg/m^3)	w/b
SN	0	194	428	0	815	996	0.45
SG30	30	194	300	128	815	996	0.45
SG50	50	194	214	214	815	996	0.45
SG70	70	194	128	300	815	996	0.45

2.2. Uniaxial Compressive Strength Test

The average compressive strength of three specimens per batch was measured according to SNI 1974:2011 method. The compressive strength was taken right before peak load begins to decrease. Concrete cylinders were tested at 7, 28, 56 and 90 days to determine the strength development of the concrete specimens.

2.3. Determination of Stress-strain Curve

The stress-strain curve of concrete was taken up to the peak strength. This curve was used to determine elastic modulus of the concrete, as well as the concrete strain at the peak strength. The test was conducted manually. Strain of the concrete specimen was measured using compressometer with digital gauge. Each strain increase was recorded, then the stress was computed for observed strain. Modulus of elasticity was determined as the ratio between concrete stresses at 40% of its peak strength and its corresponding strain. This test was carried out when concrete age reach 28 days.

2.4. Porosity Test

Volume of permeable voids of concrete specimens was measured using ASTM C642 testing standards. Porosity test using cubic concrete specimens was carried out when the concrete age was 7, 28 and 56 days. First, the specimen was dried in an oven at 100^0 C for 48 hours. Then, the specimens were cooled to room temperature and being measured for its weight. Saturation of the specimen is done by boiling the specimen for 5 hours. Tests were carried out for each GGBFS replacement levels.

3. Result and Analysis

3.1. Strength Development of Concrete

Concrete compressive strength was measured when the concrete age reached 7, 28, 56 and 90 days. At the age of 7 days, control mixture (SN) has a higher compressive strength than concrete containing 30%, 50% and 70% GGBFS (SG30, SG50 and SG70). The use of GGBFS as a partial replacement material for cement causes a decrease in compressive strength at early age, but the strength of concrete increases afterwards at later age (Figure 6).

At the age of 28 days, the compressive strength of concrete mix with 30% GGBFS (SG30) and 50% GGBFS (SG50) increased to almost equivalent to control mixture (SN), but it was still slightly lower. Concrete compressive strength containing 30% and 50% GGBFS content was respectively 2.7% and 3.8% lower than the control mixture (33.8 MPa). Concrete compressive strength was decreased further at 70% GGBFS content, which has 19.3% loss as compared to the control. Similar results were also reported by Łukowski [6]. Compressive strength test of mortar at 28 days conducted by Łukowski [6] showed a loss in strength of 4.6%; 10.3%; and 23.0% compared to control mixture (43 MPa) at GGBFS levels of 30%, 50% and 70%, respectively.

At 56 days, the compressive strength of concrete containing 30% and 50% GGBFS has exceeded the 28-day compressive strength of the control mixture. Compressive strength of concrete with 30% and 50% GGBFS content continue to increase till it exceeded the control mixture strength at 90 days.

3.2. Concrete Stress-strain Curve

It is evident from the result that in increasing GGBFS content, the strain at peak value increases (Figure 2). This indicates that concrete will fail at a higher rate of strain which also means greater durability. Similar findings was also observed by Das et.al. [7]. Das et.al. replaced partial of fine aggregate with ground granulated blast furnace slag and noticed that the rate of strain increases with the increasing of GGBFS percentage.

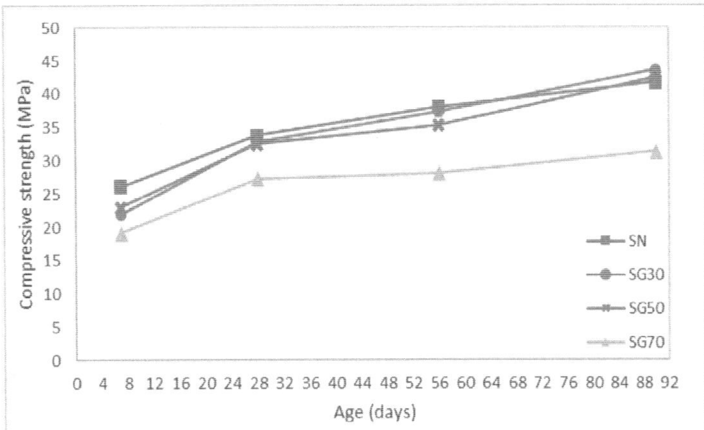

Figure 1. Development of concrete compressive strength with various replacement of GGBFS relative with age

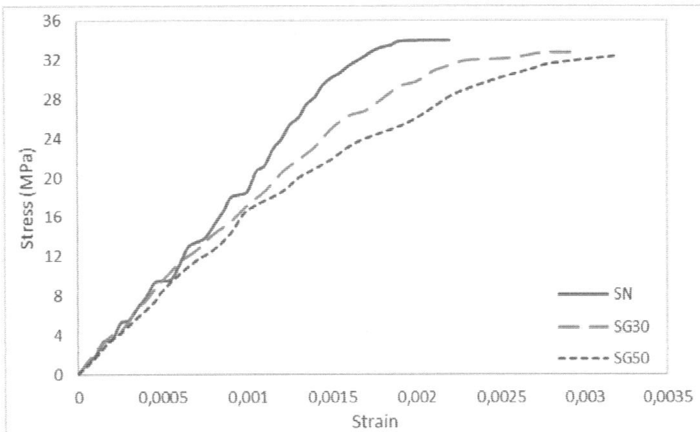

Figure 2. Relationship between stress and strain up to peak strength

Table 2. Modulus of elasticity of concrete specimens.

Specimen	Modulus of Elasticity @0,4 f_c (GPa)
SN	19.05
SG30	17.61
SG50	16.03

In this study, the modulus of elasticity is calculated at 40% of the peak stress. The results showed that the modulus of elasticity of the concrete decreased slightly due to the addition of GGBFS in the concrete mixture (Table 3). This can actually be seen from Figure 2 where the slope from 0 to 40% of peak strength was gradually becoming less steep, indicates a lower value of modulus of elasticity. Siddique and Kaur [8] observed similar findings wherein they reported 22.5%, 39.98% and 41.7% loss in modulus of elasticity as compared of the control (15.98 GPa) at 20%, 40% and 60% GGBFS content, respectively.

3.3. Concrete Porosity

Figure 3 shows the variation in volume of permeable voids in concrete with various levels of GGBFS replacement. It was shown that the voids or pore volume in concrete specimens tends to decrease along with the increases of GGBFS percentage, especially when concrete age is 28 and 56 days, where the hydration process has reduced the pores in the concrete matrix. The reduction of permeable voids in concrete means greater concrete resistance against chemical attack, when concrete is exposed to extreme environmental conditions.

Changes in the structure of the cement paste matrix due to the addition of GGBFS were also reported by Łukowski [6]. Mortars with GGBFS have better resistance to chemical attack due to the denser structure of cement paste. Research by Akram and Raza [9] also proves that the addition of GGBFS in a concrete mixture can reduce concrete permeability, with the value of concrete permeability getting smaller with increasing levels of GGBFS.

3.4. Workability of Fresh Concrete

From the results of slump test (Table 3), it was found that the concrete mixture with GGBFS has a higher slump, with the slump value increasing with increase in GGBFS content. Increased slump values due to the addition of GGBFS in concrete mixes were also reported by Turu'allo [3].

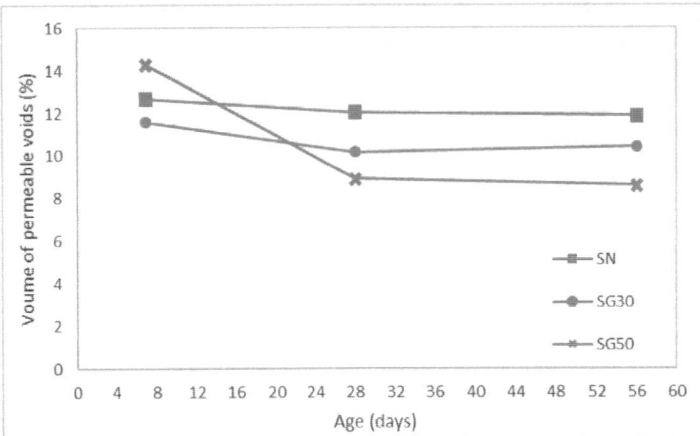

Figure 3. Volume of permeable voids of concrete with various GGBFS replacement levels

Table 3. Slump of fresh concrete mix

Specimen	Slump (mm)
SN	17.50
SG30	25.00
SG50	37.50
SG70	55.00

4. Conclusion

The effect of GGBFS as partial cement replacement on mechanical properties of concrete was investigated in this study. It was found that GGBFS can be used as partial replacement of cement up to 50% of cement weight without reducing the concrete compressive strength. However, concrete contained GGBFS has lower compressive strength at early age, so the usage of concrete with GGBFS is more suitable when high initial strength is not required. GGBFS based concrete will have slower strength development compared to conventional one, which might cause the strength to not be perfectly reached at 28 days. Incorporating GGBFS to the concrete mix could improve the fluidity of cement paste, resulting in better concrete workability. The usage of GGBFS as cement partial replacement has positive effect on the durability of concrete. The pore volume of concrete was reduced with the increase of GGBFS content, effectively raised the concrete resistance against chemical attacks. However, this result must still be verified, which is part of the next research plan.

Acknowledgements

The author would like to thank the Engineering Faculty of Muhammadiyah Malang University for the financial support for this project and PT. Krakatau Semen Indonesia for the support and recognition for this project.

References

[1] R. Vinai *et al.*, "Sustainable binders for concrete: A structured approach from waste screening to binder composition development," *Heron*, vol. 60, no. 1–2, pp. 27–57, 2015.
[2] L. Wang, P. Tian, and Y. Yao, "Application of ground granulated blast furnace slag in high-performance concrete in China," *Proc. Int. Work. Sustain. Dev. Concr. Technol.*, pp. 309–317, 2004.
[3] G. Turu'allo and M. Soutsos, "Sustainable Concrete with The Use of GGBS: The Maturity Method for Predicting Strength Development of Concrete," *International Conference on Environmentally Friendly Civil Engineering Construction and Materials,* pp. 223–232, 2014.
[4] R. Rughooputh and J. Rana, "Partial Replacement of Cement by Ground Granulated Blast furnace Slag In Concrete Corresponding Author : Reshma Rughooputh," *J. Emerg. Trends Eng. Appl. Sci.*, vol. 5, no. 5, pp. 340–343, 2014.
[5] J. V. M. Raman and V. M. Krishnan, "Partial Replacement of Cement with GGBS in Self Compacting Concrete for Sustainable Construction," *Int. J. Civ. Eng.*, vol. 4, no. 3, pp. 24–28, 2017.
[6] P. Łukowski and A. Salih, "Durability of mortars containing ground granulated blast-furnace slag in acid and sulphate environment," *Procedia Eng.*, vol. 108, pp. 47–54, 2015.
[7] S. Das, G. Singh, A. A. Ahmed, S. Saha, and S. Karmakar, "Ground Granulated Blast Furnace Slag (GGBS) based Concrete Exposed to Artificial Marine Environment(AME) and Sustainable Retrofitting using Glass Fiber Reinforced Polymer (GFRP) sheets," *Procedia - Soc. Behav. Sci.*, vol. 195, pp. 2804–2812, 2015.
[8] R. Siddique and D. Kaur, "Properties of concrete containing ground granulated blast furnace slag (GGBFS) at elevated temperatures," *J. Adv. Res.*, vol. 3, no. 1, pp. 45–51, 2012.M.
[9] R. Akram and S. Raza, "Effect of Micro Silica and GGBS on Compressive Strength and Permeability of Impervious Concrete as a Cement Replacement," *Eur. Acad. Res.*, vol. III, no. October, 2015.

Main Task and Functions Stakeholders Institutional to Improve Safety Culture in Construction Sector

R A Machfudiyanto[1,*], Y Latief[1], A Suraji[2], L Sagita[1]

[1]Department Civil Engineering, Faculty of Engineering, Universitas Indonesia, Indonesia.
[2] Department Civil Engineering, Faculty of Engineering, Andalas University, Indonesia.

[*]rossyarmyn@eng.ui.ac.id

Abstract. OSH culture cannot be carried out if the existing institutional scheme is not strong enough or the institutional authority possessed is insufficient in making improvements in the construction sector, so it can be identified Institutions / institutions / organizations / stakeholders that must be involved in OSH institutions to build a Safety culture. The purpose of this research is to find out the main tasks and functions of each stakeholder to form a strong institution in the construction sector. Analysis methods used are using benchmarking, literature studies and validation of safety experts. The results of this study are that the stakeholders identified include the Occupational Safety and Health Services Company (PJK3), DK3N, the Committee for Occupational Safety and Health (P2K3), the Ministry of Manpower, the Director General of PPK and Safety, the Directorate of OHS Norms Supervision with Main Tasks and Functions that have been identified to enhance Safety Culture in the Construction sector.

1. Introduction

Construction accidents are still one of the main concerns within the Construction Sector [1]. One of the most influential elements in order to improve safety performance is to strengthen the institutional sector [2]. In an institution there are two main components, namely the functional component and the operational component related to policy implementation [3]. The basic concept can be seen in the following Figure 1.

Policy Setting	Functional Component
Implementation of Policy	Operational Component

Figure 1. Basic Concepts of Institutional Structure

In order for an institution to run and be adhered to by its members, it is necessary to have an intensive structure that contains sanctions and rewards so that the community will obey it. [4] [5] states that institutions have three components, namely: Formal rules, including the constitution, statute, law and all other government regulations. Formal rules form political systems (governance structures, individual rights), economic systems (ownership rights in conditions of scarcity of resources, contracts), and security systems (justice, police). Information rules, including experience, traditional values, religion and all factors that influence the form of individual subjective perceptions about the world in which they live and enforcement mechanisms, all these institutions will not be effective if they are not accompanied by enforcement mechanisms.

This study aims to determine the purpose and function of the safety institutional structure in the construction sector by benchmarking safety institutions in several countries

2. Theoretical Study

Institutional can also be interpreted as a norm / rule of regulation or organization that facilitates coordination in shaping the expectations of each that might be achieved by mutual cooperation, which includes all social, economic, cultural, and other institutions, both in the form of an organization , as well as traditions and institutions contained in society consisting of elements of public, private, and non-governmental organizations [6]

From the definition of the institution it can be concluded that the institution is a process of community interaction involving the organization as the implementer to achieve common goals. In research on institutions, there are at least five fundamental questions [7] namely:

1. Who does what?
2. Who is responsible to whom?
3. Who is related to whom and in what ways?
4. What channels of communication are in the organization, how can they be used, and for what purposes?
5. What information network is in the organization?

According to [8][9], elements of institutional development can be divided into 9 (nine) elements, namely:

1. The affected community sector,
2. The needs of the program,
3. The main obstacle,
4. Possible changes,
5. The objectives of the program,
6. Benchmarks for assessing each goal,
7. Activities needed for action planning,
8. The size of the gun activity evaluates the results achieved by each activity,
9. Institutions involved in implementing the program.

OSH Institution is an independent, non-governmental national private organization that is engaged in the management of occupational safety and health, in the form of companies or business entities in Indonesia. The existing Safety institutions in Indonesia are the Committee for Occupational Safety and Health (P2K3), the National Occupational Safety and Health Board (DK3N), and the Occupational Safety and Health Services Company (PJK3).

3. Methodology

This study uses the literature study methodology from previous research, benchmarking institutional conditions in various countries and validating content and constructs to determine the main tasks and functions of stakeholders from safety institutions in the construction sector.

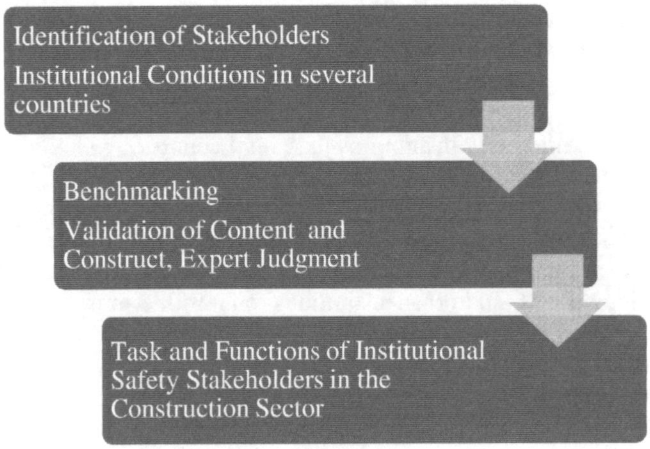

Figure 2. Research Methodology

4. Result and Discussion

4.1. Institutional Benchmarking in Thailand

The Department of Labour Protection and Welfare (DLPW) promotes OSH, including research, study and development. Underneath, the agency directly responsible is the Safety Supervisory Division which supports and facilitates law enforcement carried out by the Provincial Government and the Metropolitan Area of Bangkok (BMA) through the Office of the Protection and Welfare of the Workforce. OHS Services are carried out through 12 Service Centres in the region. The Ministry of Public Health provides occupational health services including disease prevention, health promotion, preventing and controlling the causes of illness and injuries in the workplace and providing medical care through technical departments at the centre and the public health system in the regions. The Occupational Health Program is the responsibility of the Occupational and Environmental Diseases Bureau under the Disease Control Department.

The Ministry of Industry issues factory construction and operation permits, enforces OSH regulations in the industrial sector, carries out factory inspections to ensure the application of OSH and extension of factory permits. The Department of Industrial Work (DIW) has the authority to ensure that the industry is run without endangering public health. Environmental impacts must be eliminated. DIW plays an important role in ensuring OSH practices in factories because DIW is authorized to consider extending factory permits. The agencies responsible for this matter are the Safety Technology Office, the Office of the Control of Hazardous Substances and the Industrial Estate Bureau. [10]

4.2 Institutional Benchmarking in Filipina

The Department of Labor and Employment (DOLE), the main safety institution, is responsible for the prevention of occupational diseases and occupational injuries, law enforcement, rehabilitation and compensation. The OHS Center (OSHC) is responsible for OHS research, training and dissemination of information and technical services.

The Department of Regional Government (DILG) is responsible for building and safety standards due to fire hazards in every Local Government Unit (LGU) throughout the country. DILG has 3,000 sanitation inspectors throughout the country under the supervision of the LGU. DILG also has engineers who conduct inspections.

The Department of Health (DOH) established the Agency for Environmental Health (IACEH). IACEH is responsible for protecting people from exposure to all types of occupational and environmental hazards by intervening in public and environmental health.

4.3 Institutional Benchmarking in Singapore

Occupational Safety and Health Division (OSHD) is the main national OSH institution. The OHSD strategy is to reduce mortality and work accidents. integrating Safety as part of the company, making Singapore a Center of Excellent for Safety, and building a Safety culture for Singapore.

There are four departments in OSHD namely: the Company's Service, Information and OHS Policy Department which is responsible for developing strategies and policies, effective information systems, education and training of personnel; The Safety Oversight Department is responsible for workplace inspection, supervision and law enforcement; OSH Expert Department that offers OSH expert support in developing OSH standards and best practices, investigating work accidents and occupational diseases, and conducting research; and the Work Accident Compensation Department which regulates the compensation system due to work accidents and the accident reporting system.

The Workplace Safety and Health Council (WSH) consists of 18 industry leaders, governments, trade unions and professionals. WSH is assisting the government in developing a national strategy for OSH. Its main function is to improve the industry's OHS performance, promote safety and health in the workplace, and recognize best practice in the industry, and encourage the application of OHS.

4.4 Structural Institutional Model in Indonesia

The Ministry of Manpower in this case the Director General of Manpower Supervision Development has ever arranged a Safety Coordination Mechanism in the perspective of the Ministry of Manpower and Transmigration, as follows:

1. OHS Supervision Principle
   Legal oversight is a function of the state
   OHS supervision is carried out on 2 principles: the principle of legality, and the principle of coordination (social dialogue)
   Oriented on preventive / preventive actions
   Are universal and are carried out independently and equal implementations.
2. National OHS Policy according to Law No.1 of 1970
   National Safety Policy by the Indonesian Ministry of Manpower and Transmigration
   Centralized policy
   Operational Decentralization
3. In the era of Regional Autonomy, the authority of the Central Government includes:
   National Policy Determination
   Technical standardization
   Institutional accreditation
   Personnel competency certification
4. Safety Coordination Mechanism
   Policy coordination is carried out by the Ministry of Manpower and Transmigration with the House of Representatives and the State Secretariat in terms of making laws and regulations in accordance with the hierarchy
   Operational coordination is carried out by the Ministry of Manpower and Transmigration with DK3N and related technical agencies (cross sectoral)
   Structural coordination is carried out by the Ministry of Manpower and Transmigration (Minister of Manpower and Transmigration, Secretary General of the Ministry of Manpower and Transmigration Directorate General) at the central and regional levels through Regional Coordination and Technical Coordination
   Functional coordination is carried out by the Ministry of Manpower and Transmigration with the Minister of Administrative Reform and PPNS Supervisory Coordinator (Civil Servant Investigator) regarding the functional position of investigation
5. Sectoral Operational Coordination
   Coordination of operational authority in other sectors
   Accountability to the Indonesian Minister of Manpower and Transmigration
   The responsibility of the Indonesian Minister of Manpower and Transmigration to the DPR
   Accountability to the ILO

6. Problems encountered
   Coordination between sectors has not been going well
   In the implementation / operation there are many overlaps
   Data and information flow from technical agencies is not going well
   The implementation of Safety in the era of Regional Autonomy has not been going well
   The challenges of globalization in applying OHS are getting heavier
7. Solution
   Increasing the role and empowerment of existing OSH institutions
   Intensified technical and regional coordination in the field
   Empowerment of OSH institutional and personnel functions in the region

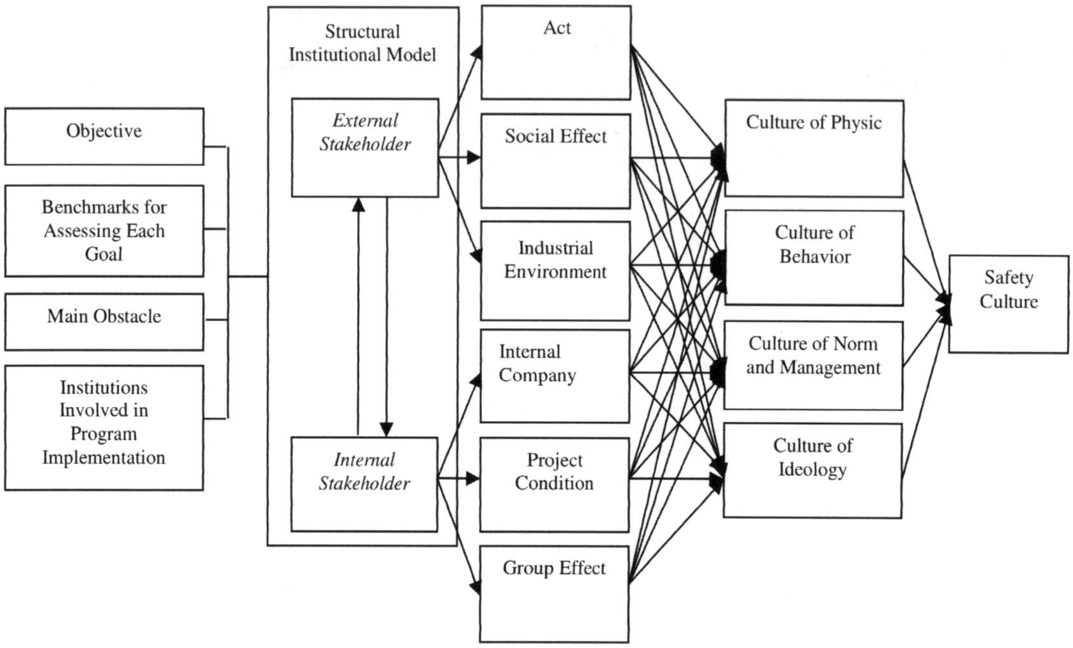

Figure 3. Institutional Model to Improve Safety Culture in Construction

4.5 Main Task and Functions of Institutional Safety

The results of the validation by the safety experts were identified that there were 6 stakeholders with the Main Tasks and Functions as in Table 1 below

Table 1. Main Task and Functions of Safety Institutional Stakeholders in the Construction Sector

No	Institutions	Institutional Task and Functions
1.	Occupational Safety and Health Services Company (PJK3)	Main Task: Assist the fulfillment of occupational safety and health requirements in accordance with applicable regulations. Function: Carry out activities related to OHS issues, starting from the consultation, fabrication, maintenance, repair, research, inspection, testing, OHS auditing and Safety guidance stages [11]
2.	National Occupational Safety and Health Council (DK3N)	Main Task: As an auxiliary body at the national level to provide national recommendations and considerations, whether requested or not, to the Government regarding national OHS Policy and assist in developing OHS towards OHS culture [12] Function: Gathering and processing all data and or occupational safety and health issues at the national and provincial levels concerned and assist the Minister in: fostering DK3W, conducting research,

		education, training, development and efforts to promote and civilize occupational safety and health
3.	Occupational Safety and Health Committee (P2K3)	Main Task: As an institution established in the company to help carry out and handle occupational safety and health efforts whose membership consists of employers and workers. Function: Improve efforts to enforce Safety provisions and their implementation in companies [13]
4.	Minister of Manpower	Main Task and Functions: Establish national policies in the field of Safety for guarantee uniform and harmonious practices throughout Indonesia [14]
5.	Director General of PPK and safety	Main Task and Functions: Conduct general supervision of the Work Safety Act [15]
6.	Safety Norms Supervision Directorate	Main Task and Functions: Through OHS supervisors and experts designated as OHS supervisors by the Ministry of Manpower and Transmigration, supervising workplaces, conducting accident investigations, conducting training courses and seminars, and promoting OHS implementation, conducting inspection and certification of heavy or dangerous machine operators [16]

5. Conclusion

The vision compiled by the government to realize OSH as a culture is a very appropriate thing. Because 80% of work accidents are triggered by unsafe behavior (unsafe behavior) of workers, unsafe behavior is caused by safety culture which is still very low of workers. The duties and functions of the stakeholders to improve the National OHS Institutional Strategy and program have followed the concepts of Occupational Health and Safety. If it is seen from the change in paradigm that Safety which has been implemented so far due to pressure and obligations from the government has changed to be the responsibility of the organization for its environment. The substance of the National OHS strategy is to want to further involve employers, workers and other social organizations in cultivating OHS so that OHS can be the responsibility of the organization for its environment. If this can be realized, then Safety will no longer be a burden on the company's costs but will be a part that contributes to improving the quality and productivity of the company.

Acknowledgment

The research was funded by the Higher Research Grant of Higher Education for Fiscal Year 2018 No: 1063 / UN2.R3.1 / PPM.00.01 / 2018 with funding source: Budget Implementation Schedule (DIPA) Kemenristekdikti No SP DIP A- 042.06.1.401516 / 2018 December 5, 2017

References

[1] Latief, Y., Machfudiyanto, R. A., Arifuddin, R., & Yogiswara, Y. (2017, March). Understanding the relationship between safety culture dimensions and safety performance of construction projects through partial least square method. In AIP Conference Proceedings (Vol. 1818, No. 1, p. 020028). AIP Publishing.

[2] Machfudiyanto, R. A., Latief, Y., Suraji, A., & Soeharso, S. Y. (2018). Improvement of policies and institutional in developing safety culture in the construction industry to improve the maturity level, safety performance and project performance in Indonesia. International Journal of Civil Engineering and Technology, *9*(10), 1022-1032.

[3] Tyas, W. P., & Ruswanti, E. Kelembagaan Implementasi Program Kredit Mikro Perumahan di Kota Semarang. *Teknik*, *29*(1), 36-50.

[4] Pejovich, S. (1999). The effects of the interaction of formal and informal institutions on social stability and economic development. Journal of Markets and Morality, 2(2).

[5] Nasution, M. (2002). Pengembangan kelembagaan koperasi pedesaan untuk agroindustri. Ipb Press.

[6] Sianipar, M. (2012). Penerapan Intrepretative Structural Modeling (ISM) dalam penentuan elemen pelaku dalam pengembangan kelembagaan sistem bagi hasil petani kopi dan agroindustri

kopi. *AGROINTEK*, *6*(1), 8-15.

[7] Firmawan, F., Othman, F., & Yahya, K. (2012). Improving Project Performance and Waste Reduction in Construction Projects: a case study of a government institutional building project. *International Journal of Technology*, *2*(1), 182-192.

[8] Saxena, J. P., Sushil, P., & Vrat, P. (1992). Hierarchy and classification of program plan elements using ISM: a case study in the Indian cement industry. *Syst. Pract*, *5*, 651.

[9] Laili Yetri (2014) The Design of Institutional Structural Models For Upgrading The Components of Upstream Oil And Gas Industry, Indonesia, Thesis, University of Indonesia, Salemba.

[10] Layanun, P., Dejthai, T., Silpasuwan, V., Buranatrevedh, S., & Pattarachat, J. (2015). The development of a health promotion management system in the workplace for the garment industry in Bangkok Metropolis.

[11] Dura C. Occupational health and safety integration in corporate social responsibility policies within BRD-GSG Romania. Annals of the University of Petroşani. Economics. 2014;14:59-70.

[12] Puplampu BB, Quartey SH. Key issues on occupational health and safety practices in Ghana: A review. International journal of business and social science. 2012 Oct 1;3(19).

[13] Mitchison N, Papadakis GA. Safety management systems under Seveso II: Implementation and assessment. Journal of Loss Prevention in the Process Industries. 1999 Jan 1;12(1):43-51.

[14] Alli BO. Fundamental principles of occupational health and safety Second edition. Geneva, International Labour Organization. 2008;15.

[15] Deakin SF, Morris GS. Labour law. Hart publishing; 2012.

[16] Choudhry RM, Fang D, Ahmed SM. Safety management in construction: Best practices in Hong Kong. Journal of professional issues in engineering education and practice. 2008 Jan;134(1):20-32.

The Effect of Iron Powder Waste as a Replacement of Amount of Fillers on HRS (Hot Rolled Sheet)

Sulfah Anjarwati[1,*] and Panji Dwi Santosa[1]

[1]Department of Civil Engineering, Faculty of Engineering, Universitas Muhammadiyah Purwokerto, 53182 Kembaran, Banyumas, Central Java, Indonesia

[*]e-mail address: sulfahanjarwati@ump.ac.id

Abstract. Hot Rolled Sheet (HRS) HRS is used on roads with moderate traffic loads, Hot Rolled Sheet is flexible and has high durability, this is due to the mixture of HRS with lame gradations having a large cavity in the mixture, so as to absorb the amount of asphalt large amounts (7-8%) without bleeding, due to the nature of the HRS above, the author chose the HRS pavement type to be modified so that the stability and flow values increased. This study was aimed to determined how much influence iron waste have on stability and flow. This study was used an experimental method that is with a laboratory experiment to get the results, thus the effect of the utilization of iron powder waste on HRS (hot rolled shett) pavement with variations of 0%, 5%, 10%, 15%, and 20% of filler weight. From the results of this study was found that: Iron powder based on gradation and specific gravity test and water absorption has fulfilled the requirements as fillers as regulated in SNI 03-4142-1996. From the Marshall test data obtained, variations that meet the requirements for the properties of Marshall parameters for HRS-WC pavement are variations of 5% and 10%. Percentage of addition of iron powder variation from filler weight which results in the highest stability and flow and fulfills all characteristics the Marshall parameter is a variation of 10%.

1. Introduction

Roads are the basic and main infrastructure in moving the wheels of the national and regional economies. Road damage in Indonesia is generally caused by overloading, heavy flow of vehicles as a result of rapid growth in the number of vehicles, especially commercial vehicles and environmental changes or due to poor drainage function, the researchers conclude the need to increase the stability and flow of a road pavement by increasing the quality of its constituent materials, one of which is the filler where the filler used is modified with variations in the number of fillers with iron powder [1].

Hot Rolled Sheet (HRS) is used on roads with moderate traffic load, Hot Rolled Sheet is flexible and has a high durability, this is due to the mixture of HRS with lame gradations having a cavity in a mixture that is large enough, so it can absorb the amount of asphalt in the amount many (7-8%) without bleeding [2]. By looking at the nature of HRS above, the authors choose the type of HRS pavement to be modified so that the stability value increases.

Broadly speaking the problems that exist are: Does the use of iron powder waste with different levels of variation as a substitute for some fillers in the HRS asphalt pavement mixture can meet the requirements of the characteristics of the Marshall parameters?. What is the composition of the best iron powder waste mixture produced from the Marshall test.

The purpose of this study is to determine how much influence the use of variations of iron powder waste on the characteristics of Marshall parameters in the asphalt pavement mixture. To get the composition of variations in the mixture of iron powder waste with the best results using Marshall parameters.

The benefits of holding this research are as follows: Can be used as a reference for asphalt mixing research by using variations of iron powder waste as a filler in asphalt pavement mixture, Can provide information about asphalt pavement innovation, Can give knowledge to readers about the use of iron powder waste, The existence of this study is expected to be able to provide insights and add insight into the effect of the use of iron powder waste as an alternative filler material in lataston mix (HRS).

Research purposes is to find out how much influence the use of variations of iron powder waste on the characteristics of Marshall parameters in the asphalt pavement mixture, and to get the composition of variations in the mixture of iron powder waste with the best results using marshall parameters.
While the benefits of research are as follows :

- Can be used as a reference for asphalt mixing research by using variations of iron powder waste as a filler in asphalt pavement mixture[3] [4]
- Can provide information about asphalt pavement innovation [5].
- Can give knowledge to readers about the use of iron powder waste[5].
- The existence of this study is expected to be able to provide insights and add insight into the effects of the use of iron powder waste as an alternative filler material in lataston mix (HRS)

2. Research Methods

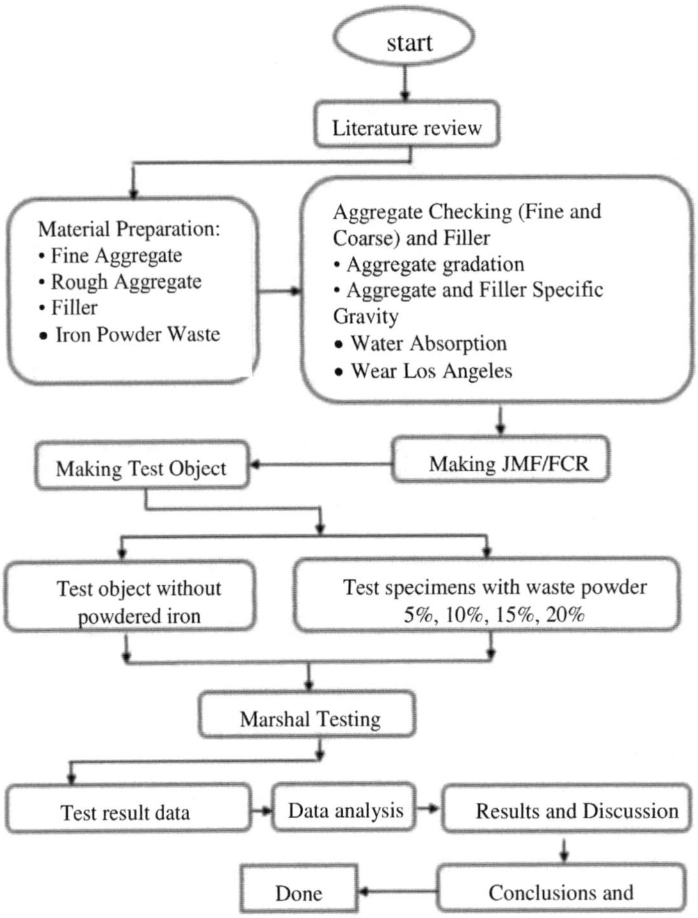

Figure 1. Flowchart Research Steps

2.1. Material Preparation

Materials used include:

a. Iron powder waste obtained from several lathe workshops in Purwokerto

b. Aggregate, including [6]:

- Coarse aggregate of size ½ Coarse aggregate comes from the Klawing river and is obtained from the results of a stone crusher from PT. AW Patikraja, Banyumas Regency, Central Java.
- Fine aggregate comes from the Kalierang river Bumiayu.
- The filler aggregate comes from the Klawing river and is obtained from the results of a stone crusher from PT. AW Patikraja, Banyumas Regency, Central Java.

c. Asphalt material using Pertamina asphalt penetration 60/70.

Material inspection is intended to determine the quality of the material to be used in research, whether by the specified requirements or not. The aggregate examination is carried out at the Jalan Raya Muhammadiyah University Laboratory in Purwokerto.

2.2. Mixed Planning

Making the Mixed Working Formula (FCK)

Mixed testing in the laboratory must be carried out in three basic steps, namely: Obtaining an appropriate aggregate gradation, making a mixture of plans, obtaining approval of mixed plans as a work plan mix[7].

Calculate the estimated asphalt level of the plan (Pb)

$$Pb = 0.035\ (\%\ CA) + 0.045\ (\%\ FA) + 0.18\ (\%\ FF) + k$$

Where :

Pb = Initial plan asphalt level is % of the weight of the mixture
CA = Coarse aggregate is% of aggregate held by filter no. 8
FA = Fine aggregate is% retained by aggregate passing filter no. 8 and retained sieve no. 200
FF = Fillers or fine aggregates pass filter no. 200
K = Constants ranging from 1-2

Manufacture of test specimens[7].

Based on SNI 06-2489-1991 testing on asphalt mixture testing methods using Marshall tools, it is planned to carry out at least 5 variations of asphalt mix testing with iron powder waste, and each variation of at least 3 test pieces is made. Testing is carried out on the asphalt and powder waste mixtures iron with a percentage variation of 0%, 5%, 10%, 15%, 20% made as many as three pieces of test specimens for each variation, so the total test pieces as many as 15 pieces.

3. Results and Discussion

Testing gradation and aggregate physical properties

a. Rough aggregate testing

Testing of coarse aggregate in the form of gravel from the Serayu river with a stone size of ½ meets SNI requirements.

b. Testing fine aggregate gradation

The testing of fine aggregate in the form of sand from the Kalierang Bumiayu river meets SNI requirements.

c. Grading gradation [8].

Table 1. gradation mix of the whole aggregate

Filter Size (ASTM)	Original Gradation			Combined Gradation	Mixed Gradation Requirements HRS-WC
	A	C	D		
¾”	100,00	100,00	100,00	100,00	100
½”	97,00	100,00	100,00	98,95	87-100

3/8"	58,00	100,00	100,00	85,30	55-88
# 4	10,50	95,40	100,00	67,99	
# 8	0,33	78,60	99,00	61,41	50-62
# 16	0	64,60	86,00	52,69	
# 30	0	49,20	62,10	38,43	20-45
# 50	0	29,20	42,70	25,73	15-35
# 100	0	7,40	22,70	12,46	
# 200	0	0,00	12,40	6,20	6, - 10
Proportion of Total Aggregates		A. Split 1-2		35%	
		C. Sand		15%	100%
		D. Filler		50%	

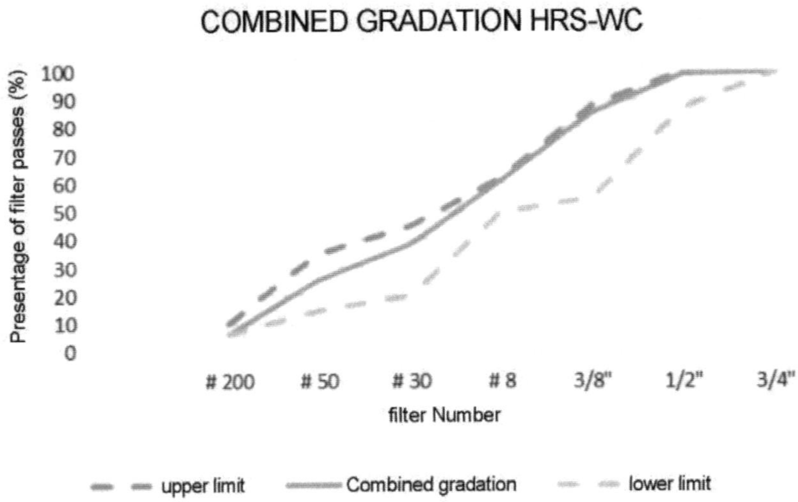

Figure 2. combined grading graph

d. Testing the aggregate physical properties [9] [10].

Table 2. Results of Aggregate Physical Examinations

No	Testing	Method	Requirement	Result	Description
		Coarse aggregate			
1	Water Absorption	SNI 03-1969-1990	≤ 30	1,7	satisfy
2	Bulk Specific Gravity	SNI 03-1969-1990	≥ 2,5	2,6	satisfy
3	Pseudo Specific Gravity	SNI 03-1969-1990	-	2,72	satisfy
4	Los Angeles Wear	SNI 03-2417-1991	-		satisfy
		Agregat Halus			
1	Water Absorption	SNI 03-1970-1990	≤ 30	2,88	satisfy
2	Bulk Specific Gravity	SNI 03-1970-1990	≥ 2,5	2,63	satisfy
3	Pseudo Specific Gravity	SNI 03-1970-1990	≤ 30	2,84	satisfy
		Filler			
1	Specific Gravity	SNI 15-2531-1991	≥1	2,82	satisfy

Determination of asphalt levels

PB = 0.035 (%CA) + 0.045 (%FA) + 0.18 (%FF) + K

= 1,357 + 2,4852 + 1,08 +1,5
= 6,42 %

e. Sample making

Table 3. Percentage of Sample Making Materials

Material	Percentage of material + iron powder waste (%)	weight of material + iron ore waste (grams)
	iron powder waste 0%	
Joint gradation	50	638,53
Filler	50	561,47
Iron powder waste	0	0
Amount	100	1200
	iron powder waste 5%	
Joint gradation	50	638,53
Filler	47,5	533,39
Iron powder waste	2,5	28,07
Amount	100	1200
	iron powder waste 10%	
Joint gradation	50	638,53
Filler	45,25	505,32
Iron powder waste	4,75	56,15
Amount	100	1200
	iron powder waste 15%	
Joint gradation	50	638,53
Filler	43,21	477,25
Iron powder waste	6,79	84,22
Amount	100	1200
	iron powder waste 20%	
Joint gradation	50	638,53
Filler	41,36	449,17
Iron powder waste	8,64	112,29
Amount	100	1200

f. Calculation of sample-specific gravity [9] [10]

1. Bulk Specific Gravity (Gmb)

Table 4. Testing of Mixed Bulk Specific Gravity (Gmb)

No	Test Objects	Iron powder	Score BJ Bulk Mix (Gmb)
1	A	0%	2,220
	B	0%	2,223
	C	0%	2,206
	Average		2,216
No	Test Objects	Iron powder	Score BJ Bulk Mix (Gmb)
2	A	5%	2,189
	B	5%	2,213
	C	5%	2,218
	Average		2,207
No	Test Objects	Iron powder	Score BJ Bulk Mix (Gmb)

3	A	10%	2,220
	B	10%	2,226
	C	10%	2,218
	Average		2,221
No	Test Objects	Iron powder	Score BJ Bulk Mix (Gmb)
4	A	15%	2,278
	B	15%	2,284
	C	15%	2,272
	Average		2,278
No	Test Objects	Iron powder	Score BJ Bulk Mix (Gmb)
5	A	20%	2,284
	B	20%	2,282
	C	20%	2,282
	Average		2,283

2. Maximum Mixed Specific Gravity (GMM)

Table 5. Calculation Results Maximum density of the mixture (Gmm)

		A	B
Bottle Weight + sample	a	2821	2825
Bottle Weight	b	1654	1654
Example Weight	c= a- b	1167	1171
Bottle + water weight	d	5782	5782
Bottle Weight + Water + Example	e	6447	6445
Fill in the Example	f= c+d-e	502	508
Specific gravity	g=c/f	2,32	2,31
Water Temperature (C))	h	25	25
Temperature Correction	1	1	1
Specific gravity	G=g x 1	2,32	2,31
BJ Average		2,315	

g. Marshall test results

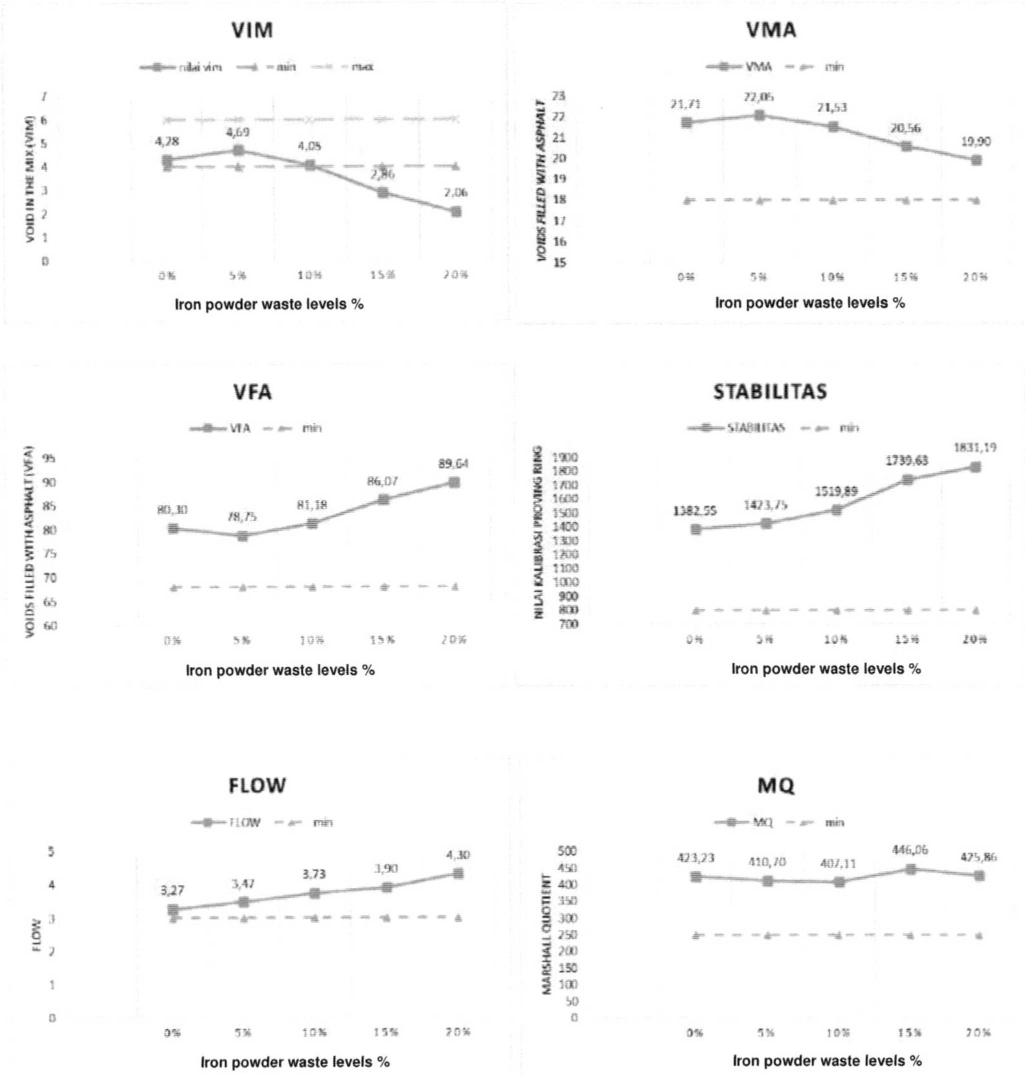

Figure 3. Graphs of Marshall Test Results (with a mixture of iron powder waste)

4. Conclusions and Suggestions

From the research results of the influence of iron powder waste as a substitute for some fillers, the following conclusions can be drawn:

a) Iron powder based on gradation test and specific gravity and water absorption have fulfilled the requirements as a filler as regulated in SNI 03-4142-1996
b) From the Marshall test data obtained, variations that meet the requirements of the characteristics of Marshall parameters for HRS-WC pavement are 5% and 10% variations.
c) The percentage increase in the variation of iron powder from the weight of the filler which produces the highest stability and meets all the criteria characteristics of the Marshall parameters is a variation of 10% content with a stability value achieved of 1519.89 kg

Some things that can be suggested in connection with the results of this study are as follows:

a) For further research, other types of pavement should be used, including AC-BC, HRS-BASE, and LATASIR.

b) Future research should consider conducting field tests so that it can prove whether the research in the laboratory is proven to improve road stability or not.

References

[1] Sugianto, Gito 2008. Study of Characteristics of Hot Rolled Asphalt Mixture Due to Addition of Used Tire Powder Waste. General Sudirman University

[2] Bahri, Samsul 2010. Effect of Iron Powder Waste as a Substitute for a Fine Aggregate of Asphalt Mixtures. Bengkulu University

[3] Ministry of Public Works. Directorate General of Highways of 2010 General Specifications HRS Revised 2. Jakarta

[4] Ministry of Public Works Directorate General of Highways, 2010. Division 6 General Specifications HRS Revised 3. Jakarta.

[5] Priambodo, Sam 2003. Laboratory studies of the effect of the use of iron sand as fine aggregate on a mixture of hot rolled asphalt HRA (HOT ROLLED ASPHALT) on marshal properties and durability. Diponegoro University

[6] Pasaribu, Iffah Fathaniah 2015 "The Effect of Use of Iron Powder Waste on Ac-Wc Hot Asphalt Mixture. University of Northern Sumatra

[7] Indonesian national standard no. RSNI m-01-2003 about the method of testing hot asphalt mixtures with marshall tools.

[8] SNI 03-1968-1990 concerning the testing method of fine aggregate and coarse aggregate sieve analysis.

[9] SNI 03-1969-1990 concerning the method for testing specific gravity and absorption of coarse aggregate water

[10] SNI 03-1970-1990 concerning the method for testing specific gravity and absorption of fine aggregate water.

Application of nanopowders in industrial production of mixed feed

M Chkalova[1,*], V Shahov[1], V Pavlidis[1], S Solovyov[2]

[1]Orenburg State Agrarian University, Chelyuskintsev str., 18, 460014, Orenburg, Russian Federation.
[2]Russian Academy of Sciences, Leninsky prosp., 14, 119991, Moscow, Russian Federation.

[*]e-mail: chkalovamv@mail.ru

Abstract. One of the main directions of the feed manufacturing industry development is the improvement of the output product quality that ensures its efficacy in modern agricultural production. Some international experience in using the essential microelements in the form of nanopowders in the diets of farm animals has been gained. The article is devoted to the development and justification of a technical solution to the problem of dispensing ultrafine (nano) powders in the industrial production of combined feeds. The authors carried out theoretical and experimental studies of the influence of the conditions of functioning of industrial-technological equipment on the properties of several nanopowders of micro metals. The parametric model of the proposed pneumatic dispenser of bulk nanomaterials allows taking into account various structural and technological changes in the dosing process during the preparation of feed mixtures. The author presents a methodology for calculating the oscillatory system of a pneumatic dispenser, including an annular membrane with fixed edges, based on the equivalent replacement of a system with distributed parameters (membrane) with a system with lumped parameters, which allows replacing the oscillatory model with a translational motion model. An engineering-mathematical model that describes forced membrane oscillations, the dispenser oscillation system and the behaviour of the ultrafine material in the working chamber was constructed on its basis. Practical implementation of the results made it possible to obtain a prototype of bulk ultrafine materials dispenser.

1. Introduction

The industrial production of mixed feed is an established industrial sector in the Russian Federation. In modern conditions, the production of mixed feeds is becoming an increasingly important link in the agro-industrial complex [1]. Qualitative and quantitative parameters of livestock production are largely determined by the efficiency of the technological processes of feed production organisation and innovative methods to improve their basic characteristics.

The development of the mixed feed industry and improving combined feeds quality is a prerequisite for the rapid and economically viable livestock production growth, as well as one of the main factors in agricultural production development [1]. The solution to the problem of increasing the efficiency of industrial production of combined feeds is becoming more relevant than ever.

The purpose of the study is to establish the possibilities and conditions for the use of ultrafine materials in a feed production line; development of a technical solution to the problem of dispensing ultrafine powders during feed preparation. To achieve this goal, it is necessary to develop a methodology for experimental studies of the influence of the operating conditions of technological equipment on the properties of ultrafine powders of micro metals; to obtain an engineering solution to the problem of dispensing ultrafine powders during feed preparation.

The total mixed ration accumulates the best qualities of the basic raw material and fully meets the animal's need for nutrient, mineral and biologically active substances. The feed nutrients will be best digested only if an equivalent amount of macro- and micronutrients, vitamins, and other essential nutrients are involved in the metabolic processes of the organism [2-5] (Table 1).

Table 1. Role of trace elements in metabolic processes of farm animals' organism

Trace elements	Importance for Metabolism
Fe	It is associated with blood haemoglobin level, included in catalase cytochromes and other enzymes and is deposited in the form of ferritin and hemosiderin. **The adult animals get it with fodder**
Cu	It belongs to strong cytoplasmatic tic poisons, but in small amounts, it is needed to convert iron into the hemo-form available for synthesis, to be transferred into the bone marrow. Copper is needed for the formation of enzymes, catalysing the transformations of tyrosine, ascorbic acid, etc. **The animals get copper with fodder**
Co	Cobalt is part of vitamin B_{12}. It activates hydrolytic enzymes, increases the synthesis of nucleic acids and muscle proteins, and contributes to higher live weight gain in young animals and increase of farm animals' performance. **The animals get cobalt with fodder**
Mn	Manganese has a beneficial effect on the growth and development of young animals, plays a significant role in the reproduction processes. **The animals get manganese with fodder**
Zn	Zinc is contained in the insulin hormone and influences the animals' growth, gain and reproduction. The zinc metabolism is connected with that of calcium, sulphur and copper. **The animals get zinc with fodder**
Cr	Chrome participates in the synthesis of proteins, cholesterol and fatty acids, activates the enzymes necessary for carbohydrate metabolism, regulates carbohydrate metabolism and the level of glucose in the blood, which normalizes the permeability of cell membranes. **The animals get chrome with fodder**

2. Materials and methods

In recent years, research aimed at exploring the possibilities of obtaining high-quality feed using the combined use of bio-additives and trace elements in nano form has been conducted [6,7]. Nanomaterials (in the form of ultrafine powders) contain structural elements whose geometric dimensions do not exceed 100 nm, and radically different in their properties and effects from substances in the form of macroscopic dispersions and continuous phases.

World science has gained some experience in the development and application of ultrafine materials in agriculture. Recent studies have shown their effectiveness in crop production, fodder production [8-10] and livestock farming.

The main methods for producing ultrafine powders of trace metals with particle diameters up to 100 nm, which can be used for feeding farm animals, are presented in Table 2.

The common disadvantages of the methods considered (excluding the Exploding Wire Method (EWM) are: low performance, significant energy consumption, a wide range of particle size distribution, poor purity of the output product and high safety requirements. The presence of these drawbacks is a serious obstacle to the production of ultrafine powders on an industrial scale. Including metal powders and trace metals for feeding farm animals including, on an industrial scale.

Table 2. Methods for producing ultrafine metal powders for feeding farm animals

Method	Methodology	Materials
	Physical methods	
Evaporation and condensation	In vacuum or inert gas	Zn, Cu, Cr

The electrical explosion of conductors	In the gas atmosphere	Cu, Fe, Zn
High-energy destruction	Grinding Detonation treatment	Fe – Cr Fe
	Chemical methods	
Synthesis	Thermal Mechanic-chemical	Fe, Cu, Co Fe – Cu
Thermal decomposition	Condensed precursors	Fe, Co, Mn

The Exploding Wire Method (EWM), among others, has a number of advantages related to the field of application of interest to us: high productivity, the stability of the output product to oxidation and sintering at room temperature, high chemical and diffusion activity when heated.

Research in the use of ultrafine materials in the preparation of animal feed for cattle and young animals was started by the researchers at the Federal State Budget Educational Establishment of Higher Education «Orenburg State Agrarian University». The results of these studies highlighted the problem of technical solutions to launch the industrial production of innovative combination feeds using ultrafine materials [9,10].

At the initial stage, we studied the conditions for the operation of technological equipment at the feed production enterprises of the Orenburg region. An analysis of the actual technological processes made it possible to single out a set of conditions that affect the physicochemical properties of ultrafine materials [11,12] (Table 3).

Table 3. Operating conditions of technological equipment affecting ultrafine material

Technological operation	**Conditions influencing the ultrafine material**		
Micronization of mixed feed components	-heating to **90** ° **C** for 45 s		
Expanding of mixed feeds	-the product passes through the expander for some seconds at the temperature from **105 to 110° C**		
Pressing of mixed feeds	- the feed moisture before it enters the press- granulator mixer is **8-15%**; -the feed moisture content after it leaves the mixer does not exceed **17.2%**; -steam pressure is up to **0.7 MPa**; -steam temperature is **140 – 180° C**; -conditions of pellets cooling: - pellet cooling conditions: increasing the temperature of feed **by 10 -11° C** as a result of conditioning in a steam mixer is equivalent to increasing its humidity **by 0.7-1.0%**		
Extrusion	-the product temperature at the outlet is **120 – 130° C**		
Separation Magnetic separators	magnet shape	sizes, mm	residual magnetic induction, mT
	bar	62*25*62	**1350**
	ring	52*27*35	**380**
	tablet	¤134*54*12	**390**

Based on the analysis of the technological conditions for the operation of the industrial feed production line, we developed a laboratory experiment technique aimed at studying the properties of nanomaterials in modelling temperature conditions, humidity, and residual magnetic and electromagnetic induction. A series of laboratory experiments showed the stability of ultrafine powders of iron, zinc and copper oxides synthesised by the electric explosion method to oxidation and sintering at room temperature. However, a high chemical and diffusion activity was revealed when heated to temperatures of the working feed mixture.

The obtained results formed the basis of the idea of many technical solutions, the implementation of which will increase the efficiency of the preparation of combined feed based on the use of ultrafine materials. The device (pneumatic dispenser of ultrafine materials) operates as follows (Figure 1).

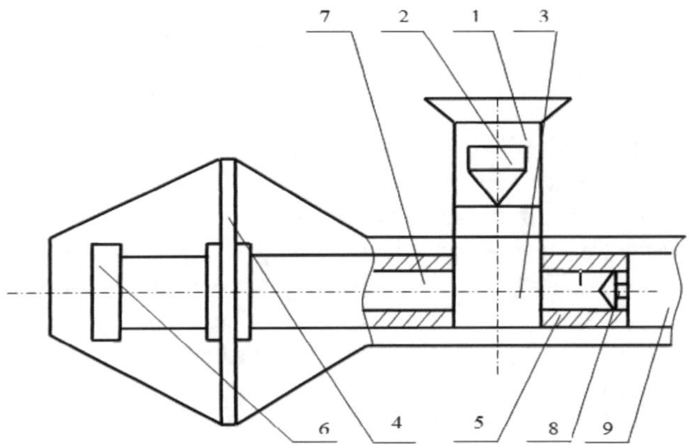

Figure 1. Principal scheme of the pneumatic dispenser of loose ultrafine materials.

The dispensing is performed by opening the solenoid-magnet valve 2 for a strictly fixed time. The loose ultrafine material from the storage tank one is moved through the calibrated opening to the dispensing chamber 3. When air is supplied into the pneumatic actuator, the membrane four is being deformed and moves the rod 5 to the right. The dispensing chamber three is overlapped by the inner spherical surface of the pneumatic actuator body. With further movement of the rod to the right, the valve 6 opens a passage for compressed air through the hollow channel 7 into the dispensing chamber 3. The compressed air presses the loose ultrafine material located in the dispensing chamber and moves it along the channel 7 to the right. Under the action of the compressed air mixture and the loose ultrafine material, the shut-off valve eight is being opened. The dispensing chamber three and the hollow channel seven are cleaned of the loose ultrafine material, which enters the receiving chamber 9 for its future use as intended when the air supply into the pneumatic drive discontinues, the shutoff valve 8 closes [13].

The pneumatic dispenser for loose ultrafine materials includes a control system located in a separate cabinet. The control system makes it possible to specify any volume (amount) of the loose ultrafine material, entering the dispensing chamber, and controls the supply of compressed air to the pneumatic drive.

3. Results and discussion

To solve the question of how to introduce ultrafine bulk materials, technical capabilities were studied, among which, in our opinion, the most promising are the design changes of the pneumatic dose of powdered materials. Due to the need to use ultrafine powders, we had to significantly change its design, based on which an engineering mathematical model was constructed that describes the forced oscillations of the membrane, the oscillatory system of the dispenser and the behaviour of the ultrafine material in the working chamber. The constructed model involves verification and therefore does not contain specific parameter values and impedance characteristics. We describe the forced oscillations of the annular membrane with fixed edges in the housing of the pneumatic actuator and calculate the oscillatory system of the pneumatic dispenser, including this membrane.

An annular membrane fixed along the contour located in the housing of the pneumatic actuator performs forced oscillatory movements under the influence of air pressure; on the other side of the membrane, the air medium remains unperturbed. Air pressure is created using the airflow pumped by the compressor, and varies according to a certain law, evenly distributed over the entire membrane area. We will assume as standard that the resistance to movement of the membrane is proportional to the speed of this movement.

The studies conducted by the authors allowed us to determine the function that describes the change in air pressure in the working chamber of the pneumatic drive as harmonic. Then using the wave differential equation of partial derivatives of the second-order of the hyperbolic type, which describes the free vibrations of the membrane, and taking into account the technical characteristics of the compressor that determine the pressure function, we obtain:

$$\rho_m \frac{\partial^2 U}{\partial t^2} + r_m \frac{\partial U}{\partial t} - T\left(\frac{\partial^2 U}{\partial r^2} + \frac{1}{r}\frac{\partial U}{\partial r}\right) = \rho_0 e^{i\omega t}, \quad (1)$$

where $u = u(r,t)$ is the deviation of the annular section of the radius r membrane at the time moment t; ρ_m– the surface density of the membrane material (constant); r_m– the resistance coefficient of the surface unit (constant); T – stress in the given annular membrane section (constant); ρ_0– the amplitude of air pressure; ω – the circular frequency of the acting driving force.

The above equation describes the established steady forced oscillations of the annular membrane with fixed edges in pneumatic drive body.

In the framework of steady oscillations, the setting of initial conditions is not required, the boundary condition has the form: $u(r,t)|_{r=R_0} = 0$, where R_0 is the outer radius of the annular membrane, the boundedness of the required function will be taken into account below. The nontrivial solutions of equation (1), with the given boundary conditions, are to be sought in the following form

$$u(r,t) = u_1(r) \cdot e^{i\omega t}, \quad (2)$$

with the coinciding boundary conditions for the amplitude and the sought functions.

When we substitute the expression (2) into equation (1) and after some transformations made, we obtain a new equation in polar coordinates

$$\frac{\partial^2 U_1}{\partial r^2} + \frac{1}{r}\frac{\partial U_1}{\partial r} + k^2 u_1 = -\frac{\rho_0}{T},$$

where $$k^2 = \frac{\rho_m \omega^2 - i r_m \omega}{T}, \quad (3)$$

From the physical point of view, the parameter k^2 is a generalised numerical characteristic of the transverse wave distribution on the membrane surface after air supply into the pneumatic drive housing of the loose ultrafine materials dispenser.

If the membrane made free oscillations, then the right side of equation (3) would be equal to zero, and the equation itself would be referred to as the Bessel's equation with the standard record of the general solution. In our case, the solution of equation (3) with allowance for the right-hand side is written as follows:

$$u_1(r) = -\frac{\rho_0}{k^2 T} + C_1 M_0(kr) + C_2 N_0(kr),$$

where $C_1 M_0(kr) + C_2 N_0(kr)$ is the general solution of the Bessel equation, $M_0(kr)$ and $N_0(kr)$are the Bessel functions of the I and II type of the zero-order of the complex variable.

In addition to the boundary condition, the equation (1) is connected with the natural requirement of the boundedness of the function in question at the centre of the membrane ring, i.e. with $r = 0{:}\forall t|u(r,t)| < +\infty$.

From this condition, which is also valid for the function $u_1(r)$, it follows that $C_2 = 0$, (the Neumann function $N_0(kr) \xrightarrow[r\to+0]{} \infty$ takes into account the damping of oscillations, linearly independent of $M_0(kr)$), and

$$C_1 = \frac{\rho_0}{k^2 T M_0(kR_0)}.$$

The solution of equation (3) for the given boundary condition and the condition for the boundedness of the function in question has the form:$u_1(r) = \frac{\rho_0}{k^2 T}\left(\frac{M_0(kr)}{M_0(kR_0)} - 1\right)$, and the required solution of equation (1) determines the function:

$$u(r,t) = \frac{\rho_0}{k^2 T}\left(\frac{M_0(kr)}{M_0(kR_0)} - 1\right) \cdot e^{i\omega t}, \quad (4)$$

The parametric solution obtained sets completely and describes the forced oscillations of the annular membrane with the predetermined parameters ($R_0\, \rho_m,\ r_m,\ T$), which is fixed along the contour in the pneumatic drive housing of the dispenser of loose ultrafine materials.
the housing of the dispenser of loose ultrafine materials.

For further studies, it may be useful to calculate the deviation of the membrane points from the equilibrium position. Reasoning inductively, we consider a special case of the membrane points' deviation from the neutral position under the influence of the static pressure uniformly distributed over the area. To do this, we write the differential equation:

$$\frac{d^2u}{dr^2} + \frac{1}{r}\frac{du}{dr} = -\frac{P}{T}, \tag{5}$$

where $u = u(r)$ –the transverse deviation of the annular membrane section of radius *r*; *T* is the tension in the given annular membrane section (constant); *P* is the value of static pressure (constant).

The special solution of equation (5), satisfying the boundary condition $u(r)|_{r=R_0} = 0$ is a function of the form:

$$u(r) = \frac{PR_0{}^2}{4T}(1 - \frac{r^2}{R_0{}^2}),$$

where $R_o$is the outer radius of the annular membrane; $\frac{PR_0{}^2}{4T}$ is the deviation value of the inner contour, which is rigidly fixed on the rod.

If we use the resulting parametric solution (4), then we can find the average deviation of the annular membrane points from the equilibrium position at any (every) time point, with the pressure being changed according to the harmonic law:

$$u_{cp}(t) = \frac{1}{\pi(R_0{}^2 - r_0{}^2)}\int_{r_0}^{R_0} u(r,t)2\pi r dr == \left[\int krM_0(kr)d(kr) = krM_1(kr)\right] =$$

$$= \frac{2\rho_0}{k^2T(R_0{}^2 - r_0{}^2)} \cdot e^{i\omega t}\left[\frac{R_0M_1(kR_0) - r_0M_1(kr_0)}{kM_o(kR_0)} - \frac{(R_0{}^2 - r_0{}^2)}{2}\right] =$$

$$= \frac{\rho_0}{k^2T}\left[\frac{2(R_0M_1(kR_0)-r_0M_1(kr_0))}{kM_o(kR_0)(R_0{}^2-r_0{}^2)} - 1\right] \cdot e^{i\omega t}, \tag{6}$$

where $M_1(kr)$-the Bessel function of the first type of the first order; R_0 –the outer radius of the annular membrane; r_0 –its inner radius.

The function $u_{cp}(t)$ sets harmonic oscillations, the amplitude and phase of which are determined by the excitation frequency ω. Moreover, $\frac{\rho_0}{k^2T}\left[\frac{2(R_0M_1(kR_0)-r_0M_1(kr_0))}{kM_o(kR_0)(R_0{}^2-r_0{}^2)} - 1\right]$– are a complex numerical characteristic, the modulus of which coincides with the amplitude of the described harmonic oscillations, and the argument coincides with the phase of these oscillations.

The authors have developed the methodology for calculating the oscillatory system of the pneumatic dispenser, which includes the annular membrane with fixed edges, based on the equivalent replacement of the system with distributed parameters (membrane) by a system with lumped parameters. Such an approach makes it possible to replace the model of vibrational motion by the model of translational motion. The substitution of the annular membrane for an equivalent system with one degree of freedom, for example, for a linear oscillator, the mass element of which is a flat round rod, greatly simplifies the calculations and avoids complex equations of mathematical physics. Therefore, under the same excitation conditions, it is possible to substitute the membrane for a piston of the same cross-sectional area [14].

The piston motion is described by a linear inhomogeneous differential equation of the second order with a special right-hand side:

$$m_p\frac{d^2x}{dt^2} + r_p\frac{dx}{dt} + k_px = p_0S_pe^{i\omega t}, \tag{7}$$

the solution of which has the form

$$x(t) = \frac{p_0S_pe^{i\omega t}}{k_p - m_p\omega^2 + ir_p\omega}, \tag{8}$$

where $x(t)$– the piston' deviation from the equilibrium position at any time moment; m_p –the piston mass; r_p – the resistance coefficient to the piston's movement; k_p – the spring stiffness of the linear oscillator; S_p –the piston area which is equal to the area of the membrane.

Then we equate the right-hand parts of equations (6) and (8), based on the identity of the laws of the membrane and the equivalent piston motion, and obtain the sought parameters of the oscillator expressed by the parameters of the annular membrane:

$$m_p = p_m S_p, \tag{9}$$

$$k_p = m_p\omega^2 + S_p T Re\left[k^2\left(\frac{2M_1(kR_0)}{kR_0 M_0(kR_0)} - 1\right)^{-1}\right], \tag{10}$$

$$r_p = \frac{S_p T}{\omega} Im\left[k^2\left(\frac{2M_1(kR_0)}{kR_0 M_0(kR_0)} - 1\right)^{-1}\right],\ k^2 = \frac{\rho_m\omega^2 - ir_m\omega}{T} \tag{11}$$

As a result, we have determined the parameters of the linear oscillator, the piston of which, on the one hand, is equivalent to the initial membrane by the response to outward action, and on the other hand, it is identical to it by the impact on the surroundings. Therefore, we can substitute the membrane with an equivalent piston, wherever it is located in the oscillatory system and carry out all further calculations with the piston of the linear oscillator [14,15].

In the course of laboratory experiments at the premises of the Engineering Department of the Orenburg State Agrarian University, the parameters of a linear oscillator (rod) were obtained, which made it possible to simulate the introduction of ultrafine powders into the finished feed mixture. It was decided to manufacture the stem from structural steel, the stem length is 230 mm, the diameter is 20 mm, the mass is 354 g, the stem stroke is 44 mm. Based on equations (9) - (11) of the mathematical model, the membrane parameters were analytically determined: effective area - 229 cm^2, thickness - not more than 8 mm, the surface density of the material - not more than 1.5 g/cm^2.

The limits of the working pressure of the sprayed mixture from 202.6 KPa to 810.4 KPa, the consumption of the sprayed mixture from 0.35 cm^3 / s to 0, 95 cm^3 / s, the spray form is a symmetrical torch with an angle of 1200 have been experimentally established. To manufacture an atomiser of ultrafine materials, the brass with good wear resistance, anticorrosion and antifriction properties was selected at the Department of Materials and Construction Materials.

4. Conclusion

A prototype dispenser of bulk ultrafine materials with overall dimensions of 322x200x260 mm and a mass of 5.48 kg is being tested during production experiments at the premises of the Orenburg State Agrarian University instructional farm. The design of bulk ultrafine materials dispenser is presented for research to the Federal Institute of Industrial Property (application № 2018114335 of 18.04.2018), where the stage of examination of the essence of the invention is completed and the patent is being prepared. The engineering solution to the problem of dosing powders of micro metals in the process of preparing the feed mixture obtained by the authors is based on:

- experimentally established the influence of the conditions of formation of the working feed mixture on the properties of ultrafine powders;
- the analysis of the set of technological conditions for the combined feed industrial production;
- the ultrafine materials dynamic properties analysis.

The mathematical model of the pneumatic dispenser of ultrafine (nano) materials constructed by the authors makes it possible to unify engineering calculations taking into account the introduced structural and technological changes and to obtain analytically some of the parameters.

References

[1] State program for the development of agriculture and control of markets of farm products, raw materials and food-stuffs 2020 *Legislative Acts of Russia* 2012 **30**

[2] BerzinYa M 1952 Importance of cobalt and copper in feeding farm animals *Microelements in the life of plants and animals: Proc. of the USSR Academy of Sci.* (Moscow) pp. 473–492

[3] Levakhin VI, Chadaeva M N 1980*Effect of feeding trace elements to young cattle* (Reserves for increasing beef production: Russia, Orenburg)

[4] Fedorov VI 1973 *Growth, development and productivity of animals* (Moscow: Kolos) 272 p.

[5] Buch L, Lrieve D 1974 Minerals for dairy cattle *Ministry of Agriculture and Food* **410/57**

[6] Sizova EA 2017 Comparative characteristics of biological effects of copper and iron nanoparticles of different size *Vestnik Rossiyskoy Nauki* **3** 13–18

[7] Miroshnikova S A 2016*Technogenicnanomaterials in agrobiocenoses: perspectives and risks* (Russia, Orenburg)320 p.

[8] SizovaYeA 2016 On the prospects of nanomaterials based on the alloys of trace elements-antagonists (on the example of Fe and Co) *Farm Biology* **51(4)** 553–562

[9] Miroshnikov S 2015 Comparative assessment of the effect of cooper and nanoparticles and microparticles in chicken *Oriental Journal of Chemistry* **31** 2327–2336

[10] Notova S 2015 Change of Elemental Composition in Muscular Tissue and Hair under Food Stress *Biosciences Biotechnology Research Asia*.**12** 25–31

[11] Rates of technological design of enterprises producing mixed fodders. [Electronic resource] RF Library of State standards and norms. Access mode: http: // infosait.ru, free. (Date of circulation: July 21, 2018)

[12] Chkalova MV, Shakhov VA, Burlutsky EM 2018 Some aspects of functioning technological lines for the production of mixed fodders by the example of enterprises of the Orenburg region *Improvement of engineering and technical support of production processes and technological systems: Proc.* (Orenburg, Russia)

[13] Chkalova MV, PavlidisVD 2018 Efficacy increase of mixed feeds preparation on the basis of using ultrafine materials *Urgent problems of personnel potential for innovative development of agroindustrial complex: Proc.*(Minsk, Belarus)

[14] Kezik VI 2013 Forced axisymmetric oscillations of a circular membrane, which is an element of an acoustic oscillatory system *Electronic J. of Technical Acoustics*. **9**

[15] Kezik VI 2014 Equivalent parameters of a circular homogeneous membrane *Electronic J. "Technical Acoustics"* **8**

Determinants factors toward the performance of the employee in the crude palm oil industry in West Sumatera, Indonesia

Hendy Tannady[1], Johanes Fernandes Andry[2], Filscha Nurprihatin[3]

[1]Department of Management, Universitas Pembangunan Jaya, Indonesia
[2]Department of Information System, Universitas Bunda Mulia, Indonesia
[3]Department of Industrial Engineering, Universitas Bunda Mulia, Indonesia

corresponding e-mail: *[1]hendy.tannady@upj.ac.id, [2]jandry@bundamulia.ac.id, [3]fnurprihatin@bundamulia.ac.id

Abstract. Human resources are the most important part in achieving organisational goals and improving organisational performance, in completing its work human resources require an adequate working environment and compensation as a reward for providing maximum work results. The purpose of this study is to examine and analyse the effect of compensation (X_1) and work environment (X_2) on employee performance (Y), especially in the crude palm oil industry in West Sumatra. Observations were made on three companies, research conducted in March to June 2019 showed that between compensation and work environment has a positive and significant effect on employee performance. With the structural equation modelling approach and Lisrel 8.8 application assistance, the coefficient of estimated of compensation toward performance is 0.37 while the estimated coefficient of work environment toward performance is 0.42. Compensation and work environment simultaneously have a positive influence on employee performance of 0.65.

1. Introduction

Human resources are the most important part in achieving the goals of an organisation, both large and small organisations, as well as companies as an organisation [4]. Equipment owned by organisations or companies with various levels of technology will be meaningless if not handled properly by appropriate human resources [3]. Human resources are an important element in improving the performance of an organisation [5], [24]. Every leader needs to ensure that every employee who works in their organisation has a high commitment toward his work and provides the best work results for the achievement of organisational goals [12]. It is easier for management to influence the work performance of production equipment because production equipment will work according to instructions and manuals, production equipment also does not require compensation or additional incentives to spur its work performance, but human resources require a combination of various factors both physical and psychologically to increased their work performance [18]. The purpose of this study is to look at the model of the relationship between compensation and work environment toward employee performance.

Research conducted by Akter & Husain on the garment industry in Bangladesh concluded that the provision of appropriate compensation has a positive impact on the work performance of employees [2]. Darma & Supriyanto, in their research at one of the telecommunication company in Indonesia, concluded that the provision of appropriate compensation could play an important role in improving employee performance and job satisfaction [7]. Regarding the impact of a supportive work environment, Malik et al. who conducted a study of 115 employees in telecommunication companies in Pakistan concluded that management could improve the work performance of employees by providing a conducive work environment [13]. In line with Malik et al., a study conducted by Samson et al. toward 173 bank employees also concluded the same thing that a good work environment can improve employee work performance [21].

2. Literature Review

2.1. Performance

Organisational performance refers to the ability of the organisation to achieve several goals set by the management team, top leader or the founder such as good product quality, large market share, high profits, and maintained business sustainability [11]. Work performance is an aggregate financial or non-financial value given by employees as a form of a contribution to the fulfilment of both directly and indirectly in their participation to succeed in achieving the targets of the organisation [6]. So that the organisation is expected to be able to fulfil [27]. Work performance of the employee is something that can be managed well to achieve organisational goals, and through proper management of human resources, the organisation will have a competitive advantage over competitors [17]. The behaviour of an employee largely determines work performance, and this behaviour can affect organisational goals so that good behaviour control over each individual in the organisation will also accelerate the achievement of organisational goals [26]. Work performance can be measured through three dimensions quantity, quality and discipline factors [25], [14].

2.2. Compensation

Compensation is various forms of payment or appreciation given by the organisation to employees which will increase according to the status of their position [8]. Compensation can play an important role in improving employee performance [10]. Appropriate compensation policies can increase employee focus on his work and reduce the level of turnover in the organisation [15]. Failure to design appropriate policies regarding compensation systems has negatively impact on productivity and job satisfaction of the employee and the overall effectiveness of the organisation [16]. Compensation can be measured through two dimensions, namely financial factors and non-financial [23].

2.3. Work Environment

The quality of work depends on safe and healthy working conditions. Working environment conditions also determine employee work behaviour [1]. The condition of the work environment will greatly affect job satisfaction and work engagement of an employee, which, of course, has the final impact on better employee performance [22]. The work environment is an internal and external condition of the organisation that can affect the working spirit and speed of task completion of an employee [3]. The work environment is all the facilities and infrastructure used and felt by an employee when working, and this greatly affects work practices. Work environment includes the workplace, funding facilities to support the smooth work, cleanliness, lighting, tranquillity, and good working relationships with colleagues or with superiors [20]. The work environment can be measured in three dimensions, namely work facilities, relationship with colleagues and work atmosphere [19].

3. Research Methodology

Analysis tools used in this research are Structural Equation Modelling (SEM). The variables used are work performance (Y), compensation (X_1) and work environment (X_2). Variable of Y uses three dimensions, namely quantity, quality and discipline factors. These three dimensions produce six statement indicators (PE1-PE6) [25], [14]. Variable of X_1 uses two dimensions, namely financial factors and non-financial factors, these two dimensions produce six statement indicators (CO1-CO6) [23]. Variable of X_2 uses three dimensions, namely work facilities, relationship with colleagues and work atmosphere. These three dimensions produce six statement indicators (WE1-WE6) [19]. The population in this study were all workers in crude palm oil in West Sumatra. Purposive sampling technique is used for sampling. The total sample of respondents is a minimum of 90 respondents using multiplication of five and 18 indicators [9]. The design of the questionnaire uses the interval measurement scale. Data quality testing uses validity and reliability tests. Data quality tests were tested on 30 respondents. After confirmed that the data are valid and reliable, it will be analysed using SEM. Validity test is conducted using Pearson correlation if the value of r-count is greater than the value of r-table, then it can be concluded that the indicators used in the study are valid, a variable is concluded to be reliable if the value of Cronbach alpha is greater than 0.6.

IOP Conf. Series: Materials Science and Engineering **771** (2020) 012066 doi:10.1088/1757-899X/771/1/012066

4. Discussion and Analysis

This section will discuss the results of a series of data quality tests which include validity and reliability tests, then proceed with confirmatory analysis using SEM. For the parameters in the validity test required the value of degree of freedom (DOF) to identify the value of r table, with the number of respondents to the initial survey of 30 people, the value of r table is 0.361 (dof : 28). Data is declared valid if it has r value greater than r table (r-value ≥ r table, α: 0.05). Table 1 shows the results of the validity and reliability tests of all indicators.

Table 1. Validity and Reliability Test Results

Compensation (Reliability: 0.74)		Work Environment (Reliability: 0.81)		Performance (Reliability: 0.76)	
Indicator	Validity Score	Indicator	Validity Score	Indicator	Validity Score
CO1	0.422	WE1	0.562	PE1	0.589
CO2	0.614	WE2	0.559	PE2	0.534
CO3	0.511	WE3	0.533	PE3	0.549
CO4	0.623	WE4	0.554	PE4	0.694
CO5	0.506	WE5	0.571	PE5	0.663
CO6	0.576	WE6	0.701	PE6	0.617

The purpose of the CFA analysis is to find the coefficient of estimation or regression of each latent variable toward the dependent variable either partially or simultaneously. Table 2 shows the results of the analysis of the coefficient of regression and t-values of the compensation and work environment on work performance.

Table 2. CFA Measurement Results

No	Variable	Estimation Coefficient	T Value
1	Compensation	0.37	0.58
2	Work Environment	0.42	0.62

The next step is to conduct goodness of fit test, known as the model feasibility test, the purpose of this test is to identify whether the model built in the study is sufficient to use data samples and instruments. The expectation of the experiments in this test is the decision to accept H0. Table 3 shows information about the final results of the goodness of fittest. Some parameters used in the goodness of fit test are GFI, RFI, NFI, IFI, CFI, NNFI, RMSEA, and AGFI.

Table 3. The goodness of fit Index

Parameters	Results	Criteria
GFI, RFI, NFI	0.88, 0.84, 0.92	Marginal fit, Marginal fit, good fit
IFI, CFI, NNFI	0.86, 0.813, 0.903	Marginal fit, Marginal fit, good fit
RMSEA	0.072	Good fit
AGFI	0.871	Marginal fit

Based on the result of analysis and data processed on the full structural model, the results show that there is an effect of compensation on work performance with a value of regression coefficient is 0.37 and t value is 0.58, these results indicate the acceptance of hypothesis 1, compensation has a positive and significant effect on work performance. The results of the analysis of the effect of the work environment on work performance, the value of regression coefficient is 0.42 and t value is 0.62, these results indicate the acceptance of hypothesis 2, the work environment has a positive and significant effect on work performance. SEM analysis results also show the simultaneous effect of latent variables in influencing the dependent variable, R^2 value is 0.65, these results indicate that simultaneously compensation and work environment has a positive and significant effect of improving the work performance of employees who were working in the crude palm oil industry in West Sumatra, Indonesia.

5. Conclusion

The variables of compensation and work environment either partially or simultaneously have a positive and significant effect in influencing the work performance of employees in the crude palm oil industry in West Sumatra. Therefore it is recommended for management to evaluate or make appropriate policies regarding compensation and work environment for employees. Based on descriptive analysis, it is found that the dimensions of non-financial factors and work facilities have an average value below the average value of the other dimensions on each variable, so efforts to improve work performance can begin with a focus on efforts to make appropriate policies related to non-financial factors and how to improve work facilities. Suggestions for further research is to estimate work performance by using other latent variables such as leadership, organisational citizenship behaviour or employee engagement. Further research can also add moderation variables or use the same model on different research objects.

References

[1] Ahmad, A. (2011). Study of work environment and employees' performance in Pakistan. *African Journal of Business Management,* 5(34), 13227-13232.

[2] Akter, N., and Husain, M. M. (2016). Effect of compensation on job performance: An empirical study. *International Journal of Engineering Technology, Management and Applied Sciences*, 4(8), 103-116.

[3] Al Omari, K., and Okasheh, H. (2017). The influence of work environment on job performance: A case study of engineering company in Jordan. *International Journal of Applied Engineering Research*, 12(24), 15544-15550.

[4] Aslam, A., Ghaffar, A., Talha, T., and Mushtaq, H. (2015). Impact of compensation and reward system on the performance of an organization: an empirical study on banking sector of Pakistan. *European Journal of Business and Social Sciences*, 4(8), 319-325.

[5] Bello, S. M. (2012). Impact of ethical leadership on employee job performance. *International Journal of Business and Sosial Science,* 3(11), 228-236.

[6] Dajani, M. A. Z. (2015). The impact of employee engagement on job performance and organisational commitment in the Egyptian banking sector. *Journal of Business and Management Sciences*, 3(5), 138-147.

[7] Darma, P. S., and Supriyanto, A. S. (2017). The effect of compensation on satisfaction and employee performance. *Management and Economics Journal,* 1(1), 69-77.

[8] Dessler,G. (2005). *Human Resource Management*. 10th edition. Prentice-Hall.

[9] Hair, J. F., Black, W. C., Babin, B. J., and Anderson, R. E. (2010). *Multivariate Data Analysis*. New Jersey: Pearson.

[10] Hameed, A., Ramzan, M., Zubair, H. M. K., Ali, G., and Arslan, M. (2014). Impact of compensation on employee performance (Empirical evidence from banking sector of Pakistan). *International Journal of Business and Social Science*, 5(2), 302-309.

[11] Koontz, H., and Donnell, C. (1993). *Introduction to Management*. New York : McGraw-Hill Inc.

[12] Maduka, C. E., and Okafor, O. (2014). Effect of motivation on employee productivity: A case study of manufacturing companies in Nnewi. *International Journal of Managerial Studies and Research* (IJMSR), 2(7), 137-147.

[13] Malik, M. I., Ahmad, A., Gomez, S. F., and Ali, M. (2011). A study of work environment and employees' performance in Pakistan. *African Journal of Business Management,* 5(34), 13227-13232.

[14] Muda, I., Rafiki, A., and Harahap, M. R. (2014). Factors influencing employees' performance: a study on the Islamic banks in Indonesia. *International Journal of Business and Social Sciences*, 5(2), 73-80.

[15] Negash, R., Zewude, S., and Megersa, R. (2014). The effect of compensation on employees motivation: In Jimma University academic staff. *Basic Research Journal of Business Management and Accounts*, 3(2), 17-27.

[16] Neo, R., Hollenbeck, J., Gerhart, B., and Wright, P. M. (2006). *Human Resource Management (Gaining Competitive Advantage)*. 5th edition. Mcgraw-Hill Companies Inc.

[17] Njoroge, S. W., and Kwasira, J. (2014). Influence of compensation and reward on performance of employees at Nakuru County government. *IOSR Journal of Business and Management*,

17(11), 87-93.

[18] Nzewi, H. N., Augustine, A., Mohammed, I., and Godson, O. (2018). Physical work environment and employee performance in selected brewing firms in Anambra State, Nigeria. *Journal of Good Governance and Sustainable Development in Africa*, 4(2), 131-145.

[19] Pawirosumarto, S., Sarjana, P. K., and Gunawan, R. (2017). The effect of work environment, leadership style, and organizational culture towards job satisfaction and its implication towards employee performance in Parador Hotels and Resorts, Indonesia. *International Journal of Law and Management*, 59(6), 1337-1358.

[20] Putra, P. P. A. A., and Sary, F. P. (2017). The influence of working environment towards employee performance in perum damri Bandung. *e-Proceeding of Management*, 4(2), 1753-1757.

[21] Samson, G. N., Waiganjo, M., and Koima, J. (2015). Effect of workplace environment on the performance of commercial banks employees in Nakuru town. *International Journal of Managerial Studies and Research,* 3(12), 76-89.

[22] Scott, K. D., Jusanne, M., and Steven, M. E. (2000). Factors influencing employee benefits beliefs that pay is tied to performance. *Journal Business Psychology,* 14, 553-562.

[23] Tannady, Hendy., Ismuhadjar., and Zami, Alex. (2017). Factors affecting the performance of driver: The experience of transjakarta bus driver. *International Journal of Research Science & Management,* 4(11), 22-28.

[24] Tannady, Hendy., Tannady, Henny., and Zami, Alex. (2019). The effect of organizational culture and employee engagement on job performance of healthcare industry in province of Jakarta, Indonesia. *Calitatea,* 20(169), 18-22.

[25] Wong, W. P., and Wong, K. Y. (2007). Supply chain performance measurement system using DEA modeling. *Industrial Management and Data Systems*, 107(3), 361–381.

[26] Zheng-Xue, L., Kan, S., Wen-Dong, L., & Dan-Min, M. (2008). Construct of job performance: Evidence from Chinese military soldiers. *Asian Journal of Social Psychology*, 11(3), 222-231.

[27] Andry, J.F., Suroso, J.S., Bernanda, D.Y. (2018). Improving quality of smes information system solution with ISO 9126. *Journal of Theoretical and Applied Information Technology*, 96(14), 4610-4620.

IOP Conf. Series: Materials Science and Engineering **771** (2020) 012067 doi:10.1088/1757-899X/771/1/012067

Social capital of local communities in the water resources management: an insight from Kepulauan Riau

W E Yudiatmaja[1,*], Yudithia[2], T Samnuzulsari[3], Suyito[4] and Edison[5]

[1] Department of Public Administration, Faculty of Social and Political Sciences, Universitas Maritim Raja Ali Haji, Jl. Raya Dompak Tanjungpinang, Kepulauan Riau 29111, Indonesia
[2] Secretary Office of the Regional House of Representatives of Kepulauan Riau, Pulau Dompak, Tanjungpinang, Kepulauan Riau, Indonesia
[3] Department of Sociology, Faculty of Social and Political Sciences, Universitas Maritim Raja Ali Haji, Jl. Raya Dompak Tanjungpinang, Kepulauan Riau 29111, Indonesia
[4] Department of Sociology, Sekolah Tinggi Ilmu Sosial dan Politik (Stisipol) Raja Haji, Jl. Raja Haji Fisabilillah No. 48 Tanjungpinang, Kepulauan Riau 29122, Indonesia
[5] Department of Public Administration, Faculty of Social and Political Sciences, Universitas Maritim Raja Ali Haji, Jl. Raya Dompak Tanjungpinang, Kepulauan Riau 29111, Indonesia

[*] Corresponding author: wayuguci@umrah.ac.id

Abstract. This paper draws the process of water resources management of local communities by using the social capital point of view. By using a comparative case study, this research was conducted in two regions in Kepulauan Riau, Senggarang and Mantang. Social capital of local communities, such as norms, trust, beliefs, and gotong-royong play a significant role in the process of governing water resources in different ways. This paper contributes to the study of water resources management by recognizing the social capital of local communities. In practical terms, this study is also valuable to the practice of water resources management in the local context in Indonesia. The theoretical, methodological, and practical implications of the study are discussed.

1. Introduction

Water has become a pivotal need in all communities around the world. Because of common interest, water has been considered as a strategic product to be sell. As long as the development of capitalism, water has transformed to be privatised by the multinational corporation, such as Aqua, Danone, and Nestle [1]. They have captured clean water resources in many places, including in Indonesia. In their hands, water has been modified as an economic entity profitable in a global market. However, in several areas in the globe, water is not only placed as commercial goods, but also social commodity maintained and preserved continuously by the local people. They culturally manage the water originated from natural resources in such a region steadily. It is oriented to the shared values, for instance, norms, beliefs, and self-regulation. The results are considerably used for the general interest of the community.

Indonesia is the country with a large number of ethnic diversities. Every ethnic community has a unique social capital contributed to the practice of communal life [2, 3], including natural resources. In Kepulauan Riau, there are ethnic communities participated in managing water resources. The community-based water resources management is more successful than government provision. It is because of the role of the social capital of the community participated in the governing process of water resources. Because of including in poor of clean water availability, the society in Kepulauan Riau attempts to maintain water resources collectively. They create the system and management of the management process of water resources through community-based orientation. In the process of governing, it also involves the social capital of the community because it has become imprescriptible of their daily life. One of the water resources managed by the local community in Kepulauan Riau is water resources in Tanjungpinang Kota Subdistrict, Tanjungpinang City, and water resource facilities in Mantang Subdistrict, Bintan Regency. It still preserves by local people until today.

The idea of social capital is widely known in social and political literature because of Bourdieu's, a French philosopher, works. Bourdieu defined social capital as the aggregation of the actual or potential resources related to the possession of a long-lasting network of more or less institutionalised interactions of mutual acquaintance and recognition or, in other words, membership of a group providing each of its members with the support of collective capital, a credential entitling them to praise in the numerous senses of the word [4]. Bourdieu's notion of social capital has been compiled and developed by the next generations, such as Putnam, Coleman, Fukuyama, and Portes. Following Siisiäinen, in sum, there is two streams in understanding social capital. They are the structural and functional perspectives [5]. The structural emphasises the influence of societal structure and environment on a set of norms, values, ideas, and practices of the humankind. On the other hand, the functional lens views social capital as a consequence of the social system.

Numerous scholars have used the framework of social capital in discussing natural resources management [6–8]. Yet, the prior studies have rarely been paid attention to the social capital in the case of water resources management. Moreover, many inquiries on social capital only focus on a single community and fewer means to compare social capital among the local people. Hence, to fill the gap in the recent literature, this paper aims to shed light on the social capital practised by local communities in Tanjungpinang City and Bintan Regency, Kepulauan Riau Province. The objectives of this paper are twofold: (1) to identify the dimension of social capital and to analyse the practical aspects of social capital among two different communities in the process of water resources management, and (2) to compare them with each other.

This paper has three contributions to the study of water resources management, respectively. First, this study enriches the literature on water resources management by adapting the concept of social capital. The previous researchers pay limited attention to utilise the social capital perspective in examining water resources management. Second, prior studies on social capital in natural resources management and water resources management as well as being dominated by single case study research. No prior scholars are employing dual or multiple-case studies and compare them. This study finally discusses the social capital of local communities in Indonesia in the process of water resources management. Through this research, we offer several recommendations to the policymakers in formulating and implementing water resources management based on community involvement.

The rest of the paper presents the methodological issue of the study. It captures the approach, design, collecting the data, and analysing the data. The results of this study are drawn in two case studies: The Chinese community in Senggarang and Malay community in Mantang. The part subsequently is followed by the discussion and implications of the study. Eventually, the conclusion and recommendations for further research are provided in the last section of the paper.

2. Methods

This research was conducted in two areas in Kepulauan Riau: (1) Senggarang urban village, Tanjungpinang Kota Subdistrict, Tanjungpinang City, and (2) Mantang Village, Mantang Subdistrict, Bintan Regency (the location can be seen in Figure 1 and 2). During March and Mei 2018, this study was structured. By implementing a qualitative strategy, this research employed a comparative case study. For two purposes, it was used. Firstly, it can draw the case deeply. Secondly, by using a comparative case study, we delivered several cases of communities' social capital.

IOP Conf. Series: Materials Science and Engineering **771** (2020) 012067 doi:10.1088/1757-899X/771/1/012067

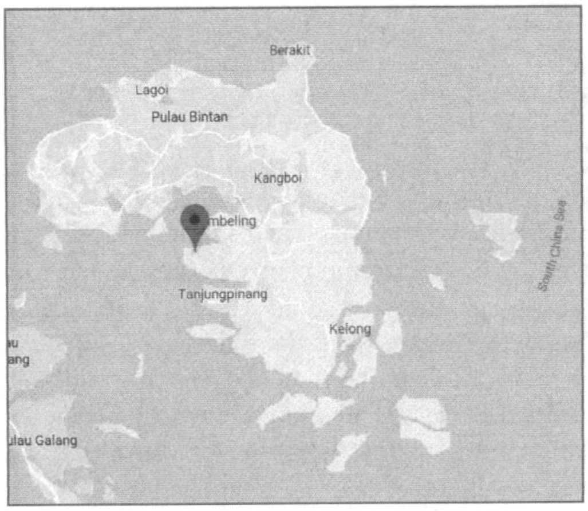

Figure 1. The Map of Senggarang
(Created with google maps, 09/12/2019)

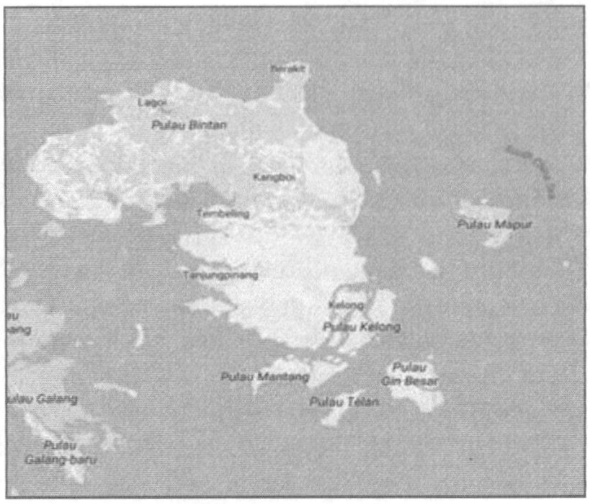

Figure 2. The Map of Mantang
(Created with google maps, 09/12/2019)

The data in this work consisted of primary and secondary data. The primary data was gathered by interviewing the participants and observing the practices of social capital in Senggarang and Mantang local communities. A series of in-depth interviews were conducted on the management of water resources management, the member, the society, community figures, and officials of the urban village and village government in Senggarang and Mantang. We used a semi-structured interview by utilizing an interview guide. The probing technique was employed to understand the meaning of informants' responds. It was coded by using open coding before the data were analyzed. To observe the participants, we joined with them in the activities, such as meeting and mutual assistance (*gotong royong*). The secondary data was collected from the government survey and scholarly journals related to the research. The data were analysed by using thematic analysis by grouped the data in specific terms of social capital, for example, networks, norms, regulations, etc.

3. Results and discussion

3.1 Case study 1: Social capital of the Chinese community in Senggarang

The first water resource managed by the local community in Kepulauan Riau discussed in this paper is Senggarang clean water facility. Senggarang is one of the urban villages in Tanjungpinang Kota Subdistrict of Tanjungpinang City. It is located near to the capital city of Tanjungpinang or approximately 24 km from the port of Sri Bintan Pura, a widely acknowledged as formerly city centre of Tanjungpinang. Generally, Tanjungpinang can be drawn as a multi-ethnic town because, in this city, there are diverse ethnic groups, such as Malay, Javanese, Chinese, Batak, and East Nusa Tenggara. The diversity of ethnic groups has established Tanjungpinang as plural and cosmopolitan city in Sumatera. Although Malay has become the largest ethnic group inhabiting this city, there are also other ethnic groups playing an important role in city development. In their societal life, Tanjungpinang's society remarkably maintains harmony in diversity.

According to the data of the Regional Bureau of Statistics of Tanjungpinang, the total of the Senggarang population is around 3.194 persons in which 222 individuals inhabit each of one km^2 of this area. From the amount, it is predicted that over 40 per cent of the population is Chinese, circa 50 per cent is Malay, and 10 per cent is other ethnic groups. The Chinese community is believed to have existed for a long period in this place. They have incorporated with local people in Tanjungpinang. Throughout our observations, the local people and the Chinese community can live together peacefully. All participants also revealed that they could wisely accept the difference.

Table 1. Demographic Conditions of Tanjungpinang Kota Subdistrict of Tanjungpinang City

Urban Village	Wide of Area (Km^2)	Population	Population Density (Person/Km^2)
Tanjungpinang Kota	0,64	5.079	7.936
Penyengat	1,11	2.275	2.050
Kampung Bugis	23,56	7.175	305

IOP Conf. Series: Materials Science and Engineering **771** (2020) 012067 doi:10.1088/1757-899X/771/1/012067

Senggarang	14,38	3.194	222
Total	39,69	17.723	447

Source: [9]

In Senggarang, there is a clean water facility established in 2002 by the Chinese community. The Chinese people attempt to address a drought issue in Senggarang, especially in the dry season. It was built by digging the soil to be a pool covering an area of 100 square meters. By using a hundred meters pipe, the water is streamed automatically to the people's homes. Because the source is in a high location, the water can flow by itself without using a machine. The water not only is benefited the Chinese people but also all ethnic groups in Senggarang. Klenteng Senggarang Foundation leads the operationalization of the facility. The foundation also handles a worship place of Chinese people in Senggarang, which is Vihara Sasana Dharma. In other words, the foundation takes place as not only a social role but also a religious function. It has been existed along with the existence of the Vihara. Although the government has built a clean water facility in Senggarang in 2015, Klenteng Senggarang Foundation still preserves and operates its clean water facility. In the recent development, the government's clean water facility cannot be performed successfully because it is not supported by the Senggarang people.

Figure 3. The Clean Water Facility Managed by Local Community in Senggarang, Subdistrict of Tanjungpinang Kota, Tanjungpinang City (Source: Research Documentation, 04/29/2018)

Figure 4. Klenteng Senggarang Foundation Managing Clean Water Facility in Senggarang, Subdistrict of Tanjungpinang Kota, Tanjungpinang City (Source: Research Documentation, 04/29/2018)

Klenteng Senggarang Foundation operates the facility autonomously. Every month they obtain from the individuals in Senggarang retribution of Rp.50,000,-/home. It is used for water plant maintenance. Interestingly, the foundation does not force the resident who is unable to pay the retribution. To keep the water clean, the Klenteng Senggarang Foundation frequently clean up the pool without any further assistance from another ethnic group. The people believe in the foundation to manage the water resources.

3.2 Case study 2: Social capital of the Malay communities in Mantang

The second case study is water resources governed by the Malay community at the Mantang Lama village, Mantang Subdistrict, Bintan Regency, Kepulauan Riau. The clean water installation is situated in Dusun (sub-village) Mantang Lama and Mantang Riau. The population of Mantang Lama is 930 persons by which the majority of them (almost 90 per cent) is the Malay ethnic group. They began to establish a clean water facility in 2007. For the first time, the development of a clean water pool is assisted by the Ministry of Public Works of the Republic of Indonesia. However, then it was turned over to the Mantang society. Currently, the freshwater facility is led by the residents in Dusun Mantang Lama and Mantang Riau. In contrast to Dusun Mantang Riau, the clean water facility in

IOP Conf. Series: Materials Science and Engineering **771** (2020) 012067 doi:10.1088/1757-899X/771/1/012067

Dusun Mantang Lama is managed by the village government of Mantang Lama. The citizens of Mantang Riau would not like to join with Mantang Lama because they not believe with the village government.

Table 2. Demographic Conditions of Mantang Subdistrict of Bintan Regency (2017)

Village	Wide of Area (Km^2)	Population	Population Density ($Person/Km^2$)
Mantang Lama	37	930	25
Mantang Besar	48	1.400	29
Mantang Baru	11	1.007	91
Dendun	18	1.062	59
Total	114	4.399	38

Source: [10]

The governance of water resources in these two places is typically similar. The management comes from the inhabitants chosen through the citizen's discussion (*musyawarah warga*). The responsibility of the administration is to ensure the availability of the water to the people in the region. The management has the mandate to collect retribution to the people. The tariff is not similar in every Dusun. If Dusun Mantang Lama assigns that the vengeance is Rp.30.000/month, Dusun Mantang Riau only inserts Rp.20.000/month. It is not the same with the Senggarang community not burdening the people with the tariff, in which in Mantang the people who cannot pay the retribution, their water facility will be withdrawn. To clean the pool, the citizens in each region always conduct mutual assistance (in Indonesia called with *gotong-royong*).

3.3 Discussion

The findings of this study go beyond prior research, indicating that the features of social capital merely limited to the network, trust, and beliefs. This study finds a new feature of social capital, typically Indonesia, which is gotong-royong. As explained above, both in Senggarang and Mantang, the local community has practised gotong-royong in governing water resources. Similar findings are also reached by Hanani and Bowen, concluding that gotong-royong in the form of social capital of Indonesian local community influencing the social cohesion and development process [11–13]. However, gotong-royong still limited to the same ethnic groups. For instance, in Senggarang gotong-royong has not involved other ethnic groups in the region whereas the clean water facility is used for all people.

Our results also show that there are many institutional issues faced by the community in managing water resources management, involving technical and non-technical problems. Dusun Mantang Lama's management of clean water is restricted to the maintenance of water suction machines because several of the necessary spare components are only accessible in the stores in the large cities in Java, such as Jakarta and Surabaya. This situation created huge expenses for both groups and a long time for a broken machine to be repaired. Water resources management practised by the groups is still managed by an unofficial system, the group name, structure, standard operating procedure (SOP), and a definite organisational unit are not yet available. Communities couldn't achieve effective, efficient, and responsible, clean water management. This is demonstrated by the absence of leadership abilities in conducting clean water management operations, uncertain organisational hierarchy, lousy knowledge of financial literacy, lack of comprehension and synchronisation of village asset management with village income, the weak capacity to manage conflicts, and inadequate ability to use contemporary technology to assist governing procedures. There are several administrative barriers, including restricted expertise in managing operations, reporting, and using the web to support management.

IOP Conf. Series: Materials Science and Engineering **771** (2020) 012067 doi:10.1088/1757-899X/771/1/012067

Figure 5. Clean water facility Managed by Local Community at the Dusun Mantang Lama, Mantang Subdistrict, Bintan Regency (Source: Research Documentation,

Figure 6. Clean water facility Managed by Local Community at the Dusun Mantang Riau, Mantang Subdistrict, Bintan Regency (Source: Research Documentation,

We analyse the difference of social capital between the Senggarang and Mantang community by using four indicators, such as trust, membership, norms of reciprocity, and network resources. Table 3. summarises the difference. In Senggarang, trust is culturally formed between Chinese and other ethnic groups, although the Chinese community performs the management of a clean water facility. However, in Mantang, it has been pushed by the economic factor. It is because of the relation between the management and the society more contractual. The society has to pay some money as compensation for the use of clean water. The membership of the community occurring in Senggarang tends to more voluntaristic than obliged because no specific norms are regulating the common interest.

On the contrary, in Mantang the society is bounded by a regulation, thereby they are more obliged. The result shows that reciprocal norms in Mantang are higher than Senggarang because the relationship constitutes among the same ethnic group. Senggarang community has high network resources than Mantang people because Chinese group in Senggarang can cooperate with other Chinese communities in Tanjungpinang to obtain their need in managing water facilities.

Table 3. The Difference of Social Capital Dimensions among Communities

Dimension	Senggarang community	Mantang community
Trust	Culturally established	Economically established
Membership	Voluntaristic	Obliged
Norms of reciprocity	Low	High
Network resources	High	Low

3.4 Theoretical, methodological, and practical implications

This study complements the research on water resources management in three categories involving theoretical, methodological, and practical features. By providing gotong-royong as social capital, this study has extended the concept of social capital. Besides, from the evidence, we argue that the concept of social capital should be strengthened with an institutionalist perspective. We can combine the notion of social capital with the institutionalism view, as suggested by Ostrom [14]. In methodological terms, this study strengthens the previous works on social capital using a qualitative approach. Finally, this study can be considered by the local government in designing water resources management in the local community. The government should prepare positive social capital and environment before implementing water resources management policy based on community governance.

4. Conclusion

In this paper, we analyse the social capital of local communities in governing water resources at two different regions in Kepulauan Riau. In summary, we argue that the different patterns of social capital established because of the distinguish of socio-cultural backgrounds of the community. Besides, social capital, including norms, networks, values, togetherness, and mutual assistance (*gotong-royong*) plays a crucial role in determining the process of water resources management. The institutional problem can inhibit the achievement of common management in long terms if the issues cannot be addressed soon. These results contribute to add our understanding of the water resources management by considering the social capital of the community. It is worthwhile to the decision-maker in Indonesian local government focussing on the issue of water resources as well as to design the effective implementation of water resources management through the recognition of local communities' social capital.

This study is not spared from several limitations, and thereby we are responsible for revealing it. These limitations are apparent in three respect. First, our study focuses on the social capital of the community. The community in two regions is the unit analysis of the study. Further works should consider the links between people across different organisations (cross-cutting ties) [15]. Second, we have used qualitative and case study research to reach some conclusion in this paper. The utilisation of qualitative research can extensively analyse the case. However, it cannot be a generalisation basis for the strengthening of our arguments.

Consequently, the next studied can adopt quantitative [16] or mixed-methods (quantitative and qualitative simultaneously). Third, this study has not considered gender viewpoint in investigating the social capital of the community. It tends to assume that gender is not a significant part of forming social capital. Therefore, further research should look into the gender aspect in the process of water resources management [17].

References

[1] N. Heynen and P. Robbins, "The neoliberalization of nature: Governance, privatization, enclosure and valuation," *Capitalism Nature Socialism*, vol. 16, no. 1, pp. 5–8, 2005.

[2] P. Barron, K. Kaiser, and M. Pradhan, "Understanding variations in local conflict: Evidence and implications from Indonesia," *World Development*, vol. 37, no. 3, pp. 698–713, 2009.

[3] D. Mavridis, "Ethnic diversity and social capital in Indonesia," *World Development*, vol. 67, pp. 376–395, 2015.

[4] P. Bourdieu, "The forms of capital," in *Handbook of theory and research for the sociology of education*, J. G. Richardson, Ed., New York: Greenwood Press, 1986, pp. 241–258.

[5] M. Siisiäinen, "One concept, two approaches: Bourdieu and Putnam on social capital," *International Journal of Contemporary Sociology*, vol. 40, no. 2, pp. 183–204, 2003.

[6] F. Cleaver, "Reinventing institutions: Bricolage and the social embeddedness of natural resource management," *European Journal of Development Research*, vol. 14, no. 2, pp. 11–30, 2002.

[7] J. Pretty, "Social capital and the collective management of resources," *Science*, vol. 302, no. 5652, pp. 1912–1914, 2003.

[8] T. Claridge, "Social capital and natural resource management: An important role for social capital?," M.A. Thesis, The University of Queensland, Queensland, 2004.

IOP Conf. Series: Materials Science and Engineering **771** (2020) 012067 doi:10.1088/1757-899X/771/1/012067

[9] Regional Bureau of Statistics of Tanjungpinang, *Tanjungpinang Kota Subdistrict in figure 2018*. Tanjungpinang: Regional Bureau of Statistics of Tanjungpinang, 2018.
[10] Regional Bureau of Statistics of Bintan, *Mantang Subdistrict in figure 2018*. Bintan: Regional Bureau of Statistics of Bintan, 2018.
[11] V. A. Beard, "Individual determinants of participation in community development in Indonesia," *Environment and Planning C: Government and Policy*, vol. 23, no. 1, pp. 21–39, 2005.
[12] S. Hanani, "Studi negosiasi kultural yang mendamaikan antaretnik dan agama di Kota Tanjungpinang [Study on reconciled cultural negotiation inter-ethnics and religion in Tanjungpinang city]," *Episteme*, vol. 12, no. 1, pp. 201–230, 2017.
[13] J. R. Bowen, "On the political construction of tradition: Gotong Royong in Indonesia," *Journal of Asian Studies*, vol. 45, no. 3, 545-561, 1986.
[14] E. Ostrom, *Understanding institutional diversity*. Princeton, New Jersey: Princeton University Press, 2005.
[15] P. Paxton, "Is social capital declining in the United States? A multiple indicator assessment," *American Journal of Sociology*, vol. 105, no. 1, pp. 88–127, 1999.
[16] D. Narayan and M. F. Cassidy, "A dimensional approach to measuring social capital: Development and validation of a social capital inventory," *Current Sociology*, vol. 49, no. 2, pp. 59–102, 2001.
[17] R. Silvey and R. Elmhirst, "Engendering social capital: Women workers and rural-urban networks in Indonesia's crisis," *World Development*, vol. 31, no. 5, pp. 865–879, 2003.

IOP Conf. Series: Materials Science and Engineering **771** (2020) 012068 doi:10.1088/1757-899X/771/1/012068

Regional Food Hubs For Distribution of Regional Food Logistics (Case Study on the Establishment of a Food Distribution Center in Banten Province, Indonesia)

H Winarno[1,*], T Perdana[1], Y Handayati[2], D Purnomo[3]

[1]Agribusiness, Padjadjaran University, Indonesia.
[2]Business and Management School, Bandung Institute of Technology[3], Indonesia.
[3]Agricultural Industrial Technology, Padjadjaran University, Indonesia.

[*]heruwinarno42@gmail.com, tomyperdana@gmail.com, yuanita@sbm-itb.ac.id, dwipurnomo@gmail.com

Abstract. Agriculture has a strategic role in sustaining the economy of a region, especially in improving food security and production of agricultural products. Food Hubs plays the role of building a food system that is capable of aggregating, distributing and connecting farmers with consumers directly. This requires the role and management of effective distribution methods by integrating the roles of farmers and consumers in a distribution center. The Distribution Center (DC) is managed by a Regional-Owned Enterprise (*BUMD*) that assists farmers in producing, selling and distributing food. Distribution Centers are able to overcome the problem of stock scarcity, price disparity, and price fluctuations in goods - especially staples. The Food hubs are supported by the determination of the number of warehouses, warehouse locations, and warehousing standards for the management of food staple inventory warehouses. Besides determining the level of inventory of safe food staples (buffer stock) is based on the main parameters: lead time and demand requirements. Food Hubs Operations embody the role of farmers, traders, in supporting the activities of distribution centers managed by Regional owned enterprises. Food hubs interactions involve institutions that play a role in sustaining these activities.

1. Introduction

The agricultural sector is one sector that has an important role in supporting the Indonesian economy. The agricultural sector in Indonesia needs to be developed in line with population growth and technological developments in order to increase agricultural production. Agriculture plays an important role in the development and as a measure of the welfare level of a country's population, especially in meeting the needs, supplies, and food consumption of its population. One of the commodities that is very important and in line with the government's program in food self-sufficiency is food crops. Food crops are one of the commodities and are currently being developed and are of concern to the government. Problems with food security (rice) such as lack of rice availability and rising rice prices can trigger social insecurity, economic and political instability and can thoroughly disrupt national stability. Aspects that must be met to achieve food security conditions are the adequacy and stability of food availability, affordability/distribution of food and food quality/safety. Distribution is the process flow from one process to the next. Supply Chain Management includes management information systems, purchasing, customer service, resources, transportation, production schedules, demand fulfilment processes, inventory management, warehousing and marketing [15]. The food hubs is an emerging food distribution model that can link small and medium farmers with other institutions (consumers, distribution agents), by creating new value chains that strengthen local and regional food systems [13]. A Regional Food hub is a business or organization that actively manages the aggregation, distribution and marketing of food products whose sources are identified primarily from local and regional producers to strengthen their ability to meet wholesale, retail and institutional demand [11]. Food hubs as a network and intersection of grassroots, community-based organizations and individuals who work together to build a more socially, economically and environmentally friendly food system that connects farmers with consumers directly "[5]. Food Knives contribute to strengthening local and regional food systems and broader community goals of sustainability and health"[4].

The Food Hubs builds effective information flows and transparency among value chain partners, thus enabling each partner in the supply chain to fully understand the costs of production, processing, transportation and marketing operations, all of which help ensure that supply chain partners can negotiate acceptable returns [20]., revealed that the supply chain is a dynamic that involves a constant flow of information, products, and funds between different stages [6] There are 3 main components in the supply chain, namely the upstream supply chain that includes the activities of manufacturing companies with distribution and linking with retailers, internal supply chain management / internal supply chain management which includes storage, and finally is a downstream supply chain / downstream supply which includes shipping to end customers with a chain with an agricultural supply chain company. Commodity supply chain management of agricultural commodities and products is different from supply chains of non-agricultural commodity businesses, because: (i) commodities and agricultural products are perishable, (ii) the process of planting, growth and harvesting depends on climate and season, (iii) results harvests have varying shapes and sizes. (iv) agricultural commodities and products are cages making agricultural products difficult to handle [1]. All of these factors must be considered in the management design of commodity supply chains and agricultural products in order to obtain a comprehensive, effective, efficient, responsive and sustainable supply chain system.

2. Related Research

This section provides guidance on what and how the scope of regional food hubs is in realizing regional food logistics. Describe the appropriate methods for applying food hubs for regional food logistics. Innteraction involved in the food hubs.

2.1. Scope of Regional Food hubs

The recent emergence of the concept of the food node and its development throughout the country have attracted much attention. Food hub operations have been framed as a strategy to improve local and regional food systems, specifically by increasing market access to small and medium scale agriculture.

Several case studies have commented on expected and observed operational characteristics of food hubs, including the ability of food hubs to differentiate their products throughout the supply chain as well as their potential to engage in transparent and collaborative relationships with business partners and other stakeholders. Product differences refer to the ability to identify and market certain product attributes. This can include information about producers and their locations, emerging practices, or group branding.

The emergence of alternative food networks (AFN) over the past few decades, including farmers' markets and Community Supported Agriculture (CSA), has signaled a direct response to the failures of industrial agriculture, and presented new opportunities for regional / rural communities to facilitate development and change [14]. Through AFN, food production, consumption and distribution ideally inspires social justice, improves ecological health, and prioritizes the needs of the community while increasing economic benefits.

Regional Food Hubs / (RFHs) represent the latest type of AFN, and are most related to the results of developing regional / rural development for producers, consumers, and communities over the past five years [8]. Policy makers, practitioners, and funders have actively embraced and financed RFHs which is defined as an organization that collects, stores, markets and distributes locally and regionally sourced food from several producers to many consumers [3], especially with regard to the capacity of such organizations, namely regional / rural revitalization and agricultural development [20].

A significant gap, however, exists in what is known about RFHs (Blay-Palmer et al., 2013; Fischer et al). To stem the loss of family agriculture and stabilize rural communities in decline. Fischer et al. With RFHs, provide some support for theorization, but with limited empirical investigations of RFHs to date, such ideas are still untested and debated. What research has been done for RFHs offers little insight into regional / rural development, and is often at odds in a narrow result within the scope, lack of input from members of the local community, especially from consumers, without respecting site conditions, or applied outside the village, in the context of high prices where the ability to engage with claims related to food products.Many initiatives such as community-supported agriculture and farmers' markets exist as alternatives to global food systems, industrialization, conventional . This initiative broadens infrastructure and market opportunities for "middle agriculture" and promotes more sustainable food systems and food value chains. One concept that is quickly gaining recognition and attention from various stakeholder groups - from non-profit organizations and urban designers to universities and the United States Department of Agriculture - is food hubs. Food Hubs seem to offer many benefits, including expanded market opportunities for farmers, job creation, and increased access to healthy food by consumers [16].

The Food hubs / food center concentrates on aggregation, distribution and marketing, connecting producers with buyers in the supply chain. USDA [18], maintains a list of parent food centers that can be classified as follows:

1). Agriculture to the consumer model: producers who sell directly to consumers through farmers' markets, enforce agriculture, purchasing clubs and CSA
2). Agricultural to business / institutional model: producers selling to supermarket chains, independent grocery stores, schools, hospitals, institutions, restaurants and other businesses
3). Hybrid model: a combination of the agricultural to consumer and agricultural to business / institutional models

Determine the characteristics of Regional Food Hubs [2], including :

1) Carry out or coordinate the aggregation, distribution and marketing of food produced locally or regionally from various producers to various markets.
2) Regard producers as valuable business partners, not interchangeable suppliers and commit to buying from small to medium scale local producers whenever possible.
3) Work closely with producers, especially small scale operations, to ensure they can meet buyer requirements by providing technical assistance or finding partners who can provide this technical assistance.
4) Use product differentiation strategies to ensure that producers get good prices for their products. Examples of product differentiation strategies include preservation of identity (knowing who produced it and where it came from), group branding, special product attributes (such as heirlooms or unusual varieties), and sustainable production practices (such as certified organics, minimum pesticides, or "independently naturally "grown or raised).
5) Aiming to be financially viable while also having positive economic, social, and environmental impacts in their communities, as demonstrated by conducting certain production, community, or environmental services and activities.

Important factors in supporting the success of the Regional Food Hubs :

1) Thought and practice of leadership and entrepreneurship.
   This includes visionary, committed and experienced individuals or teams, supported by strong and capable Board who are willing to innovate, take risks and come up with creative solutions to meet business growth and consumer demand for local and regional products, as well as design and provide services that meeting the needs of vulnerable people in the local community [3], Partnerships and networks are key, especially with producers [19]. Opportunities to develop in and / or utilize resources from established organizations, different from building entirely new organizations, are very important, as they start small and simple without the need to focus on physical infrastructure.
2) Engagement in local communities is very important to ensure relevance and credibility as well as a strong understanding of consumer motivation that drives demand for identified sources and values [12].
3) Reaching the minimum sales threshold and / or non-sales income is essential for long-term survival.

IOP Conf. Series: Materials Science and Engineering **771** (2020) 012068 doi:10.1088/1757-899X/771/1/012068

2.2. Regional Food Hubs Distribution Methods

The distribution method in the food hubs in the handling of the Provincial distribution center is described as follows:

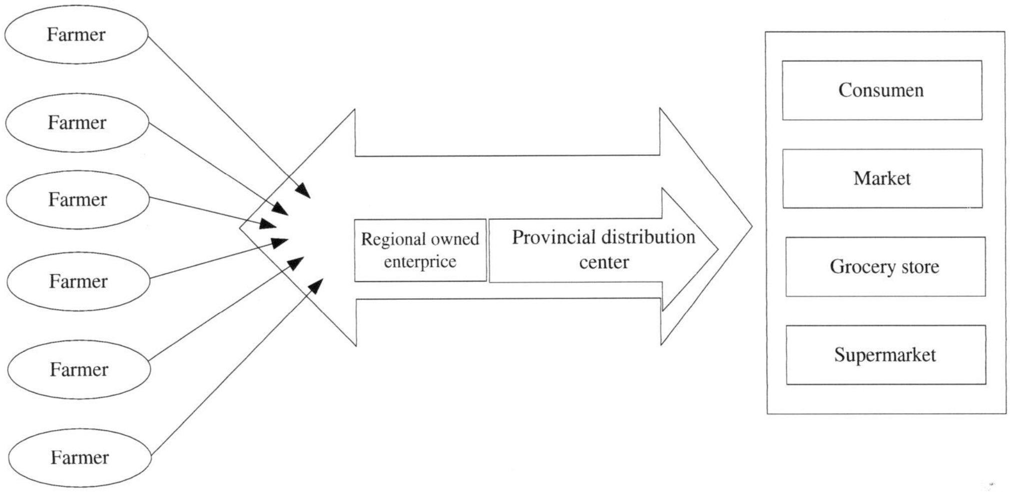

Figure 1. Flow distribution method in food hubs

2.3. Farmers

The Food Hubs overcomes the challenge of improving alternative food systems to serve as aggregators, processors, distributors and marketers and/or sales brokers in alternative food supply chains. Every node in the supply chain is important to enhance its development. Some nodes facilitate sales from small and medium-sized producers to institutional buyers by combining products from various sources. This provides buyers with the high volume and variety of product lines they need and allows small and medium-sized producers to compete with larger operations [9].

The role of farmers providing food (rice) which is spread throughout the regions in Banten Province is aggregated in each district which is spread in various regions, accommodated in warehouses owned by individuals or regions through warehouse receipts. Purchases through warehouse receipts are funded through regional banks to collect food/rice before being distributed to distribution centers. Usually, when it enters the rice warehouse receipts it has gone through stages from rice to milling. In this process, there has been a flow of sales from farmers to Regionally Owned Enterprises with appropriate prices and controlled rice.

Some nodes offer processing services that can cover everything, this allows the node to make food available throughout the year in areas with limited planting seasons; add value to farmers' income; meet consumer demand for comfort [17].

Aggregation/collection, food products before and after harvest plans made between farmers and the food/food node are mechanisms to anticipate what products will be available together with expectations for when they will be sent [10]. Pre-harvest talks with potential buyers provide guidance on consumer trends, desired products, quantity needed, and preferred packaging. Successful sales of food products depend on providing quality/inventory quality and consistency. Farmers must produce products with market recognized standards.

2.4. Regionally Owned Enterprises (BUMD)

This agency has the role of managing the Distribution Center because it is calculated to have the advantage of accommodating all aspects of Distribution Center activities to achieve physical, economic and trade development.

From physical development, BUMD has a role in managing infrastructure assets/facilities and infrastructure of distribution centers including buildings, facilities, transportation and distribution.

Economic development, namely creating purchase prices at the producer level and selling prices at the consumer level, is more stable. On the other hand, farmers get the consumption of goods and Saprotan needs at lower prices. Thus farmers are expected to be prosperous and consumers get a guaranteed supply of rice commodities at stable prices.

The Regional Government through BUMD has an interest in the solution to the problem of stock scarcity, price disparity, and fluctuations in the price of goods - especially staples, especially rice.

Mapping of food needs (demand side) is carried out to identify the types of food staples and the estimated needs of each type of food staples in an area. Food supply mapping (supply side) is intended to identify the supply of each type of food and the estimated volume of food production in an area.

.

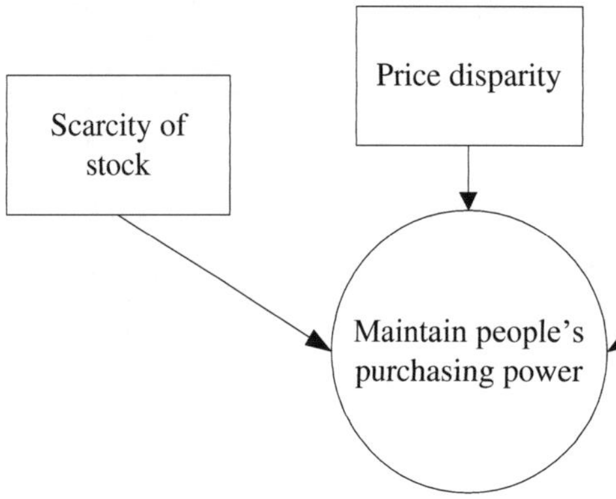

Figure 2. Maintaining people's purchasing power

Identification of Commodity Availability,
price and supply/demand controls in several regions in the district/city are in accordance with the figure

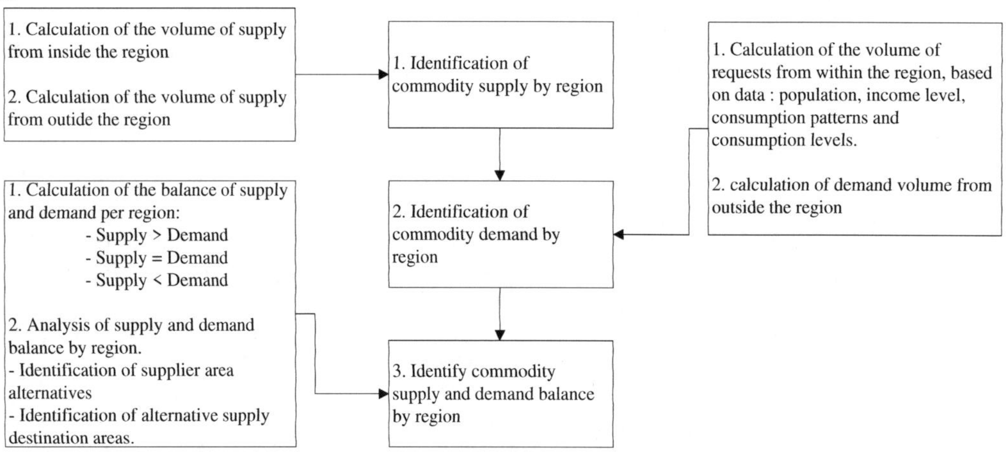

Figure 3. Identifying material availability, price control and supply-demand

The smooth distribution will guarantee the accuracy of the stock and be able to control the disparity and fluctuations in commodity prices of basic commodities. Fluctuations in prices of staples contribute significantly to regional and national inflation rates. Fluctuating inflation rates, moreover tend to be high, have an impact on overall economic stability [6]. Warehousing plays an important role in overcoming fluctuations in surpluses and shortages of food staples. Determination of the number of warehouses, warehouse locations, and warehousing standards for the management of food staple inventory warehouses. Determination of the level of inventory of safe food staples (buffer stock) is based on the main parameters: lead time and demand requirements. Food packaging and handling processes need to be done safely to protect from damage, loss and degradation of food quality from primary producers (farmers), industrial producers (food processing manufacturing companies), wholesalers, retailers to consumers. In managing food staple warehouses, food expiration needs to be considered. Food staples are perishable goods. Warehouse standards and cold chain transportation modes are required for certain types of staples. To maintain the limit of food supply, the application of FIFO (first-in-first-out) in the management of staple food supplies is absolutely necessary. The use of an inventory card will ensure the supervision of food inventory in and out. In warehouses with SKUs (stock-keeping units), a large and complex supply of food staples requires the use of a WMS (warehouse management system) application system.

With this WMS application, it can be monitored the stock of food staples, receipts, storage, picking, and expenditure of food staples from the warehouse. From the description of the method of distribution of food hubs, operational activities in the distribution center can be described as follows :

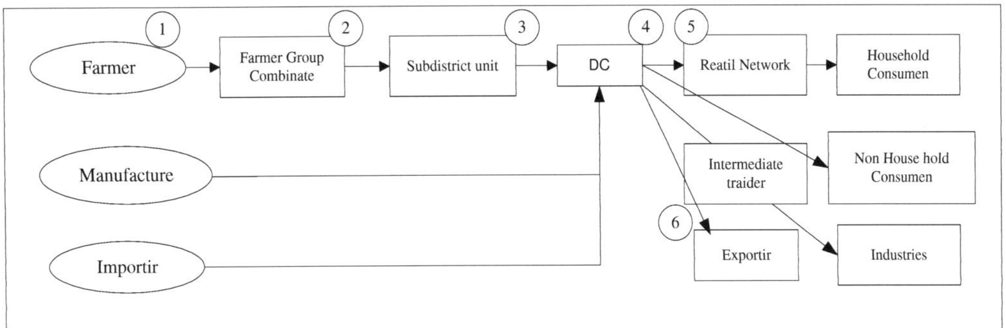

Area	Business that can be developed	Actor
Farmer	Trading: Buyer, Seller	BUMD, Logistical Affairs Agency, Regional
Outline & Sub-distric unit	Trading, Processing, (Collect, Rise Milling Sort, Grading, Packing)	BUMD, Logistical Affairs Agency, FGC
DC	Trading, Processing, (Collect, Sort, Re-Grading, Re-Packing) Warehousing, warehouse receipt Support business: surveying collateral management	State-Owned Enterprises, BUMD BUMD, Logistical Affairs Agency Surveying : PT. Socofindo/FI Finance : Bank, Micro Finance Inssurance
Intermediate Trader	Trading in Intermediate Market	Central Market, BUMD, Finance
Retail Network	Retail business Microfinance	Public Market, Modern Market, Bank Pt Pos
1 to 6	Transportation	PT. Pos, PT BGR, PT KA, PT Pelni, Land Transportation Organitation

Figure 4. Food Hubs operations in a distribution center

2.5. Interactions involved in the food hubs

Distribution Centers make purchases and sales directly through established markets, for specified commodities.

1). Purchase

The Distribution Center has a procurement division that is responsible for purchasing food commodities to various suppliers and from various regions. The purchase of food commodities is aimed at businesses upstream in the supply chain of each commodity. With notes, the upstream part is included in the criteria, as follows:

a) Manufacturers incorporated in groups/organizations.
b) Groups/individuals who have rice mills.
c) Small and Medium Enterprises that have the ability to process agricultural produce.

Quantity of purchases is carried out with reference to the level of demand (market size) in a certain area and the trade deficit of commodities in other regions. That is, there will be a calculation for the proportion of the purchase scale of each commodity managed by the Distribution Center. That, as a solution to meet food needs in areas which are the marketing areas of every Distribution Center presence. If the strategy continues and is stable, it will have a slow impact on the movement of disparity between regions.

2). Sales

The Distribution Center has a marketing and business division that is responsible for carrying out commodity sales. The payment system for commodity sales is done in 2 ways: cash and tempo.

Payments from consumers are made in cash. Payment methods can be in cash or transfer. Payments from consumers that are made on time can be in cash or transfer and the maximum payment duration is 45 days after the sale and purchase transaction.

The various Sales Centers for Distribution Centers are as follows:

a) People's market.
b) Traders (non-market people).
c) Modern markets.
d) Horeca (Hotels, restaurants and cafes).
e) Catering company.
f) Processing industry.
g) Shops.

The sales targets set are the nodes of the supply chain of basic commodities that will approach the final consumer. Thus, ideally, the supply chain of commodities managed and traded by the Distribution Center does not exceed 4 links.

One of the causes of the frequent fluctuations in prices of basic commodities, due to the supply chain length is too long, even as happened in early 2017, prices soared sharply in public markets that have a negative impact on high inflation. For this reason, the support of Government regulations concerning the trade system for essential commodities is essential. The role of the distribution center's control range which is controlled by the Government is also needed to oversee the operation of the Distribution Center's operational systems and procedures so that the development objectives of the Distribution Center can be realized quickly.

Commodities come from the coverage area and outside the scope of the suppliers managed by the Distribution Center management.

(1) Supply area

The commodities managed and traded by the Distribution Center are food products, whichever make purchases to suppliers from various regions of the production center. Each Distribution Center has a different coverage area with the presence of Distribution Centers in other regions. That is, there will be a determination of regional objects as the scope of each Distribution Center. For example, the Makassar Distribution Center will have different marketing coverage areas than the Papua Distribution Center. However, the Makassar Distribution Center and the Papua Distribution Center in obtaining a supply of commodities can be sourced from the same supplier partners originating from the commodity production centers.

(2) Mechanism, empowerment and procurement

Distribution Centers in obtaining a various supply of commodities through the production center areas. The Distribution Center procurement division is a division that is closely related to suppliers in the area of production centers. The production center of each commodity has a different distribution. For example, centers of rice production with shallots and red chillies. One of the centers of rice production is in Banten and East Java, shallots in Brebes Regency, Central Java, while red chili is in Garut Regency, West Java. The relationship between the Distribution Center and commodity production centers is closely related to several aspects, as follows:

(a) Empowerment

Related to the activities of coaching, growing and developing suppliers.

(b) Procurement.

Associated with the activity of realizing the purchase of suppliers' production results from the activities of guidance, growth and development of suppliers. With the empowerment and procurement system by the Distribution Center procurement division, it will reduce the level of the price disparity between regions. Distribution Centers will work together with suppliers through the principle of partnership.

The realization of the interaction of food hubs produces a picture of the distribution center as follows:

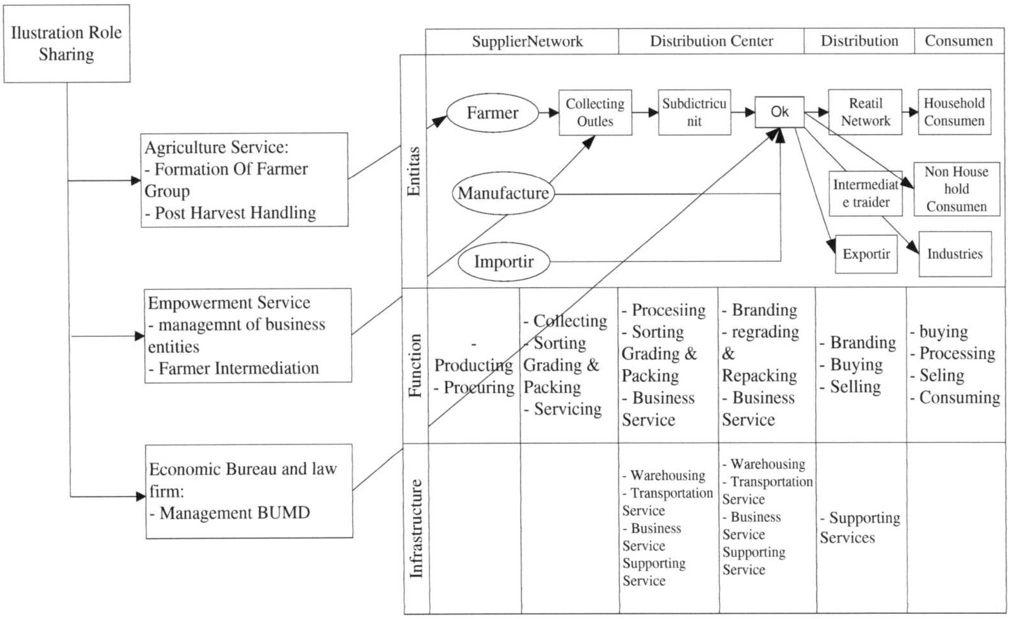

Figure 5. Interaction of food hubs in realizing the Distribution Center

3. Conclusions

(a) Regional Food Hubs is a business or organization that actively manages the aggregation, distribution and marketing of food products whose sources are identified mainly from local and regional producers

(b) Food Hubs distribution method shows distribution that gives producer/farmer capable of supplying products to distribution centers and the role of Regionally Owned Enterprises that can handle and manage all activities at distribution centers so as to produce adequate supply distribution.

(c) Interaction Food Hubs divide the role of institutions in realizing distribution management in the activities of well-managed distribution centers.

Reference

[1] Austin,(1994), Brown, W.J. (2003). Agribusiness Cases in Supply Chain Management. Paper. IFMA Congress.

[2] Barham, (2012), Regional Food Hubs Resource Guide, U.S. Dept. of Agriculture, Agricultural Marketing Service. Washington, DC. April 2012. http://dx.doi.org/10.9752/MS046.04-2012

[3] Barham, et All, (2011), Selling to Food Hubs in North Carolina *RegionalFood Hub Resource Guide*. U.S. Dept. of Agriculture, Agricultural Marketing Service

[4] Barham, et All, (2014), Selling to Food Hubs in North Carolina *RegionalFood Hub Resource Guide*. U.S. Dept. of Agriculture, Agricultural Marketing Service

[5] Blay Palmer A, Landman K (2013), dissertation with Attitudes and interest in the development of Northern New York food hubs, Corner University

[6] Business Concept and management of The Provincial Distribution Center (2018), Banten Provincial Trade Office

[7] Chopra S, Meindl, P, (2010), Supply Chain Management Strategy, Planning and Operation, New Jersey: Prentice-Hall

[8] Cleveland, DA, et al, (2014), Local food hubs for alternative food systems: A case study from Santa Barbara County URPL711, California, Journal of Rural Studies, http://dx.doi.org/10.1016/j.jrurstud.2014.03.008

[9] Day Farnsworth and Morales, (2011), Marketplaces and Food Systems,Sustainable Agriculture http : // www.nal.usda.gov /afsic/pubs/terms/srb 9902 terms.shtml Food Safety http://www.allfoodbusiness.com/ foodsafety_glossary.phpUniversity of Wisconsin-Madison

[10] Day Farnsworth and Morales, (2009), Marketplaces and Food Systems, Sustainable Agriculture

[11] Fischer M, Pirog R, Hamm MW (2015) Food Hubs: Definitions, Expectations, and Realities, Journal of Hunger & Environmental Nutrition, DOI: 10.1080/19320248.2015.1004215

[12] Krejci CC, Stone RT, Domeich MT, Gilbert SB,(2015), Analysis of Food Hub Commerce and ParticipationUsing Agent-Based Modeling: IntegratingFinancial and Social Drivers, Iowa State University, Ames

[13] LadonaKnigge, Jacob N, Brimlow and Sara S Metcalf,(2015), Food hubs connecting farm with local and regional Market

[14] Martsden T (2012)AlternativeAgrifood network and short supply chain: literature review, Journal economia agro alimentareDOI**:** 10.3280/ECAG2013-002002

[15] Nabhani F, and Sholeri A, (2009), Reducing the delivery lead time in a six sigma methodology, Journal of Manufacturing Technology Management.

[16] National Good Food Network,2011

[17] Ohberg*, et all (2012*) Food sovereignty in Canada: Creating just and sustainable food systems. Journal Agricultural , Food System and community development, Volume 8, Issue 2Spring–Summer 2018

[18] Rogoff, J, (2014), Improving Systems of Distribution and Logistics for Regional Food Hubs, Jurnal Rural Development USDA

[19] Severson RM and Schmit TM,(2015)Building Success of Food Hubs Through the Cooperative Experience- A Case Study Perspective

[20] USDA. (2009). USDA Results: Local and Regional Food Systems. *Results.*Retrievedfromhttp://origin.www.usda.gov/wps/portal/usda/usdahome?contentidonly=true&conteAgriculture and Human Values,1-15.5,393-141

The effect of composition on hardness and wear resistance of rice plant fiber reinforced composite as a material of brake lining

Agung Setyo Darmawan[1,*], Pramuko Ilmu Purboputro[1] and Bambang Waluyo Febriantoko[1]

[1]Department of Mechanical Engineering, Faculty of Engineering, Universitas Muhammadiyah Surakarta, Surakarta, Central Java, Indonesia

[*]Corresponding author: Agung.Darmawan@ums.ac.id

Abstract. Asbestos type brake lining is widely circulating at a cheap price and guarantees the durability of brake shoes, but the result of friction powder in the form of small particles is very dangerous for human health. The study is based on the investigation of hardness and wear resistance in rice plant fiber reinforced composite which is fabricated with polyester resin, fiberglass, and addition of aluminium powder. The composite is hardness tested by using durometer shore D hardness tester. The composite is also wearing resistance tested. The result showed that the mechanical properties were improved due to the increasing of the rice plant fiber composition. The more composition of the rice plant fiber, the harder and the higher wear resistance of the composite. An increase in the composition of the rice plant fiber from 20% to 40% resulted in an increase in hardness of 23% and a decrease in wear by 9%

1. Introduction

Composite is a material that is composed of a mixture of two or more materials with different chemical, physical properties and mechanical, and produces a new material that has different properties from the constituent materials. Composite materials are composed of two types of materials namely matrix and fiber (reinforcement). Both have different functions, fiber functions as the frame material that composes composites, while the matrix functions to glue the fiber and keep it from changing positions. A mixture of the two will produce a hard, high strength, but lightweight material [1, 2].

Fiber has properties that are easy to change its shape by cutting or also forming according to the needs of the design. In addition, differences in the arrangement of fiber structures will also change the properties of the resulting composite. This can be used to obtain composite properties according to the parameters needed. The reinforcing material in the composite has the role of holding the load received by the composite material. The nature of the reinforcing material is usually rigid and tough. Strengthening materials commonly used so far are carbon fiber, glass fiber and ceramics [3, 4].

The matrix is generally made of resin. It functions as an adhesive of fiber material so that the stack of fibers can adhere strongly. The resin will bind to the fiber material so that the burden imposed on the composite will spread evenly. In addition, the resin also serves to protect fiber from chemical attack or also extreme weather conditions that can damage it. Matrix materials can also generally be in the form of metals, polymers, ceramics, and carbons. The nature of the matrix is usually ductile. The matrix in the composite serves to distribute the load into all composite reinforcing materials [5].

In addition to the ease of designing composites in any form, one of the main reasons for using composite materials is that high strength materials are obtained with a much lighter weight than conventional materials.

Composite materials are widely used. Its uses include the fields of aircraft and military, aerospace applications [6, 7], the automotive industry [8], sports equipment [9], packaging materials [10] and others

Reinforcement material to be able to produce composite materials that are lightweight, high strength, environmentally friendly and economical [11-15]. One of them is natural fiber ingredients. Natural fiber as a type of fiber that has advantages began to be applied as a reinforcing material in polymer composites. Natural fiber composites combine good mechanical properties with a low specific mass and seem to be an alternative material to glass-fibre-reinforced plastics in some technical applications [16].

Elastic, high strength, abundant, environmentally friendly and lower production costs are advantages possessed by natural fibers [17]. In addition, there are also disadvantages of this type of fiber, especially strength that is not always evenly distributed. Types of natural fibers such as jute, coconut, sisal, hemp and flax began to be used as reinforcing materials for polymer composites [18].

This composite material is mainly used as a car exterior material [19]. Based on the results of their development can be obtained polymeric composite materials - natural fibers with a strength of 40% stronger and lighter than glass-fiber polymer composites. Lighter means it can reduce the total weight of the vehicle, which in turn can contribute to fuel savings. The process of making natural fiber based composites is relatively cheaper and more environmentally friendly. Its manufacture consumes about 70% less energy than glass-fiber polymer composites. From an ecological point of view, this natural fiber based polymer composite material can be recycled for reuse, although in terms of performance it has dropped considerably [20-22].

So far, several educational and research institutions have begun to conduct research on the use of natural fiber as a reinforcing material for composites. Starting from coconut fiber, water hyacinth fiber and palm fiber. Besides that, rice plant fiber is a material that has the potential to be used as reinforcement in natural composite.

When materials are used as components that rub against each other, hardness and wear resistance play an important role [23]. Therefore, this study investigates the effect of fiber composition on the hardness and wear resistance of composites. The composite is used for brake lining as a component of automotive.

2. Materials and Methods

The research was carried out in five stages. The first stage is preparing tools and materials that will be used in the research process. The material was weighed using a digital weighing according to the composition shown in table 1.

Table 1. The composition of composite.

Rice Plant Fiber (%)	Polyester Resin (%)	Fiberglass (%)	Al Powder (%)
20	60	10	10
30	50	10	10
40	40	10	10

The second stage is the process of mixing dry materials, mixed using a blander machine so that the results of mixing the material can be mixed evenly

Then in the third stage, the brake lining molds are prepared and put the brake lining plates into the mold. Furthermore, the brake lining plate was given an epoxy resin that is useful for materials attached to the surface of the brake lining. After that, the material that has been mixed is put into the mold. The next step is to press the mold using a press machine. Pressing is done by giving heat with a temperature of 90 °C. Then the sintered canvas with a temperature of 180 °C with duration of 15 minutes. The fourth stage is to do microphotographs, hardness test using shore D durometer hardness test equipment, and wear test. The final stage is analyzing data obtained from testing and making conclusions.

3. Results and Discussions

Microphotographs of composites are shown in Figure 1. Composites were made with variations in rice plant fiber's composition of 20%, 30% and 40%.

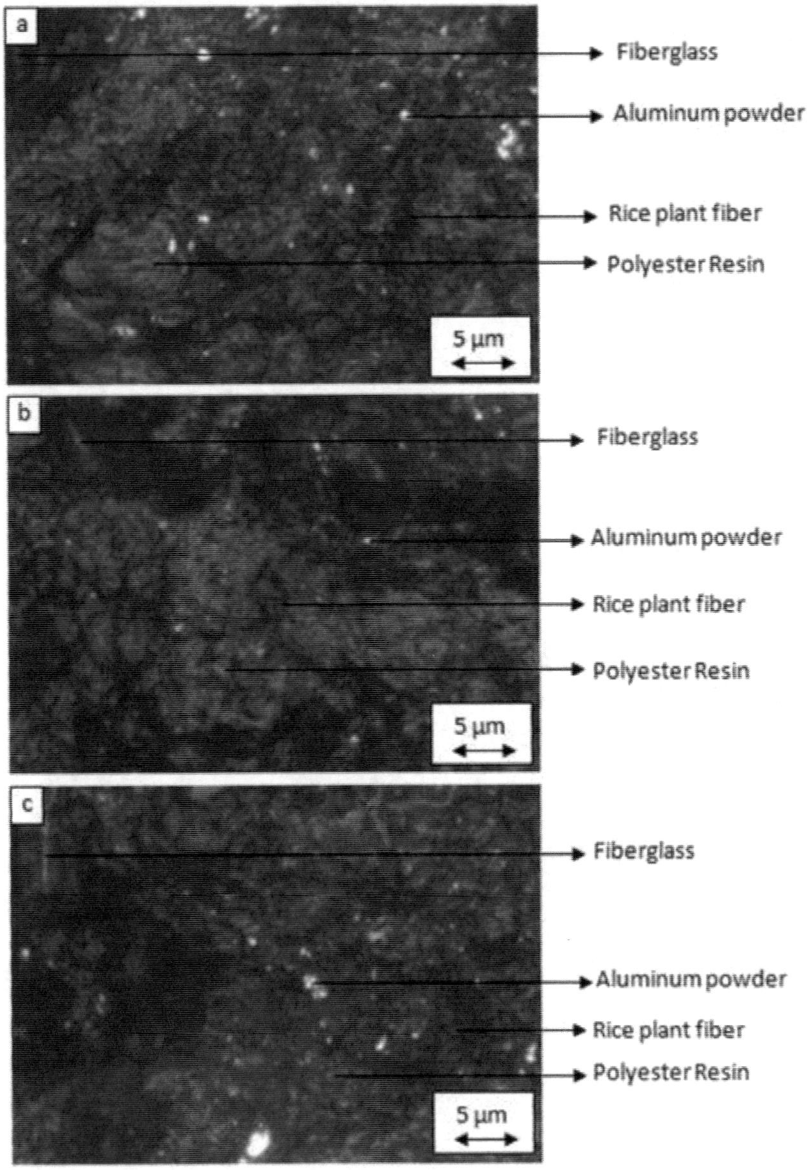

Figure 1. Microphotograph of composite, (a) composition of rice plant fiber is 20 % (b) composition of rice plant fiber is 30 % (c) composition of rice plant fiber is 40 %.

Figure 2 and figure 3 show the results of hardness and wear testing of the composite. Hardness testing is performed with a durometer shore D hardness tester while wear testing is carried out using a friction-testing machine.

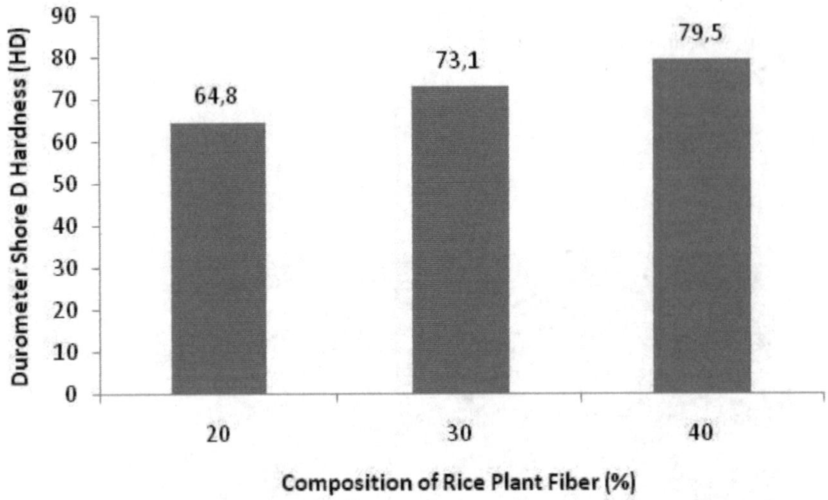

Figure 2. The effect of rice plant composition on the composite's hardness.

From the results of this hardness test (figure 2), the hardness value with 20% composition of rice plant fiber was 64.8 HD, the value of hardness with 30% composition of rice plant fiber was 73.1 HD, and the hardness value with 40% composition of rice plant fiber was 79.5 HD. Figure 2 shows that the highest hardness value was 79.5 HD when the composition of the rice plant fiber was 40%. This result shows that the more the composition of rice fiber plants, the hardness will increase if the composition of fiberglass and aluminium powder remains unchanged.

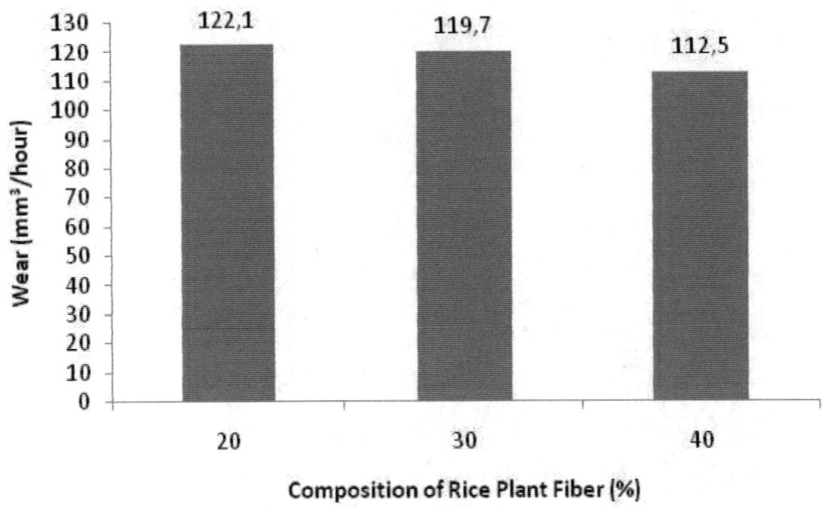

Figure 3. The effect of rice plant composition on wear.

Figure 3 shows the results of wear testing with dry conditions, wear results obtained on composites with 20% fiber composition was 122.1 mm^3/hour, wear on composites with 30% fiber composition was 119.7 mm^3/hour, and wear on composites with 40% fiber composition was 112.5 mm^3/hour.

The results of this test show a tendency to increase composite wear resistance when fiber composition increases. This is due to surface hardness that also increases when fiber composition increases.

4. Conclusion

From the results and discussions, it is concluded that the more composition of rice plant fiber, the harder and the higher wear resistance to the composite. An increase in the composition of rice plant fiber from 20% to 40% results in an increase in hardness from 64.8 HD to 79.5 HD and a decrease in wear from 122.1 mm^3/hour to 112.5 mm^3/hour.

Acknowledgement

The authors would like to acknowledge the Directorate of Research and Community Service, The Ministry of Research, Technology and Higher Education of the Republic of Indonesia [Project Number = 199.24/A.3-III/LPPM/V/2019]. The author would also like to thank the Mechanical Engineering Laboratory of Universitas Muhammadiyah Surakarta for mechanical test facilitations.

References

[1] E. Salernitano and C. Migliaresi, Composite Materials for Biomedical Applications: A Review, *Journal of applied biomaterials & biomechanics* **1(1)**, 3-18 (2003).

[2] D. Verma, P.C. Gope, A. Shandilya, A. Gupta and M.K. Maheshwari, Coir Fibre Reinforcement and Application in Polymer Composites: A Review, *J. Mater. Environ. Sci.* **4(2)**, 263-276 (2013).

[3] Z. Ding, Y. Li, C. Lu and J. Liu, An Investigation of Fiber Reinforced Chemically Bonded Phosphate Ceramic Composites at Room Temperature, *Materials* **11**, 858 (2018).

[4] G.I. Williams and R.P. Wool, Composites from Natural Fibers and Soy Oil Resins, *Applied Composite Materials* **7**, 421–432 (2000).

[5] M. Surappa, Aluminium Matrix Composites: Challenges and Opportunities, *Sadhana* **28(1)**, 319-334 (2003).

[6] D.K.K. Cavalcanti, M.D. Banea, J.S.S. Neto, R.A.A. Lima and R.J.C. Carbas, Mechanical characterization of intralaminar natural fibre-reinforced hybrid composites, *Composites Part B: Engineering* **175**, 107149 (2019).

[7] J. Zhou, Y. Li, L. Cheng and L. Zhang, Indirect Microwave Curing Process Design for Manufacturing Thick Multidirectional Carbon Fiber Reinforced Thermoset Composite Materials, *Applied Composite Materials* **26(2)**, 533–552 (2019).

[8] E. Sarikaya, H. Çallioğlu and H. Demirel, Production of epoxy composites reinforced by different natural fibers and their mechanical properties, *Composites Part B: Engineering* **167**, 461-466 (2019).

[9] J. L. Wang, Application of Composite Materials on Sports Equipments, *Applied Mechanics and Materials* **155-156**, 903-906 (2012).

[10] S.A.A. Mohamed, M. El-Sakhawy, E.H.A. Nashy and A.M. Othman, Novel natural composite films as packaging materials with enhanced properties, *International Journal of Biological Macromolecules* **136**, 774-784 (2019).

[11] K.I. Alzebdeh, M.M.A. Nassar and R. Arunachalam, Effect of fabrication parameters on strength of natural fiber polypropylene composites: Statistical assessment, *Measurement* **146**, 195-207 (2019).

[12] S. Goutianos, T. Peijs, B. Nystrom and M. Skrifvars, Development of Flax Fibre based Textile Reinforcements for Composite Applications, *Applied Composite Materials* **13(4)**, 199–215 (2006).

[13] M. Ramesh, K. Palanikumar, and K. H. Reddy, Plant fibre based bio-composites: Sustainable and renewable green materials, *Renewable and Sustainable Energy Reviews* **79**, 558-584 (2017).

[14] R. Sivagurunathan, S.L.T. Way, L. Sivagurunathan and M.Y. Yaakob, The Effects of Triggering Mechanisms on the Energy Absorption Capability of Circular Jute/Epoxy Composite Tubes under Quasi-Static Axial Loading, *Applied Composite Materials* **25(6)**, 1401–1417 (2018).

[15] Y. Zhou, M. Fan, and Li. Chen, Interface and bonding mechanisms of plant fibre composites: An overview, *Composites Part B: Engineering* **101**, 31-45 (2016).

[16] J. Gassan and A. K. Bledzk, Possibilities to Improve the Properties of Natural Fiber Reinforced Plastics by Fiber Modification – Jute Polypropylene Composites –, *Applied Composite*

Materials **7(5–6)**, 373–385 (2000).
[17] K. Oksman, Mechanical Properties of Natural Fibre Mat Reinforced Thermoplastic, *Applied Composite Materials* **7(5–6)**, 403–414 (2000).
[18] F. Tian and Z. Zhong, Modeling of load responses for natural fiber reinforced composites under water absorption, *Composites Part A: Applied Science and Manufacturing* **125**, 105564 (2019).
[19] M.M. Davoodi, S.M. Sapuan, A. Aidy, N.A. Abu Osman, A.A. Oshkour and W.A.B. Wan Abas, Development process of new bumper beam for passenger car: A review, *Materials and Design* **40**, 304–313 (2012).
[20] E. Asmatulu, J. Twomey and M. Overcash, Recycling of fiber-reinforced composites and direct structural composite recycling concept, *Journal of Composite Materials* **48(5)**, 593–608 (2014).
[21] T.Q. Truong Hoang, F. Lagattu and J. Brillaud, Natural Fiber-Reinforced Recycled Polypropylene: Microstructural and Mechanical Properties, *Journal of Reinforced Plastics and Composites* **29(2)**, 209-217 (2010).
[22] V. Srebrenkoska, G. Bogoeva-Gaceva and D. Dimeski, Biocomposites based on polylactic acid and their thermal behavior after recycling, *Macedonian Journal of Chemistry and Chemical Engineering* **33(2)**, 277–285 (2014).
[23] A.S. Darmawan, W.A. Siswanto and T. Sujitno, Comparison of Commercially Pure Titanium Surface Hardness Improvement by Plasma Nitrocarburizing and Ion Implantation, Advanced Materials Research 789, 347-351 (2013).

Characteristic of Vertical Mixed Flow Dryer in Coffee Bean Drying Process

Sutrisno[1,*], D Ariwibowo[1], M E Yulianto[2], R Sitawati[3]
[1]Department of Mechanical Engineering, Vocational School, Diponegoro University
[2]Department of Chemical Engineering, Vocational School, Diponegoro University
Jl. Prof Soedarto SH, Tembalang, Semarang 50239
[3]Department of Accounting, STIE Dharmaputra

[*]e-mail: didik.ariwibowo@live.undip.ac.id

Abstract. This paper proposes a new type of vertical mixed flow dryer and evaluates the characteristic of the dryer for coffee bean drying. The specific design of vertical mixed flow dryer was in the drying chamber that equipped with a perforated-sloped tray to retard grain flow while drying. The dryer characteristic was presented in a form of humidity and temperature behaviour during the drying process. This characteristic could be a database to get coffee with standard moisture content (MC). Parameter stated were air temperature (°C), humidity (kg/kg dry air), coffee bean mass flow rate (kg/s), and air velocity (m/s). For air velocity of 2 m/s, this dryer was able to dry coffee bean from 52% initial MC to final 14% MC for 11,5 hours.

1. Introduction

Drying takes a role in a post-harvest treatment and is important from the energy consumption point of view. In coffee bean drying, the moisture content is decreased from 55% *w.b.*to 12% *w.b.* If the drying process is poor, microorganism could grow on coffee bean surface, lead to biochemical reactions that could be toxic and unconsumable. It could also affect sugar content [1], and changes in organoleptic properties such as taste, color, and aroma [2]. Besides the biodegradation, coffee trading requires a standard dried coffee bean.

Coffee has a moisture content (MC) of 50-60% (wet basis, w.b.) and unevenness in maturity. Therefore, drying should be performed immediately to reduce the high moisture content of the bean and to avoid unwanted fermentation [3]. However, thermal stress following with mechanical damage can occur if drying not performed correctly [4].

The most study of drying mechanism is based on an empirical approximation model. The model can give information to predict drying time in an experimental condition. A phenomenological-based model has been developed to get a whole feature of coffee bean drying. This type of model can be generalized to a different condition. A process modeling methodology [5,6] has been presented by several authors. The methodology that used empirical approximation model and phenomenological-based model give to a simple and intuitive procedure to be used in engineering.

2. Method

The investigation of a new type of vertical mixed flow dryer behaviour was performed experimentally. This dryer has the ability to keep contact between air and coffee bean in a specific drying condition that facilitates heat and mass transfer occurs simultaneously. The heat transfer was governed by introducing hot air, whereas mass transfer was controlled by the humidity of the air. The new design was referred to some prototypes of the dryer, whereas its behaviour based on both dryer phenomenological model and empirical model. Some drying model were used to approach the drying process of coffee bean in the drying chamber. The feature of the drying chamber related to relative humidity and temperature was performed. In this study, the coffee drying process was based on the specific design of vertical mixed flow dryer at Mechanical Engineering Laboratory Vocational School Diponegoro University. The coffee dryer consists of three main elements namely drying chamber, heat exchanger, and blower as seen in Figure 1.

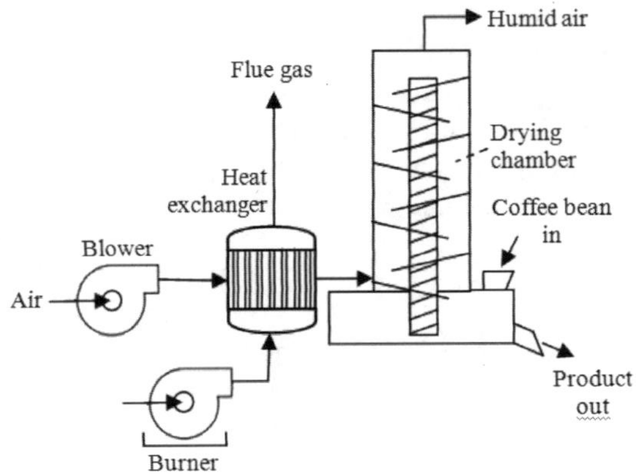

Figure 1. Schematic of vertical mixed flow dryer

Coffee bean fed into the bin and conveyed to the upside of the drying chamber by conveyor. The bean flows downward by gravity and flow on sloped-tray. Meanwhile, hot-dried air from heat exchanger flow upward in the drying chamber. This hot air contact with a coffee bean in a mixed flow pattern. The contact between coffee bean and air occurs in three-stage as seen in Figure 2. The first stage, hot air keep contact with bean in the coffee drying zone. The hot air capture moisture from coffee bean and become humid. In the second stage, the humid airflow passes pre-drying zone, retiring more moisture from the bean in this zone. Finally, in the third stage, humid airflow passes the wet coffee bean zone before going out of the chamber.

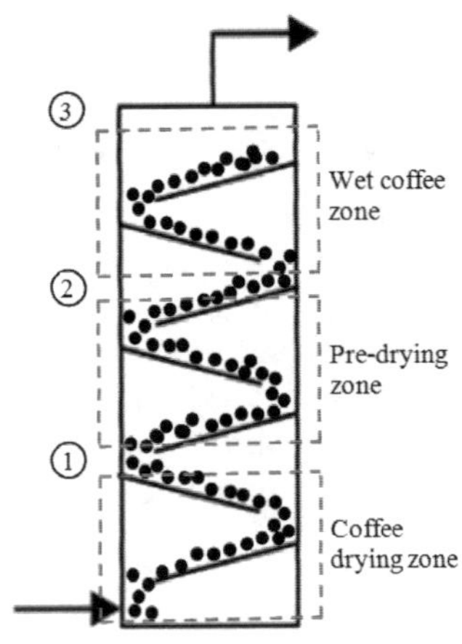

Figure 2. Schematic of drying process

3. Materials and Devices Research

Raw materials to be used in this study is coffee cherry obtained from Ambarawa, Semarang District. The raw material is peeled in order to get a coffee bean. The 1 kg of the bean was prepared for every run. Liquified Petroleum Gas – LPG was used as fuel for heating the heat exchanger.

The main equipment used in this research is the vertical mixed flow dryer which is used to dry coffee bean by the mechanism of heat and mass transfer simultaneously. Supporting instruments needed include thermometer, hygrometer, grain moisture meter, stopwatch, a digital scale with 0,01 precision, and electric oven.

4. Variable Experiment

There are 3 variables in this experiment those are fixed variable, dependent variable and independent variable. The fixed variables including air temperature that stated at 60 °C, and air velocity introduced to the drying chamber at 2 m/s. The dependent variables were temperature and humidity of the drying chamber, whereas the independent variable were mass flow rate of coffee bean that vary at 1 kg, 2 kg, and 3 kg per minutes.

5. Experimental procedure

In this experiment, the coffee cherries was peeled using a peeler machine to get a coffee bean. One kilogram bean was prepared to dry. The moisture content of the stock was measured by grain moisture meter before fed it into the drying chamber. The stock was conveyed upward by using screw conveyor. The coffee bean flows downward by gravity and flow on a perforated-slope tray in the drying chamber. While bean flowing downward, hot air was introduced into drying chamber. Contact between coffee bean and hot air occurs in drying chamber. The coffee bean was recirculated upward to experience frequent contact with hot air. Every 15 minutes, the moisture content of coffee bean was measured. Temperature and humidity in drying chamber were also measured and logged every 15 minutes. The drying process was stopped when the moisture content of coffee bean reach 14%.

6. Result and Discussion

Coffee bean drying using vertical mixed flow pattern with sloped-tray performed contact between hot air and bean individually. This contact leads to increase in the coefficient of convection heat transfer and increase the mass transfer coefficient. Higher convection heat transfer coefficient affects to heat energy flowing easily. The higher water reduction rates occur early in the process, as seen in Figure 1. This cause that small energy needed to evaporates free water content. The next course of drying seems that water reduction rates slower. In this stage, more energy needs to be required for evaporating more strongly bonded water. This drying profile was similar to studies from some authors [7][8][9].

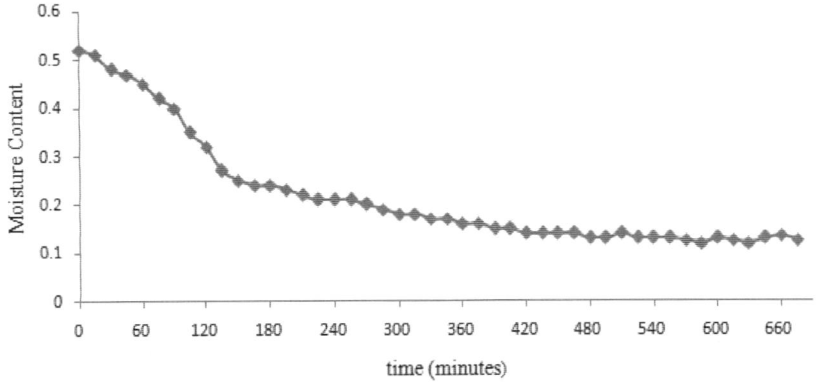

Figure 3 Profile of decreasing moisture content during coffee bean drying

As shown in Figure 3, the water reduction curve was steep in the first 130 minutes of drying period. This occupied 0,25 percent of drying period.

IOP Conf. Series: Materials Science and Engineering **771** (2020) 012070 doi:10.1088/1757-899X/771/1/012070

The drying chamber temperature and humidity take a role to the drying process. This couple parameter, that shown at Fig. 4, determine water reducing rate. The profile of Fig.4 can be a database of this typical dryer to get water reduction profile at Fig.3.

The parameter are governed by an equation that is introduced by [10], as seen at equation 1 and 2.

$$\boldsymbol{h} = \frac{\left(-\frac{dX}{dt}\right)BL}{A(T_a - T_c)} \quad (1)$$

$$\boldsymbol{k} = \frac{h(T_a - T_c)}{L(Y_{sup} - Y_a)} \quad (2)$$

Where *X* refers to the average moisture content of coffee beans in kg water kg dry coffee^{-1}; *h* the coefficient of convective heat transfer W m^{-2} °C^{-1}; *k* the mass transfer coefficient in kg m^{-2} s^{-1}; *B* is the mass flow of coffee beans in kg s^{-1}; L is the latent heat of evaporation in kJ kg^{-1}; *A* the surface area of the coffee beans in m^2; *Ta* the air temperature in °C; *Tc* the temperature of the coffee beans in °C; *Ysup* the absolute moisture content on the surface of the coffee beans in kg water kg dry air^{-1} and *Ya* the absolute humidity of air in kg water kg dry air^{-1}.

As can be seen, the equations proposed are issued from general balances used to calculate the mass and heat transfer coefficients in a particular dryer configuration. Moreover, the evolution of the air temperature is not considered but measured which is part of the simplification of the proposed equations.

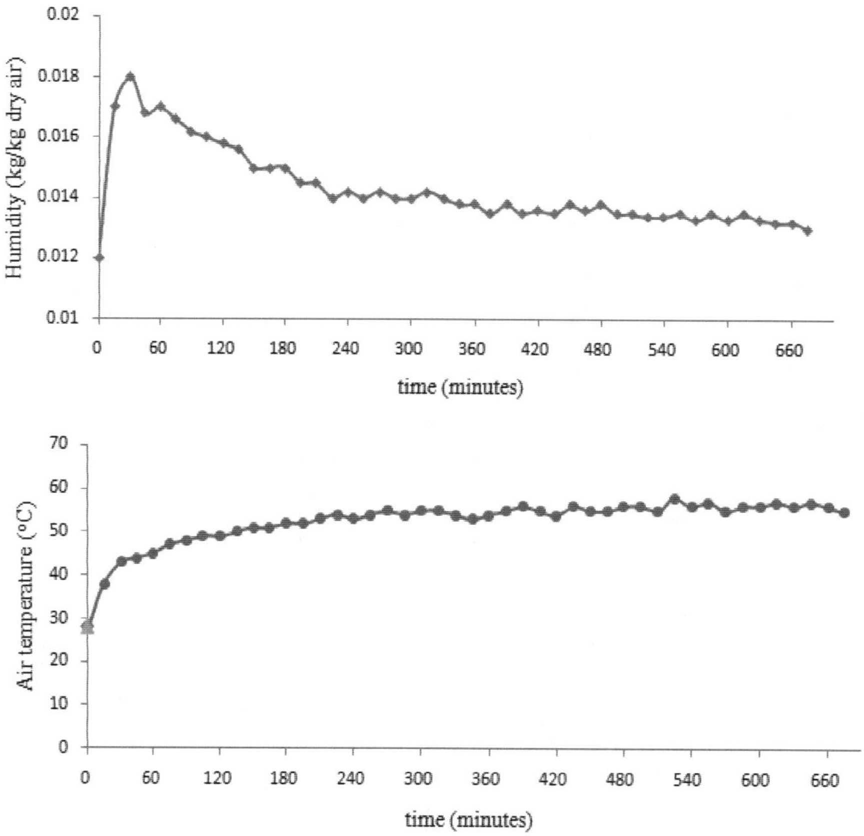

Figure 4 Profile of air temperature and humidity in drying chamber

7. Conclusion

At air velocity and temperature of 2 m/s and 60 °C, subsequently, the coffee bean can be dried from 52% initial MC to final 14% MC for 11,5 hours in this dryer. The drying rate can be increased by increasing air temperature and velocity, and humidity. The tray installed in the drying chamber has a role to retard drying contact time between air and coffee bean so that temperature of bean raise and leads to increase water diffusivity in the endosperm.

References

[1] M. Kleinwachter and D. Selmar, "Influence of drying on the content of sugars in wet processed green Arabica coffees," Food Chemistry, vol. 119, no. 2, pp. 500–504, Mar. 2010

[2] C. Reh, A. Gerber, J. Prodolliet, and G. Vuataz, "Water content determination in green coffee. Method comparison to study specificity and accuracy, "*Food Chemistry*, vol. 96, no.3, pp.423-430, Jun. 2006.

[3] Banga J. R. and R. P. Singh (1994). Optimization of the air drying of foods. Journal of. Food Engineering. 23:189-211

[4] Estiasih, Teti dan Kgs Ahmadi, 2009. *Teknologi Pengolahan Pangan.* Bumi Aksara. Malang.

[5] S. Haberman and A. E. Renshaw, "Generalized Linear Models and Actuarial Science,"*Journal of the Royal Stastistical Society. Series D (The Statistician),* vol. 45, no. 4, pp. 407-436, Jan 1996.

[6] K. Hangos and I. T. Cameron, Process Modelling and Model Analysis. Academic Press,2001,vol. 4

[7] Widyotomo, Sukrisno & Mulato Sri, "Penentuan Karakteristik Pengeringan Kopi Robusta Lapis Tebal", Buletin Ilmiah INSTIPER Vol 12 No 1, pp 15-37, 2005.

[8] Hendarson, S. M. and R. L. Perry. 1976. *Agricultural Process Engineering. 3 rd ed.* The AVI publ. Co., Inc, Wesport, Connecticut, USA

[9] Ruiz-Lopez, I.I., Córdova, A.V., Rodríguez-Jimenes, G.C., & Garcia-Alvarado, M.A. (2004). Moisture and temperature evolution during food drying: effect of variable properties. Journal of Food Engineering. 63(1): 117-124

[10] Sfredo, .M.A., Finzer, J.R.D., Limaverde, J.R. (2005). Heat and mass transfer in coffee beans drying. Journal of Food Engineering. 70: 15-25.

IOP Conf. Series: Materials Science and Engineering **771** (2020) 012071 doi:10.1088/1757-899X/771/1/012071

Environmental effects on hardness and wear resistance of composites reinforced by rice plant fiber as a material of brake lining

Pramuko Ilmu Purboputro[1,*], Agung Setyo Darmawan[1] and Bambang Waluyo Febriantoko[1]
[1]Department of Mechanical Engineering, Faculty of Engineering, Universitas Muhammadiyah Surakarta, Pabelan, Surakarta 57162, Indonesia

[*]e-mail: pip272@ums.ac.id

Abstract. The environment has an important role in determining the mechanical properties of composites. This study aims to investigate the effect of in-service environment on composite wear resistance in its use as a brake lining. Composites made from a mixture of rice plant fiber, fiberglass, aluminum powder and polyester resin are produced into brake linings. The brake lining is then tested for wear resistance using a friction machine. The test environment is dry, water, salt-water and brake oil. The highest wear is shown by the brake lining in the salt-water environment, while the lowest wear is shown when the brake lining is working in the brake oil environment.

1. Introduction

In the field of material technology, natural fiber materials are candidates as reinforcement materials to be able to produce composite materials that are lightweight, strong, environmentally friendly and economical. Nature has provided many human needs ranging from food to shelter. One of them is natural fiber ingredients [1-6].

Composite is a material that is formed from a combination of two or more materials to produce a composite material that has different mechanical properties and characteristics from the forming material. Composite consists of two main parts, namely: Matrix, serves as an adhesive or binder and protective filler (fillers) from external damage. Commonly used matrices such as carbon, glass and Kevlar [7]. Composite materials are widely used, such as in the fields of automotive industry [8], aerospace [9], aircraft and military [10], the sports equipment [11] and packaging materials [12].

In general, composite materials consist of two kinds, namely particulate composite materials and fiber composite materials. Particle composite material consists of particles that are bound by a matrix. The shape of these particles can vary as round, cubic, tetragonal or even irregularly shaped randomly. While the fiber composite material consists of fibers that are bound by a matrix. There are two kinds of forms, namely long fiber and short fiber [13].

In a composite structure, particle composite materials are composed of particles called particulate composite by definition; these particles are of various shapes such as round, cubic, tetragonal or even randomly shaped, but on average of the same dimension. Particle composite materials are generally used as fillers and reinforcement of ceramic composite materials (ceramic matrix composites). Particle size affects the strength of the composite. Particle composite materials are generally weaker than fiber composite materials. Particle composite materials have advantages, such as resistance to wear, do not small crack and have a binding force with a good matrix [14].

The main element of the composite is fiber that has many advantages; therefore, fiber composite materials are the most widely used. Fiber composite materials consist of fibers - as well as those bound by interconnected matrices. This fiber composite material consists of two types, namely long fiber (continuous fiber) and short fiber (short fiber and whisker). The use of fiber composite materials is very efficient in accepting loads and forces. Because of that, the fiber composite material is very strong and stiff when loaded in the direction of the fiber; on the contrary, it is very weak when burdened in the perpendicular direction of the fiber [15, 16].

Fiber composites in the industrial world began to be developed rather than using particle material. Fiber composite materials have the main advantages that are strong (strong), stiff (tough), and more resistant to heat when inside the matrix. In the development of fiber processing technology, making fiber is now more favored than the materials used. The method used to combine high tensile strength fiber and high elastic modulus with lightweight mass, lace tensile strength, and low elastic modulus is increasingly being developed in order to obtain maximum results. Composites generally use plastic material which is the material most often used as a fiber-binding material besides that the plastic is easy to obtain and easier to treat, rather than metal materials that require their own material [17,18].

Along with innovations made in the material field, natural fibers are re-examined to be used as composite reinforcing materials. Natural fibers have elastic, strong, abundant and environmentally friendly properties [19-22]. In addition, there are also disadvantages of this type of fiber, especially strength that is not always evenly distributed. The types of natural fibers that began to be used as reinforcement for polymers include Hemp, sisal, Coconut, Flex, and Jute [23].

Furthermore, because the in-service environment also has an important role in determining the mechanical properties of composites, an investigation is carried out to determine the effect of the environment on composite wear

2. Materials and Methodology

Composite material is made of a mixture of rice plant fiber, fibreglass, aluminum powder and polyester resin. The composition of this composite can be seen in table 1.

Table 1. The composition of composite.

Rice Plant Fiber (%)	Fiberglass (%)	Al Powder (%)	Polyester Resin (%)
20	10	10	60
30	10	10	50
40	10	10	40

To achieve the research objectives, this research was conducted in three stages. The first step is to make the brake lining from the composite. At this stage, the composite material that has been mixed then pressed with a compacting load of 4500 kg with a heated temperature of 80 °C and held for 15 minutes. After reaching the desired holding time, the brake lining is removed from the mold then put into the oven and the sintering process is carried out at a temperature of 180 °C for 15 minutes

The second stage is the wear test. In this stage, the wear test is carried out on the brake lining using a friction machine based on the China National Standard (CNS) GB 5763 standard. The test is carried out in a dry, air, saltwater and brake oil, respectively.

The final stage of this research is to analyze the data obtained from the test and draw conclusions of the research.

3. Results and Discussion

Photograph of brake lining made of rice fiber reinforced composites shown in figure 1. Meanwhile, figure 2 shows microphotographs of rice plant fiber reinforced composites. Figure 2a, figure 2b and figure 2c show the rice plant fiber reinforced composites with the composition of the rice plant fiber are 20%, 30% and 40%, respectively.

IOP Conf. Series: Materials Science and Engineering **771** (2020) 012071 doi:10.1088/1757-899X/771/1/012071

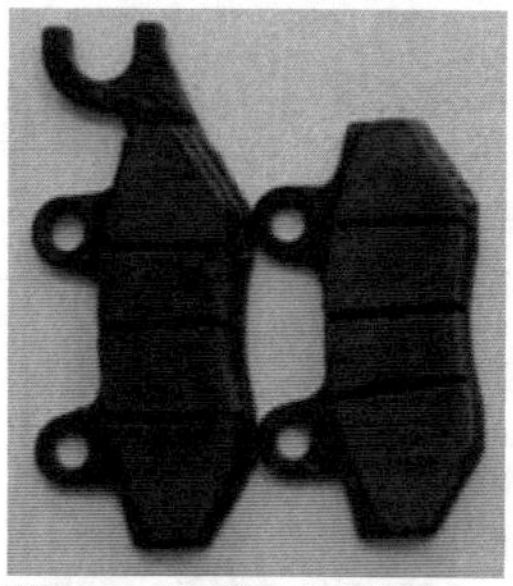

Figure 1. Brake lining.

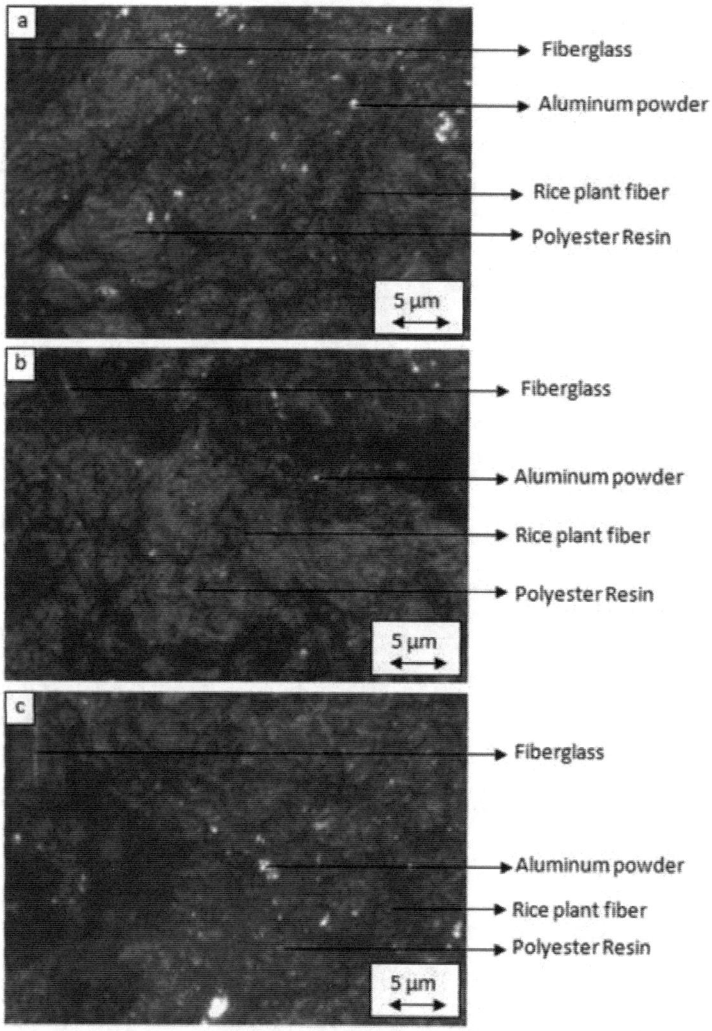

Figure 2. The microphotographs of rice plant fiber reinforced composites (a) the rice plant fiber is 20%, (b) the rice plant fiber are 30% (c) the rice plant fiber are 40%

Figure 3 shows the results of wear testing in different environments. Wear is seen reaching the highest levels of 126.4 mm^3/hour in the salt-water environment with a composition of 20 % rice plant fiber. This is because the 20 % rice plant fiber is the smallest composition of the variation of fiber composition tested, while the presence of Na^+ ions and Cl^- ions in salt solution will make the Al^+ ions escape into the environment. Meanwhile, the lowest wear of 110 mm^3/hour experienced by composites in the brake oil environment with 40% composition of rice plant fiber. This is because in addition to the fiber composition is the highest composition; the friction that occurs in the specimen is also reduced by the lubrication by brake oil.

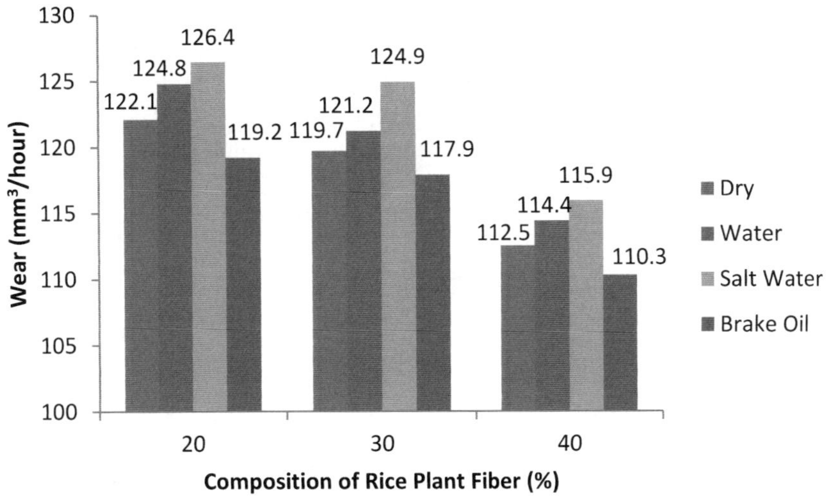

Figure 3. The effect of environment on the wear of the composites.

4. Conclusion

After testing for wear on several types of environments, it can be concluded that the environment has an effect on composite wear resistance. The highest wear experienced by composites in the salt-water environment and the lowest wear experienced by composites in the brake oil environment

Acknowledgement

This research is funded by the Directorate of Research and Community Service, The Ministry of Research, Technology and Higher Education of the Republic of Indonesia [Project Number = 199.24/A.3-III/LPPM/V/2019]. The authors would also like to acknowledge the Mechanical Engineering Laboratory of Universitas Muhammadiyah Surakarta for material preparations and mechanical test facilitations.

References

[1] Elanchezhian C, Ramnath B V, Ramakrishnan G, Rajendrakumar M, Naveenkumar V and Saravanakumar M K 2018 Review on mechanical properties of natural fiber composites *Materials Today: Proceedings* **5** 1785–1790.

[2] Salernitano E and Migliaresi C 2003 Composite Materials for Biomedical Applications: A Review *Journal of applied biomaterials & biomechanics* **1(1)** 3-18.

[3] Chegdani F and Mansori M E 2018 Friction scale effect in drilling natural fiber composites *Tribology International* **119** 622–630.

[4] dos Reis J M L 2012 Effect of temperature on the mechanical properties of polymer mortars *Materials Research* **15(4)** 645–649.

[5] Dittenber D B and Rao H V S G 2012 Critical review of recent publications on use of natural composites in infrastructure *Composites: Part A* **43** 1419–1429.

[6] Verma D, Gope P C, Shandilya A, Gupta A and Maheshwari M K 2013 Coir Fibre Reinforcement and Application in Polymer Composites: A Review *J. Mater. Environ. Sci.* **4(2)** 263-276.

[7] Hassan M L, Rowell R M, Fadl N A, Yacoub S F and Christainsen A W 2000 Thermoplasticization of Bagasse. II. Dimensional Stability and Mechanical Properties of Esterified Bagasse Composite *Journal of Applied Polymer Science* **76** 575–586.

[8] Sarikaya E, Çallioğlu H and Demirel H 2019 Production of epoxy composites reinforced by different natural fibers and their mechanical properties *Composites Part B: Engineering* **167** 461-466.

[9] Zhou J, Li Y, Cheng L and Zhang L 2019 Indirect Microwave Curing Process Design for Manufacturing Thick Multidirectional Carbon Fiber Reinforced Thermoset Composite Materials *Applied Composite Materials* **26(2)** 533–552.

[10] Cavalcanti D K K, Banea M D, Neto J S S, Lima R A A and Carbas R J C 2019 Mechanical characterization of intralaminar natural fibre-reinforced hybrid composites *Composites Part B: Engineering* **175** 107149.

[11] Wang J L 2012 Application of Composite Materials on Sports Equipments *Applied Mechanics and Materials* **155-156** 903-906.

[12] Mohamed S A A, El-Sakhawy M, Nashy E H A and Othman A M 2019 Novel natural composite films as packaging materials with enhanced properties *International Journal of Biological Macromolecules* **136** 774-784.

[13] Zhang C, Li F, Li J, Wang L, Xie Q, Xu J and Chen S 2017 A new biodegradable composite with open cell by combining modified starch and plant fibers *Materials & Design* **120** 222-229.

[14] Ahmed S and Jones F R 1990 A review of particulate reinforcement theories for polymer composites *Journal of Materials Science* **25** 4933-4942.

[15] Ding Z, Li Y, Lu C and Liu J 2018 An Investigation of Fiber Reinforced Chemically Bonded Phosphate Ceramic Composites at Room Temperature *Materials* **11**, 858.

[16] Williams G I and Wool RP Composites from Natural Fibers and Soy Oil Resins *Applied Composite Materials* **7** 421–432.

[17] Oksman K 2000 Mechanical Properties of Natural Fibre Mat Reinforced Thermoplastic *Applied Composite Materials* **7(5–6)** 403–414.

[18] Gassan J and Bledzk A K Possibilities to Improve the Properties of Natural Fiber Reinforced Plastics by Fiber Modification – Jute Polypropylene Composites – *Applied Composite Materials* **7(5–6)** 373–385.

[19] Ramesh M, Palanikumar K and Reddy K H 2017 Plant fibre based bio-composites: Sustainable and renewable green materials *Renewable and Sustainable Energy Reviews* **79** 558-584.

[20] Alzebdeh K I, Nassar M M A and Arunachalam R 2019 Effect of fabrication parameters on strength of natural fiber polypropylene composites: Statistical assessment *Measurement* **146** 195-207.

[21] Goutianos S, Peijs T, Nystrom B and Skrifvars M 2006 Development of Flax Fibre based Textile Reinforcements for Composite Applications *Applied Composite Materials* **13(4)** 199–215.

[22] Zhou Y, Fan M, and Chen L 2016 Interface and bonding mechanisms of plant fibre composites: An overview, *Composites Part B: Engineering* **101** 31-45.

[23] Tian F and Zhong Z 2019 Modeling of load responses for natural fiber reinforced composites under water absorption, Composites Part A *Applied Science and Manufacturing* **125** 105564.

AUTHOR INDEX